ACS SYMPOSIUM SERIES **583**

Taxane Anticancer Agents

Basic Science and Current Status

Gunda I. Georg, EDITOR
University of Kansas

Thomas T. Chen, EDITOR
University of Tennessee

Iwao Ojima, EDITOR
State University of New York at Stony Brook

Dolatrai M. Vyas, EDITOR
Bristol-Myers Squibb PRI

Developed from symposia sponsored
by the Divisions of Chemical Health and Safety,
Medicinal Chemistry, and Organic Chemistry
at the 207th National Meeting
of the American Chemical Society,
San Diego, California,
March 13–17, 1994

American Chemical Society, Washington, DC 1995

Library of Congress Cataloging-in-Publication Data

Taxane anticancer agents: basic science and current status: developed from symposia sponsored by the Divisions of Chemical Health and Safety, Medicinal Chemistry, and Organic Chemistry at the 207th National Meeting of the American Chemical Society, San Diego, California, March 13–17, 1994 / Gunda I. Georg, editor ... [et al.].

p. cm.—(ACS symposium series, ISSN 0097–6156; 583)

Includes bibliographical references and indexes.

ISBN 0–8412–3073–0

1. Taxol—Congresses.

I. Georg, Gunda I. II. American Chemical Society. Division of Chemical Health and Safety. III. American Chemical Society. Division of Medicinal Chemistry. IV. American Chemical Society. Division of Organic Chemistry. V. American Chemical Society. Meeting (207th: 1994: San Diego, Calif.) VI. Series.

[DNLM: 1. Taxol—chemistry—congresses. 2. Taxol—pharmacology—congresses. QV 269 T235 1994]

RC271.T38T38 1994
616.99'4061—dc20
DNLM/DLC
for Library of Congress 94–37226
CIP

The paper used in this publication meets the minimum requirements of American National Standard for Information Sciences—Permanence of Paper for Printed Library Materials, ANSI Z39.48–1984. ∞

PRINTED IN THE UNITED STATES OF AMERICA

Foreword

THE ACS SYMPOSIUM SERIES was first published in 1974 to provide a mechanism for publishing symposia quickly in book form. The purpose of this series is to publish comprehensive books developed from symposia, which are usually "snapshots in time" of the current research being done on a topic, plus some review material on the topic. For this reason, it is necessary that the papers be published as quickly as possible.

Before a symposium-based book is put under contract, the proposed table of contents is reviewed for appropriateness to the topic and for comprehensiveness of the collection. Some papers are excluded at this point, and others are added to round out the scope of the volume. In addition, a draft of each paper is peer-reviewed prior to final acceptance or rejection. This anonymous review process is supervised by the organizer(s) of the symposium, who become the editor(s) of the book. The authors then revise their papers according to the recommendations of both the reviewers and the editors, prepare camera-ready copy, and submit the final papers to the editors, who check that all necessary revisions have been made.

As a rule, only original research papers and original review papers are included in the volumes. Verbatim reproductions of previously published papers are not accepted.

M. Joan Comstock
Series Editor

Contents

Preface

THE ANTICANCER TAXANES PACLITAXEL AND DOCETAXEL are the most promising new chemotherapeutic agents developed for cancer treatment in the past decade. Early concerns about paclitaxel's lack of activity against certain leukemia models, limited availability, and problems with its formulation almost prevented its development as a clinical agent. But then, clinical trials conducted in the late 1980s and early 1990s demonstrated impressive clinical activities against advanced ovarian and breast cancer. The U.S. Food and Drug Adminstration has approved the clinical use of paclitaxel for these two types of cancer. Recent clinical trials have shown that paclitaxel and docetaxel may also be useful agents for the treatment of non-small-cell lung cancer, head and neck cancer, and other types of cancers.

The promising clinical activities of the anticancer taxanes combined with potential problems due to limited supply generated a multitude of programs worldwide, such as broader clinical trials and in-depth clinical evaluation of anticancer taxanes, efforts to provide an adequate supply of paclitaxel for clinical trials and for cancer patients, development of better drug formulations, biochemical studies to elucidate the precise mechanism of action, chemical studies to obtain structure–activity information, development of second-generation paclitaxel analogues, elaboration of semisynthetic methods for the large-scale production of the anticancer taxanes, and efforts directed at the total synthesis of paclitaxel.

Chemistry plays a pivotal role in many of these endeavors, and so the American Chemical Society held its first symposium on paclitaxel in 1992. Because of continued excitement about the anticancer taxanes, three divisions of the ACS held a series of symposia on taxane research in 1994. The Division of Chemical Health and Safety asked Thomas T. Chen to organize their symposium entitled "Symposium on the Cellular Mechanism of Action and Toxicity of Taxol." He is grateful for the support of this symposium from Glaxo, SmithKline Beecham, TPL Phytogen, the University of Tennessee Physicians' Medical Education and Research Foundation and the Graduate School. For the Division of Medicinal Chemistry, Gunda I. Georg and Dolatrai (Dinesh) M. Vyas organized a symposium entitled "Advances in the Medicinal Chemistry of Taxol and Taxoids." They acknowledge the financial contributions from Bristol-Myers Squibb to make the symposium possible. Iwao Ojima organized

NOTE: Paclitaxel is the generic name for Taxol, which is now a registered trademark.
Docetaxel is the generic name for Taxotere, which is also a registered trademark.

the symposium "Advances in the Chemistry of Taxane and Taxoid Anticancer Agents," for the Division of Organic Chemistry. He gratefully acknowledges support from Bristol-Myers Squibb, Rhône-Poulenc Rorer, Indena, and Abbott for this symposium.

The publication of the symposium contributions provides comprehensive coverage of the newest research results in a timely fashion to interested chemists, biochemists, biologists, clinicians, and other scientists in the United States and abroad. This book contains 24 chapters, ranging in content from recent clinical results, paclitaxel formulation problems, metabolism studies, taxane biosynthesis, tubulin biochemistry to medicinal chemistry, semisynthesis, and total synthesis of anticancer taxanes. It is the first multi-author, comprehensive overview of important areas of taxane research, containing a multitude of previously unpublished research results. With the publication of this book, the authors hope to further stimulate the interest and the ongoing research activites in the exciting area of anticancer taxanes.

We dedicate this book to people who live with cancer.

GUNDA I. GEORG
Department of Medicinal Chemistry
University of Kansas
Lawrence, KS 66045–2506

THOMAS T. CHEN
Departments of Zoology and
Obstetrics and Gynecology
University of Tennessee
Knoxville, TN 37996

IWAO OJIMA
Department of Chemistry
State University of New York at Stony Brook
Stony Brook, NY 11794–3400

DOLATRAI (DINESH) M. VYAS
Bristol-Myers Squibb PRI
5 Research Parkway
Wallingford, CT 06492–7660

September 29, 1994

Brief History of the Discovery and Development of Taxane Anticancer Agents *(1-3)*

1962 Collection of *Taxus brevifolia* in the Gifford Pinchot National Forest in the State of Washington by USDA botanist Arthur Barkley as a result of a collaborate effort between the USDA and the NCI Plant Program under the direction of Jonathan Hartwell.

1964 Cytotoxicity of bark extracts to KB cells established. Shipment of 30 lbs. of bark to Monroe Wall at the Research Triangle Institute (RTI).

1966 Isolation of taxol (K172) and cytotoxic activity of the pure compound against KB cells by the Wall group.

1967 Disclosure of taxol isolation at the American Chemical Society National Meeting in Miami Beach in Florida by the Wall group.

1971 Publication of the structure and biological activity of taxol in the *Journal of the American Chemical Society* by M. Wall with M. Wani and H. Taylor (RTI) and A. McPhail and P. Coggon (Duke University).

1974 Activity in the *in vivo* B16 (ip, ip) melanoma model (NCI).

1977 Championed by Matthew Suffness, taxol becomes an NCI development candidate.

1978 Activity against the NCI LX-1 lung xenograft, MX-1 breast xenograft and CX-1 colon xenograft.

1978 Publication by D. A. Fuchs and R. K. Johnson on the anti-mitotic activity of taxol (*Cancer Treat. Repts.*).

1979 Susan Horwitz with P. Schiff and J. Fan disclose that taxol is a promoter of microtubule assembly (*Nature*).

1979 Pierre Potier's laboratory (Institut des Chimie des Substances Naturelles (ICSN), CNRS) in France begins bioassay-guided purification of the European yew (*Taxus baccata*).

1980 At NCI, Cremophor formulation selected and route and schedule dependency studies completed. Toxicology studies with taxol begin.

1980 10-Deacetylbaccatin III, isolated from *Taxus baccata*, identified as readily accessible starting material for the semisynthesis of anticancer taxanes by the Potier group (CNRS).

1981 Rhone-Poulenc and CNRS sign a research agreement concerning anticancer taxanes.

1982 Toxicologie studies completed. Taxol approved by NCI for Investigational New Drug Application (INDA) filing.

1983 INDA filed for taxol with the Food and Drug Administration (FDA).

1984 INDA approved by FDA. Phase I clinical trials begin.

1984 Taxotere synthesized by Francoise Gueritte-Voegelein and Daniel Guenard (ICSN-CNRS).

1985 NCI approves taxol for phase II clinical trials.

1985 *In vitro* assays demonstrate cytotoxicity of taxotere.

1986 The first semisynthesis of taxol via the oxamination process (ICSN-CNRS and Rhone-Poulenc patent).

1986 First asymmetric synthesis of the C-13 phenylisoserine side chain of taxol by the Green group (*J. Org. Chem.*).

1987 Rhone-Poulenc selects taxotere for clinical development.

1988 *J. Am. Chem. Soc.* publication of the semisynthesis of taxol published via direct esterification of 10-deacetylbaccatin III by the groups of Potier and Greene. CNRS - Rhone-Poulenc patent on the semisynthesis of taxol and taxotere.

1989 Johns Hopkins group publish on the activity of taxol in advanced ovarian cancer (*Ann. Intern. Med.*).

1989 NCI issues a request for applications for a Cooperative Research and Development Agreement (CRADA) and selects Bristol-Myers Squibb as the CRADA partner.

1989 Semisynthesis of taxol by Holton via *N*-acyl beta-lactams. Patent on the semisynthesis of taxol.

1989 Asymmetric synthesis of beta-lactams as phenylisoserine precursors by the Georg group and the Ojima group.

1990 Phase I clinical trials of taxotere begin in Europe and the US.

1991 M. D. Anderson group publishes finding of activity of taxol in metastatic breast cancer (*J. Natl. Cancer Inst.*).

1991 *J. Org. Chem.* publication by the Ojima group and the Georg group on the asymmetric synthesis of phenylisoserine and beta-lactam intermediates.

1991 *J. Org. Chem.* publication by the Green group on the semisynthesis of taxol and taxotere from phenylglycine and 10-deacetylbaccatin III.

1992 New Drug Application (NDA) for taxol filed by Bristol-Myers Squibb.

1992 6 Month after filing, the FDA approves taxol for the treatment of refractory ovarian cancer.

1992 Phase II clinical trials for taxotere begin worldwide.

1992 Semisynthesis of taxol and taxotere via the oxazolidinone route by Commercon and collaborators at Rhone-Poulenc Rorer (*Tetrahedron Lett.*)

1992 Semisynthesis of taxol and taxotere through coupling of metalated baccatin III with β-lactam intermediates by the Holton group (patent) and the Ojima group (*Tetrahedron*).

1993 Taxol is marketed by Bristol-Myers Squibb Co.

1993 SNDA (Supplemental NDA) for the semisynthesis of taxol from 10-deacetylbaccatin III filed by Bristol-Myers Squibb Co.

1994 Supplemental FDA approval of taxol for the treatment of metastatic breast cancer.

1994 NDA and MAA (Manufacture Approval Authorization) for the use of taxotere in the treatment of breast and lung cancers filed.

1994 *J. Am. Chem. Soc.* publication by the Holton group and *Nature* publication by the Nicolaou group on total syntheses of taxol.

References and Notes:

(1) Many other individuals, not listed in this short history, have made very significant contributions to the development of taxol and to our current understanding of the biology and chemistry of the anticancer taxanes. Most of their names and contributions can be found in the chapters and references of this book.

(2) For a detailed account of the history of the development of taxol see: Suffness, M. Wall, M. In *Taxol: Science and Applications*; Suffness, M., Ed.; CRC: Boca Raton, FL, 1994 (in press).

(3) Taxol® is a registered trademark of Bristol-Myers Squibb Company. The generic name is paclitaxel. Taxotere® is a registered trademark of Rhone-Poulenc. Its generic name is docetaxel.

Chapter 1

Overview of Paclitaxel Research

Progress on Many Fronts

Matthew Suffness

Grants and Contracts Operations Branch, Developmental Therapeutics Program, Division of Cancer Treatment, National Cancer Institute, Bethesda, MD 20892

Great progress has been made in research on taxol production and on future generation taxol drugs since the report of important activity in human ovarian cancer in 1989. Recent results have established that: there is strong long term potential in plantations of either *Taxus baccata* or several ornamental *Taxus* cultivars, for either direct production of taxol or production of its precursor, 10-desacetylbaccatin III (10-DAB); plant cell culture, especially utilizing advances in biosynthetic understanding and genetic engineering, is a likely future drug source; semisynthesis of taxol from 10-DAB is well established through major advances in side chain synthesis and coupling reactions; the binding site on microtubules involves both the α- and β-subunits of tubulin, with the N-terminal 31 amino acids of the β-subunit being critical; effects of single substitutions of substituents at nearly all positions on the taxol molecule have been established; and there is a strong opportunity for new generations of taxol-like drugs through peptidomimetic approaches.

For the first 28 years of knowledge of taxol, from the first paper in 1971 (*1*) until the first report of the activity in Phase II clinical trials in refractory ovarian cancer in 1989, (*2*) taxol progressed at a rather slow pace and its development was fraught with many obstacles (*3,4*) including solubility, supply, and toxicities (*5*). There were many candidate compounds being developed at the National Cancer Institute (NCI) and taxol was a particularly difficult and expensive one. The clinical activity in ovarian cancer was a great stimulus to research, and the purpose of this article is to provide a perspective of where research on taxol was in 1989, where it is today, and how it got there. The activity in refractory ovarian cancer brought to the fore two main classes of problems, the supply of the drug both short term and long term, and the possibilities to improve on taxol through understanding its properties well enough to design better

NOTE: Paclitaxel is the generic name for Taxol, which is now a registered trademark.

delivery systems and analogues with better potency, solubility, and specificity.

The immediate problem of supply was beyond the scope of the NCI's resources as the yield of drug was only about 100 mg per kg of dried bark of *Taxus brevifolia* (Western or Pacific yew), and the only pilot plant facility contracted to the NCI, Polysciences, Inc., could only handle about 5,000 kg of *Taxus* bark per year to produce about 500 g of pure taxol (*6*). It was apparent that a massive program was needed for collection, extraction, and purification of taxol and that a commercial partner was needed for taxol development. A request for proposals for enter into a Cooperative Research and Development Agreement (CRADA) was issued by the NCI in August, 1989 and after careful review of proposals, Bristol-Myers (soon to become Bristol-Myers Squibb; BMS) was selected in November, 1989. The final CRADA was signed in January, 1991. Surveys of abundance of Pacific yew were conducted by the United States Department of Agriculture, (USDA) and by the Bureau of Land Management (BLM) on public lands, and while it rapidly became apparent that there was a lot more yew available than had been expected, it was also clear that due to the slow growth of the trees and the killing of the trees by harvesting bark, Pacific yew bark was not a sustainable resource for taxol production. Alternate sources or alternative compounds were needed for the near future (4-8 years) and for the long term (greater than 8-10 years). The agreement between the NCI and BMS was that BMS would take responsibility for short term production of taxol and would sponsor such studies as might be expected to rapidly effect taxol supply, while the NCI would sponsor the basic research which would deal with the long term goals.

The NCI in 1989 had a very small portfolio of grants related to taxol research and it was felt essential to stimulate research in the area. This involved a two faceted approach, the stimulation of collaborations through a workshop, the National Cancer Institute Workshop on Taxol and *Taxus*, held in Bethesda in June, 1990, and the setting

Table I. Understudied areas in Preclinical Taxol and Taxus research, 1990

Production:

Biological Studies
- biosynthesis
- genetics
- tissue culture
- agronomics
- genetic engineering
- strain selection
- strain improvement
- chemical ecology

Chemical Studies
- total synthesis
- improved semisynthesis
- improved assays

Future generation drugs:

Biological studies
- taxol binding site

Chemical studies
- synthesis of analogues
- structure-activity relationships

aside of funds for grants in taxol research, which took the form of a Request for Applications (RFA) issued in July, 1990 with a set aside of one million dollars per year for up to five years. Through use of additional funding sources, 16 research grants with first year costs of approximately two million dollars were funded from applications received in response to this RFA and most of the work reviewed in this paper was supported through NCI research grants. The summary of 1990 preclinical research needs in the taxol area is found in Table I, which is reproduced from a slide presented at the workshop. This paper looks briefly at the areas of production and future generation drugs as they were then and where they are now in order to demonstrate: 1) the progress that has been made; 2) the new directions that have opened; 3) research directions which are now less important; and 4) areas where increased efforts are needed.

PRODUCTION OF TAXOL - BIOLOGICAL STUDIES

Biosynthesis and Genetic Engineering. Understanding of the biosynthetic pathway for taxol in *Taxus* has many implications for drug production. These include: 1) increased yields through feeding of precursors present in limiting quantities; 2) increased yields through manipulation of the genes coding for the enzymes involved in biosynthesis, through activation of gene transcription, increase in copy number of genes, introduction of promoter sequences to enhance expression, or ultimately transfer of the whole biosynthetic pathway to a rapidly growing bacterium such as *E. coli*; 3) enhancement of production in plant cell culture; and 4) the use of mutational programs to alter key genes and enhance the efficiency of their derived enzymes. The results most needed for taxol production are to define the rate limiting steps in the biosynthetic pathway and apply genetic technologies to enhance rates.

Prior studies of biosynthesis of terpenoids show complex pathways of many steps, nearly every one catalyzed by a separate enzyme, and the elucidation of such a pathway is a truly daunting task, made even more difficult in the case of taxol by the very low yield of the compound, and hence the putative intermediates, in the source plant and the possibility that discrete subsets of the pathway might be taking place in different tissues (compartmentalization). On the positive side there was the initial presumption based on structures of previously isolated taxanes that biosynthesis of the side chain and nucleus of taxol were separate so that if the side chain was limiting, the problem was much simpler. Equally important is the conceptual placement in biosynthesis of the key C-20 precursor for all diterpenes, geranylgeranyl pyrophosphate, which can lead to diterpenes or tetraterpenes and which arises from the centrally placed mevalonate pathway. Since mevalonate is a precursor for an extremely wide set of terpenoids, sterols, and prenylated compounds, and is readily produced, the availability of geranylgeranyl pyrophosphate is unlikely to be rate limiting. Studies in biosynthesis of other diterpenoids indicate that the initial folding and cyclization which gives rise to the specific diterpenoid nucleus is often a slow step for the whole pathway leading to a diterpenoid family. Thus if the diterpenoid cyclase enzyme in the taxol pathway could be isolated, characterized, and cloned it could be manipulated to result in enhancement of the entire pathway.

Progress in taxol biosynthesis has been quite remarkable. The entire pathway for the side chain has been elucidated (*7,8*), the incorporation of both mevalonate and acetate has been reported in *T. canadensis* (*9*) and in bark of *T. brevifolia* (*10*) and very recently Croteau and collaborators have isolated an enzyme which appears to be taxadiene cyclase (R. Croteau, Wash. State U. unpublished).

Genetics. Genetic variation among *Taxus* species and *Taxus* populations within a species may not only relate directly to biosynthesis of taxol but also to growth rates, nutrient utilization, disease resistance, and many other factors that would influence cultivation of *Taxus* for either taxol or precursors such as 10-desacetylbaccatin (10-DAB). The current production method for both taxol and taxotere of isolation of 10-DAB and conversion by semisynthesis relies mainly on wild *Taxus* or *Taxus* which has been cultivated for ornamental purposes. The yields of 10-DAB from such material is obviously not optimized and future needs are expected to be met from plantations of *Taxus* specifically selected for high yield of taxol or precursors. The time frame of tree genetics studies precludes results at this time but important studies have been set up by the Weyerhaeuser Company taking diverse genetic populations, cultivating them under identical conditions in one location, and performing taxane analyses to correlate genetic markers with taxane production.

Tissue Culture (Plant Cell Culture). The NCI had sponsored contracts for plant cell culture production of antitumor agents including taxol in the late 1970s with no success, but significant advances in plant cell culture in the 1980s enabled new work including the results of Christen (*11,12*). Much success has been achieved in the last four years and taxol is now apparently being produced with reasonable titers by several organizations including Phyton Catalytic and ESCAgenetics in the US. There are at least 30 and perhaps as many as 50 laboratories worldwide working on production of taxol or other taxanes by plant cell culture. While the yields are closely held commercial data, it can be surmised from the expansion into large fermentation equipment that the data are quite positive. Culture manipulation to promote secretion of taxol as an extracellular product offers the possibility of semi-continuous or continuous fermentation and recent results from the biosynthetic studies cited above may enable further yield enhancements through precursor feeding. Plant cell culture would be the logical place to exploit advances in biosynthesis and genetic engineering and for example, the introduction of an activated gene for taxadiene cyclase could have dramatic effects on yield by funneling nutrients down the mevalonate pathway into taxanes. There is no doubt that taxol can be produced by cell culture; the only question is whether the cost will be competitive with semisynthesis.

Another important opportunity with plant cell culture is the production of analogues by directed fermentation, i.e. the feeding of unnatural substrates to produce new analogues. It has recently been established that *Taxus* cell cultures produce taxanes which have not previously been identified from plant extracts (*13*).

Agronomics. Taking a plant from the wild and cultivating it as a crop may not seem like a great feat until one thinks about crops like corn, wheat, and rice that have been under cultivation for thousands of years and are still undergoing improvement. Questions that must be answered for a medicinal plant include: what conditions of soil, water, light, temperature, climate, and altitude are best for growth; are these the same conditions that are optimal for metabolite production; can populations be expanded by clonal propagation or from seed; how does pollination occur and is it controllable for breeding purposes; what conditions are needed for seed germination; what traits are environmental and which are inherited; what populations exist and what is the extent of genetic variation among them? With regard to the Pacific yew (*T. brevifolia*) the information was very limited as this small, spindly, slow growing, tree had almost no commercial value, and there had been no reason to conduct studies to answer these many questions. The NCI had previously initiated discussion with the Weyerhaeuser Company but until the confirmed clinical activity in refractory ovarian cancer was reported in 1989, it was not possible to generate significant commercial interest. On the other hand, there were a large number of yews being cultivated for ornamental purposes; their hardiness to heat and cold, shade tolerance, disease resistance, and relatively slow growth made them ideal for foundation plantings and yews are the most popular evergreen for this purpose in the northern and central US. The main species cultivated as ornamentals are the Japanese yew *(T. cuspidata)*, the European yew (*T. baccata*), and the cross of these two *T. x media* which includes about 40 currently cultivated varieties (cultivars) of which about a dozen are common in nurseries. A survey in 1991 (*14*) found seven *T. x media* cultivars 'Densiformis', 'Hicksii', 'Brownii', 'Dark Green Spreader', 'Runyan', 'Wardii', and 'Tauntonii') and one *T. cuspidata* cultivar ('Capitata') each with more than one million plants in the field in nurseries in the US, mainly in Michigan, Ohio, Pennsylvania, and Rhode Island. This was a tremendous blessing because the nurseries engaged in the production of these ornamental yews had developed an extremely valuable knowledge base in how to germinate, propagate, transplant, fertilize, and cultivate them. However, before 1990 there had been no systematic effort to analyze the ornamental yews for content of taxol and other taxanes. A major problem for all natural products work involving taxol was the lack of accurate and reliable assays that could detect specific taxanes in crude extracts. This is discussed in more detail below in the chemistry section but the two main points are that taxol is present in very low concentrations in complex mixtures, and the only significant chromophoric groups present are the benzene rings which are common to many natural products. Both chemical and immunological assays are now well developed and there are numerous reports (15-20) of analysis of both natural and cultivated *Taxus* for taxol content as well as content of other taxanes.

A very critical and as yet unresolved question in selection of which *Taxus* to grow is the issue of whether it is more desirable to select for high taxol content or high content of 10-DAB. Production for taxol has the obvious advantage of directly producing the final marketed compound, but there are three significant disadvantages, purity, flexibility, and cost. The purity issue is one of reproducibility of minor compounds that are co-eluted with taxol, i.e. the impurity profile. This profile can differ with climatic changes, location of plantations, time of harvest, etc. This would not be an issue if the end product of cultivation was 10-DAB and the last several steps in taxol production were synthetic steps with more defined and reproducible chemical byproducts. The question of how long taxol will remain on the market before it is displaced by newer analogues is a critical one and the more flexible approach is to cultivate for 10-DAB which can act as an intermediate for a wide variety of compounds altered in the C-13 side chain, including taxotere. At present, cost analysis seems to favor production of 10-DAB since the literature yields cited are from 200 to 1,000 milligrams per kg of fresh leaves of *T. baccata* (*21-23*) while the yields cited for taxol from *T. brevifolia* bark are 50 to 200 milligram per kg of dried bark (*1,17, 20, 24*). Taxol yields from select ornamental *Taxus* cultivars have been cited as high as 800 mg per kg for *T. x media* 'nigra' (*25*). Examination of cultivars has shown wide differences in content of both taxol and 10-DAB with good consistency within a cultivar but wide variations in different cultivars of the same species (*17,25,26*).

Another agronomic issue is whether it is better to grow *Taxus* as a hedge crop taking clippings two or three times a year, or to periodically harvest whole plants and maintain a constant flow of plantings. Studies of variation of content by season have shown that Spring collections of bark had considerably higher taxol content than Autumn collections (*27*), and that shade-grown trees had higher taxol content than sun-exposed trees (*20*). Finally, there have been quite important results on the stability of taxol content in leaves of *Taxus*. The genesis of this question goes back to the early studies conducted in 1969 by the NCI and USDA on the best sources of taxol (*28*) in which it was found that the order of taxol content in plant parts of Pacific yew was bark > roots > leaves >> wood, but that there was a large variability in taxol content of leaves compared to other plant parts. Recent studies of both Pacific yew and ornamental yews found substantial variation in taxol content in leaves that could not be attributed to the analytical methods used; the strong inference is that taxol is unstable in leaves and begins to degrade on harvest and much of the variation in analytical results observed is a function of harvest conditions and post-harvest treatment (drying, freezing, etc.), and the time from harvest to assay (*20*). Taxol yield from processing 100,000 pounds of leaves of *T.* x *media* 'Hicksii' could be maintained at an acceptable level through use of a suitable drying protocol (*29,30*).

Strain Selection and Improvement. This is a general problem which can directly have major effects on yields and production costs whether one is considering growing *Taxus* in plantations, or in cell cultures. While the most obvious parameter of interest is yield of taxol or 10-DAB, selection for rapid growth, resistance to infectious organisms, and suppression of other metabolites which are either toxic, or which are difficult to separate from the desired end product also should be considered. Hardiness in terms of resistance to heat, cold, and drought are important in plantations, while for cell cultures, selection for excretion into media, rapidity of production after growth phase, and stability of the end products in the media are key parameters. By 1990, extensive work on strain selection had been carried out on ornamental *Taxus* by nurseries to select for hardiness, disease resistance, and growth forms (height, foliage color, spreading, shape) but there was no work reported on selection for content of taxol or other taxanes. Much progress has been made in the last four years, with reports of selection of plant cell cultures for taxol content from several laboratories (13, 31-34). As described above, many analyses of both wild and ornamental *Taxus* for various taxanes have been completed, and higher yielding strains have been identified but most of this work has been on biomass combined from multiple plants of individual species or cultivars. It is now necessary to examine individual plants of the highest yielding populations, to grow them under optimal conditions, and to clone the best of these using tissue culture methodology as starting stock for plantations. Studies are needed to be certain that the desired traits are stable and hereditable.

Chemical Ecology. The chemical ecology of Pacific yew refers specifically to interactions with the organisms in its environment, and the chemicals that produce those interactions. For Pacific yew, this would include deer, moose, and elk which use it as browse, microorganisms, particularly fungi, which can infect the tree, insects which want to use yew as food, and other higher plants that can compete by blocking seed germination, pollination or otherwise interfere with the survival and expansion of yew in its ecological niche. The most critical question in 1990 was whether there was any evidence for taxol being a stress metabolite, because if so, application of the proper physical or chemical conditions that induce a particular type of stress in *Taxus* might markedly enhance production. Since taxol is present at low concentrations, is not highly potent, is largely destroyed when administered orally, and from its mechanism as a tubulin interactive agent should not be toxic to procaryotic organisms, it seemed unlikely that it was a stress metabolite and there has been no evidence brought forth to the contrary. Taxol content in Pacific yew has been shown to vary several fold with the time of year, the population studied, light versus shade conditions, the age of the plant, etc. (20,27), and this data may be important for plantations of other *Taxus* species.

One of the most fascinating developments in taxol research in the last several years has been the finding that a new genus of fungus isolated from the inner bark of *Taxus*, *Taxomyces*, can produce taxol, albeit at very low levels (*35,36*). There is no obvious answer to why both a plant and an associated microorganism would produce the same complex secondary metabolite, especially given the complex biosynthesis proposed for taxol; possible hypotheses would include co-evolution or gene transfer from the plant to the fungus. Perhaps once key enzymes of the biosynthetic pathway for taxol in *Taxus* have been isolated, the probes can also be used to isolate corresponding enzymes from *Taxomyces* for comparison of sequence homologies to shed light on this question.

The recent biological developments in production of taxol can be summarized by stating that the original source of taxol, the Pacific Yew (*T. brevifolia*), does not appear to be a promising source for the long term and indeed is already being replaced by semisynthesis using 10-DAB produced from leaves of *T. baccata* and its close ally *T. wallichiana* (considered by some authorities as a sub-species of *T. baccata*). Further studies on Pacific yew are likely to have limited impact on future production. Plantations of either *T. baccata* or one or more of the ornamental cultivars, selected for high 10-DAB content should give maximum flexibility and yield for future production. The potential contribution of plant cell culture is significant, particularly in combination with likely future developments in biosynthesis and cloning of genes of the biosynthetic pathway. Finally, the discovery of taxol production in the fungus *Taxomyces* offers interesting possibilities if titers can be markedly enhanced, although given the relatively advanced state of progress in plant cell culture of *Taxus* species, *Taxomyces* may not have much impact on production.

PRODUCTION OF TAXOL - CHEMICAL STUDIES

Total Synthesis. The total synthesis of taxol has been a challenging goal to organic chemists since the publication of its structure in 1971 (*1*) and many investigations have been reviewed (*37-40*). The most dramatic development in synthesis in recent years has been the almost simultaneous completion of the total synthesis by both the Nicolaou and Holton groups (*41-43*). Both of these syntheses appear too long to have commercial value. The most attractive approach thus far reported appears to be Wender pinene pathway (*44*) but this approach has not yet been elaborated to the final product. With major advances in semisynthesis and greater availability of taxol via 10-DAB, total synthesis is less likely to be important. The development of the critical chemistry for the elaboration of the taxane skeleton and introduction of functional groups is nonetheless likely to have great value in subsequent synthesis of simpler analogues, and in making congeners which are not accessible from natural materials.

Semisynthesis. The production of taxol and related compounds through semisynthesis from 10-DAB is now well established, starting with early work in Europe to show the feasibility of side chain coupling to protected 10-DAB (*23*) and with subsequent work from many laboratories to improve the coupling reaction by use of condensed side chain equivalents including oxazolines or β-lactams to reduce steric bulk (*45-48*). The final major improvement in semisynthesis has been the specific activation of the C-13 hydroxyl by metals to create an anion which enhances reactivity and attack on the condensed side chain equivalent (*49*). This combination of new methodologies yields a highly efficient conversion of 10-DAB to taxol and little fundamental research remains to be done in the semisynthesis area; further utility of this process is dependent on improving 10-DAB yield through better biological sources as discussed above.

Assays. The ability to detect and quantitate taxol and other taxanes is the crux of a great deal of the work done in the last five years. Every biological approach to taxol research and production as discussed above including isolation, agronomics, tissue culture, strain selection, and ecological studies is dependent on sensitive assays with good throughput. Assays for taxol as a pure chemical or in pharmaceutical formulations were satisfactory, but a truly major problem in the 1980s was detection and quantitation of taxol and other taxanes in complex biological matrices such as crude extracts of plant samples or cell cultures, or in patient samples. The difficulty was three-fold: the lack of a distinctive chromophore (only phenyl groups which are present in many natural products); the low concentration of taxol in its plant sources; and the inability to achieve clean separation from the hundreds of other components present in plant extracts. Evaluation of taxol content of plant samples thus required several preliminary extraction, solvent partition and in some cases preliminary chromatography steps before the samples could be analyzed on an hplc column, severely limiting the number of assays that could be performed and also requiring large starting samples. Due to the complexity of the chemistry of taxol, it was not possible to cleanly enhance sensitivity by either pre-column or post-column derivitization. Since 1989 improved physico-chemical methods including tlc, hplc, new columns, more selective extractions, super-critical fluid chromatography, mass spectrometry, and ms-ms have been published (*16,*

17, 34, 50-59) but most of these remain marginally adequate for rapid evaluation of large numbers of samples, particularly of crude extracts or fractions which have low taxane content. Nonetheless these improved assays have had a lot of impact on assays of purer fractions, formulations, and final products, and some of the solvent partition and chromatography methods developed for analytical purposes have been applied to enhancement of bulk isolation procedures for taxol and 10-DAB resulting in higher yields of purer materials and sparing trees. One method that deserves particular mention is supercritical fluid extraction (SFCE) and chromatography (*52,60*); this is still at a pilot stage for taxol isolation but the use of recyclable gasses in place of organic solvents tremendously minimizes organic solvent waste and it is likely that SFCE will become the major industrial method for bulk processing of most types of natural products in future years.

The key that opened the door to large numbers of assays of crude natural products for content of various taxanes was immunoassays. The first paper in this area was from Jaziri and co-workers in 1991 (*61*) but the work that had the most impact was the development of assays by Raybold and Grothaus at Hawaii Biotechnology, Inc., supported under an NIH Small Business Innovation Research (SBIR) grant. These workers were able to make three important taxane antibodies, one with high specificity for taxol, one with high specificity for baccatin III, and one which recognizes most compounds having the taxane skeleton (*18*). These antibodies have been used to develop competitive inhibition enzyme immunoassays (CIEIAs) which have been made available in kits for easy use by many investigators. More recently, the Erlanger group has developed assays based on taxol antibodies which have been shown useful in analysis of plant extracts and plasma samples (*19,62*). At this point, fully adequate analytical methods are available for taxol and other taxanes, and development of new assays is not a high priority area for future research.

FUTURE GENERATION DRUGS - BIOLOGICAL STUDIES

Taxol Binding Site. It has been pointed out many times in this volume and elsewhere that taxol is unique in its mechanism as an antimitotic agent, in that it shifts the equilibrium between soluble tubulin dimers and polymerized microtubule structures through stabilization of microtubules and suppression of depolymerization. All other antimitotic drugs including vincristine, colchicine, podophyllotoxin, maytansine, and rhizoxin bind to various sites on tubulin and inhibit polymerization, essentially the opposite reaction. Given the efficacy of taxol in treatment of human cancer, the unique binding site for taxol on microtubules is an attractive target for drug discovery and improvements on the taxol motif. Current thinking in drug design is that the most promising leads will come from examination of ligand-receptor complexes in conjunction with studies on receptor and ligand alone so that changes in conformation on binding can be taken into account in drug design. Taxol is a black box in this regard. Details of the solution conformation of taxol are known as is the high resolution crystal structure of the closely related taxotere (*63-66*) but in 1989 virtually nothing was known about the taxol binding site on microtubules. Tubulin is a soluble dimeric protein with non-equivalent α- and β-subunits, each of molecular weight about 50 kd. To date no high resolution structure of tubulin has been established and clearly the high resolution structure of the microtubule polymer derived from tubulin is beyond current technology. There are a few assumptions that can be made about the taxol binding site based on available data; first, the site must involve a highly conserved sequence of tubulin since taxol is active in many species across many phyla. (*67*) but this is only modestly helpful since numerous sequences in both the α- and β- subunits are highly conserved; second, the site might be near the ends of one of the chains or overlapping the joining of α- and β- subunits on microtubules since the binding site is not present in soluble tubulin. The most available approach to trying to define the binding site in 1989 was through photolabelling experiments. The idea was to attach photolabile groups to taxol at positions not involved in the binding to microtubules, or at positions which were involved in binding but with labels that didn't cause much loss of tubulin binding; then to allow the modified taxol to bind to its microtubule receptor site and then to photolyze to give highly reactive functional groups that would form covalent bonds to tubulin. The tubulin could then be partially digested and the locations of covalently modified amino acids determined, since the sequences of human, bovine and porcine tubulins are well known. By introducing such labels at various positions around the taxol nucleus and on the side chain, it would be possible to develop a picture of which amino acids in the receptor bound to which parts of the taxol molecule and to get a useful model of the receptor. There are several reports of syntheses of photolabelled taxol derivatives (*68-71*).There are experimental difficulties with this approach, including obtaining good inding of the modified taxols to the receptor, getting a good yield of the photolysis roduct, selectively photolyzing the bound taxol derivative without affecting uninvolved ortions of tubulin, and being able to sequence the modified tubulin without loss or igration of labelled amino acids. Direct labelling experiments by the Horwitz group owed that when taxol was allowed to bind to microtubules and was then photolyzed, d the microtubules depolymerized, the covalently bound taxol was localized exclusively the beta chain (*72*). Subsequent studies using a *p*-azido label on the 3'-benzamido oiety in the taxol side chain showed localization of the label to the N-terminal 31 amino acid units of β-tubulin (*73*). Future work by this group will involve analysis of peptide fragments of the 31 N-terminal amino acids of β-tubulin to determine the smallest binding sequence. Another very recent study utilized an [(azidophenyl) ureido] taxoid derivative where the label was located on what would be the 3′ position on taxol. Photolysis of this analogue resulted in incorporation of radiolabel into both the β- and α-tubulin subunits in a ratio of 2.5:1 (*74*). These data indicate that the taxol binding site is at the interface of the α- and β-subunits and either subunit can be labelled depending on the precise position of the photolabel on the analogue, and the particular orientation at the moment of photolysis.

A second approach to the taxol binding site is to screen for molecules of diverse structure which have taxol-like properties in stabilization of microtubules with the idea of finding molecules that bind at the same site and which will then give key information about the binding site based on modelling the commonalities. Several pharmaceutical companies are currently screening for taxol-like activity searching broadly through libraries of natural products and synthetic compounds.

A more direct approach which is quite imaginative and elegant is the anti-idiotype antibody strategy used by the Erlanger group. The concept of defining a binding site in this way is that an antibody to taxol has a binding site complementary to taxol and expressed in a sequence of amino acids representing a portion of the F_{AB} (antigen-binding fragment) of the antibody. If one can then make an antibody to the portion of the F_{AB} representing the taxol binding site, this anti-idiotype antibody will represent a double inversion of the taxol structure, and thus a mimic of taxol, but now in a peptide sequence. Erlanger's laboratory has produced just such an anti-idiotype antibody named 82H which is able to stoichiometrically polymerize tubulin into microtubules, shows competitive kinetics with taxol binding on microtubules, and stabilizes microtubules to calcium chloride and to cold, hallmarks of taxol activity. Partial cleavage of the 82H antibody shows that the F_{AB} portion retains the activity of the full antibody and sequencing of the F_{AB} has revealed a peptide sequence with taxol activity (*75*). Several smaller peptides have been found which retain activity and further work is in progress to get an optimal small peptide sequence. These peptides can be modelled against taxol to look for similarities in shape, charge, functionality, etc., which should provide a significant contribution to our understanding of the binding site.

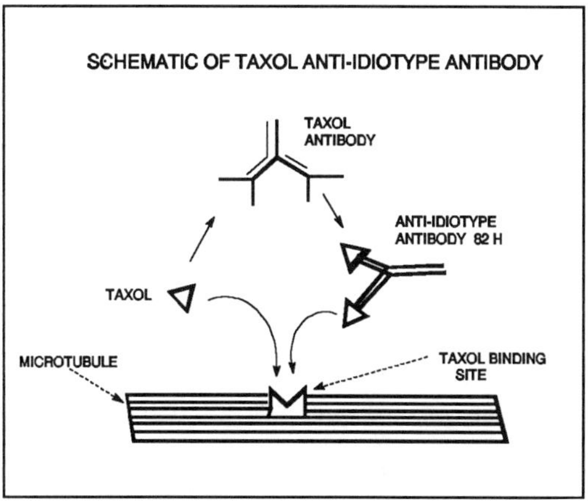

While the peptides discovered through the anti-idiotype approach are unlikely to become drugs themselves due to the typical problems of peptides in transport and in stability to peptidases and proteases, they can be used as starting materials for peptidomimetic agents. This is likely to lead to a whole new area of analogues which have the same biological function as taxol but completely unrelated chemical structures. The anti-idiotype work thus opens a major opportunity for new cancer therapeutics mechanistically related to taxol.

FUTURE GENERATION DRUGS - CHEMICAL STUDIES

Synthesis of Analogues. There are four types of analogues that may be considered: type 1) close chemical analogues prepared from naturally occurring taxanes; type 2) close chemical analogues with substituents not accessible from natural taxanes; type 3) markedly simplified analogues prepared by total synthesis and retaining only critical shapes, groups, and electronics; and type 4) biological analogues binding at the same site on microtubules but with fundamentally different chemical structures. Thus far the great majority of analogues prepared have been of type 1, starting from 10-DAB, baccatin III (*76-79*), 14-hydroxy taxoids (*80*), and A-nor-taxols (*81*). There are a few reports of type 2 analogues including work of Blechert and Nicolaou (*82,83*) but the biological data on these is inadequate to draw firm conclusions about the utility of these compounds. There is not enough data on SAR as yet to enable design of type

3 analogues and this may need another few years. Type 4 analogues are likely to arise from the work on anti-idiotype antibodies and work on defining the binding site which is giving rise to peptide sequences with taxol-like activity.

Structure Activity Relationships (SAR). The state of the art in 1989 was that the C-13 side chain of taxol was recognized as critical to activity but details of what changes could be made were incomplete. Extensive studies in several laboratories have almost totally clarified the substitution patterns and structural requirements of the side chain. (*47,79,84-86*). The effects of substituents at positions 2,4,5,7,9,10, and 14 have all been reported (*87-95*) and are discussed in other papers in this symposium as well as in earlier reviews (*3,86,96*) and the data is far too extensive to review in detail here. A broad summary of the SAR to date is shown in the figure, and can be stated as follows: 1) SIDE CHAIN: the presence of the C-13 side chain is an absolute requirement for activity and the length, hydroxyl substituent at C-2′ and the phenyl group at C-3′ are all needed for activity. Variation of substituents on the nitrogen at C-3′ is tolerated including aromatic or aliphatic groups as amides or carbamates. The natural taxol steroechemistry of 2′R, 3′S is markedly more active than the other possibilities. Minor changes in the side chain can have profound effects on activity and the side chain therefore is a key recognition element; 2) POSITIONS 7,8,9, AND 10: variations in these positions across the top of the molecule can be made without destroying activity and even fairly large substituents are tolerated. Modest enhancements of activity have been achieved by modification at these positions, and the 7- and 10- positions have been used as handles to attach polar groups in attempts to enhance solubility. These positions do not appear to be critical to receptor binding; 3) POSITIONS 2,4,5: removal of either the 2-benzoate or the 4-acetate causes dramatic loss of activity and these positions appear necessary

STRUCTURE - ACTIVITY RELATIONSHIPS OF TAXOL

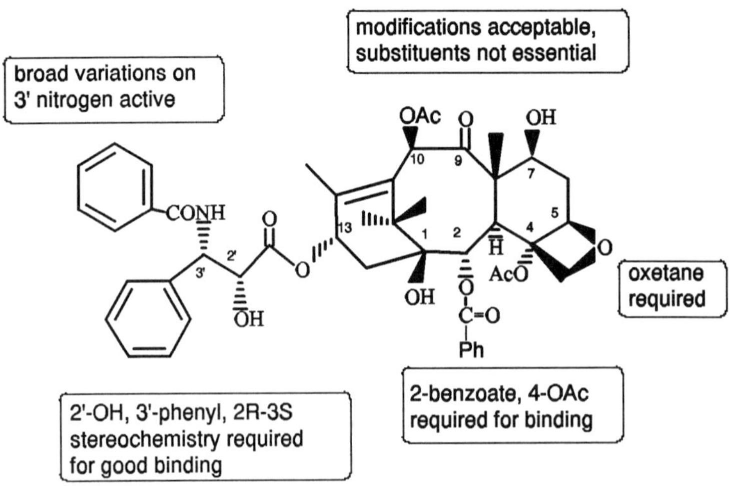

for interaction with the binding site; the presence of the 4,5-oxetane is necessary for activity but it is not clear whether it is involved in binding or serves as a "conformational lock" on the preferred binding conformation. Most of the many compounds prepared for analysis of structure activity relationships were deliberately modified at a single position to obtain baseline data and the next step will be to combine multiple favorable changes to create more advanced analogues.

While the above structural changes have been very important in SAR, modelling studies have an equally important role in our current understanding. The taxol molecule is shaped like a cup and in either organic or polar solutions of taxol, there is clustering of the side chain with the 2-benzoate and the 4-acetate (*47,63,65,79,86*). However, in polar solutions there is a further hydrophobic clustering of these substituents resulting in significant conformational change from organic solution to polar solutions (*64*) which is likely quite important to activity of taxol analogues. It will be fascinating to see these results compared with results of modelling of active peptides derived from the anti-idiotype antibody work.

REFERENCES

1. Wani, M. C.; Taylor, H. L.; Wall, M.E.; Coggon, P.; McPhail, A.T. *J. Am. Chem. Soc.* **1971**, 93, 2325-2326.
2. McGuire, W. P.; Rowinsky, E. K.; Rosenheim, N. B.; Grumbine, F. C.; Ettinger, D. S.; Armstrong, D. K.; Donehower, R. C. *Ann. Intern. Med.,* **1989**, 111, 273-279.
3. Suffness, M. *Ann. Rep. Med. Chem.*, **1993**, 28, 305-314.
4. Suffness, M.; Wall, M.E. In *Taxol: Science and Applications*; Suffness, M., Ed.; CRC Press: Boca Raton, Fl, 1995, In Press.
5. Rowinsky, E. K.; Cazenave, L. A.; Donehower, R. C. *J. Natl. Cancer Inst.*, **1990**, 82, 1247-1259.
6. Huang, C. H. O.; Kingston, D. G. I.; Magri, N. F.; Samaranayake, G.; Boettner, F.E. *J. Nat. Prod.,* **1986**, 49, 665-669.
7. Fleming, P. E.; Mocek, U.; Floss, H. G. *J. Am. Chem. Soc.*, **1993**, 115, 805-807.
8. Lansing, A.; Haertel, M.; Gordon, M.; Floss, H. G. *Planta Med., Suppl. 2*, **1991**, 57, A83.
9. Zamir, L. O.; Nedea, M. E.; Garneau, F. X. *Tetrahedron Lett.*, **1992**, 33, 5235-5236.
10. Strobel, G. A.; Stierle, A.; van Kujik, F.J.G.M. *Plant Sci.*, **1992**, 84, 65-74.
11. Christen, A. A.; Gibson, D. M.; Bland, J. *Proc. Amer Asso. Cancer Res.*, **1989**, 30, 566.
12. Christen, A. A.; Gibson, D. M.; Bland, J. **1991**, U.S. Patent 5,019,504.
13. Ma, W.; Park, G. L.; Gomez, G. A.; Nieder, M. H.; Adams, T. L.; Aynsley, J. S.; Sahai, O. P.; Smith, R. J.; Stahlhut, R. W.; Hylands, P. J. *J. Nat. Prod.*, **1994**, 57, 116-122.
14. Croom. E. M. Jr. In, *Taxol: Science and Applications*; Suffness, M., Ed.; CRC Press: Boca Raton, Fl, 1995, In Press.
15. Harvey, S. D.; Campbell, J. A.; Kelsey R. G.; Vance, N. C. *J. Chromatog.*, **1991**, 587, 300-305.
16. Vidensek, N.; Lim, P.; Campbell, A.; Carlson C. *J. Nat. Prod.,* **1990**, 53, 1609-1610.

17. Witherup, K. M.; Look, S. A.; Stasko, M. W.; Ghiorzi, T. J.; Muschik, G. M.; Cragg, G. M. *J. Nat. Prod.*, **1990**, 53, 1249-1255.
18. Grothaus, P. G.; Raybould, T. J. G.; Bignami, G. S.; Lazo, C. B.; Byrnes, J. B. *J. Immunol. Meth.* **1993**, 158, 5-15.
19. Leu, J.-G.; Jech, K. S.; Wheeler, N. W.; Chen, B.-X.; Erlanger, B. F. *Life Sci.* **1993**, 53, PL 183-187.
20. Kelsey, R. G.; Vance, N. *J. Nat. Prod.* **1992**, 55, 912-917.
21. Chauvière, G.; Guénard, D.; Picot, F.; Sénilh, V.; Potier, P. *C.R. Acad. Sci., Paris Ser. II*, **1981**, 293, 501-503.
22. Colin, M.; Guénard, D.; Guéritte-Voegelein, F.; Potier, P. **1989**, U.S. patent 4,857,653
23. Denis, J.-N.; Greene, A.; Guénard, D.; Guéritte-Voegelein, F.; Mangatal, L.; Potier, P. *J. Am. Chem. Soc.*, **1988**, 110, 5917-5919.
24. Stull, D. P.; Jans, N. A. *Abstracts: Second National Cancer Institute Workshop on Taxol and Taxus,* **1992**, Alexandria, VA.
25. Mattina, M. J. I.; Paiva, A. A. *J Environ. Hort.*, **1992**, 10, 187-191.
26. ElSohly, H. N.; El-Kashoury, E. A.; Croom, E. M. Jr.; ElSohly, M. A.; McChesney, J. D. *Abstracts: Second National Cancer Institute Workshop on Taxol and Taxus*, **1992**, Alexandria, VA, Poster D-10.
27. Wheeler, N. C.; Jech, K.; Masters, S.; Brobst, S. W.; Alvarado, A. B.; Hoover, A. J.; Snader, K. M. *J. Nat. Prod.,* **1992**, 55, 432-440.
28. Suffness, M. *Abstracts: Pacific Yew: A Resource for Cancer Treatment*, **1992**, Corvallis, OR.
29. Hansen, R. C.; Holmes, R. G.; Shugert, R. B. Jr.; Croom, E. M. Jr.; El Sohly, H.; Keener, H. M.; Cochran, K. D. *Abstracts: Second National Cancer Institute Workshop on Taxol and Taxus*, **1992**, Alexandria, VA.
30. Croom, E. M. Jr. *Abstracts: Second National Cancer Institute Workshop on Taxol and Taxus*, **1992**, Alexandria, VA, Poster D-9.
31. Fett-Neto, A. G.; Melanson, S. J.; Sakata, K.; DiCosmo, F. *Bio/Technology*, **1993**, 11, 731-734.
32. Smith, R. J.; Adams, T. L.; Barton, C. R.; Stahlhut, R. W. **1991**, U.S. Patent Appl. 794,711.
33. Bringi, V.; Kadkade, P. G. **1993**, International Patent Application WO 93/17121.
34. Bitsch, F.; Ma, W.; Macdonald, F.; Nieder, M.; Shackelton, C. H. L. *J. Chromatog.*, **1993**, 615, 273-280.
35. Stierle, A.; Strobel, G.; Stierle, D. *Science*, **1993**, 260, 214-216.
36. Strobel, G.; Stierle, A.; Stierle, D.; Hess W. M. *Mycotaxon*, **1993**, 47, 71-80.
37. Blechert, S.; Guénard, D. *The Alkaloids*, **1990**, 39, 195-238.
38. Swindell, C. S. *Org. Prep. Proc. Intl.*, **1991**, 23, 465-543.
39. Paquette, L. A. In: *Studies in Natural Products Chemistry Vol. 11*, Atta-ur-Rahman, Ed.; Elsevier, Holland, 1992, pp. 3-69.
40. Wender, P. A.; Natchus, M. G.; Shuker, A.J. In: *Taxol: Science and Applications*; Suffness, M., Ed.; CRC Press: Boca Raton, Fl, 1995, In Press.
41. Nicolaou, K. C.; Yang, Z.; Liu, J. J.; Ueno, H.; Nantermet, P. G.; Guy, R. K.; Claiborne, C. F.; Renaud, J.; Couladouros, E. A.; Paulvannin, K.; Sorensen, E. J. *Nature (London)*, **1994**, 367, 630-634.

42. Holton, R. A.; Somoza, C.; Kim, H. B.; Liang, F.; Biediger, R. J.; Boatman, P. D.; Shindo, M.; Smith, C. C.; Kim. S.; Nadizadeh, H.; Suzuki, Y.; Tao, C.; Vu, P.; Tang, S.; Zhang, P.; Murthi K. K.; Gentile, L. N.; Liu, J. H. *J. Am. Chem. Soc.*, **1994**, 116, 1597-1598.
43. Holton, R. A.; Kim, H. B.; Somoza, C.; Liang, F.; Biediger, R. J.; Boatman, P. D.; Shindo, M.; Smith, C. C.; Kim. S.; Nadizadeh, H.; Suzuki, Y.; Tao, C.; Vu, P.; Tang, S.; Zhang, P.; Murthi K. K.; Gentile, L. N.; Liu, J. H. *J. Am. Chem. Soc.*, **1994**, 116, 1599-1600.
44. Wender, P. A.; Mucciaro, T. P. *J. Am. Chem. Soc.*, **1992**, 114, 5878-5879.
45. Ojima, I.; Habus, I.; Zhou, M.; Zucco, M.; Park, Y. H.; Sun, C. M.; Brigaud, T. *Tetrahedron*, **1992**, 48, 6985-7012.
46. Georg, G. I.; Cheruvallath, Z. S.; Harriman, G. C. B.; Hepperle, M.; Park, H. *Bioorg. Med. Chem. Lett.*, **1993**, 3, 2467-2470.
47. Swindell, C.; Krauss, N. E.; Horwitz, S. B.; Ringel, I. *J. Med. Chem.*, **1991**, 34, 1176-1184.
48. Holton, R. A. **1991**, U.S. Patent 5,015,744.
49. Holton, R. A. **1991**, International Patent Appl. WO 93/06079.
50. Auriola, S. O. K.; Lepistö, A.-M.; Naaranlahti, T.; Lapinjoki, S. P. *J. Chromatog.*, **1992**, 594, 153-158.
51. Witherup, K. M.; Look, S. A.; Stasko, M. W.; McCloud, T. G.; Issaq, H.J.; Muschik, G. M. *J. Liq. Chromatog.*, **1989**, 12, 2117-2132.
52. Heaton, D. M.; Bartle, K. D.; Rayner, C. M.; Clifford, A. A. *J. High Res. Chromatog.*, **1993**, 16, 666-670.
53. Blay, P. K. S.; Thibault. P.; Thiberge, N.; Kiecken, B.; Lebrun, A.; Mercure, C. *Rapid Commun. Mass Spec.*, **1993**, 7, 626-634.
54. Hoke, S.H. II.; Cooks, R. G.; Chang, C.-J.; Kelly, R. C.; Qualls, S. J.; Alvarado, B.; McGuire, M. T.; Snader, K. M. *J. Nat. Prod.*, **1994**, 57, 277-286.
55. Ketchum, R. E. B.; Gibson, D. M. *J. Liq. Chromatog.*, **1993**, 16, 2519-2530.
56. Vanhaelen-Fastre, R.; Diallo, B.; Jaziri, M.; Faes,M.-L.; Homes, J.; Vanhaelen, M. *J. Liq. Chromatog.*, **1992**, 15, 697-706.
57. Willey, T. A.; Bekos, E. J.; Gaver, R. C.; Duncan, G. F.; Tay, L. K.; Beijnen, J. H.; Farmen, R. H. *J. Chromatog.*, **1993**, 621, 231-238.
58. Castor, T. P.; Tyler, T. A. *J. Liq. Chromatog.*, **1992**, 16, 723-731.
59. Harvey, S. D.; Campbell, J. A.; Kelsey, R. G.; Vance, N. C. *J. Chromatog.*, **1991**, 587, 300-305.
60. Jennings, D. W.; Deutsch, H. M.; Zalkow, L. H.; Teja, A. S. *J. Supercrit. Fluids*, **1992**, 5, 1-6.
61. Jaziri, M.; Diallo, B. M.; Vanhaelen, M. H.; Vanhaelen-Fastre, R. J.; Zhiri, A.; Becu, A. G.; Homes, J. *J. Pharm. Belg.*, **1991**, 46, 93-99.
62. Leu, J.-G.; Chen, B.-X.; Schiff, P. B.; Erlanger, B. F. *Cancer Res.*, **1993**, 53, 1388-1391.
63. Williams, H. J.; Scott, A. I.; Dieden, R. A.; Swindell, C. S.; Chirlian, L. E.; Francl, M. M.; Heerding, J. M.; Krauss, N. E. *Tetrahedron*, **1993**, 49, 6545-6560.
64. Vander Velde, D. G.; Georg, G. I.; Grunewald, G. L.; Gunn, C. W.; Mitscher, L. A. *J. Am. Chem. Soc.*, **1993**, 115, 11650-11651.
65. Dubois, J.; Guénard, D.; Guéritte-Voegelein, F.; Guedira, N.; Potier, P.; Gillet, B.; Beloeil, J.-C. *Tetrahedron*, **1993**, 49, 6533-6544.

66. Guéritte-Voegelein, F.; Mangatal, L.; Guénard, D.; Potier, P.; Guilhem, J.; Cesario, M.; Pascard, C. *Acta Cryst.*, **1990**, C46, 781-784.
67. Suffness, M. *In Vivo*, **1994**, 8, In Press.
68. Chatterjee, A.; Williamson, J. S.; Zjawiony, J. K.; Peterson, J. R. *Bioorg. Med. Chem. Lett.*, **1992**, 2, 91-94.
69. Georg, G. I.; Harriman, G. C. B.; Himes, R. H.; Mejillano, M. R.; *Bioorg. Med. Chem. Lett.*, **1992**, 2, 735-738.
70. Rimoldi, J. M.; Kingston, D. G. I.; Chaudhary, A. G.; Samaranayake, G.; Grover, S.; Hamel, E. *J. Nat. Prod.*, **1993**, 56, 1313-1330.
71. Carboni, J. M.; Farina, V.; Rao, S.; Hauck, S. I.; Horwitz, S. B.; Ringel, I. *J. Med. Chem.*, **1993**, 36, 513-515.
72. Rao, S.; Horwitz, S. B.; Ringel, I. *J. Natl. Cancer Inst.*, **1992**, 84, 785-788.
73. Rao, S. R.; Krauss, N. E.; Heerding, J. M.; Swindell, C. S.; Ringel, I.; Orr, G. A.; Horwitz, S. B. *J. Biol. Chem.*, **1994**, 269, 3132-3134.
74. Combeau, C.; Commerçon, A.; Mioskowski, C.; Rousseau, B; Aubert, F.;Goeldner, M. *Biochemistry*, **1994**, 33, 6676-6683.
75. Leu, J.-G.; Chen, B.-X.; Diamanduros, A. W.; Erlanger, B. F. *Proc. Natl. Acad. Sci. USA*, **1994**, In Press.
76. Parness, J.; Kingston, D.G.I.; Powell, R.G.; Harracksingh, C.; Horwitz, S. B. *Biochem. Biophys. Res. Commun.*, **1982**, 105, 1082-1089.
77. Magri, N. F.; Kingston, D. G. I. *J. Nat. Prod.*, **1988**, 51, 298-306.
78. Lataste, H.; Senilh, V.; Wright, M.; Guénard D.; Potier P. *Proc. Natl. Acad. Sci. USA*, **1984**, 81, 4090-4094.
79. Guéritte-Voegelein, F.; Guénard, D.; Lavelle, F.; Le Goff, M.-T.; Mangatal, L.; Potier, P. *J. Med. Chem.*, **1991**, 34, 992-998.
80. Kant, J.; Farina, V.; Fairchild, C.; Kadow, J. F.; Langley, D. R.; Long, B. H.; Rose, W. C.; Vyas, D. M. *Bioorg. Med. Chem. Lett.*, **1994**, 4, 1565-1570.
81. Georg, G. I.; Cheruvallath, Z. S.; Vander Velde, D.; Ye, Q. M.; Mitscher, L. A.; Himes, R. H. *Bioorg. Med. Chem. Lett.*, **1993**, 3, 1349-1350.
82. Blechert, S.; Kleine-Klausing, A. *Angew. Chem. Int. Ed. Engl.*, **1991**, 30, 412-414.
83. Nicolaou, K. C.; Claiborne, C. F.; Nantermet, P. G.; Couladouros, E. A.; Sorensen, E. J. *J. Am. Chem. Soc.*, **1994**, 116, 1591-1592.
84. Sénilh, V.; Gueritte, F.; Guénard, D.; Colin, M.; Potier, P. *C.R. Acad. Sci. Paris*, **1984**, 299, 1039-1043.
85. Georg, G. I.; Cheruvallath, Z. S.; Himes, R. H.; Mejillano, M. R.; Burke, C. T. *J. Med. Chem.*, **1992**, 35, 4230-4237.
86. Guénard, D.; Guéritte-Voegelein, F.; Potier, P. *Acc. Chem. Res.*, **1993**, 26, 160-167.
87. Chen, S.-H.; Wei, J.; Farina, V. *Tetrahedron Lett.*, **1993**, 34, 3205-3206.
88. Chaudhary, A. G.; Gharpure, M. M.; Rimoldi, J. M.; Chordia, M. D.; Kingston, D. G. I.; Grover, S.; Lin, C. M.; Hamel, E.; Gunitilaka, A. A. L. *J. Am. Chem. Soc.*, **1994**, 116, 4097-4098.
89. Chen, S.-H.; Farina, V.; Wei, J.; Long, B. H.; Fairchild, C.; Mamber, S. W.; Kadow, J. F.; Vyas, D. M.; Doyle, T. W. *Bioorg. Med. Chem. Lett.*, **1994**, 4, 479-482.
90. Chen, S.-H.; Huang, S.; Wei, J.; Farina, V. *Tetrahedron*, **1993**, 2805-2828.
91. Chen, S.-H.; Huang, S.; Kant, J.; Fairchild, C.; Wei, J.; Farina, V. *J. Org. Chem.*, **1993**, 58, 5028-5029.

92. Chaudhary, A. G.; Rimoldi, J. M.; Kingston, D. G. I. *J. Org. Chem.*, **1993**, 58, 3798-3799.
93. Chen, S.-H.; Fairchild, C.; Mamber, S. W.; Farina, V. *J. Org. Chem.*, **1993**, 58, 2927-2928.
94. Chaudhary, A. G.; Kingston, D. G. I. *Tetrahedron Lett.*, **1993**, 34, 4921-4924.
95. Klein, L. L. *Tetrahedron Lett.*, **1993**, 34, 2047-2050.
96. Hepperle, M.; Georg, G. I. *Drugs Future*, **1994**, 19, 573-584.

RECEIVED September 16, 1994

Chapter 2

Paclitaxel: From Discovery to Clinic

Monroe E. Wall and Mansukh C. Wani

Research Triangle Institute, P.O. Box 12194, Research Triangle Park, NC 27709–2194

Taxol, a potent antitumor agent, was isolated from a tree, *Taxus brevifolia*, by procedures guided by bioactivity. Final isolation in pure form was accomplished using repeated Craig Countercurrent Distribution. Taxol is a diterpene ester with unique structural features and many asymmetric centers. The compound displayed considerable cytotoxic and antitumor activity, particularly toward B-16 melanoma. Taxol binds strongly to tubulin by a unique mechanism. Early clinical trials with patients with ovarian cancer showed remarkable efficacy resulting in complete and partial remissions. Taxol has received much chemical study, particularly in regard to structure/activity relationship (SAR), semi- and total synthesis. It is currently regarded as one of the best new anticancer agents.

Natural products chemists and phytochemists have always been impressed by the fact that compounds found in nature display an almost unbelievable range of diversity in terms of their structures and physical and biological properties. Most of these compounds are secondary metabolites whose functions in plants, fungi, and marine organisms are still not widely understood. Currently, it is believed that many of these compounds act in defense against the harmful effects of toxins, carcinogens, or mutagens found in the plant (*1,2*) or attack by external predators (*3*).

Some secondary metabolites may have excellent therapeutic potential. Examples are camptothecin (*4*) and taxol (*5*), found in low concentrations, respectively, in the bark of *Camptotheca acuminata* and *Taxus brevifolia*. This chapter will describe in detail the events and research leading to the discovery of taxol (cf. Figure 1). A brief summary will be given of subsequent developments culminating in the clinical trial and the general availability of the drug for treatment of patients with ovarian, breast, and other solid tumors.

NOTE: Paclitaxel is the generic name for Taxol, which is now a registered trademark.

0097–6156/95/0583–0018$08.00/0

Discovery of taxol

Since 1988, the remarkable clinical efficacy of taxol (Figure 1), resulting in numerous observations of partial and complete remission of advanced ovarian cancer in women, and, more recently, reports of the drug's efficacy in breast, lung, and prostate cancer, have aroused great interest in this antitumor compound which was discovered at the Research Triangle Institute (RTI) many years ago (*5,6*). Like so many other investigations of this type, a combination of serendipity followed by much hard work led to the discovery of this very active antitumor agent.

Initial Procurement

A screening program for antitumor agents in the plant kingdom was initiated in 1960 under Dr. Jonathan L. Hartwell. In this program, plant samples collected at random were supplied by the U.S. Department of Agriculture under an interagency agreement with The National Cancer Institute (NCI). In August 1962, USDA botanist, Arthur S. Barclay, and three college student field assistants collected 650 plant samples in California, Washington, and Oregon, including bark, twigs, leaves and fruit of *Taxus brevifolia* in Washington state (*7*).

T. brevifolia is a slow growing tree which is found primarily localized in the coastal areas of the above-mentioned West Coast states. It had never received any chemical investigation until it was assigned to RTI by Dr. Hartwell. The assignment of the plant was not entirely serendipitous. Some of these samples had been shown to have cytotoxicity against 9KB cell culture derived from a human cancer of the nasopharynx. We had noted an excellent correlation in our camptothecin studies between L1210 (lymphoid leukemia in mice) in vivo activity and the 9KB cytotoxicity. Accordingly, we had requested Dr. Hartwell to assign to us as many 9KB actives as possible. From this arose the assignment to RTI of *T. brevifolia.* A number of other plants also highly active in 9KB were also assigned to our group, and several highly active novel compounds were found in these cases also, including colubrinol, a maytansine analog (*8*), carminomycin, related to daunomycin (*9*), and an active quassinoid (*10*).

Extraction and Isolation

Initial samples of *T. brevifolia* arrived at RTI by 1964. By 1966, at least a year prior to our isolating the pure material, in a letter dated April 15, 1966 to Jonathan Hartwell, we requested that the extracts we had sent "receive a special priority with the biological screeners" as we regarded it as one of the most important samples we had seen in a long time. By May 1966, in RTI Progress Report #18 (The Screening of Fractionated Plants for Antitumor Inhibitory Substances), we stated "At present, a major effort by our group is being placed on this plant (Taxus brevifolia)." By November 1966 (RTI Progress Report #20) we were able to report the isolation of a purified fraction and presented some physical constants. The actual isolation was completed by June 1967. The method finally adopted after several unsuccessful trials

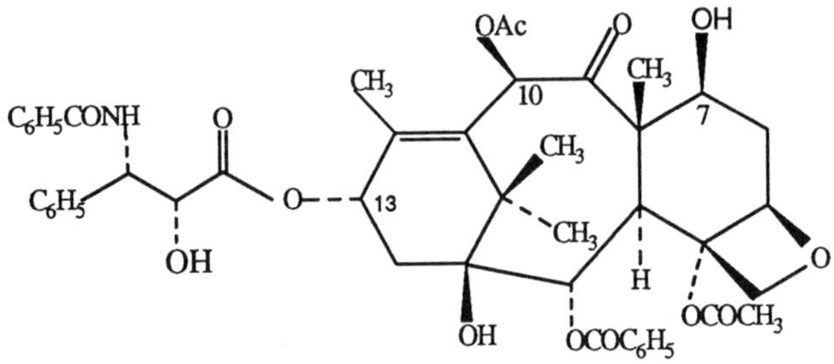

Figure 1. Structure of Taxol

Stem Bark - 12 kilos, air dried

Extract 95% Ethanol

Concentrate

Partition between H2O and chloroform-methanol (4:1)

Aqueous layer (discard)

Chloroform (146 g solids)
Activity 5WM,*
T/C** = 31% at 100 mg/kg

*5WM is a solid tumor known as Walker-256 intramuscular rat carcinosarcoma.

**T/C = mean tumor weight of treated animals ÷ mean tumor weight of control animals x 100

Chart 1. Fractionation of *Taxus brevifolia*

is shown in Charts 1 and 2. Extraction was carried out by our standard procedure—ethanol extraction and partition of the ethanolic residue between water and chloroform. Purification and isolation utilized a large number of Craig countercurrent distribution treatments, the last of which involved a 400-tube Craig countercurrent distribution. In this manner, approximately 0.5 g of taxol was isolated starting with 12 kg of air dried stem and bark from *T. brevifolia*. The yield was about 0.004%. All the various steps were monitored by an in vivo bioassay which, at that time, involved the inhibition of the solid tumor known as Walker-256 intramuscular rat carcinosarcoma (5WM). As is shown in Chart 2, increased purification was accompanied by increased antitumor activity at lower doses. The isolation steps were laborious, but because of the mild countercurrent distribution methodology, losses or alterations of the active constituent were avoided. Much simpler procedures have been subsequently developed both at RTI and elsewhere. Thus by 1966 we were able to compare the activity of crude chloroform extracts from various samples of *T. brevifolia* collected in Alaska, California, Washington, Idaho, Oregon, and Montana and present the data on their cytotoxicity and 5WM inhibition. In 1967 we presented the first report on taxol to the American Chemical Society (*6*).

The Structural Determination of Taxol

As soon as we had isolated taxol in pure form, the structure of the compound was investigated using available spectroscopic methods. Although methods for ultraviolet, infrared, and mass spectrometry were at a reasonably advanced stage in the late 1960s, NMR was relatively primitive compared to the sophisticated instrumentation and procedures now available. Some of the physical and chemical properties of taxol are shown in Table I.

Table I. Physical and Chemical Properties of Taxol

1.	Needles from 50% aqueous methanol or ether.
2.	M.P. 213-216°C (with previous shrinking around 204°C)
3.	$[\alpha]_D^{20}$ -49.6° (MeOH)
4.	Unstable towards mineral acid and base
5.	Forms mono and diacetate
6.	Analysis Calcd. for $C_{47}H_{51}NO_{14}$: C, 66.11; H, 6.20, N, 1.64. Found: C, 65.98; H, 6.10; N, 1.57. Required m/z 853. Found m/z 853
7.	UVλ_{max}^{MeOH} 227 nm (ε 29,800)

It was evident by this time that taxol probably contained the taxane skeleton. A number of taxane derivatives had been reported in previous literature (*11-13*). It was evident that taxol was more complex than previously reported taxanes since its molecular weight from high resolution mass spectrometry was $C_{47}H_{51}NO_{14}$, cor-

Chloroform (146 g solids)

|

11 Tube Craig CCD
System: hexane, acetone, t-butanol, water (5:4:4:2)

|

Tubes 6, 7, 8
41 g solids, activity 5WM*
T/C* 30% at 45 mg/kg

|

30 Tube Craig CCD
System: methanol, water, CCl_4, $CHCl_3$ (8:2:7:3)

|

Tubes 8-18
14 g solids
T/C 30% at 23 mg/kg

|

400 Tube Craig CCD
System: methanol, water, CCl_4, $CHCl_3$ (8:2:7:3)

|

Tubes 170-189
2.4 g solids, T/C 16% at 15 mg/kg

|

Trituration with benzene
0.5 g Taxol, T/C 16% at 10 mg/kg
24% at 5 mg/kg
Yield circa 0.004%

*T/C in 5WM. For definitions of 5WM and T/C, see footnotes of Chart 1.

Chart 2. Purification of Crude Extract of *T. Brevifolia* by Craig Countercurrent Distribution (CCD) and Isolation of Taxol

$RCOOR^1$		RCOOMe		$R^1(OH)x$		CH_3COOCH_3
Taxol	MeOH, NaOMe, 0°C →	Methyl Ester	+	Hydroxy Compound	+	Methyl Acetate
$C_{47}H_{51}NO_{14}$		$C_{17}H_{17}NO_4$		$C_{29}H_{36}O_{10}$		
		Major Components				Not Isolated

Figure 2. Methanolysis of Taxol

responding to a molecular weight of 853. The evidence then indicated that taxol was comprised of a taxane nucleus to which an ester was attached, as preliminary experiments indicated that an ester moiety was easily cleaved from the rest of the molecule. Attempts were made to prepare crystalline halogenated derivatives of taxol. However, none had properties suitable for x-ray analysis. Taxol was therefore subjected to a mild base catalyzed methanolysis at 0°C, which yielded a nitrogen containing α-hydroxy methyl ester, $C_{17}H_{17}NO_4$, a tetraol, $C_{29}H_{36}O_{10}$, and methyl acetate (Figure 2). Full details are presented in our paper dealing with the structure of taxol (*5*). The methyl ester thus obtained by the mild methanolysis procedure was converted to a parabromobenzoate ester and characterized by the x-ray analysis as $C_{24}H_{20}BrNO_5$ with the structure shown in Figure 3. The ester may be regarded as an N-benzoyl derivative of (2R,3S)-3-phenylisoserine. The tetraol formed by the methanolysis of taxol was converted to a bisiodoacetate, $C_{33}H_{38}O_{12}I_2$, which again received x-ray analysis (*5*). The structure is shown in Figure 4.

Since the ester could have originally been joined to hydroxyl groups at either C_7, C_{10}, or C_{13} (Figure 4), it was necessary to establish at which of these hydroxyl moieties the ester had originally been located. When taxol was oxidized with MnO_2 under neutral conditions, no reaction occurred. However, MnO_2 oxidation of taxol under alkaline conditions smoothly yielded a reaction product with the structure shown in Figure 5. It is evident that MnO_2 oxidation of taxol under neutral conditions did not effect the hydroxyl groups available for oxidation at C_7 and $C_{2'}$. Under alkaline conditions, the C_{13} ester and C_{10} acetate of taxol can be hydrolyzed yielding a tetraol with structure I, Figure 4. When taxol was oxidized with alkaline MnO_2, an analog of Baccatin III with a conjugated carbonyl moiety as shown in Figure 5 was obtained (*5*). It is evident that under the alkaline conditions that both the 10-acetyl moiety and the ester were hydrolyzed yielding the tetraol (Figure 4, I). It is well known that MnO_2 oxidation of allylic hydroxyl groups under alkaline conditions smoothly forms the corresponding conjugated ketone. This reaction in conjunction with the x-ray structure determination of the structures of the ester and taxane moieties established the structure of taxol.

Prior to the isolation of taxol, no natural taxane derivative was reported to have antitumor activity. There are many interesting features of the molecule, particularly the 4-membered oxide ring at C_4 and C_5, which is not found in any of the other natural taxanes. The ester moiety itself is of interest, containing two phenyl groups, one of which is attached as part of an amide function.

For antitumor activity, it is essential that the ester moiety at C_{13} be present. We have shown that the ester and the tetraol formed by low temperature cleavage of taxol are each essentially inactive (*5*).

Biological Activity of Crude And Purified Taxol

We assigned the name "taxol" to this compound before we really knew its complete structure, but it was evident that it did contain some hydroxyl groups, and the name had a nice ring to it. Apart from the actual isolation of the pure material, the crude extracts were subjected to a number of assays in rodent leukemias and solid tumors.

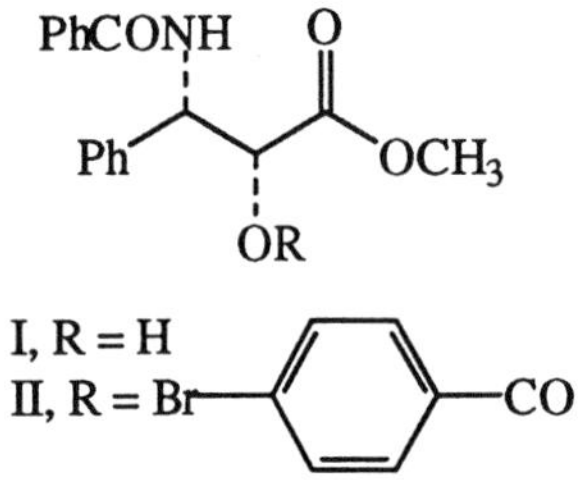

Figure 3. Structures of RCOOMe (I) and Its P-Bromobenzoate (II)

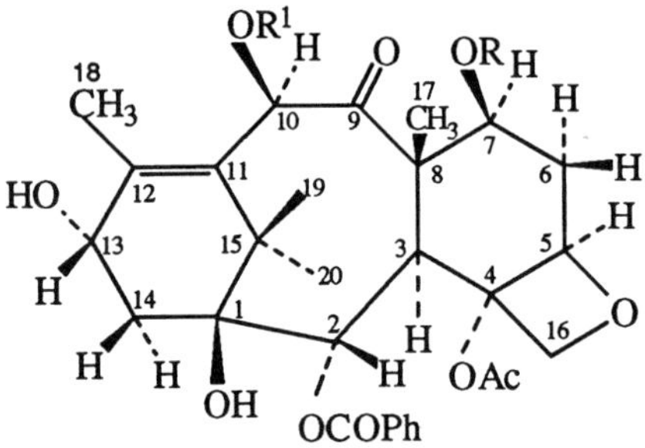

I, R = R^1 = H

II, R = R^1 = ICH_2CO

Figure 4. Structures of Tetraol (I) and Tetraol Bis-Iodoacetate (II)

In early work we found that the crude extracts were active not only in the Walker tumor inhibition assay, but also had shown modest activity in L1210 leukemia and particularly high activity in the 1534 (P4) leukemia assay. The latter assay was a life prolongation assay in mice, and it had been used previously by scientists at Eli Lilly during the isolation of the vinca alkaloids which showed high activity in life prolongation in this system. The same was noted by us for taxol with T/C values in the P4 system in excess of 300, even with crude extracts. The activity of pure taxol in a number of in vivo rodent assays is shown in Table II. Particularly high activity was shown in the B-16 melanoma assay. Years later it was one criteria by which taxol was moved to clinical trial (*14*).

Table II. Cytotoxic and Antitumor Activity of Taxol

Cytotoxic Activity
KB (human carcinoma of the nasopharynx):
$ED_{50} = 3.5 \times 10^{-5}$ µg/mL
(ED_{50} = conc. required for 50% inhibition of growth)

Antitumor Activity

System Tested	Administration	Activity (% T/C)
i.p. P388 Leukemia	i.p.	+ (164)
i.p. B16 Melanoma	i.p.	++ (283)
i.p. L1210 Leukemia	i.p.	+ (139)
S.R.C.* CX-1 Colon Xenograft	s.c.	++(3)
S.R.C. LX-1 Lung Xenograft	s.c.	+ (8)
S.R.C. MX-1 Mammary Xenograft	s.c.	++(-77)

*Sub-renal capsule

Subsequent Developments 1971-1974

During the period 1971-1974, we made repeated efforts to interest the NCI administrators, who were at that time involved in the procurement of new antitumor agents, in obtaining larger quantities of taxol from *T. brevifolia.* The response was that since the compound was present in the bark in low concentration, extraction and isolation on a large scale would be difficult and expensive. Moreover, the activity of taxol against L1210 leukemia or other rodent leukemia and solid tumors was modest. Hence for some years taxol remained literally "on the shelf" in the NCI repository.

Discovery of Activity in B-16 Melanoma

It became apparent in the early 1970s that the most intensively studied models, such as L1210 leukemia, while valuable in discovering clinical agents for human leukemia and lymphomas, were less useful for prediction of activity against solid tumors. By

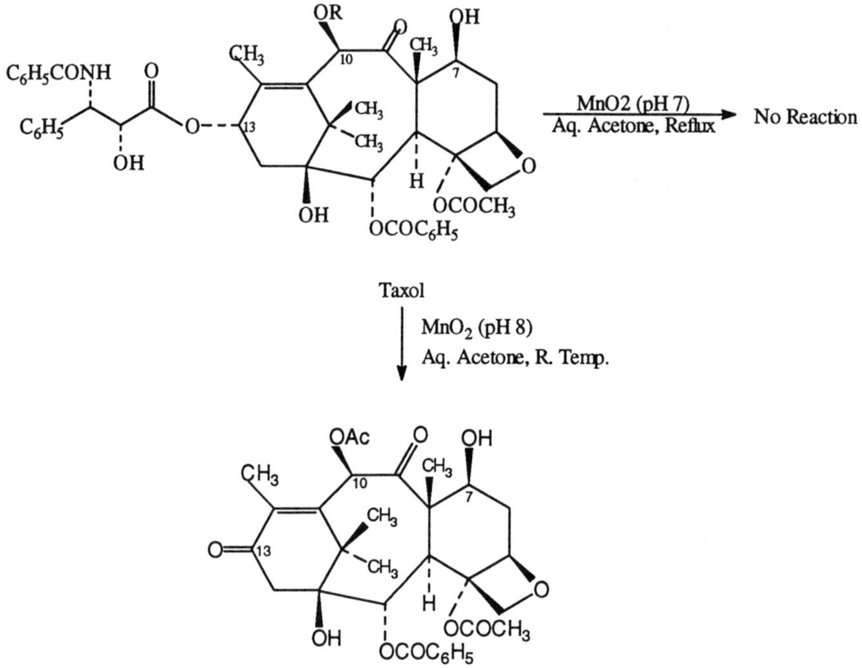

λ_{max}^{MeOH} 272 nm (ε 4,800)

$\gamma_{max}^{CHCl_3}$ 1680 cm^{-1}

^{1}H-NMR d 6.46 (s, ^{1}H, H-10)

Figure 5. Selective Hydrolysis and Oxidation of Taxol
(Reproduced with permission from reference 5. Copyright 1971.)

1974, taxol was found to have good activity against this animal model (Table II). By 1977, because of its activity against B16, taxol was finally selected as a development candidate. Thus taxol was finally removed from the NCI storage shelves. A recent review presents a detailed account of this development (*14*).

Tubulin Binding

For some time there had been some interest in the mechanism of action of taxol. Was the unique structure accompanied by an unusual mechanism of action? The initial studies by Fuchs and Johnson (*15*) indicated that taxol inhibited proliferation at the G2-M phase in the cell cycle and blocked mitosis. Thus it appeared to be one in a series of naturally occurring spindle poisons such as vincristine and vinblastine, colchicine, podophyllotoxin, maytansine, and others. The finding that taxol was a spindle poison was not necessarily encouraging. Shortly thereafter a more detailed investigation by the Horwitz group (*16*) established that, while taxol was a mitotic inhibitor, the mechanism was unique in that it stabilized microtubules and inhibited depolymerization back to tubulin; this was the opposite effect of the other antimitotic agents cited above, which all bind to soluble tubulin and inhibit the polymerization of tubulin to form microtubules (*17-19*). This information was important in making the argument that by virtue of uniqueness of both structure and mechanism, taxol was a worthy candidate for development.

Our early observations that taxol extracts and pure taxol were highly active in leukemia P1534 (P-4) assay now became explicable. Tubulin is a protein involved in the process of mitosis. The vinca alkaloids were noteworthy for their activity against P-4 leukemia. Taxol is also highly active against P-4. Both compounds bind to tubulin but, as described above, by completely different mechanisms.

Era of Rapid Progress 1982-1994

Rapid progress both in chemical synthesis of taxol and in clinical developments occurred in the decade 1982-1994 (Table III). The events in this decade have been covered in detail by a number of reviews (*14, 19, 20*). Hence this section will present only a brief outline of the tremendous progress and extensive research conducted during this period.

Animal toxicology and formulation were in place by 1982-1984 (*14*). Clinical Phase I and Phase II trials were conducted over the period 1983 to 1986. Great interest was generated in taxol, both in scientific circles and by the general public, by the announcement of the remarkable efficacy of taxol against ovarian cancer (*21,22*). Subsequently, taxol and an analog, Taxotere, have been studied for application in many solid tumors (breast, lung) (*14, 19, 20*).

In an effort to obtain adequate supplies of taxol, the National Cancer Institute issued a Cooperative Research and Development Award (CRADA) which was open to competition. The award was then issued to Bristol-Myers Squibb in 1991. The company moved rapidly to obtain FDA approval for the marketing. An

NDA was filed in July 1992 and approved in December 1992. The initial shortage of taxol for general clinical use has been at this time (May 1994) greatly alleviated.

Table III. Chronology of Taxol Development

• Isolation and Structure/P388 and L1210 Activity	1971
• Activity in A Panel of Tumor Systems	1975-1976
• Preclinical Development	1977
• Mechanism of Action	1979
• Animal Toxicology	1982
• Phase I Clinical Trials	1983-1984
• Phase II Clinical Trials/Activity in Ovarian Cancer	1985-1986
• Synthesis of Taxol Side Chain	1986
• Semisynthesis of Taxol	1988
• Improved Syntheses of Taxol Side Chain	1990-1993
• Bristol-Myers Squibb Receives CRADA from NCI	January 1991
• NDA Filed with FDA	December 1992
• Total Synthesis of Taxol/Holton-Nicolaou	1994

Chemical Developments

Structure Activity. There has been great interest in carrying out various chemical transformations, both with taxol and taxotere, an analog. Such studies have been reviewed in detail by Kingston (*23*), in whose laboratory extensive SAR studies on taxol have been conducted. Recently, Kingston and colleagues have reported the discovery of a very active analog with considerably greater potency than taxol. This analog is formed by removing the 2-benzoate group of taxol and subsequently reacylating with certain meta-substituted benzoic acids (*24*).

Synthesis of the Taxol Sidechain. During the period 1990-1994, extensive progress has been made on the synthesis of the ester moiety of taxol (and these researchers have been reviewed in detail) (*19, 20*). Moreover, the side-chain has been synthesized in a manner which permits facile esterification of the taxane portion of taxol (Figure 4) so that there is now a practical semisynthesis of taxol available. As a consequence, Bristol-Myers Squibb has been quoted as stating that in the relatively near future, taxol will not be produced from the bark of *Taxus brevifolia*, but from baccatin III or 10-deacetylbaccatin III which occur in other *Taxus* species in much greater quantity. Moreover, these are smaller shrub-like plants such as *Taxus baccata*, a European yew. This plant is a renewable source. The component used in the semisynthesis is found in the needles (19). As a consequence, the availability of taxol in a continuing supply is now assured.

Total Synthesis. Because of the enormous interest in taxol, a number of groups, particularly those of Wender at Stanford, Holton at Florida State, and Nicolau at Scripps Institute, have made intensive efforts to synthesize taxol. Recently, both the Holton and Nicolau groups have announced the total synthesis of taxol (*25, 26*), truly an epochal event. Although this will not be a practical supply source, the availability of synthetic methodology may lead to important new information.

Concluding Remarks

The authors have been delighted that their initial discovery more than twenty-five years ago of a novel natural product with excellent activity in a number of animal models has presently reached the stage where taxol is now available in adequate quantity for therapeutic use. Undoubtedly, there are other highly-active natural products from plant, marine, and fungal sources as yet unknown which, when discovered, will have therapeutic utility. Cancer is not one, but several hundred diseases and will require many different types of agents.

Acknowledgments

The authors are appreciative of the National Cancer Institute for continuous support of their work. The discovery, isolation, and structure elucidation of taxol was carried out under Contract SA-43-ph-4322.

Literature Cited

1. Mitscher, L. A.; Drake, S.; Gollapudi, S. R.; Harris, J. A.; and Shankel, D. M. In *Antimutagenesis and Anticarcinogenesis Mechanisms;* Shankel, D. A., Hartman, P. A., Cotta, T., Hollaender, A., Eds; Plenum Press: New York, 1986: 153-165.
2. Williams, D. H.; Stone, M. J.; Hauck, P. R.; Rahman, S. K. *J. Nat. Prod.* **1989**, *52*, 1189-1208.
3. Woodbury, A. M.; Wall, M. E.; Willaman, J. J. *J. Econ. Bot.* **1961**, *15*, 79-86.
4. Wall, M. E.; Wani, M. C.; Cook, C. E.; Palmer, K. H.; McPhail, H. T.; Sim, G. A. *J. Am. Chem. Soc.* **1966**, *88*, 3888-3890.
5. Wani, M. C.; Taylor, H. L.; Wall, M. E.; Coggon, P.; McPhail, A. T. *J. Am. Chem. Soc.* **1971**, *93*, 2325-2327.
6. Wall, M. E.; Wani, M. C. Abstracts of Papers, 153rd National Meeting of the American Chemical Society, Miami, Beach, FL 1967; No. M-006.
7. *Washington Insight*, Persinos, E., Ed.; Sept. 15, 1990.
8. Wani, M. C.; Taylor, H. L.; Wall, M. E. *Chem. Commun.* **1973**, *Com. 470*, 390.
9. Wani, M. C.; Taylor, H. L.; Wall, M. E.; McPhail, A. T.; Onan, K. D. *J. Am. Chem. Soc.* **1975**, *97*, 5955-5956.
10. Wani, M. C.; Taylor, H. L.; Thompson, J. B.; Wall, M. E. *Lloydia* **1978**, *41*, 578-583.
11. Woods, M. C.; Nakanishi, K.; Bhacca, N. S. *Tetrahedron* **1966**, *22*, 243-258.

12. Harrison, I. W.; Scrowston, R. M.; Lythgoe, B. *J. Chem. Soc.*, **1966**, 1933-1945.
13. Della Casa de Marcano, D. P.; Halsall, T. G. *Chem. Commun.*, **1970**, 1382-1383.
14. Suffness, M.; Wall, M. E. In *Taxol—Science and Application*; Suffness, M., Ed.; CRC Press: Boca Raton, FL, in press 1994.
15. Fuchs, D.A.; Johnson, R.K. *Cancer Treat. Repts.* **1978**, *62*, 1219-1222.
16. Schiff, P. B.; Fant, J.; Horwitz, S. B. *Nature*, **1979**, *22*, 665-667.
17. Horwitz, S. B.; Parness, J.; Schiff, P. B.; Manfredi, J. J. *Cold Spring Harbor Symp. Quant. Biol.* **1982**, *46*, 219-226.
18. Horwitz, S. B. *Trends Pharmacol. Sci.* **1992**, *13*, 134-136.
19. Suffness, M. In *Ann. Repts. Med. Chem.*; Bristol, J. A., Ed.; Academic Press: San Diego, CA, 1993: 305-314.
20. Wall, M. E.; Wani, M. C. In *Alkaloids*; Pelletier, S. W., Ed.; Pergamon Press: Oxford, U.K., Vol. 9, in press 1994.
21. McGuire, W. P.; Rowinsky, E. K.; Rosenshein, N. B.; Grumbine, F. C.; Ettinger, D. S.; Armstrong, D. K.; Donehower, R. C. *Ann. Intern. Med.* **1989**, *111*, 273-279.
22. Slichenmyer, W. J.; VonHoff, D. D. *Anticancer Drugs*, **1991**, *2*, 519-530.
23. Kingston, D.G.I. *Pharmac. Therap.* **1991**, *52*, 1-34.
24. Chaudhary, A. G.; Gharpure, M. M.; Rimoldi, J. M.; Chordia, M. D.; Gunatilaka, A.A.L.; Kingston, D.G.I. *J. Am. Chem. Soc.* **1994**, *116*, 4097-4100.
25. Holton, R. A.; Somoza, C.; Kim, H.-B.; Liang, F.; Biediger, R. J.; Boatman, P. D.; Nadizadeh, M.; Suzuki, Y.; Tao, Ch.; Vu, P.; Tang, S.; Zhang, P.; Murthi, K. K.; Gentile, L. N.; Liu, J. H. *J. Am. Chem. Soc.* **1994**, *116*, 1597-1600.
26. Nicolaou, K. C.; Yang, Z.; Liu, J. J.; Ueno, H.; Nantermet, P. G.; Guy, R. K.; Claiborne, C. F.; Renaud, J.; Couladouros, E. A.; Paulvannan, K. *Nature* **1994**, *367(6464)*, 630-634.

RECEIVED August 31, 1994

Chapter 3

Current Status of Clinical Trials with Paclitaxel and Docetaxel

F. A. Holmes[1,4], A. P. Kudelka[1], J. J. Kavanagh[1], M. H. Huber[2], J. A. Ajani[3], and V. Valero[1]

Departments of [1]Breast and Gynecologic Medical Oncology, [2]Thoracic/Head and Neck Medical Oncology, and [3]Gastrointestinal Medical Oncology, The University of Texas M. D. Anderson Cancer Center, Houston, TX 77030

The unique mechanism of action of the antimicrotubule agent paclitaxel suggested that it would be a potent antineoplastic agent. However, even after multiple preclinical problems with paclitaxel were surmounted, a number of unique clinical problems still required resolution. Despite the existence of over 50 active antineoplastic agents, drug resistance and patient tolerance limit the number of effective agents in specific tumor types. Paclitaxel has shown antitumor activity in multiple clinical trials in cancers of the ovary, breast, head and neck, lung, and gastrointestinal tract. In many of these trials, paclitaxel was active despite evidence of the tumors' resistance to other important drugs. Paclitaxel may be the first of a series or family of drugs: in preliminary trials, an analogue, docetaxel, has also shown significant antineoplastic activity. However, many issues regarding the optimal use of both of these drugs remain unresolved.

Overview: Paclitaxel and Chemotherapy

Schiff and Horowitz's description of paclitaxel's unique mechanism of action was the catalyst for paclitaxel's clinical development *(1)*. The difficulties in translating paclitaxel from the forest to the pharmacy were described in chapters 1 and 2. Additional difficulties awaited clinicians who began using paclitaxel in clinical trials. This chapter will address those clinical problems and the results of therapeutic trials in patients with tumors of the ovaries, breast, head and neck, lungs, or gastrointestinal tract. The trials in ovarian and breast cancers confirmed that paclitaxel has major clinical activity, leading to its approval by the Food and Drug Administration (FDA) for commercial use. The clinical results of trials of a semisynthetic taxane (taxoid), docetaxel (Taxotere), will be also discussed here.

[4]Current address: Department of Breast and Gynecologic Medical Oncology, Box 56, The University of Texas M. D. Anderson Cancer Center, 1515 Holcombe Boulevard, Houston, TX 77030–4009

0097–6156/95/0583–0031$09.26/0

Clinical Problems Unique to Paclitaxel. The early clinical trials of paclitaxel faced three specific problems.

Hypersensitivity Reactions. The purpose of the first series of trials of a new drug in humans (phase I trials) is to determine the maximum tolerated dose by treating consecutive cohorts of patients with escalating doses of the drug. The maximum tolerated dose is defined by the occurrence of toxic effects, called dose-limiting toxicity, whose severity or permanence limit further dose escalation. In addition to the expected problems of myelosuppression (low white blood cell counts) and neuropathy encountered in the initial phase I trials of paclitaxel, an acute allergic reaction occurred that was fatal in one patient *(2)*. This reaction was similar to the severe reactions experienced by some patients who receive iodinated intravenous contrast medium for radiographic procedures. The unpredictability of this reaction terminated clinical trials until it was discovered that it was rapid infusion of the diluent, Cremophor EL, that caused the reaction. To prevent this, the National Cancer Institute (NCI) recommended infusing paclitaxel over 24 hours and premedication with corticosteroids and antihistamines *(3)*. These strategies have reduced the incidence of serious hypersensitivity reactions to 1% or less. Since both the premedication regimen and the slow infusion duration were developed simultaneously, it was initially unclear which was more important. Further studies have shown that each is effective independently.

Cardiac Toxic Effects. When clinical trials were resumed, all patients were treated in an intensive care unit with cardiac monitoring. Nearly 30% of patients were observed to have an abnormal but generally benign slowing of the cardiac rhythm (sinus bradycardia) while receiving paclitaxel. In a few cases, however, this rhythm was so slow that a pacemaker was required to continue treatment. A few patients with severe but undiagnosed coronary artery occlusions died of myocardial infarctions (heart attacks) or had life-threatening rhythm abnormalities *(4)*. Cardiac monitoring was required in all clinical trials and extensive data were collected. Analysis of 3400 patients revealed that the incidence of life-threatening events was less than 0.5%. Review of trials conducted before these heart problems were observed and of historical data from other drug development studies revealed that multiple benign rhythm abnormalities are common in patients receiving chemotherapy. It was recommended that patients who had known disease of or took drugs affecting the conduction system be given paclitaxel only with cardiac monitoring.

Drug Supply. After phase I trials have determined an effective and safe dose, most active new drugs are tested simultaneously in several trials in specific tumor types (phase II) and at different research centers, with exploration of different infusion durations and retreatment intervals (administration schedules). The early scarcity of paclitaxel meant that the results of each trial had to be carefully evaluated before planning the subsequent trial, that the numbers of patients treated would be limited, that the duration of treatment would be curtailed, that only limited types of tumors would be tested, and that the time frame for the start of these trials was delayed. To determine the level of activity with narrow confidence levels, the usual numbers of patients treated in a trial of a drug with such high activity would be 35-50 patients. In the early trials in breast cancer, for example, only 25 patients could be treated. Similarly, in a drug with such a high level of activity, most patients would be given at least six treatments before their tumor was judged unresponsive. In the lung cancer studies, patients whose tumors did not evince a 50% or greater shrinkage in the perpendicular diameters of bidimensionally measurable lesions (i.e., an objective response), were removed from study even though clinical information, such as shortness of breath or pain control, suggested they were having clinical benefit. The initial phase II trials were limited to only three tumor types; renal (kidney), melanoma, and ovarian. Of these, only ovarian cancer is fairly common. Finally, although trials in breast cancer were planned in 1985, sufficient supplies of the drug did not become available until 1990. As noted above, improved extraction methods and formulation of a semisynthetic drug have alleviated the supply problem.

Clinical Problems of Chemotherapy. Inherent limitations in the current practice of clinical oncology make paclitaxel a needed addition to the therapeutic armamentarium.

Drug Resistance. Although there are nearly 50 different antineoplastic drugs in use, only a dozen or fewer are effective in the treatment of each specific tumor type because of intrinsic or primary resistance. The initial and subsequent regimens of chemotherapy allow development of secondary or acquired resistance by selecting cells that survive. Ultimately, by a variety of mechanisms, a multiply drug-resistant tumor evolves, and further chemotherapy induces only toxic effects without tumor kill *(5)*. Tumors are classed by their degree of chemosensitivity. Breast and ovarian cancers are moderately sensitive. However, when the tumor becomes resistant to doxorubicin (breast) or cisplatin (ovarian), few other drugs are effective.

Patient Tolerance. As the tumor grows, the patient becomes increasingly debilitated by the accumulation of secondary effects from the tumor (cachexia, pain) or previous treatments (bone marrow or heart muscle failure from irradiation or prior chemotherapy or both) as well as primary effects, which depend on the site of involvement (shortness of breath, liver failure, bone fractures). Thus, with the exception of those with very indolent tumors, most patients have the physical reserve to endure only a limited number of aggressive chemotherapy regimens.

Ovarian Cancer

Background. Ovarian cancer is the leading cause of death among gynecologic malignancies in the USA, surpassing the mortality from cervical and endometrial cancer combined. Approximately one woman in 70 will develop ovarian cancer. In American women it is the sixth most common cancer and the fourth most common cause of death *(6)*. In 1994, 24,000 cases and 13,600 deaths are attributed to ovarian cancer *(7)*. The peak incidence is in the seventh decade; it is uncommon below 50 years of age *(8)*. The incidence is high in North America and Northern Europe, and low in Japan *(9)*.

The cause of ovarian cancer is unknown, but it is associated with consumption of animal fat, and is more common in patients with a history of breast cancer *(6)*.

Only about 5% of cases are hereditary. Childbearing and use of birth control pills reduce the risk of developing ovarian cancer by 30-60%, but use of replacement estrogen has no effect on the incidence *(10)*.

The two most important features of the disease which determine outcome are the extent of the disease (stage) and the aggressiveness of the tumor as determined by microscopic evaluation (histologic grade). Unfortunately, 75-85% of patients are diagnosed with advanced disease which has metastasized from the ovaries in the pelvic cavity to the abdominal cavity, because symptoms are often absent until the disease involves other organs in the abdominal cavity. These patients are rarely cured *(11-14)*, but treatment with chemotherapy may reduce symptoms and prolong life.

Standard therapy consists of cyclophosphamide with either cisplatin or its newer analogue, carboplatin. This causes tumor regression in 60-80% of patients of which 30-50% are complete responses (CR). The median duration of survival is 18 to 24 months, and 5- and 10-year survival for ovarian cancer metastatic to the abdominal cavity and elsewhere is 5-20%, and 0-10%, respectively. Carboplatin has less neurologic, kidney, and auditory toxic effects than cisplatin, causes less nausea and vomiting, and provides a better quality of life than cisplatin. However, it is more expensive and causes more depression of the white blood cell and platelet counts (myelosuppression and thrombocytopenia) *(15)*. Hormonal therapies are transiently effective in 10-20% of patients.

Patients whose tumors previously responded to chemotherapy and who have a treatment-free interval of at least six months are defined as potentially "platinum-" or "platin-sensitive," as 30-50% of these patients will have tumor regression if retreated with either carboplatin or cisplatin *(16,17)*. Patients with tumors that are platin-resistant, defined as worsening disease during treatment, persistent disease after four to six treatments, or recurrent disease within 6 months after completing therapy *(18,19)*, have a median survival of 12 months or less, and no currently available drugs have been shown to prolong these patients' life span. It is in this group of patients that new drug treatments are urgently needed and in whom they are first tested.

Paclitaxel Trials in Ovarian Cancer

The initial clinical trials of paclitaxel in humans treated patients with multiple tumor types that had failed all standard therapies. The intent of these trials was to determine the safest and most effective dose (maximum tolerated dose, MTD) and infusion duration (schedule). Evidence of tumor regression is uncommon in such trials. However, unexpectedly, tumor regression was seen in patients with platin-resistant ovarian cancer *(20)*.

Single-agent trials. These above results were the basis for a series of trials in patients with ovarian cancer designed to evaluate antineoplastic activity (phase II trials) of paclitaxel. A total of 111 patients were treated with doses of 100-250 mg/m^2 infused over 24 hours every three weeks *(21-24)*. Overall, 20-37% of patients had tumor regression, and in seven patients it was complete. Analysis by platin-sensitivity revealed responses in 40-50% of patients with potentially platin-sensitive tumors, and in 24-30% of patients with platin-resistant tumors, with at least two of the patients achieving a CR. The median duration of response was 6 months (range, 2-30). The overall median survival was 11 months. It was 17 months for patients with potentially platin-sensitive tumors and 9 months for those with platin-resistant tumors *(24)*. The major toxic effect in these studies was granulocytopenia (lowered granulocyte count, a subset of the white blood cells responsible for preventing bacterial infection) which necessitated dose-reduction to prevent infection. These studies showed the potential range of antineoplastic activity but did not define whether the response rate was a function of dose.

High-Dose Trials. To determine whether antineoplastic activity was higher at higher doses, the National Cancer Institute (NCI) and M. D. Anderson each tested paclitaxel in a single-arm study at 250 mg/m^2 over 24 hours in patients with platin-resistant ovarian cancer. Granulocyte-colony stimulating factor (G-CSF), a biosynthetic form of the endogenous chemical that stimulates production and maturation of granulocytes, was given to prevent severe granulocytopenia *(25-28)*. More than 50% tumor regression (a partial response) was seen in 48% of patients in each study. The duration of response was 6 months, and the median survival was 12 months.

Trials in Patients with Multiple Prior Therapies. Although paclitaxel was still not approved by the FDA and thus not widely available, this striking evidence of antineoplastic activity suggested that many patients could potentially benefit from paclitaxel. Therefore, the NCI supplied paclitaxel to approved cancer centers for compassionate use (Treatment Referral Center, TRC, mechanism) for patients whose tumor had progressed despite three or more prior chemotherapy regimens. The dose was lower, 135 mg/m^2 over 24 hours, because of the extensive previous treatment. The overall response rate was 22%; the median survival was 9 months *(29)*.

Dose- and Schedule-Comparison Trial. In a European-Canadian (NCI-Canada) study of paclitaxel in patients who failed one or two platin regimens, 60% had platin-resistant disease, and all had measurable or evaluable disease *(30)*. The study was performed to answer two questions. First, is the response rate dependent on dose? Two dose levels were tested, 175 mg/m^2 and 135 mg/m^2. Second, is the 3-hour infusion schedule as safe and effective as the standard, but more cumbersome, 24-hour infusion? Thus, there were four treatment arms, because the high- and the low-dose arms were each given by 3- or 24-hour infusion. Responses were more frequent with the larger dose (20% versus 15%, respectively) and with the longer infusion schedule (19% versus 16%, respectively). However; neither of these differences achieved statistical significance. The response rate for platin-resistant patients was 13%, but the survival was not reported for this subgroup *(30)*.

TABLE I. Studies of Paclitaxel in Advanced and Refractory Ovarian Cancer

Institution (Reference)	*No. of Patients*	*Dose, mg/m²*	*Overall Response*	*CR % (No.)*	*Median Survival, mo.*
Single Agent					
JHOC *(21)*	40	135 (110-170)	30%	2.5(1)	8.2
GOG *(23)*	41	170 (↓)	37%	12 (5)	15.9
Einstein *(22)*	30	180-250	20%	3 (1)	6.5
NCI-TRC *(29)*	619	135	22%	3	9
European-Canadian *(30)*	195 187	135 175	15% 20%	1 (2) 2 (4)	11.0 11.5
High Dose (With G-CSF)					
NCI *(24,25)*	44	250	48%	14	11.5
MD Anderson *(27, 28)*	48	250	48%	4	12

Conclusions. Single-agent paclitaxel infused over 24 hours produces antineoplastic responses of 10-22% and 48-50%, respectively, for doses of 135 mg/m^2 and 250 mg/m^2, suggesting a benefit to higher doses. However, two additional points are necessary to interpret these data correctly. First, only a concurrent, randomized trial comparing the 135 mg/m^2 and 250 mg/m^2 doses can prove this hypothesis. Second, even though more patients responded at the higher dose, the median survival was comparable in both trials. The apparent lack of benefit in terms of increased survival for the high-dose group, with its attendant increase in incidence of toxic effects and cost (owing to both the paclitaxel and the G-CSF), suggests to us that the higher doses may be most useful in alleviating the severe cancer-induced symptoms of some patients with advanced platin-resistant ovarian carcinoma. The dose and schedule approved by the FDA is 135 mg/m^2 over 24 hours given every 21 days.

Combination Trials. The next obvious step to optimize paclitaxel's antineoplastic activity was to combine it with platin and to treat patients who had not previously received chemotherapy. A trial comparing the standard combination of cisplatin with cyclophosphamide to the same dose of cisplatin with paclitaxel 135 mg/m^2 was initiated. Only preliminary data has been reported *(24)*. Patients on the paclitaxel experienced significantly more numbness, hair loss, allergic reactions, and episodes of fever during periods of granulocytopenia. The overall response rate was 64% for the standard arm and 77% for the paclitaxel-containing arm, a statistically significant difference with $p = 0.02$ *(24)*. Still unreported is information on survival or comparison of the patients' quality of life or cost of treatment. For all of these reasons, the use of paclitaxel combinations in previously untreated ovarian cancer patients should be considered investigational and not a standard of care. A number of other combination trials with other agents are ongoing, but the data are too preliminary for conclusions.

Breast Cancer

Background. In 1994, 182,000 patients will be diagnosed with breast cancer; 1% of these will be men. Of the 46,000 patients who will die of this disease, most will have been diagnosed previously *(7)*. Breast cancer is the most common malignancy affecting women. For unexplained reasons, the incidence of breast cancer has stabilized recently. Exposure to pesticides at a young age has recently been implicated in breast cancer because of the estrogen-like effect of DDT (*31*). The current trend for delayed or deferred childbearing may cause the incidence of breast cancer to increase *(32)*, but even if the rate of breast cancer does not increase, the absolute numbers of patients will rise with the "graying" of America.

The most important variable determining a patient's outcome is the extent of disease, or stage *(33)*. Tumors that are confined to the locoregional area comprising the breast and draining lymph nodes in the axilla (armpit), stages I - III, are potentially curable with local therapy (total mastectomy or lumpectomy and axillary lymph node dissection followed by irradiation of the remaining breast tissue) with or without systemic therapy, that is, chemotherapy or hormones or both (adjuvant or prophylactic therapy for micrometastases). Tumors that have metastasized beyond the locoregional area (stage IV) are incurable. However, even in patients with incurable tumors, palliative therapy with chemotherapy, hormones, and/or radiation may prolong good quality of life. Before the use of chemotherapy, the average life expectancy after the diagnosis of stage IV (metastatic) breast cancer was 9 months *(34)*. With chemotherapy, the median life expectancy is 2 to 3 years. The first chemotherapy regimen is generally effective at reducing tumor size by at least half in 50-70% of patients, and this effect lasts for a median of 9 to 12 months. Subsequent treatments are effective in about 25-40% of patients for 3 to 6 months. The most effective antineoplastic drug available for breast cancer is doxorubicin. Tumors that develop resistance to doxorubicin usually have only very brief responses to other agents. It is in these patients, who have stage IV disease and who have received a variety of previous treatments, that new chemotherapy agents are first tested. With the increased awareness of the multiple mechanisms of drug resistance, however, many new drugs are tested in patients who have received only one or fewer prior therapies in order to prevent false-negative results from tumors that are resistant to all drugs.

In contrast to some other tumors, in breast cancer the histologic type, with few exceptions, has little impact on outcome. The degree of differentiation of the tumor, or grade, as determined by a variety of methods, is more important. Poorly differentiated, or anaplastic, tumors grow rapidly and develop early resistance to therapy. In stage IV disease, the extent and sites of metastases correlate with outcome. Patients with metastases to visceral organs generally have a more rapidly fatal outcome. A unique feature of breast cancer is its responsiveness to hormonal treatments in a subset of patients. Eventually, however, all hormone-sensitive tumors become insensitive.

Paclitaxel Trials in Breast Cancer Patients. The initial investigative trials tested paclitaxel alone in patients who had received various numbers of prior chemotherapy regimens. Later trials investigated paclitaxel in combination with other standard antineoplastic drugs.

Single-Agent Trials in Patients with Limited Prior Chemotherapy. The initial trial of the antineoplastic activity of paclitaxel in breast cancer was at M. D. Anderson in patients with stage IV breast cancer who had received one prior regimen of chemotherapy *(35)*. Paclitaxel was given at a dose of 250 mg/m^2 over 24 hours to 25 patients of whom 66% had visceral disease. In 56% of these patients, tumors shrank 50% or more, and this objective response lasted for a median of 9 months (range, 5-27). In three patients, tumor regressed completely. The treatments were generally well tolerated; the median number of treatments per patient was 13 courses (range, 2-21). The toxic effects were those expected: low blood counts in 88% of patients (but complications of that in only 6% of the 297 courses), total hair loss, and muscle aches which were severe in only a minority. Long-term effects included a sensory neuropathy manifested primarily by impairment of fine motor functions such as buttoning, sewing, fastening jewelry, etc. However, no patient experienced an allergic reaction. This unexpected result was confirmed by review of all x-rays and patient charts at the NCI.

A trial at the Memorial Sloan-Kettering Cancer Center in New York confirmed these results in a similar group of patients except that only 16 had received prior chemotherapy *(36)*. The dose and duration of infusion were similar, except that G-CSF was administered to prevent the severe drop in white blood cell counts, with the hope of decreasing the incidence of infection. Of 26 patients, 62% had a decrease in tumor size of 50% or more. The use of G-CSF reduced the incidence of infectious complications by half.

Finally, when more paclitaxel became available, a large trial commenced in Europe and Canada (Canadian-European Taxol Study Group, CETSG) which focused on women who had received only one prior chemotherapy treatment for metastatic disease regardless of whether they had received preventive therapy (adjuvant) after mastectomy *(37)*. The objectives of the trial were two: to determine the efficacy in breast cancer of the 3-hour schedule, which had been developed for ovarian cancer, and to compare doses of 175 mg/m^2 and 135 mg/m^2. A total of 471 women were treated. Nearly a third had received adjuvant treatment, another third had received treatment for metastatic disease, and the remaining third had received both adjuvant and metastatic treatment. The response rates, 29% for the high dose and 22% for the low dose, were not significantly different. Although these response rates were not as high as in the previous two trials, the doses used were lower and the infusion schedule was shorter. Both of these changes diminished the effective given dose. One measure of this is the determination of the incidence of granulocyte counts below 500 cells/mm^3 (grade 4 granulocytopenia). In the original M. D. Anderson study, 100% of patients (88% of courses) experienced grade 4 granulocytopenia. In the Memorial Sloan-Kettering study, which used G-CSF to prevent or diminish this effect, 66% of courses had grade 4 granulocytopenia. In the European- Canadian study, only 27% and 21% of courses, respectively, had grade 4 granulocytopenia. The data from this trial were the basis for approval of paclitaxel as "salvage" therapy for metastatic breast cancer which worsened despite doxorubicin treatment in December 1993. The approved dose and schedule is 175 mg/m^2 by 3-hour infusion given every 21 days.

Single-Agent Trials in Patients with Multiple Prior Therapies. Both M. D. Anderson and Memorial Sloan-Kettering tested the 24-hour infusion schedule in patients who had received two or more prior treatment regimens. The Memorial Sloan-Kettering trial used G-CSF which allowed treatment at higher doses. The planned dose level in the Memorial Sloan-Kettering study was 200 mg/m^2 *(38)*. In patients who had received two or three or more prior regimens, respectively, response rates were 31% and 20%. At M. D. Anderson, doses of 175 mg/m^2 and 150 mg/m^2, respectively, without G-CSF, were used for patients with two or three or more prior regimens *(39)*. Responses were seen in 23% and 20% of patients, respectively. Approximately 35 patients were treated in each of the four trials. Nearly 15% of patients on each trial had received high doses of chemotherapy on a regimen that required transplant of bone marrow or stem cells.

Using data from cell cultures, which showed that prolonged infusion of drugs retards the development of drug resistance, investigators at NCI tested paclitaxel by 96-hour infusion in patients who had received two or more prior chemotherapy regimens and were resistant to doxorubicin, a group of patients with a dismal prognosis *(40)*. Objective responses were seen in 48% of patients.

TABLE II. **Phase II Trials of Paclitaxel in Metastatic Breast Cancer**

Institution (Reference)	*No. of Patients*	*No. Prior Chemorx.*	*Dose/infusion schedule (hr)*	*Response*
MD Anderson *(35)*	25	1	200-250/24	56%
Memorial *(36)*	26	0 - 32% pts 1 - 68% pts	200-250/24 + G-CSF	62%
CETSG *(37)*	471	1 - 69% 2 - 31%	randomized 135/3 175/3	 22% 29%
Memorial *(38)*	22	2	200/24 + G-CSF	36%
Memorial *(38)*	24	≥3	200/24 + G-CSF	21%
MD Anderson *(39)*	33	2	175/24	20%
MD Anderson *(39)*	35	≥ 3	150/24	18%
NCI-96 *(40)*	33	2	140/96	48%

Abbreviations: *No.*, number; *Chemorx.*, chemotherapy treatments

Combination Trials with Doxorubicin. Because doxorubicin is the most effective drug for breast cancer, it was the logical choice for a combination trial. The scarcity of paclitaxel limited trials to two centers, M. D. Anderson and the NCI. The centers used different schedules, but both used G-CSF to diminish effects on the white blood cell counts. Based on the cell culture data with prolonged drug infusions, the NCI gave each drug by 72 hours concurrently *(41)*. M. D. Anderson used a sequential approach, giving paclitaxel first over 24 hours followed by doxorubicin over 48 hours *(39)*. The purpose of these phase I trials was to determine the maximum tolerated dose. Since the trials were conducted simultaneously, data on the incidence and type of side effects associated with various dose levels were shared by the investigative teams. The sequential combination with paclitaxel preceding doxorubicin (paclitaxel-doxorubicin) produced much more severe mucous membrane toxicity (mucositis) which precluded intake of solid food at paclitaxel doses which were 30% lower than those in the simultaneous combination. This suggested that paclitaxel potentiated the effects of doxorubicin. For this reason, the reverse sequence (doxorubicin-paclitaxel) was then studied at M. D. Anderson *(42)*. With this sequence, the maximum tolerated doses of paclitaxel and doxorubicin were very similar to those reached in the NCI concurrent infusion schedule. To further define the nature of this drug interaction, pharmacokinetic studies were performed in a third group of patients to evaluate doxorubicin levels in patients who had received either sequential regimen. In the first cohort of patients doxorubicin preceded paclitaxel for course 1 and was reversed in course 2. In the second cohort of patients the sequences were reversed for courses 1 and 2. The results showed that when paclitaxel preceded doxorubicin, the peak doxorubicin concentration at the end of infusion as well as the area under the concentration x time curve (AUC) was 30% higher. Conversely, doxorubicin clearance was decreased by 70% when paclitaxel preceded it. The pharmacodynamic effects were similarly striking: the median granulocyte count, in the sequence paclitaxel-doxorubicin, was 0.2 granulocytes/mm^3 versus 1.3 in the reverse sequence. Also, the incidence and severity of mucositis was greater when paclitaxel preceded doxorubicin. The conclusion was that if paclitaxel by 24-hour infusion is to be given in sequence with doxorubicin, doxorubicin should be given first.

Although phase I studies are not designed to determine response rates with a narrow confidence interval, response data are evaluated. Independent of schedule, the combination of doxorubicin with paclitaxel by 24-hour or longer infusion resulted in 70% objective responses. Only 10% of these responses were complete, a condition necessary for development of a curative regimen. This is similar to standard aggressive combinations of doxorubicin with 5-fluorouracil and cyclophosphamide ("CAF" or "FAC"). Additionally, the NCI trial produced an unusual toxic effect, typhlitis, inflammation of the intestines *(43)*. This is uncommon in patients with solid tumors and may be lethal.

Another phase I trial also tested this combination but gave doxorubicin by a rapid intravenous injection ("bolus"), which is the more common method of administration in most physicians' offices. Cognizant of the M. D. Anderson sequence data, Sledge et al. interposed a 4-hour delay between the drugs *(44)*. Their trial evaluated the questions of sequence and dose. The sequence of drug administration was alternated both between and across patients. Two dose levels were tested: paclitaxel 150 mg/m^2 with doxorubicin 50 mg/m^2 and paclitaxel 160 mg/m^2 with doxorubicin 60 mg/m^2. This trial also confirmed that the sequence of paclitaxel-doxorubicin produced more severe mucositis; not unexpectedly, the higher dose produced more granulocytopenia. Response data have still not been reported in final form.

Preliminary data have only recently been reported by investigators in Milan who treated a similar patient population but used the bolus infusion of doxorubicin with the 3-hour schedule for paclitaxel *(45)*. These investigators did not observe any sequence effect. Moreover, the response rate was over 90%, with nearly 40% complete responses. These encouraging early data must be verified, but do suggest a potentially synergistic schedule for this combination.

Other Combination Trials. As the supply of paclitaxel increased due to more efficient extraction procedures and its approval for use in patients with ovarian cancer made it commercially available, a host of other combination trials have begun. One trial tested the combination of paclitaxel with cisplatin but used lower than standard doses of each drug to allow retreatment of patients every 14 days instead of the more standard 21-day interval *(46)*. Responses were seen in almost 75% of patients, nearly all of whom had received doxorubicin as an adjuvant therapy. Two trials combined paclitaxel with cyclophosphamide, arguably the second most effective drug for breast cancer. Both Johns Hopkins Oncology Center (JHOC) *(47)* and the NCI tested various schedules and found the combination active. Hopkins also evaluated the sequence question and found more severe granulocytopenia when paclitaxel preceded cyclophosphamide. However, pharmacokinetic studies have not shown any difference in the standard pharmacokinetic parameters, so the nature of this interaction is unexplained.

Novel Approaches. Three important trials representing a new direction in research are ongoing. At Memorial Sloan-Kettering, paclitaxel is administered in combination with monoclonal antibodies directed to growth factors located on the cell surface *(48)*. Preliminary data from studies of anti-epidermal growth factor receptor in combination with low doses of paclitaxel are positive. Another trial involves gene therapy for patients who will undergo high-dose chemotherapy with bone marrow transplantation and takes advantage of one of the mechanisms of resistance used by tumor cells, the p-glycoprotein multidrug resistance pump, which actively pumps any number of structurally unrelated but naturally occurring compounds out of the tumor cell *(49)*. This gene will be inserted into the transplanted white blood cell precursors of patients who will later receive high doses of paclitaxel and will allow these cells to, in effect, eject the paclitaxel before it impairs cell function and, importantly, proliferation. The third approach uses paclitaxel in the adjuvant setting *(50)*.

Unresolved issues. The most effective schedule for paclitaxel in breast cancer has not been defined. An ongoing North American trial is comparing the 3- and 96-hour infusions. Synergistic combinations are being developed. The trials already described have shown that the sequence of any combination must be carefully studied to minimize toxic effects *(42)*. Paclitaxel is excreted primarily by the liver. Use of paclitaxel in patients whose liver function has been impaired because of tumor has exacerbated all the standard toxic effects and been lethal in some. A safe but effective dose-schedule for these patients must be established. Finally, a better understanding of the mechanisms of resistance and methods to circumvent it is needed.

Lung Cancer

Lung cancer is the leading cause of cancer death in the United States in both men and women: more than 170,000 cases are diagnosed each year, and more than

140,000 people die of the disease in the same period *(51)*. Lung cancer is divided into two major biological and clinical subtypes: non-small cell lung cancer, which is relatively insensitive to chemotherapy, and small cell lung cancer, which is very sensitive to chemotherapy *(52)*.

Non-Small Cell Lung Cancer. Non-small cell lung cancer accounts for more than three quarters of all lung cancer cases. Early stage non-small cell lung cancer is frequently curable with surgical resection, but three quarters of patients present with either locally advanced disease, which is only infrequently curable with surgery, or distant metastasis, which is almost always fatal. The treatment for individuals with metastatic disease consists of chemotherapy; however, the best available single agents, such as cisplatin, ifosfamide and mitomycin C, induce major responses in only 20-30% of patients. Furthermore, the duration of these responses is frequently only a few months, and almost all patients will relapse. Combinations of active agents may yield higher response rates, but the overall outcome remains dismal. Large randomized trials have shown that chemotherapy may prolong survival in patients with metastatic non-small cell lung cancer, but the benefit in overall survival duration is still measured in months. Therefore, it is obvious that new agents are needed for the treatment of non-small cell lung cancer *(52)*.

Phase II Trials. Paclitaxel was studied in two phase II trials in patients with non-small cell lung cancer based on its activity in preclinical models. In one trial at M. D. Anderson, 27 patients with non-small cell lung cancer who had not received prior chemotherapy were treated with paclitaxel 200 mg/m^2 as a 24-hour infusion *(53)*. Six of the 25 evaluable patients (24%) had major responses, one of which was a complete remission. The median response duration was 27 weeks, and the median survival of all patients enrolled on the study was 40 weeks.

Another phase II study reported by the Eastern Cooperative Oncology Group (ECOG) treated patients with metastatic non-small cell lung cancer with paclitaxel 250 mg/m^2 as a 24-hour infusion *(54)*. Five of 24 evaluable patients (21%) had major responses, and the median survival was 24.1 weeks; however, 42% of the patients were alive at one year, which is higher than is usually reported in phase II studies in lung cancer. The authors commented that this result made paclitaxel the most active single agent for non-small cell lung cancer evaluated in the past ten years of phase II studies by the ECOG.

While the overall response rates in these two studies may not appear impressive, they clearly demonstrate that paclitaxel is among the most active agents for non-small cell lung cancer identified to date. Furthermore, in additional trials non-small cell lung cancer patients who had not responded to prior cisplatin-based therapy had partial responses to paclitaxel, which is unusual with any other agent *(55,56)*.

Investigational Approaches: Combination Studies and Locally Advanced Disease. Some high-profile investigations on the use of paclitaxel in the treatment of non-small cell lung cancer involve 1) combination with other agents and radiotherapy and 2) treatment for patients with tumors that are not resectable but are limited to the lung only (locally advanced disease).

Data from cell lines and animal models suggest that paclitaxel may have synergy with cisplatin, one of the most active single agents in the treatment of non-small cell lung cancer *(57)*. A large randomized cooperative trial is now underway evaluating the efficacy of this combination in patients with metastatic non-small cell lung cancer.

For patients with stage III non-small cell lung cancer, which is not resectable but has no documented evidence of distant metastasis, paclitaxel may have a major impact on survival. This group comprises 25-30% of all patients with non-small cell lung cancer; therefore, a therapeutic impact in this population, even if relatively small, may prolong the survival of thousands of patients. Surgery and/or radiotherapy have usually been used in this patient population, but the majority of patients so treated develop either distant metastasis or local recurrence. Recent data indicate that chemotherapy prior to radiotherapy prolongs survival in this population of patients *(58)*; the development of combinations with greater activity may lead to even greater improvements in survival.

Concurrent use of chemotherapy and radiotherapy may provide an additional benefit in patients with locally advanced disease by avoiding delay of either modality and exploiting the radiosensitizing capacity of some agents. Paclitaxel is an ideal candidate for this strategy because it is among the most active agents for metastatic non-small cell lung cancer and it is a potential radiosensitizer *(59,60)*. Clinical trials are now exploring the combination of paclitaxel with radiotherapy in the treatment of locally advanced non-small cell lung cancer *(61)*.

Small Cell Lung Cancer. Although the role of chemotherapy in the management of patients with non-small cell lung cancer remains unclear, chemotherapy is essential in the treatment of patients with small cell lung cancer *(62)*. Numerous agents have been found to be active in small cell lung cancer, and combination therapy can yield response rates of more than 80%. However, responses are frequently incomplete, and the duration is short in the majority of patients. Therefore, the development of new agents is essential in order to prolong overall survival of these patients.

Phase II Trials. The first paclitaxel trial produced major responses in 34% of patients with previously untreated small cell lung cancer *(63)*. This was probably an underestimate of the true response rate, however, as patients were allowed only two cycles of treatment unless they had at least a partial response. The activity of paclitaxel in small cell lung cancer was confirmed in a second phase II study which identified a response rate of 68% *(64)*. Current studies are exploring the combination of paclitaxel with other agents with known activity in small cell lung cancer.

Head and Neck Cancer

Background. Head and neck cancer comprises a broad range of tumors of the oral cavity, pharynx, and larynx, and accounts for more than 40,000 cases of cancer in the United States each year. The vast majority of these cancers are squamous cell histologic type, and early stage lesions can be cured with surgery and/or radiotherapy. Unfortunately, patients frequently present with disease that has either invaded adjacent local structures or metastasized to local lymph nodes. Either of these markedly shortens survival to a median of 6 months; death is usually due to local-regional or distant metastasis.

The most active chemotherapeutic agents in this setting include cisplatin, carboplatin, methotrexate, and ifosfamide. However, these agents rarely produce complete responses, and none have been found to have a significant impact on overall survival in this population. Combinations of agents such as cisplatin and 5-fluorouracil yield higher response rates than single agents but have no significant impact on survival *(65)*.

Phase II Trials. A recent phase II study found that paclitaxel may be among the most active agents in the management of squamous cell carcinoma of the head and neck: 47% of patients had responses following paclitaxel alone, 11% complete responses *(66)*. Another phase II study has reported a similar high response rate in patients with recurrent disease *(67)*. A large trial is currently underway in Europe under the auspices of the EORTC to confirm the efficacy of paclitaxel. The study is comparing two paclitaxel schedules, 175 mg/m^2 over 24 hours versus a standard arm with methotrexate 40 mg/m^2/week. A second large trial is ongoing in the Eastern Cooperative Oncology group evaluating the efficacy of paclitaxel 250 mg/m^2 over 24 hours with G-CSF versus 135 mg/m^2, both in combination with cisplatin. Current trials will define the role of paclitaxel in combination with other standard active agents.

Gastrointestinal Tract Cancers

Background. The gastrointestinal system is the most frequent site of malignant diseases in the U.S. population. Colon and rectal carcinomas are the most frequent gastrointestinal malignancies, and nearly 50% of patients with these diseases are cured by surgical excision of the primary tumor. Although the incidence of carcinomas of the pancreas, stomach, and esophagus is much less than that of colorectal carcinoma, these malignancies are generally not curable by surgery. Nearly 100,000 patients are diagnosed with incurable gastrointestinal malignancies each year in the U.S. *(68)*. We must develop new chemotherapeutic and biological agents to improve response rate, quality of life, and survival duration.

Carcinomas of the esophagus and gastroesophageal junction account for approximately 1% of all malignancies in the U.S. *(68)*. Approximately 11,000 new cases and 10,400 deaths are expected in 1994 *(68)*. For reasons yet to be explained, the incidence of adenocarcinoma of the esophagus and proximal stomach has increased dramatically in the past 15 years, particularly in the U.S. *(69)*. An increase in adenocarcinoma of the upper gastrointestinal tract has also been reported in Europe *(70-72)*. The median survival duration of patients with advanced disease is 4 to 8 months.

Phase II Trial in Carcinoma of the Esophagus. A phase II NCI-sponsored study of paclitaxel by a 24-hour infusion was conducted in previously untreated patients with unresectable local-regional or metastatic carcinoma of the esophagus. The protocol, whose purpose was to evaluate response rate, duration of response, and toxicity, accrued patients simultaneously at M. D. Anderson and Memorial Sloan-Kettering *(73)*. Patients with histologic proof of esophageal carcinoma and good performance status were premedicated and received paclitaxel at a starting dose of 250 mg/m^2 infused intravenously over 24 hours; this was repeated every 3 weeks. Patients also received G-CSF. Fifty-three patients were enrolled, 33 with adenocarcinoma and 20 with squamous cell carcinoma. Fifty-one patients were evaluable, and 16 (32%) achieved a complete or partial response. Complete response of liver metastases and local recurrence was observed in a patient with adenocarcinoma. Eleven patients achieved a minor response, 15 had no change in their disease, and 19 had progressive disease while receiving paclitaxel. The median duration of partial response was 17 weeks (range, 7 to 58+ weeks). Among the 33 patients with adenocarcinoma, 12 (36%) achieved either a complete (one patient) or partial response (11 patients) and six had a minor response. Four (22%) of 18 patients with squamous cell carcinoma had a partial response and 4 (22%) had a minor response.

Toxic Effects. The combination of paclitaxel followed by G-CSF was well tolerated. No treatment-related deaths occurred in this group of patients. There were no unexpected toxic effects. No grade 4 (life-threatening) toxic effects were observed in any patient. Three patients developed grade 3 (severe) neuropathy (nerve injury which causes numbness), but the severity was reduced in two patients as of this writing. Grade 3 or 4 granulocytopenia occurred in 107 (66%) courses. There were 11 hospitalizations in 9 (18%) patients for the management of fever during granulocytopenia.

Phase II Trial in Carcinoma of the Pancreas. Carcinoma of the pancreas afflicts nearly 25,000 people each year in the U.S. The prognosis of patients with this disease is extremely poor. Most patients develop widely metastatic disease shortly after the establishment of diagnosis. There is no effective palliative chemotherapy.

Paclitaxel has been studied at the starting dose of 250 mg/m^2 infused over 24 hours with G-CSF *(74)*. Among 44 patients registered, 30 were evaluable for response, but only two achieved a partial response. The duration of response was brief. There were no unexpected toxic effects. Overall, paclitaxel was considered ineffective in patients with carcinoma of the pancreas. This agent has not been pursued in this group of patients.

Paclitaxel seems to have definite activity in patients with adenocarcinoma or squamous cell carcinoma of the esophagus. A trial combining paclitaxel with 5-fluorouracil and cisplatin is now underway. However, paclitaxel appears to be ineffective in patients with carcinoma of the pancreas. Studies of paclitaxel in patients with colorectal carcinoma, biliary carcinoma, and gastric carcinoma are underway.

Docetaxel (Taxotere)

Background. Docetaxel is a new taxoid with a broad and promising antitumor profile. This semisynthetic compound was developed by French researchers from the Institut de Chimie des Substances Naturelles and Rhône-Poulenc in 1981 *(75)*. Docetaxel is prepared from a noncytotoxic precursor, 10-deacetyl baccatin III. The lateral chain at C-13, which is responsible for its cytotoxic effect, is added chemically. Docetaxel differs structurally from paclitaxel at the 10-position on the baccatin ring and at the 3'-position on the lateral chain. Docetaxel has a molecular weight of 807.9 grams, is practically insoluble in water but freely soluble in alcohol, and is currently formulated in polysorbate 80 *(20,75)*.

The mechanism of action and the effects of docetaxel on the cell cycle are unique. Docetaxel promotes the polymerization of tubulin into stable microtubules as the other taxoid paclitaxel does, but it does not alter the number of protofilaments in normal microtubules. It also appears to alter the structure of tubulin polymers in other classes of microtubules, including those that are Tau dependent. On the basis of an *in vitro* tubulin assay, docetaxel is twice as potent an inhibitor of microtubule depolymerization as paclitaxel. Docetaxel thus acts as a mitotic spindle poison and induces a mitotic block *(20,76)*.

The *in vitro* cytotoxicity of docetaxel in murine and human tumor cell lines and its *in vivo* preclinical antitumor activity in murine and human xenografts were very impressive *(20,76-79)*. Docetaxel showed higher cytotoxic activity than other antineoplastic agents such as paclitaxel, cisplatin, cyclophosphamide, and doxorubicin against the same tumor models *(20)*. The higher cytotoxic potency relative to paclitaxel may result from docetaxel's high affinity for microtubules, high intracellular concentrations, and slow cellular efflux. Synergistic antitumor effects have been seen with cyclophosphamide, fluorouracil, etoposide, and vinorelbine but not with cisplatin or doxorubicin *(20,80)*. Incomplete cross-resistance between docetaxel and paclitaxel has been seen in human cell lines *in vitro (20)*. Docetaxel is inactive against the doxorubicin-resistant P388 cell line, but it is active in other cell lines expressing the multidrug resistance phenotype *(20)*. Some preliminary data suggest that docetaxel could be a radiation sensitizer *(20)*. The preclinical toxicologic studies in mice and dogs found that docetaxel exerted its toxicity in tissues with high cell turnover. The toxic effects were greater with a 5-day schedule than with a single-dose schedule and were dose dependent *(20)*.

Phase I Trials. Six phase I trials of docetaxel at different schedules, comprising a total of 238 patients, were conducted throughout the world starting in 1990 *(81)*. The highest maximum tolerated dose and highest dose intensity were obtained using a short 1-hour infusion every 3 weeks. The dose-limiting toxic effect was grade 4 granulocytopenia, which was schedule independent, noncumulative, and dose dependent. Its duration was brief (less than 7 days), and it was rarely associated with significant morbidity. There was no significant anemia or thrombocytopenia. Mucositis was schedule dependent, being more frequent and severe with longer or repeated infusions (6-hour, 24-hour, or 5 consecutive-day schedules). Other toxic effects included alopecia (>50% of patients); mild nausea, vomiting, and diarrhea (<30% of patients); reversible paresthesias (<20% of patients); and hypersensitivity reactions (20% of patients; 4% severe). Other minor side effects were conjunctivitis, asthenia, and myalgias. The short infusion schedules were unexpectedly associated with skin toxic effects. An unusual fluid retention was noted in association with cumulative dosing in some patients who responded to the drug. Premedication was not administered in these trials. Major clinical activity was documented in 17 patients: eight objective responses were seen in breast cancer, two in ovarian cancer, three in lung cancer, and two in carcinoma of unknown origin *(81)*.

Clinical Pharmacology. Clinical pharmacology studies of docetaxel showed linear pharmacokinetics and no schedule-dependent disposition *(20)*. Docetaxel is significantly bound to plasma proteins (>90%). It is metabolized in the liver, with more than 90% of the drug being eliminated in the feces during the first 7 days (mainly in the first 2 days); renal excretion is minimal (5%). Mean pharmacokinetic values following dosing of 100 mg/m^2 as a 1-hour infusion are: peak concentration, 3.6 μg/ml; alpha, beta, and terminal gamma half-lives: 0.1, 0.6, and 12.2 hours (triphasic model); plasma clearance, 21.2 l/h/m^2. The dose recommended for phase II trials was 100 mg/m^2 infused over 1 hour every 21 days *(81)*.

Phase II. Phase II studies have been conducted in Europe, Japan, and North America. Preliminary results are shown in Tables III-V.

Metastatic Breast Cancer. In the first three trials, conducted by the European trials group (EORTC-ECTG), NCI-Canada, and Memorial Sloan-Kettering *(82-84)*, patients without prior chemotherapy for metastatic disease received 100 mg/m^2. Ninety-five patients were assessed for drug efficacy. The preliminary data show an overall objective response rate of 60%, with 9% complete responses. The response rate was similar whether or not patients had received prior adjuvant chemotherapy. The results from four other trials that used lower drug doses showed a lower rate of objective response. The response rate was 48% in 46 patients treated with 75 mg/m^2 and 44% in 155 patients treated with 60 mg/m^2 *(85,86)*.

The EORTC-ECTG then treated 32 patients who had received one prior chemotherapy regimen, 23 for metastatic disease and 9 for adjuvant, or preventive, therapy after mastectomy, with a dose of 100 mg/m^2 *(87)*. The overall objective response was 53%, with 6% complete responses.

Two studies have been performed in patients whose disease had worsened during treatment with an anthracycline (e.g., doxorubicin) or anthracenedione *(88,89)*. As noted earlier, once a tumor develops resistance to doxorubicin, it progresses rapidly and patients die. Fifty-nine patients were evaluable for response. The median number of prior chemotherapeutic regimens was two. The preliminary overall response rate was 55%, with 4% complete responses.

TABLE III. **Phase II Studies of Docetaxel (100 mg/m^2 every 3 weeks) in Patients With Metastatic Breast Cancer Without Prior Chemotherapy**

Institution (Reference)	*No. Patients*	*Total Responses*	*Complete Responses*
EORTC-CSG *(82)*	32	73%	6 (16%)
NCI-Canada *(83)*	34	65%	3 (9%)
Memorial Sloan-Kettering *(84)*	29	76%	2 (7%)
EORTC-ECTG[1] *(87)*	32	53%	2 (6%)
MD Anderson[1] *(88)*	33	55%	0 (0%)
UT San Antonio[1] *(89)*	26	60%	3 (11%)

[1]Patients in these trials had prior chemotherapy.
Abbreviations: *No.*, number

Lung Cancer, Non-Small Cell. Docetaxel 100 mg/m^2 *(91-93)* was given to 130 patients in the U.S. and Europe; in Japan 185 patients received 60 mg/m^2 as initial chemotherapy *(86,94)*. The preliminary overall objective response rates were 30% and 22%, respectively. Docetaxel was also tested as second or "salvage" chemotherapy in 69 patients treated with 100 mg/m^2 *(91,95)*. The preliminary overall objective response rate was 26%.

TABLE IV. **Phase II Studies of Docetaxel (100 mg/m^2 every 3 weeks) in Metastatic Non-Small Cell Lung Cancer without Prior Chemotherapy**

Institution	*No. of Patients*	*Responses*
Memorial Sloan-Kettering	29	38 %
EORTC	32	31 %
MD Anderson	39	33 %
UT San Antonio	30	30 %
MD Anderson[1]	39	27 %
UT San Antonio[1]	30	23 %

[1]Patients in these trials had received prior chemotherapy.
Abbreviations: *No.*, number

Lung Cancer, Small Cell. Twenty-eight patients were treated, and the preliminary overall response rate was 25% *(96)*.

Ovarian Cancer. Several studies have been conducted in which docetaxel was used as salvage therapy in a total of 271 platinum-exposed patients *(97-100)*. The preliminary overall objective response rate was 26%. The granulocytopenia is comparable in severity to that induced by paclitaxel 200 mg/m^2 infused over 24 hours. The depth, duration, and associated complications have been markedly reduced by the use of prophylactic G-CSF in our institution. At this time paclitaxel and docetaxel seem to be therapeutically equivalent in platinum-refractory of ovarian cancer. The toxicity profile and cost may differentiate the two. However, a randomized trial will be required to confirm this.

Gastrointestinal Cancers. Docetaxel was evaluated in 33 patients with gastric cancer *(101)* and 18 patients with pancreatic cancer *(102)*. The preliminary overall objective response rates were 24% and 28%, respectively. However, only one objective response was documented in 64 patients with colorectal cancer *(103-105)*. The preliminary activity in pancreatic cancer is very encouraging. A confirmatory phase II study is under way in our institution in untreated patients with pancreatic cancer.

TABLE V. **Phase II Studies of Docetaxel in Other Tumor Types**

Tumor Type	*Institutions (Reference)*	*Prior Therapy*	*No. of Patients*	*Responses*
Small Cell Lung	ECTG *(96)*	Pretreated	28	25 %
Ovarian	ECTG, CSG, MD Anderson *(97-100)*	Pretreated	271	26 %
Gastric	ECTG *(101)*	Untreated	33	24 %
Pancreatic	Institut Gustave Roussy *(102)*	Untreated	18	28 %
Melanoma	ECTG *(106)* MD Anderson *(107)*	Untreated	44	17 %
Head/Neck	EORTC *(108)*	Pre/untreated	37	32%
Sarcoma	EORTC *(109)*	Pretreated	29	17 %

Other Cancers. Objective responses were documented when docetaxel was used as initial chemotherapy for melanoma (17%) *(106,107)* and head and neck cancer (32%) *(108)* and in second- or third-line therapy for sarcomas (17%) *(109)*. There were no responses in patients with renal cell cancer *(110)*.

Docetaxel is one of the most active antineoplastic drugs in patients with metastatic breast, ovarian, or lung cancers, whether untreated or previously treated. Patients whose tumors were resistant to anthracyclines or cisplatin responded to docetaxel. The dose of 100 mg/m^2 produced higher rates of objective responses than lower doses such as 60 mg/m^2 or 75 mg/m^2. Duration of response and overall survival data are still not available for most studies.

Safety Profile. Many of the side effects of docetaxel are similar to those of paclitaxel. However, docetaxel has its own unique toxic effects.

Hematologic (Blood) Toxic Effects. Anemia has been mild in patients receiving docetaxel, with fewer than 20% of the courses associated with grade 2 or greater. Thrombocytopenia was infrequent (<5% of patients). Moderate to severe granulocytopenia was seen across all studies. The median nadir granulocyte count was approximately 0.3/mm^3. The granulocytopenia usually occurs early, between days 6 and 15 with a median starting day of 7 or 8. The median duration of granulocytopenia has been approximately 7 to 8 days. In patients with breast cancer, EORTC-CSG reported no episodes of neutropenic fever/infection in 168 courses in patients given docetaxel as initial chemotherapy for metastatic disease, but this was seen in 9% of 437 courses in patients at M. D. Anderson and in San Antonio. However, these latter patients had more aggressive disease which was resistant to doxorubicin.

Nonhematologic Toxic Effects. Hypersensitivity reactions were seen in the early clinical trials when no premedication was given. These reactions are usually manifested as rash, urticaria, dyspnea, or back pain. Patients experiencing severe hypersensitivity reactions may rarely develop bronchospasm or angioedema (swelling which can be lethal if the main air passage becomes occluded), sometimes with lowering of the blood pressure. Mild or moderate hypersensitivity reactions were noted in approximately 20% of patients and 10% of courses, but this varied significantly among studies *(111-113)*. Premedication with an antihistamine and a corticosteroid has nearly abolished this complication. The incidence decreased from 50% to 5% of patients in the M. D. Anderson trial *(114)*.

Other side effects of docetaxel are common to many other chemotherapy agents. Hair loss has been total and nearly universal. Gastrointestinal effects, including mild to severe nausea, vomiting, diarrhea, and mucositis, occurred in 10-25% of the courses; in most, however, these have been of moderate severity. Peripheral neuropathy was documented but was uncommon, mild, and mainly sensory. Fatigue and asthenia (a generalized sense of debility) of moderate to high severity (grades 2-4) were seen in 30-60%; the incidence of myalgia (muscle aches) varied greatly among studies, ranging from <5% to 20-40% of the courses.

Two unique toxic effects were also noted. Skin reactions were documented in 25-60% of the patients and in 20-35% of the courses prior to the administration of premedication. The skin manifestations included: 1) scattered maculopapular eruptions with desquamation (peeling and flaking of skin), usually localized to the extremities; 2) fingernail and toenail changes, including thinning, subungual (below the nail) erythema (redness), ridging of the nail plates, shedding of the nail either from above or below, and subungual hemorrhage; 3) hand-foot syndrome (redness, swelling, and tenderness of the hands and feet); 4) photodermatitis (skin redness and thickening in response to light, an enhanced sensitivity to sunlight); 5) reactive dermatitis; and 6) inflammation of the hair follicles *(20)*. In some patients the lesions appeared in areas of prior injury or areas of pressure. The use of corticosteroids has significantly decreased the cutaneous reactions, including the hand-foot syndrome *(88,113)*. Biopsies showed nonspecific collections of lymphocytes around blood vessels in the skin *(20)*.

The second unique toxic effect was fluid retention. This initially presented as a progressive increase in weight. Later, patients developed edema (swelling), mainly in the lower extremities; pleural effusions (fluid collections in spaces between the lungs and the chest wall); pericardial effusion (fluid in the space around the heart); and ascites (accumulation of fluid in the abdominal cavity) were also manifestations of this fluid retention. Without premedication, fluid retention was associated with a cumulative dose of more than 300 mg/m^2 *(111-113)*. The incidence of fluid retention was greater than 80% in long-term responders. It is not associated with abnormal function of the kidneys, heart, liver, lungs, or thyroid or adrenal glands. Vascular (blood vessel) permeability has been shown to increase in patients taking docetaxel *(114)*.

Two approaches have been taken to overcome these effects in a small number of patients. First, the dose of docetaxel was decreased to 75 mg/m^2, but this produced no definite evidence of improvement *(86,111)*. When doses were lowered to 60 mg/m^2, however, as in the Japanese studies, the incidence of these effects was under 10%. Second, premedication with steroids was used. This appeared to have an impact on the onset of fluid retention in the preliminary data of some studies *(113,115)*. The

NCI-Canada used a short-term premedication regimen similar to the one used for paclitaxel without any change in the severity or frequency of these effects *(111);* however, multiple-day administration of steroids appeared to delay the onset and the severity of significant fluid retention in the M. D. Anderson trial. The median number of courses the patient could receive before the onset of moderate fluid retention increased from three to eight after steroids were routinely used, but the numbers of patients in these trials were small. The EORTC-ECTG reported a decrease in fluid retention from 36% to 8% when a short course of methylprednisolone, cetirizine, and ketotifen was administered *(115).* Other studies showed a shift in the median onset of fluid retention to a cumulative dose of 500 mg/m^2.

Future Issues. Three areas are under extensive laboratory and clinical research in the development of docetaxel. First, synergistic combinations with other active antineoplastic agents such as cisplatin, anthracyclines, cyclophosphamide, 5-fluorouracil, and vinorelbine are being developed. Randomized studies to compare docetaxel with standard first- or second-line therapies in lung and breast cancer are ongoing or planned. Studies to assess the pattern of cross-resistance between docetaxel and paclitaxel must be done (a pilot trial of docetaxel in paclitaxel-resistant metastatic advanced breast cancer is under way). Second, more comprehensive studies in other tumor types for which docetaxel has demonstrated preliminary antitumor activity, including head and neck, pancreatic carcinoma, and sarcomas, are needed. Third, efforts to understand the pathophysiology of the skin reactions and fluid retention, as well as the development of effective preventive and therapeutic measures to enhance the safety profile, are needed. Several ongoing pilot studies and a major randomized trial with a different premedication regimen will address this issue.

Conclusions

After surmounting multiple preclinical and clinical problems, paclitaxel has demonstrated potent antineoplastic activity in ovarian, breast, lung, head and neck, and esophageal cancers. Paclitaxel is currently approved for the treatment of ovarian and breast cancer. Side effects have been acceptable. However, many issues regarding the optimal use of paclitaxel are unresolved: determination of the most effective and least toxic infusion duration; development of guidelines for use in patients whose liver function is compromised from tumor invasion or prior therapy; understanding of how tumors develop resistance to paclitaxel and how to prevent or reverse this; and development of synergistic combinations. Unlike many currently available drugs, development of synergistic combinations will require attention to the sequence in which the drugs are given as well as the infusion duration. The sequence and infusion duration of paclitaxel in combination with other drugs affect the type and severity of side effects. An analogue, docetaxel, appears to have even greater antineoplastic activity and may be safely and effectively given by a convenient 1-hour infusion. However, significant fluid retention occurs in most patients who have received five or more courses. Paclitaxel and docetaxel may be the first of a new family of potent antineoplastic drugs, and they are fertile areas of ongoing clinical and basic research.

Acknowledgments

The authors acknowledge the editorial assistance of Katy Hale, MLS, and the secretarial assistance of Pat Mayers, Melinda Garza, Laura Butanda, Esther Garza, and Judy Dillon.

Literature Cited

1. Schiff, P. B.; Horowitz, S. B. *Nature* **1979,** *22*, 665-667.
2. Rowinsky, E. K.; Cazenave, L. A.; Donehower R. C. *J. Natl. Cancer Inst.* **1990,** *82*, 1247-1259.
3. Weiss, R.; Donehower, R. C.; Wiernik, P. H.; et al. *J. Clin. Oncol.* **1990,** *8*, 1263-1268.
4. Arbuck, S. G.; Christiansen, M. C.; Fisherman, J. S.; Cazenave, L., A.; Sarosy, G.; Suffness, M.; Adams, J.; Canetta, R.; Cole, K. E.; Friedman, M. A. *Monogr. Natl. Cancer Inst.* **1993,** *15*, 11-24.
5. DeVita, V. T. In *Cancer: Principles and Practice of Oncology 4th ed.;* DeVita V. T.; Hellman, S.; Rosenberg, S. A., Eds.; J. B. Lippincott Co.: Philadelphia, PA, 1993; pp. 276-292.
6. Greene, M. H.; Clark, J. W.; Blaynew, D. W. *Semin. Oncol.* **1984,** *11*, 209-226.
7. Boring, C. C.; Squires, T. S.; Tong, T.; Montgomery, S. *CA* **1994,** *1*, 7-26.
8. Yancik, R.; Ries, L. G.; Yates, J. W. *Am. J. Obstet. Gynecol.* **1986,** *154*, 639-647.
9. In *Cancer Incidence in Five Continents, Volume 5;* Muir, C.; Waterhouse, J.; Mack, T.; Powell, J.; Whelan, S., Eds.; IARC Scientific Publication 88: Lyon, France, 1987.
10. Kaufman, D. W.; Kelly, J. P.; Welch, W. R.; Rosenberg, L.; Stolley, P. D.; Warshauer, M. E.; Lewis, J.; Woodruff, J.; Shapiro, S. *Am. J. Epidemiol.* **1989,** *130*, 1142-1151.
11. Hoskins, P. J.; O'Reilly, S. E.; Swenerton, K. D.; Spinelli, J. J.; Fairey, R. N.; Benedet, J. L. *J. Clin. Oncol.* **1992,** *10*, 1561-1568.
12. Neijt J. P.; ten Bokkel Huinink, W. W.; van der Bug, Mel; van Oosterom, A. T.; Willemse P. H. B.; Vermorken, J. B.; van Lindert, A. C. M.; Heintz, A. P. M.; Aartsen, E.; van Lent, M.; Trimbos, J. B.; de Meijer, A. J. *Eur. J. Cancer* **1991,** *27*, 1367-1372.
13. Sutton, G. P.; Stehman, F. B.; Einhorn, L. H.; Roth, L. M.; Blessing, J. A.; Ehrlich, C. E. *J. Clin. Oncol.* **1989,** *7*, 223-229.
14. Advanced Ovarian Cancer Trials Group. *BMJ* **1991,** *303*, 884-893.3
15. ten Bokkel Huinink, W. W.; Rodenhuis, S.; Simonetti, G.; Dubbelman, R.; Franklin, H.; Dalesio, O.; Vermorken, J. B.; McVie, J. G. In *Carboplatin (JM-9): Current Perspectives and Future Directions;* Bunn, P. A.; Canetta, R.; Ozols, R. F.; Rozencweig, M., Eds.; W. B. Saunders: Philadelphia, PA, 1990; pp. 165-173.
16. Markman, M.; Rothman, R.; Hakes, T.; Reichman, B.; Hoskins, W.; Rubin, S.; Jones, W.; Almadrones, L.; Lewis, Jr., J. L. *J. Clin. Oncol.* **1991,** *9*, 389-393.
17. Blackledge, G.; Lawton, F.; Redman, C.; Kelly, K. *Br. J. Cancer* **1989,** *59(4)*, 650-653.
18. Markman, M. *J. Clin. Oncol.* **1992,** *10*, 513-514.
19. Thigpen, J. T.; Vance, R. B.; Khansur, T. *Cancer* **1993,** *71*, 1559-1664.

20. Pazdur, R.; Kudelka, A. P.; Kavanagh, J. J.; Cohen, P. R.; Raber, M. N. *Cancer Trt. Rev.* **1993,** *99(4)*, 351-386.
21. McGuire, W. P.; Rowinsky, E. K.; Rosenshein, N. B.; Grumbine, F. C.; Ettinger, D. S.; Armstrong, D. K.; Donehower, R. C. *Ann. Intern. Med.* **1989,** *111*, 273-279.
22. Einzig, A.I.; Wiernik., P. H.; Sasloff, J.; Runowicz, C. D.; Goldberg, G. L. *J. Clin. Oncol.* **1992,** *10*, 1748-1753.
23. Thigpen, T.; Blessing, J.; Ball, H.; Hummel, S.; Barret, R. *Proc. Am. Soc. Clin. Oncol.* **1990,** *9*, 604.
24. McGuire, W. P. In *Am. Soc. Clin. Oncol. Educational Book, 30th Annual Meeting*, Dallas, TX, 1994; pp. 204-220.
25. Sarosy, G.; Kohn, E.; Stone, D. A.; Rothenberg, M.; Jacob, J.; Adamo, D. O.; Ognibene, F. P. ; Cunnion, R. E.; Reed, E. *J. Clin. Oncol.* **1992,** *10*, 1165-1170.
26. Sarosy, G.; Kohn, E.; Link, C.; Adamo, D.; Davis, P.; Ognibene, F.; Goldspiel, B.; Christian, M.; Reed, E. *Proc. Am. Soc. Clin. Oncol.* **1992,** *11*, 716.
27. Kavanagh, J. J.; Kudelka, A. P.; Edwards, C. L.; Freedman, R. S.; Gibbs, H.; Gonzalez de Leon, C.; Canetta, R.; Harper, K. J.; Kopplin, S.; Mante, R.; Krakoff, I. H. *Proc. Am. Soc. Clin. Oncol.* **1993,** *13*, 259.
28. Kavanagh, J. J.; Kudelka, A. P. Unpublished results; survival data updated October 1994.
29. Trimble, E. L.; Adams, J. D.; Vena, D.; Hawkins, M. J.; Friedman, M. A.; Fisherman, J. S.; Christian, M. C.; Canetta, R.; Oneteto, N.; Hayn, R.; Arbuck, S. G. *J. Clin. Oncol.* **1993,** *11,* 2405-2410.
30. Eisenhauer, E. A.; ten Bokkel Huinink, W. W.; Swenerton, K. D.; Gianni, L.; Myles, J.; van der Burg, M. E. L.; Kerr, I.; Vermorken, J. B.; Buser, K.; Colombo, N.; Bacon, M.; Santabarbara, P.; Onetto, N.; Winograd, B.; Canetta, R. *J. Clin. Oncol.* **1994,** in press.
31. Wolff, M. S.; Toriolo, P. G.; Lee, E. W.; Rivera, M.; Dubin, N. *J. Natl. Cancer Inst.* **1993,** *85*, 648-652.
32. White E. *Am. J. Public Health* **1987,** *77*, 495-497.
33. Harris, J. R.; Morrow M.; Bonadonna, G. In *Cancer: Principles and Practice of Oncology;* DeVita V. T.; Hellman, S.; Rosenberg, S. A., Eds.; J. B. Lippincott Co.: Philadelphia, PA, 1993; pp. 1278-1280.
34. In *Cancer: Principles and Practice of Oncology;* DeVita V. T.; Hellman, S.; Rosenberg, S. A., Eds.; J. B. Lippincott Co.: Philadelphia, PA, 1993; pp. 1315-1332.
35. Holmes, F. A.; Walters, R. S.; Theriault, R. L.; Forman, A. D.; Newton, L., K.; Raber, M. N.; Buzdar, A. U.; Frye, D. K.; Hortobagyi, G. N. *J. Natl. Cancer Inst.* **1991,** *83*, 1797-1805.
36. Reichman, B. S.; Seidman, A. D.; Crown, J. P. A.; Hoolan, R.; Hakee, T. B.; Lebwohl, E. E.; Gilewski, T. A.; Surbono, A.; Currie, V.; Hudis, C. A.; Yoo, T. J.; Klecker, R.; Jamis-Dow, C.; Collins, J.; Quninllvan, S.; Berkery, R.; Toomasi, F.; Canetta, R.; Fisherman, J.; Arbuck, S.; Norton, L. *J. Clin. Oncol.* **1993,** *11*, 1943-1951.
37. Naboholtz, J. M.; Gelmon, K.; Bontenbal, M.; et al. *Proc. Am. Soc. Clin. Oncol.* **1993,** *12*, 61 (abstract 42).
38. Seidman, A. D.; Crown J. P. A.; Reichman, B. S.; et al. *Proc. Am. Soc. Clin. Oncol.* **1993,** *12*, 63 (abstract 53).

39. Holmes, F. A.; Valero, V.; Walters, R. S.; Theriault, R. L.; Booser, D. S.; Fraschini, G.; Buzdar, A. U.; Frye, D.; Gibbs, H. R.; Hortobagyi, G. N.; *Monogr. Natl. Cancer Inst.* **1993,** *15*, 161-170.
40. Wilson, W. H.; Berg, S.; Bryant, G.; Wittes, R. E.; Bates, S.; fojo, A.; Steinberg, S. M.; Goldspiel, B. R.; Herdt, J.; O'Shaughnessy, J.; Balis, F. M.; Chabner, B. A.; *J. Clin. Oncol.* **1994,** *12*, 1621-1629.
41. Fisherman J. S.; McCabe, M.; Hillig M.; Ognibene, F.; Goldspiel, B.; Venzon, D. J.; Cowan, K. H.; O'Shaughnessy, J. A. *Monogr. Natl. Cancer Inst.* **1993,** *14*, 189-194.
42. Holmes, F. A.; Newman R. A.; Madden, T.; et al. *Ann. Oncol.* **1994,** *5* (Suppl. 5), 197 (abstract 489).
43. Pestalozzi, B. C.; Sotos, G. A.; Choyke, P. L.; Fisherman, J. S.; Cowan, K. H., O'Shaughnessy, J. A. *Cancer* **1993,** *71*, 1797-1800.
44. Sledge, G. W.; Goldstein, R. N.; Sparano, J.; Cobleigh, M.; Baughman, C.; Neuberg, D.; Rowinsky, E. *Eur. J. Cancer* **1993,** *29A (Suppl. 6)*, S79 (abstract 413).
45. Gianni, L.; Straneo, M.; Capri, G.; Villani, E.; Munzone, E.; Bonadonna, G. *Proc. Am. Soc. Clin. Oncol.* **1994,** *13*, 74 (abstract 97).
46. Gelmon, K. A.; O'Reilly, S., Plenderleith, I. H.; Bryce, C.; Ragaz, J.; Coppin, C.; Campbell, C.; Healey, D.; Onetto, N. *Proc. Am. Soc. Clin. Oncol.* **1994,** *13*, 71 (abstract 87).
47. Kennedy, M. J.; Armstrong, D.; Donehower, R.; Noe, D.; Sartorius, S.; Chen, T-L.; Bowling, K.; Rowinsky, E. *Proc. Am. Soc. Clin. Oncol.* **1994,** *13*, 137 (abstract 342).
48. Baselga, J.; Norton, L.; Coplan, K.; et al. *Proc. Am. Assoc. Cancer Res.* **1994,** *35*, 380 (abstract 2262).
49. Deisseroth, A. B.; *Ann. Oncol.* **1994,** *5* (Suppl. 5), 68 (abstract 8).
50. Hudis, C.; Seidman, A.; Baselga, J.; Raptis, G.; Lebwohl, D.; Gilewski, T.; Moynaham, M.; Sklarin, N.; Fennelly, D.; Crown, J.; Uhlenhopp, M.; Riedel, E.; Yao, T. J.; Norton, L. *Proc. Am. Soc. Clin. Oncol.* **1994,** *13*, 65 (abstract 62).
51. Boring, C. C.; Squires, T. S.; Tong, T. *CA Cancer J. Clin.* **1992,** *42*, 19-38.
52. Ginsberg, R. J.; Kris, M. G.; Armstrong, J. G. In *Cancer: Principles and Practice of Oncology;* DeVita V. T.; Hellman, S.; Rosenberg, S. A., Eds.; J. B. Lippincott Co.: Philadelphia, PA, 1993; pp. 673-722.
53. Murphy, W. K.; Fossella, F. V.; Winn, R. J.; Shin, D. M.; Hynes, H. E.; Gross, H. M.; Davilla, E.; Leimert, J.; Dhingra, K.; Raber, M. N.; Krakoff, I. H.; Hong, W. K. *J. Natl. Cancer Inst.* **1993,** *85*, 384-388
54. Chang, A. Y.; Kim, K.; Glick, J.; Anderson, T.; Karp, D.; Johnson, D. *J. Natl. Cancer Inst.* **1993,** *85*, 388-394.
55. Ruckdeschel, J.; Wagner, H.; Williams, C.; Heise, M.; Hilstro, J. *Proc. Amer. Soc. Clin. Oncol.* **1994,** *13*, 357 (abstract).
56. Murphy, W. K.; Winn, R. J.; Huber, M.; Fossella, F. V.; et al: *Proc. Amer. Soc. Clin. Oncol.* **1994,** *13*, 1224 (abstract).
57. Citardi, M.; Rowinsky, E. K.; Schaefer, K. L.; et al. *Proc. Am. Assoc. Cancer Res.* **1991,** *31*, 2431 (abstract).
58. Dillman, R. O.; Seagren, S. L.; Propest, K. J.; et al. *N. Engl. J. Med.* **1990,** *323*, 940-945.

59. Tishler, R. B.; Geard, R. B.; Hall, E. J.; Schiff, P. B. *Cancer Res.* **1992,** *52,* 3495-3497.
60. Liebmann, J. E.; Cook, J. A.; Lipschultz, C.; Teague, D.; Fisher, J.; Mitchell, J. B. *Br. J. Cancer* **1993,** *68,* 1104-1109.
61. Choy, H.; Akerley, W.; Safarin, H.; Brown, M.; Rege, V.; Papa, A.; Glantz, M.; Capistrano, M.; Puthawal, Y.; Soderberg, C.; Leone, L. *Proc. Am. Soc. Clin. Oncol.* **1994,** *13,* 1210 (abstract).
62. Ihde, D. C.; Pass, H. I.; Glatstein, E. J. In *Cancer: Principles and Practice of Oncology;* DeVita V. T.; Hellman, S.; Rosenberg, S. A., Eds.; J. B. Lippincott Co.: Philadelphia, PA, 1993; pp. 723-758.
63. Ettinger, D. S. *Semin. Oncol.* **1993,** *20,* 46-49.
64. Kirschling, R. J.; Jung, S. H.; Jett, J. R. *Proc. Am. Soc. Clin. Oncol.* **1994,** *13,* 1076 (abstract).
65. Wolf, G.; Lippman, S. M.; Laramore, G.; Hong, W. K. In *Cancer Medicine* 3rd ed.; Holland, J. F.; Frei, E.; Bast, R. C. Jr.; Kufe, D. W.; Morton, D. L.; Weichselbaum, R., Eds.; Lea & Febiger: Philadelphia, PA, 1993; pp. 1211-1275.
66. Forastiere, A. A. *Semin. Oncol.* **1993,** *20,* 56-60.
67. Thornton, D.; Singh, K..; Putz, B.; Gams, R.; Schuller, D.; Smith, R. *Proc. Am. Soc. Clin. Oncol.* **1994,** *13,* 933 (abstract).
68. Boring, C. C.; Squires, T. S.; Tong, T. *CA Cancer J. Clin.* **1993,** *43,* 7-26.
69. Blot, W. J.; Devesa, S. S.; Kneller, R. W.; Fraumeni, J. F. *J. A. M. A.* **1991,** *265,* 1287-1289.
70. Levi F.; La Vecchia, C. *J. A. M. A.* **1991,** *265,* 2960 (letter).
71. Powell, J.; McConkey, C. C. *Br. J. Cancer* **1991,** *62,* 440-443.
72. Reed, P. I. *Lancet* **1991,** *338,* 178.
73. Ajani, J. A.; Ilson, D. H.; Daugherty, K.; Pazdur, R.; Lynch, P. M.; Kelsen, D. P. *J. Natl. Cancer Inst.* **1994** (in press).
74. Brown, T.; Tangen, C.; Fleming, T.; Macdonald, J. *Proc. Am. Soc. Clin. Oncol.* **1993,** *12,* 200 (abstract).
75. Guéritte-Voegelein, F.; Guénard, D.; Lavelle, F.; Le Goff, M. T.; Mangatal, L.; Potier, P. J. *Med. Chem.* **1991,** *34,* 992-998.
76. Ringel, I.; Horwitz, S. B. *J. Natl. Cancer Inst.* **1991,** *83,* 288-291.
77. Riou, J. F.; Naudin, A.; Lavelle, F. *Biochem. Biophys. Res. Comm.* **1992,** *187,* 164-170.
78. Bissery, M. C.; Guénard, D.; Guéritte-Voegelein, F.; Lavelle, F. *Cancer Res.* **1991,** *51,* 4845-4852.
79. Harrison, S. D., Jr.; Dykes, D. J.; Shepherd, R. V.; Griswold, D. P., Jr.; Bissery, M. C. *Proc. Am. Assoc. Cancer Res.* **1992,** *33,* 526 (abstract).
80. Bissery, M. C.; Vrignaud, P.; Bayssas, M.; Lavelle, F. *Proc. Am. Assoc. Cancer Res.* **1994,** *35,* 327 (abstract).
81. Piccart, M. J. In *Am. Soc. Clin. Oncol. Educational Book, 29th Annual Meeting;* Bostrom Corporation: Orlando, FL, 1993; pp. 25-32.
82. Fumoleau, P.; Chevallier, B.; Kerbrat, P.; Dieras, V. Le Bail, N.; Bayssas, M.; Van Glabbeke, M. *Proc. Am. Soc. Clin. Oncol.* **1993,** *12,* 56 (abstract).
83. Trudeau, M. E.; Eisenhauer, E.; Lofters, W.; Norris, B.; Muldal, A.; Letendre, F.; Vandenburg, T.; Verma, S. *Proc. Am. Soc. Clin. Oncol.* **1993,** *12,* 64 (abstract).

84. Seidman, A. D.; Hudis, C.; Crown, J. P. A.; Balmaceda, C.; Lebwohl, D.; Currie, V.; Gilewski, T.; Hakes, T.; Robles, M.; Klem, K.; Lepore, J.; Norton, L. *Proc. Am. Soc. Clin. Oncol.* **1993,** *12,* 63 (abstract).
85. Dieras, V.; Fumoleau, P.; Chevallier, B.; Kerbrat, P.; Krakowski, Y.; Roche, H.; Misset, J. L.; Lentz, M. A.; Azli, N.; Pouillart, P. *Proc. Am. Soc. Clin. Oncol.* **1994,** *13,* 78 (abstract).
86. Data on file; Rhône-Poulenc Rorer, Antony, France and Collegeville, PA, USA.
87. Ten Bokkel Huinink, W. W.; Prove, A. M.; Piccart, M.; Steward, W.; Tursz, T.; Wanders, J.; Franklin, H.; Clavel, M.; Verweij, J.; Alakl, M.; Bayssas, M.; Kaye, S. B. *Ann. Oncol.* **1994,** *5(6),* 527- .
88. Valero, V.; Walters, R.; Theriault, L.; Esparza, L.; Holmes, F.; Fraschini, G.; Plasse, T.; Bellet, R.; Buzdar, A.; Hortobagyi, G. *Proc. Am. Soc. Clin. Oncol.* **1994,** *13,* 470 (abstract).
89. Burris, H. A.; Ravdin, P. M.; Fields, S. M.; Smith, L. S.; Peacock, N. W.; Elledge, R. M.; Von Hoff, D. D. *Breast Cancer Res. Treat.* **1993,** *27,* 132.
90. Burris, H.; Eckardt, J.; Fields, S.; Rodriguez, G.; Smith, L.; Thurman, A.; Peacock, N.; Kuhn, J.; Hodges, S.; Bellet, R.; Bayssas, M.; Le Bail, N.; Von Hoff, D. *Proc. Am. Soc. Clin. Oncol.* **1993,** *12,* 335 (abstract).
91. Francis, P. A.; Rigas, J. R.; Kris, M. G.; Pisters, K. M. W.; Orazem, J. P.; Woolley, K. J.; Heelan, R. T. *J. Clin. Oncol.* **1994,** *12,* 1232-1237.
92. Fossella, F. V.; Lee, J. S.; Murphy, W. K.; Lippman, S. M.; Calayag, M.; Pang, A.; Chasen, M.; Shin, D. M.; Glisson, B.; Benner, S.; Huber, M.; Perez-Soler, R.; Hong, W. K.; Raber, M. *J. Clin. Oncol.* **1994,** *12,* 1238-1244.
93. Watanabe, K.; Yokoyama, A.; Furuse, K.; Nakai, Y.; Ohnoshi, T.; Kudo, S.; Niitani, H. *Proc. Am. Soc. Clin. Oncol.* **1994,** *13,* 331 (abstract).
94. Fossella, F. V.; Lee, J. S.; Shin, D. M.; Calayag, M.; Perez-Soler, R.; Huber, M.; Murphy, W. K.; Benner, S.; Lippman, S.; Glisson, B.; Raber, M.; Hong, W. K. *Proc. Am. Soc. Clin. Oncol.* **1994,** *13,* 336 (abstract).
95. Smyth, J. F.; Bowman, A.; Smith, I.; Sessa, C.; Kaye, S. *Eur. J. Cancer.* **1993,** *29A (Suppl. 6),* S154.
96. Piccart, M. J.; Gore, J.; ten Bokkel Huinink, W.; Van Oosterom, A.; Verweiji, J.; Kaye, S.; Wanders, J.; Franklin, H.; Le Bail, N.; Bayssas, M. *Proc. Am. Soc. Clin. Oncol.* **1993,** *12,* 258.
97. Kavanagh, J.; Kudelka, A.; Freedman, R.; Edwards, C.; Bayssas, M.; Bellet, R.; Harper, K.; Newman, B.; Raber, M. *Proc. Am. Soc. Clin. Oncol.* **1994,** *13,* 237 (abstract 732).
98. Aapro, M. S.; Pujade-Lauraine, E.; Lhommé, C.; Fumoleau, P.; Kerbrat, P.; Lentz, M. A.; Azli, N.; Chevallier, B. *Ann. Oncol.* **1994,** *5(Suppl. 5),* 202.
99. Francis, P.; Hakes, T.; Schneider, J.; Spriggs, D.; Fennelly, D.; Hann, L.; Barakat, R.; Curtin, J.; Jones, W.; Lewis, Jr., J.; Phillips, M.; Hoskins, W. *Proc. Am. Soc. Clin. Oncol.* **1994,** *13,* 260 (abstract 825).
100. Sulkes, A.; Cavalli, F.; van Oosterom, A.; Kaye, S.; Wanders, J.; Franklin, H.; Le Bail, N.; Bayssas, M. *Eur. J. Cancer.* **1993,** *29A (Suppl. 6),* S101.
101. Rougier, P. H.; De Forni, M.; Adenis, A.; Ducreux, M.; Djazouli, K.; Adams, D.; Bonneterre, J.; Clouet, P.; Blanc, C.; Bayssas, M.; Armand, J. P. *Proc. Am. Soc. Clin. Oncol.* **1994,** *13,* 200 (abstract).

102. Clark, T.; Kemeny, N.; Conti, J. A.; André, M. *Proc. Am. Soc. Clin. Oncol.* **1994,** *13,* 212 (abstract).
103. Sternberg, C. N.; Ten Bokkel Huinink, W.; Kaye, S.; Bruntsch, U.; van Oosterom, A. T.; Pavlidis, N.; Robinson, E.; Clavel, M.; Franklin, H.; Wanders, J.; Le Bail, N.; Bayssas, M. *Eur. J. Cancer.* **1993,** *29A (Suppl. 6),* S100.
104. Pazdur, R.; Lassere, Y.; Soh, L. T.; Ajani, J. A.; Bready, B.; Soo, E.; Sugarman, S.; Patt, Y.; Abbruzzese, J. L.; Levin, B. *Ann. Oncol.* **1994,** *5(5),* 468- .
105. Aamdal, S.; Verweij, J.; Piccart, M.; Bruntsch, U.; Paridaens, R.; Sulkes, A.; Kaye, S.; Wanders, J.; Franklin, H.; Bayssas, M.; Le Bail, N. *Eur. J. Cancer.* **1993,** *29A (Suppl. 6)* S180.
106. Bedikian, A.; Legha, S.; Eton, O.; Buzaid, A.; Benjamin, R. *Proc. Am. Assoc. Cancer Res.* **1994,** *35,* 86 (abstract).
107. Catimel, G.; Verweij, J.; Mattijssen, V.; Hanauske, A.; Piccart, M.; Wanders, J.; Franklin, H.; Le Bail, N.; Clavel, M.; Kaye, S. B. *Ann. Oncol.* **1994,** *5(6),* 533- .
108. van Hoesel, Q. G. C. M.; Verweij, J.; Catimel, G.; Clavel, M.; Kerbrat, P.; Bui, N. B.; Kerger, J.; van Oosterom, A. T.; Kerger, J.; Tursz, T.; van Glabbeke, M.; van Pottelsberghe, C.; Le Bail, N.; Mouridsen, H. *Ann. Oncol.* **1994,** *5(6),* 539- .
109. Mertens, W. C.; Eisenhauer, E.; Jolivet, J.; Ernst, S.; Moore, M.; Muldal, A. *Ann. Oncol.* **1994,** *5(Suppl. 5),* S203.
110. Eisenhauer, E. A.; Lu, F.; Muldal, A.; Trudeau, M.; Latreille, J.; Mertens, W. *Ann. Oncol.* **1994,** *5 (Suppl. 5),* S202.
111. Wanders, J.; Schrijvers, D.; Bruntsch, U.; Gore, M.; Verweij, J.; Hanauske, A. R.; Franklin, H.; Roelvink, M.; Bayssas, M.; Kaye, S. B. *Proc. Am. Soc. Clin. Oncol.* **1993,** *12,* 73 (abstract).
112. Galindo, E.; Kavanagh, J.; Fossella, F.; Valero, V.; Bedikian, Y.; Pazdur, R. *Proc. Am. Soc. Clin. Oncol.* **1994,** *13,* 164 (abstract).
113. Oulid-Aissa, D.; Béhar, A.; Spielmann, M.; Kayitalire, L.; Chau, A.; Plasse, T.; Le Bail, N. *Proc. Am. Soc. Clin. Oncol.* **1994,** *13,* 465 (abstract).
114. Schrijvers, D.; Wanders, J.; Dirix, L.; Prove, A.; Vonck, I.; van Oosterom, A.; Kaye, S. *Ann. Oncol.* **1993,** *4,* 610-611.

RECEIVED September 19, 1994

Chapter 4

The Pacific Yew Environmental Impact Statement

S. J. Campbell and S. A. Whitney

Natural Resources Group, Pacific Northwest Region, U.S. Department of Agriculture Forest Service, Portland, OR 97208

Pacific yew, an evergreen tree found in the forests of the Pacific Northwest, contains the cancer-fighting drug taxol. The need to harvest Pacific yew from federal lands to obtain taxol for research and commercial sale precipitated a number of federal actions in the early 1990's. These included passage of a law protecting Pacific yew, a three-state inventory of the species, cooperative agreements between three federal agencies and a major pharmaceutical company, and an analysis of the impacts of Pacific yew harvest on the environment. The analysis is documented in an environmental impact statement, published in September of 1993.

Pacific yew, *Taxus brevifolia,* is native to western North America. It ranges from southeast Alaska to northern California and from the Pacific coast to interior Idaho and Montana (Figure 1). A rather inconspicuous member of the forest, it usually grows slowly beneath the forest canopy. In the drier, more easterly portions of its range it grows in the open as a shrub, rarely reaching more than five feet in height. Pacific yew is a member of the Taxaceae family, distinguished by evergreen foliage and seed enclosed in arils (red, fleshy, berry-like structures) rather than cones. The bark of Pacific yew is scaly, reddish-brown to purple, and extremely thin (2 to 6 mm in thickness). Most parts of the Pacific yew plant - the bark, wood, and needles - contain the compound taxol. Taxol is now a registered trademark name. However, due to greater familiarity with the word taxol, we use it in this article in lieu of "paclitaxel".

The ecological role of Pacific yew in the forest is not well understood, especially in comparison to our knowledge of many other tree species. We know, though, that Pacific yew is utilized as food or habitat by many other forest species, ranging from insects to spotted owls (Figure 2). Traditional and contemporary uses

0097–6156/95/0583–0058$08.00/0

Figure 1. Geographical Range of Pacific Yew, *Taxus brevifolia*. Reproduced from Pacific Yew Environmental Impact Statement, 1993, USDA Forest Service.

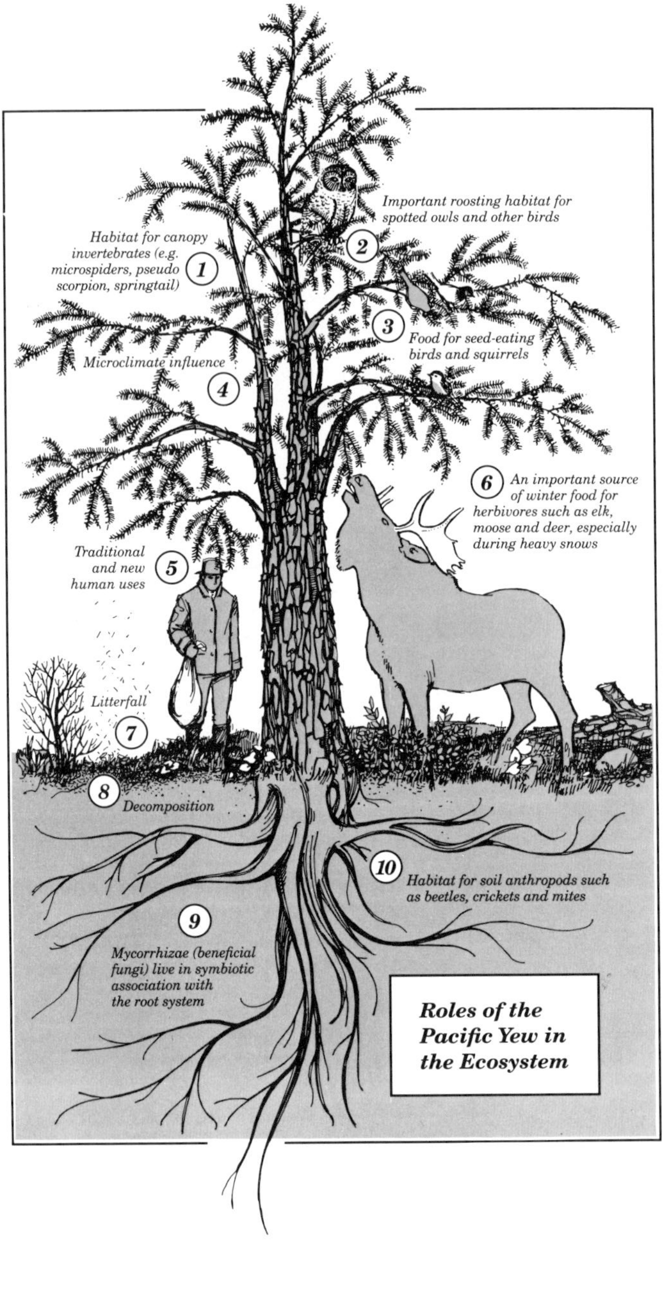

Figure 2. Ecological Roles of Pacific Yew. Reproduced from Pacific Yew Environmental Impact Statement, 1993, USDA Forest Service.

by humans are numerous as well: Native Americans valued and used the extremely hard and decay-resistant wood for tools, weapons, and ceremonial and decorative items. Foliage, fruits, and bark were used medicinally. Today, bow-makers and various other craftsmen seek out Pacific yew for its beauty and the same practical qualities that earlier peoples did.

But apart from firewood, fence wood, and wood for craftsmen, Pacific yew historically has not been used by the forest industry. It has not been harvested for pulp or lumber; it was not normally inventoried along with the other major tree species on federal or private forest land; and it often was piled and burned along with the logging slash and other vegetation remaining after timber harvest.

Forest Service and BLM Participation in the Taxol Program

In 1962, the National Cancer Institute (NCI) collected Pacific yew bark while collecting new plant material to screen for cancer fighting properties. Extracts from the bark showed *in vitro* activity against cancer cells. Isolation of the active principle, determination of the structure and mechanism of action, and toxicology studies were carried out through the 1960's and 70's. Clinical trials started in 1983 and, in 1989, NCI began working with the two major federal land management agencies in the west - the Forest Service and the Bureau of Land Management (BLM) - to obtain a supply of yew bark for expanded clinical trials.

Following the selection of Bristol-Myers Squibb Company by NCI in January of 1991 to develop taxol for commercialization, a Memorandum of Understanding was signed in June of the same year by the Secretaries of Health and Human Services (the parent agency for NCI), Department of Agriculture (parent agency for the Forest Service), and the Department of the Interior (parent agency for the BLM) to formalize the working partnership between the three Departments. Separate cooperative agreements between Bristol-Myers Squibb and each land management agency were developed to facilitate the transfer of Pacific yew bark to Bristol-Myers Squibb.

Bark Harvest on Federal Lands. In 1989, approximately 60,000 pounds of dry bark were collected from federal lands and sent to NCI. In 1990, 74,000 pounds were harvested. In 1991, the year the MOU was signed, almost 850,000 pounds were harvested. Demand for bark from national forests and BLM lands was projected to be at least 750,000 pounds per year until other sources of taxol were developed by Bristol-Myers Squibb. Private landowners were also harvesting bark for Bristol-Myers Squibb.

Federal yew bark harvest occurred primarily in areas where timber harvest (of other commercial tree species) had already taken place or was scheduled to occur later. Harvest usually consisted of cutting the yew tree down, peeling the bark off the trunk and larger limbs with the aid of chisels or hatchets, and then bagging the bark. Once all the large yew in an area were peeled, the bagged bark was taken to one of several processing plants in the northwest, chipped and dried, and then sent on for further processing for taxol extraction.

Events Leading to an Environmental Impact Statement

While bark peelers were harvesting yew from northwest forests for the purpose of producing taxol, many questions remained unanswered about the impacts of the harvest on forest ecosystems. The public spoke out. Some people believed the Forest Service and BLM were proceeding with yew harvest without adequately assessing the impacts and without complying with the law (National Environmental Policy Act); others thought yew bark was being wasted due to less than thorough peeling of each yew tree or because some yew trees in timber sale areas were left unpeeled. Poachers cut and peeled yew trees or even stripped the bark from living trees in federal and private woodlands, public parks, and citizens' yards and many people viewed bark stealing as an issue the agencies needed to control quickly. Some groups and individuals feared that Bristol-Myers Squibb would monopolize the collection, production, and sale of taxol. Many concerns surfaced about yew harvest, the impacts on the environment, and the production of taxol for research and cancer treatment. Information and guidelines were needed.

Interim Guide Team. Before the 1991 season of yew bark collection, the Forest Service and BLM established an interageny team of scientists (Forest Service, BLM, State of Washington, and Oregon State University) to assess the impacts of yew harvest and develop guidelines for the conservation of the species. This team produced "The Interim Guide to the Conservation and Management of Pacific Yew" (*1*), a 72-page book that outlines Pacific yew harvest and management on a long-term basis while ensuring the survival of the species. It was based on the known biological and ecological information that existed for Pacific yew as well as on general ecological and genetic principles. The Guide was completed in 1992; it governed the 1992 yew harvest season.

The Yew Act, 1992. Concurrent with the preparation of the Pacific Yew EIS, Congress, responding to public concern about the need for taxol and the waste of yew trees, drafted the Yew Act, H.R. 3836. This Act ensures that federal lands will be managed to provide for the sustainable harvest and long-term conservation of the Pacific yew. While the Act ensures yew and ecosystem protection, it also ensures that Bristol-Myers Squibb or other qualified companies could obtain the yew. The purpose of the Act is: to provide for the efficient collection and utilization of those parts of the Pacific yew that can be used to manufacture taxol for the treatment of cancer; to provide for the sale of Pacific yew from federal lands and the subsequent sale of taxol at a reasonable cost to cancer patients; ensure the long-term conservation of Pacific yew; and prevent the wasting of Pacific yew while successful and affordable alternative methods of manufacturing taxol are being developed.

NEPA. The National Environmental Policy Act of 1969 (as amended), the basic national charter for protection of the environment, requires preparation of an environmental impact statement (EIS) whenever a major project or action that might affect the environment is proposed for federal lands. An environmental impact statement serves to ensure policies and goals of the National Environmental Policy Act are infused into the programs and actions of the federal government. It provides full and fair disclosure of significant environmental impacts and provides reasonable proposals for the decision-maker and the public that avoid or minimize adverse impacts to the environment or enhance the quality of the human environment.

Because yew harvest could potentially impact the environment, Forest Service officials, in 1991, named a team to analyze the possible impacts and prepare the environmental impact statement. The Pacific Yew EIS was a cooperative effort between three federal agencies: the two land management agencies (Forest Service and BLM) and the Food and Drug Administration (FDA). The Forest Service and the BLM used the results of the EIS analysis to decide how to harvest yew on federal lands; the decision was documented in a joint Record of Decision. The FDA used the draft EIS to aid in the decision to approve the New Drug Application (submitted by Bristol-Myers Squibb) for sale and use of taxol from Pacific yew bark in treating patients with refractory ovarian cancer.

The Environmental Impact Statement

Environmental impact statements follow a format recommended in Forest Service and BLM implementing regulations. Four major chapters contain the heart of the environmental analysis: Chapter 1, the proposed action and the purpose and need for it; Chapter 2, the issues and alternatives; Chapter 3, the affected environment; and Chapter 4, the environmental consequences.

Proposed Action and Need. Harvest of Pacific yew (*Taxus brevifolia*) for taxol from public lands administered by the Forest Service and the Bureau of Land Management (BLM) is the proposed federal action addressed in the Pacific Yew Environmental Impact Statement (EIS) (*2*).

The need for harvest of federal yew was driven by the need for taxol for cancer research and treatment. Federal lands were thought to contain most of the yew population in the Pacific Northwest. And, at the time, only taxol from Pacific yew bark had FDA approval for use in clinical trials in the United States and the only FDA-approved process for taxol production was extraction of taxol from the bark of Pacific yew.

Use of Pacific yew bark for taxol production was viewed as a transition source by both FDA and Bristol-Myers Squibb. Other methods of taxol production were being developed at a rapid pace: semi-synthesis using *Taxus* needles, nursery production of *Taxus* spp, cell-culture, and full synthesis. The expectation was that one or more of these alternate sources would be developed and operational within five years. As these alternate sources "scaled up," harvest of Pacific yew would "scale down". Pacific yew taxol was considered a temporary, but vital, link in the development of taxol from sustainable sources.

Issues. The EIS process begins with public involvement to determine the issues and culminates in a decision based primarily on the results of the analysis and public input. Three main issues were distilled from comments from the general public, members of interest groups, and government employees who participated in early public involvement ("scoping"): provide material from the Pacific yew for taxol; protect the ecosystem; and protect the Pacific yew and maintain its genetic diversity. The main items commentors wanted the team to consider included: establish a sustainable level of collection -- analyze a range of minimum to maximum levels; establish areas of collection; utilize the yew completely -- all bark, twigs, needles, and wood; establish collection methods; regenerate yew -- plant and manage for natural regeneration; stop theft and illegal harvest; protect the Pacific yew gene pool -- establish reserve areas; protect the ecosystem -- riparian areas, wildlife, other plants, soil, fire cycles, old growth forests; develop other sources of taxol as soon as possible; consider the economic impacts of yew collection on timber production, local employment, sustained forest ecology to ensure the future supplies of taxol and other possible drugs, and monopolistic agreements for taxol production. Many people expressed a desire to see some level of yew harvest as long as it did not harm the ecosystem and did not decimate the population of Pacific yew. These issues and comments helped form the alternatives and raise questions for analysis.

Alternatives. Seven alternate programs of yew harvest were proposed and analyzed in the EIS. All except the "No action" alternative respond to the need for Pacific yew for taxol. The alternatives range from no yew harvest, to yew harvest in timber sale units only, to varying degrees of harvest outside timber sales, including spotted owl habitat. The main differences between alternatives are the amount of yew that would be harvested, where it would be harvested, and the amount of protection it would be given. The alternatives are graphically displayed in Figure 3.

Analysis. Determining the impacts of each alternative on the environment depended on two vital pieces of information: "The Interim Guide to the Conservation and Management of Pacific Yew" (*1*) and the Pacific yew inventory.

The Guide was completed shortly after the Pacific Yew EIS was begun and provided the EIS team with state-of-the-art information about Pacific yew. Many of the conclusions about the impacts of the alternatives on various aspects of the environment and on Pacific yew itself are based on the Guide. In summary, it stated: harvest no yew if sites of 20 to 100 acres within local management areas of about 20,000 acres do not contain at least 500 mature yew trees; establish genetic reserve areas of 20 to 100 acres for every 2,000 foot elevation band in the management area; harvest no yew within 75 feet of perennial streams; incorporate current Port-Orford cedar management guidelines where Port-Orford cedar is present; incorporate current owl management guidelines within 1/4 mile of northern spotted owl nests; follow guidelines for moose winter range; leave a 12 inch stump with bark on it to ensure resprouting; regenerate yew with seedlings, cuttings, layering, or sprouting; leave either 50 percent of the yew per acre or five yew trees, whichever is greater.

The Pacific yew inventory was the other piece of information necessary to determine the effects of each alternative. Until the inventory was completed, there was little data telling the Forest Service and BLM how much Pacific yew was present in the forest, where it was located, and how big it was. Yew was not consistently inventoried in the past because it was not a valuable commercial species. With the increasing demand for Pacific yew for taxol, it became imperative that the scope of the yew resource be known - both for the sake of conserving the yew species and for determining the amount of taxol that this source could provide. An inventory method was designed specifically to measure Pacific yew and field work on six national forests began in the fall of 1991. A year later, a total of eight national forests and several BLM districts in Oregon, Washington, and Idaho had been inventoried for Pacific yew. The information obtained from the inventory allowed the EIS team to estimate the number of yew trees and the amount of bark and needles potentially available under each alternative. It also gave the team an idea of the relative abundance, distribution, and habitat of the species.

The analysis focused on the impacts of each alternative on the Pacific yew, the ecosystem, and on people and their need for taxol. The results of the analysis are summarized in Table 1.

Preferred Alternatives. The National Environmental Policy Act allows a preferred alternative to be identified in the draft EIS, based on the results of the analysis. Preferred alternatives are recommendations by the EIS team; the final decision rests with the responsible official(s). Following public comments on the draft, the preferred alternative may be changed to another alternative, be modified, or remain the same.

In the case of the Pacific Yew EIS, Alternative D was identified as the preferred alternative in the draft EIS; it attempts to strike a balance between the demand for yew bark for taxol and the need to protect and conserve the species and the ecosystem. Prior to completion of the final EIS, Bristol-Myers Squibb announced it would no longer need Pacific yew bark due to its success in developing an alternate source of taxol. In keeping with these new developments and public comments, Alternative B was identified as the preferred alternative in the final EIS. Alternative B would maximize yew and ecosystem protection while providing a small quantity of yew products for taxol research and development efforts if needed.

Record of Decision and EIS Implementation. The Pacific Yew EIS Record of Decision was signed by the Forest Service Regional Forester for the Pacific Northwest Region and the BLM Director for the State of Oregon, the responsible officials for this EIS. It documents their decision for Pacific yew harvest for taxol on lands administered by the Forest Service and BLM. Both selected the preferred alternative, Alternative B, identified in the final EIS.

Alternative B has never been implemented. By the time the final EIS was published (September of 1993), Bristol-Myers Squibb had made significant progress

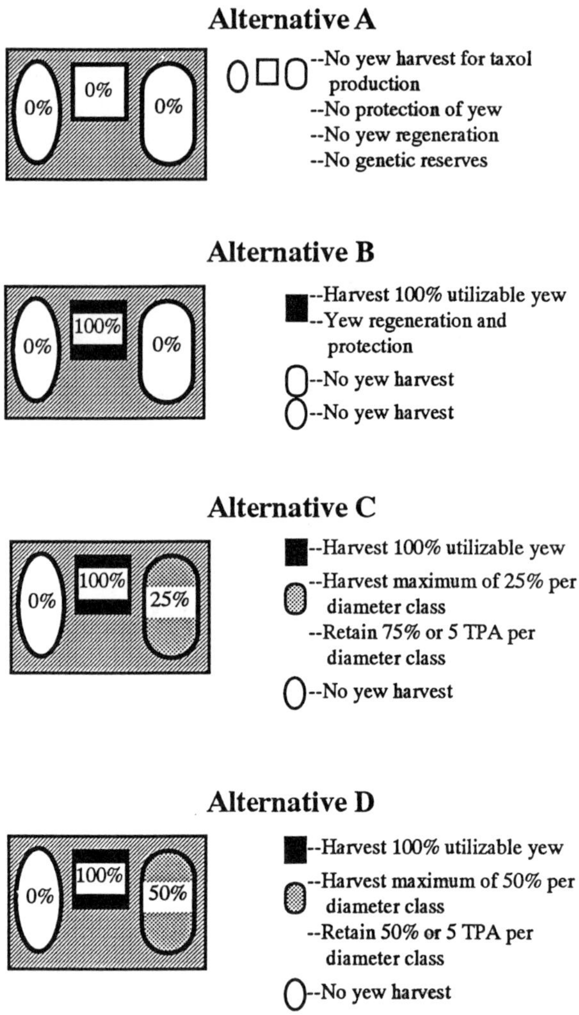

Figure 3. Graphic Representation of the Pacific Yew Harvest Alternatives. Adapted from Pacific Yew Environmental Impact Statement, 1993, USDA Forest Service.

Continued on next page

Alternative F

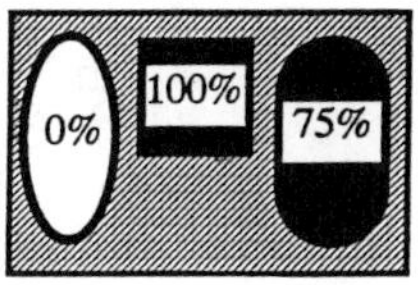

--Harvest 100% utilizable yew

--Harvest maximum of 75% per diameter class
--Retain 25% or 2 TPA per diameter class

--No yew harvest

Alternative G1

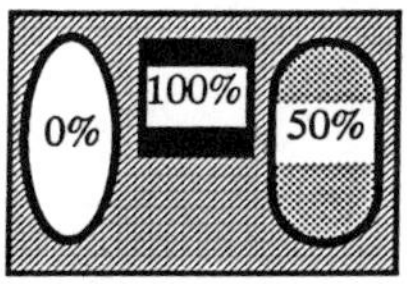

--Harvest 100% utilizable yew

--Harvest maximum of 50% per diameter class
--Retain 50% or 1 TPA per diameter class

--No yew harvest

Alternative G2

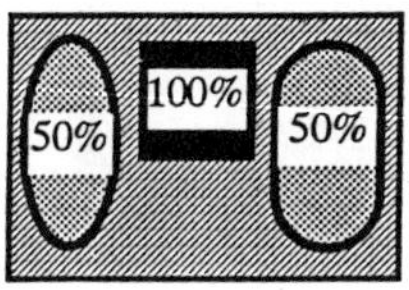

--Harvest 100% utilizable yew

--Harvest maximum of 50% per diameter class
--Retain 50% or 1 TPA per diameter class

--Harvest maximum of 50% per diameter class
--Retain 50% or 5 TPA per diameter class

Key

Owl Conservation Area

Timber Sale Units—clearcut, shelterwood, or seedtree harvest units

Partial-cut Sale Units such as thinning or uneven-aged cuts and non-sale areas where yew harvest is allowed.

Figure 3. *Continued*

Table 1. Comparison of Pacific Yew Harvest Alternatives

	Alt A	*Alt B*	*Alt C*	*Alt D*	*Alt F*	*Alt G1*	*Alt G2*
Issue: Protect Pacific Yew							
Impact on Distribution of Pacific Yew over the Landscape	Low Risk	Low Risk	Low Risk	Low Risk	Low to Moderate Risk	Low to Moderate Risk	Low to Moderate Risk
Impact on Pacific Yew Reproduction	Moderate	Minor	Minor	Minor	Minor to Moderate	Minor to Moderate	Minor to Moderate
Impact on Genetics of Pacific Yew	Minor	None	Minor	Minor	Moderate to High	Minor to Moderate	Minor to Moderate
Issue: Protect the Ecosystem							
Impact on Ecosystem Structure and Function	Low Risk	Low Risk	Low Risk	Low to Mod Risk	High Risk	Moderate Risk	Moderate Risk
Loss of Diversity	Some	Negligible	Negligible	Negligible	Negligible	Negligible	Negligible
Impact on Water and Aquatic Habitat	None	None	Negligible to Minor	Negligible to Minor	Negligible to Minor	Negligible to Minor	Negligible to Minor
Impact on Wildlife Species	Minor	Minor	Minor	Minor	Minor to High	Minor to Moderate	Minor to Moderate
Impact on Northern Spotted Owl	Minor	Minor	Minor to Moderate	Minor to Moderate	Moderate to High	Minor to Moderate	Moderate to High
Impact on Forest Health	Minor	Minor	Minor	Minor	Minor	Minor	Minor

Continued on next page

Table I. *Continued*

Issue: Provide Pacific Yew for Taxol							
MM Yew Trees Available for Taxol Harvest	None	0.26-0.39	1.51-2.27	2.36-3.94	4.23-6.35	3.14-4.71	4.22-6.33
Available Bark (MM lbs/yr)	None	0.3-0.4	1.1-1.7	1.9-2.8	3.2-4.8	3.2-4.6	3.8-5.7
KG Taxol Produced (15,000 lbs bark = 1 KG)	None	17.3-26.0	73.3-113.3	126.6-186.7	213.3-320.0	213.3-306.7	253.3-380.0
Potential Patients Treated per Year (1 KG/480 patients)	None	8,300 - 12,400	35,184 - 54,384	60,768 - 89,616	102,384 - 153,600	102,384 - 147,216	121,584 - 182,400

1 . Adapted from Table II-1 in USDA Forest Service 1993.

2. Based on the 1991-92 Pacific yew inventory on eight national forests and six BLM districts. Availability is reduced from the total estimated yew population (about 50 million trees on inventoried land) by Forest Service and BLM land use restrictions and restrictions unique to each alternative.

in the development of alternate approaches to taxol production and was working with FDA to obtain approval of production of taxol from sources other than bark. The summer of 1993 was the last time that yew bark was collected from federal lands for Bristol-Myers Squibb. Pacific yew on federal lands was then available to other qualifying companies or individuals.

The Future of Yew and Taxol

Alternate Sources of Taxol. The Bristol-Myers Squibb Company found alternative sources of taxol-supplying material well within the predicted five years. According to the cooperative agreement with NCI, Bristol-Myers Squibb was required to investigate and develop other sources; by 1993 several of the research and development programs were showing positive signs of producing enough taxol to meet their needs.

Several alternate sources other than wild Pacific yew trees from federal lands could provide material for taxol: semi-synthesis using *Taxus* needles, nursery production of *Taxus* spp, cell-culture, and full laboratory synthesis. Bristol-Myers Squibb found its supply of taxol producing material could be met by yew bark from private lands and from other yew sources abroad. One of the most promising sources appeared to be semi-synthesis of taxol from *Taxus* needles harvested in Europe and Asia. The process of semi-synthesis begins with extraction of the taxol precursor, baccatin-III, from needles of a number of *Taxus* species; the remainder of the taxol molecule is attached synthetically to form the complete taxol molecule. This process is similar to that used by Rhone-Poulenc Rorer, a French pharmaceutical firm that developed Taxotere, a taxol analog.

Other alternatives included a contract with the Weyerhaueser Company to propagate millions of yew seedlings in Oregon and Washington nurseries; Phyton Catalytic Inc. of Ithaca, New York and ESCA Genetics of San Carlos, California's cell culture production of taxol and taxol-like compounds from yew cells grown in culture; and total synthesis of the complex taxol molecule, which has proven difficult.

Following the development of alternate sources, Bristol-Myers Squibb Company filed a supplement to the New Drug Application (NDA) with the FDA, applying for approval to market taxol derived from needles of other *Taxus* species.

Lessons. Although wild Pacific yew was unheralded for years and was treated more as a weed than a valuable resource, its brief stardom changed it status. The Forest Service and BLM, in the process of harvesting the bark and preparing the Guide and the EIS, gathered what information existed and then added substantially to it. Information was gained through the inventory, compilation of scattered reports and data, and through various observations and research projects. Ecologists, botanists, wildlife biologists, geneticists, pathologists, and entomologists investigated various aspects of Pacific yew biology, genetics, and ecological relationships.

In addition, the Pacific yew EIS gave several federal agencies an opportunity to collaborate on an analysis of a major federal action (yew harvest) that crossed agency boundaries and included a non-natural resource agency, the FDA.

Further, the need for large quantities of wild Pacific yew bark for taxol gave the Forest Service and BLM a unique opportunity to provide to the public a medicinal product rather than a traditional forest product such as lumber. The need appeared at a time when both agencies were changing from a timber production orientation to an emphasis on stewardship and ecosystem conservation. The problem was how to both use Pacific yew and conserve it; how to be stewards of the ecosystems where yew was a component and, at the same time, provide human beings with a cancer-fighting drug.

Pacific yew, after its short fame, will most likely never be ignored again by forest land managers. Knowledge of its medicinal properties and its ecological niche has given it the status that it, and all forest species, deserve for their unique roles in forest ecosystems and human communities.

Literature Cited

1. USDA Forest Service. 1992. An Interim Guide to the Conservation and Management of Pacific Yew. Pacific Northwest Region. Portland, OR. 72 pp.

2. USDA Forest Service. 1993. Pacific Yew Environmental Impact Statement. Pacific Northwest Region. Portland, OR.

RECEIVED September 13, 1994

Chapter 5

Paclitaxel Biosynthesis

The Early Steps

Rodney Croteau, Mehri Hezari, Jerry Hefner, Alfred Koepp, and Norman G. Lewis

Institute of Biological Chemistry, Washington State University, Pullman, WA 99164–6340

The biosynthesis of taxol and related taxoids in yew (*Taxus*) species is thought to involve the cyclization of the common isoprenoid intermediate geranylgeranyl diphosphate to a taxadiene followed by extensive, largely oxidative, modification of this diterpene olefin precursor. This sequence of reactions leading to taxol likely requires in excess of a dozen distinct enzymatic steps, exclusive of the construction of the *N*-benzoyl phenylisoserine side chain moiety, which has been shown by H.G. Floss and his colleagues to originate from phenylalanine and to involve esterification at C13 of the very advanced intermediate baccatin III. No early steps of the pathway have been defined and none of the enzymes involved in taxol biosynthesis have been described until quite recently. A cell-free extract of yew saplings was shown to catalyze the conversion of [1-^{3}H] geranylgeranyl diphosphate to a cyclic diterpene olefin that was identified as taxadiene by spectroscopic means and was demonstrated to serve as a precursor of taxoids *in vivo*. The cyclization enzyme that catalyzes this first committed step of taxol biosynthesis is described, as are preliminary studies on the initial oxygenation of taxadiene catalyzed by a presumptive cytochrome P450 hydroxylase. Demonstration of the first two specific enzymatic steps of taxol biosynthesis suggests that the complete pathway can be defined by a systematic, stepwise approach at the cell-free enzyme level.

Although the total synthesis of taxol has been achieved (*1-3*), this approach to supplying the drug is not now commercially viable (*4*). For the foreseeable future, taxol will continue to be produced by biological means, including isolation from the renewable foliage and other tissues of *Taxus brevifolia* (Pacific yew) and related *Taxus* species, semi-synthesis of the product and its analogs from baccatin III (see Scheme 1) and similar, late-stage taxoid (taxane diterpenoid) metabolites that are more readily available, and, potentially, by generation of the drug in *Taxus* cell cultures (*5*). Therefore, it is important to elucidate the biosynthesis of taxol, particularly the slow steps of the pathway, the manipulation of which may be expected to lead to increased yield and to the production of the drug in large quantities at reasonable cost (*6,7*).

NOTE: Paclitaxel is the generic name for Taxol, which is now a registered trademark.

0097–6156/95/0583–0072$08.00/0

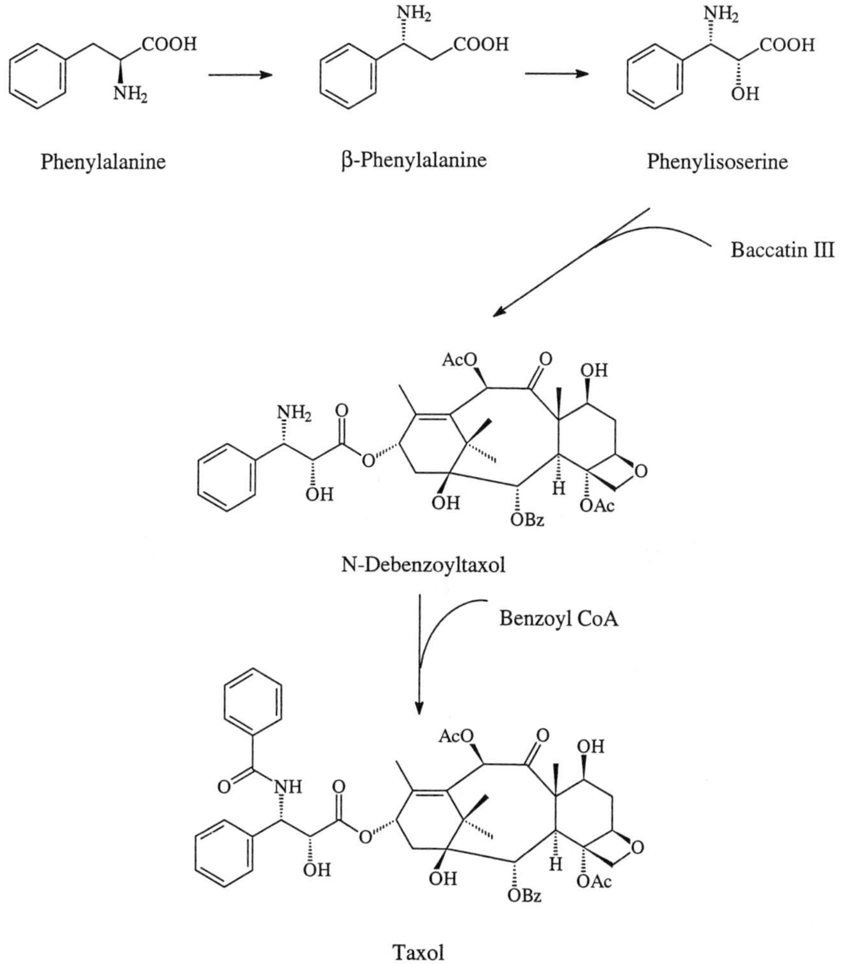
Phenylalanine
β-Phenylalanine
Phenylisoserine
Baccatin III
N-Debenzoyltaxol
Benzoyl CoA
Taxol

Scheme 1.

The structure of taxol (*8*) is quite complex; the metabolite possesses the unusual taxane diterpene carbon skeleton (pentamethyl tricyclo[9.3.1.0.^{3,8}] pentadecane) that bears eight oxofunctional groups and an assortment of side-chains (including the novel *N*-benzoyl phenylisoserine) for a total of eleven stereocenters. The pathway to taxol from the universal diterpenoid precursor geranylgeranyl diphosphate (*9*) must involve well over a dozen discrete enzymatic steps. Deciphering the origin of this metabolite thus represents a formidable biosynthetic challenge. Although the biogenesis of taxol has been the subject of considerable speculation (*10-14*), very little is known about the origin of taxol and related taxoids with the exception of very preliminary *in vivo* studies with basic precursors such as acetate, mevalonate and phenylalanine (*15,16*), and the detailed feeding studies by Floss and colleagues with advanced metabolites that have demonstrated the origin and timing of assembly of the *N*-benzoyl phenylisoserine ester (*17,18*). Thus, baccatin III was shown to be a specific precursor of taxol with the side chain added most likely as phenylisoserine (via phenylalanine) followed by *N*-benzoylation as the last step in taxol formation (Scheme 1); this same sequence with ultimate *N*-tigloylation would afford cephalomannine. A description of this work and an excellent overview of previous speculations on the biogenetic pathway(s) leading to taxol have been provided by Floss and Mocek (*19*). No information on the enzymology of taxol biosynthesis was available until quite recently. In this chapter, we describe these new studies, carried out at the cell-free enzyme level, that are focused on the very early, slow steps of the pathway, including the first committed step of taxol biosynthesis.

The Olefin Precursor of Taxol

The biosynthesis of taxol has been suggested (*10,13,14*) to involve cyclization of geranylgeranyl diphosphate to taxa-4(20),11-diene (Scheme 2), since taxoids bearing the 4(20)- and 11(12)-double bond pair are very common (*20*), and the reaction can be readily formulated based on accepted mechanisms for the construction of related terpenoids (*21*). Such a cyclization to establish the taxane skeleton could then be followed by oxidative elaboration of this, as yet unknown, diterpene olefin progenitor (*13*) by enzymatic processes similar to those responsible for the formation of other oxygenated terpenoid natural products (*21*).

To examine the putative cyclization step, *T. brevifolia* saplings were chosen as an experimental system because the taxoid content of immature tissue is relatively high and because sapling stems contain a high proportion of phloem parenchyma cells in which taxol is thought to be produced (*22,23*). Additionally, by raising yew saplings in the greenhouse, the effects of environmental and developmental variation on taxoid content (*24,25*) can be minimized. Rigorous pulverization of frozen stem sections, and extraction of the resulting tissue powder under conditions designed to minimize the deleterious effects of endogenous phenolics and resinous materials (*26*), yielded an operationally soluble enzyme preparation that catalyzed the Mg^{2+}-dependent conversion of [1-^{3}H]geranylgeranyl diphosphate to a labeled hydrocarbon fraction isolated by column chromatography. Radio-GLC analysis of this material indicated the presence of essentially a single product (~ 1% yield) with chromatographic properties consistent with that of a polycyclic diterpene olefin.

To determine if the presumptive diterpene olefin could serve as a precursor of taxol and related taxoids, preparative enzyme incubations were carried out to generate several μCi of the chromatographically purified olefin that was suspended in buffer and vacuum-infiltrated into *T. brevifolia* stem discs. Following incubation for a week, the labeled products were extracted and separated by combination of normal and reversed-phase column chromatography into a taxoid fraction containing metabolites more polar than (+)-taxusin (the tetraacetate of taxa-4(20),11(12)-dien-5,9,10,13-tetraol). This material, representing about 10% incorporation of the labeled olefin, was fractionated by reversed-phase HPLC to reveal at least a dozen labeled taxoids of polarity between 10-deacetylbaccatin III and taxol, including the major bark taxoids baccatin III, cephalomannine and taxol, and their 10-deacetyl derivatives (*25*). The coincidental migration of radioactivity with the principal taxoids was also demonstrated by normal phase TLC, and several of these metabolites were crystallized to constant specific activity to confirm that the diterpene olefin product of the cell-free system could serve as a precursor of taxol and closely related taxoids of yew stem. These results suggested that the cyclization product of geranylgeranyl diphosphate was a taxadiene.

To identify the biosynthetic diterpene olefin product, it was necessary to obtain a sufficient amount of this material for spectroscopic analysis. An extract of 750 kg of dried *T. brevifolia* bark powder was therefore diluted with the tritium-labeled biosynthetic material and the natural product was isolated by combination of normal phase, argentation and reversed-phase chromatography, while monitoring the fractionation by aliquot counting and capillary GLC/GLC-MS, to yield about 1 mg of radiochemically pure olefin. The trace levels of this olefin in bark clearly indicate that this intermediate does not significantly accumulate and, thus, that the cyclization of geranylgeranyl diphosphate to the taxoid precursor is a very slow step of the reaction sequence relative to subsequent oxygenations. The mass spectrum of the product was consistent with a cyclic diterpene olefin of molecular weight $C_{20}H_{32}$ and the structure was determined to be taxa-4(5),11(12)-diene (Scheme 3) by 1D and 2D ^{1}H- and ^{13}C-NMR spectrometry. The structure of the biosynthetic product, obtained by preparative-scale enzyme incubations, was subsequently confirmed directly by GLC-MS analysis, and the authentic standard was recently prepared in low yield from (+)-taxusin obtained from yew wood.

The cyclization of geranylgeranyl diphosphate to taxa-4(5),11(12)-diene, as the first dedicated step in the biosynthesis of taxol and related metabolites, is consistent with earlier suggestions that the pathway involves preliminary formation of a parent taxane olefin followed by oxidative modification (13). However, the identification of taxa-4(5),11(12)-diene as the precursor of taxoids, rather than taxa-4(20),11(12)-diene as originally proposed on the basis of metabolite co-occurrence (*10,13,14*), was unexpected. Nevertheless, the cyclization (Scheme 3) can be readily formulated as involving ionization of the geranylgeranyl diphosphate ester with closure of the A-ring and deprotonation to afford 1*S*-verticillene. Protonation at C7 can then initiate transannular cyclization to provide the taxenyl cation, which upon deprotonation at C5 yields the endocyclic double bond of the taxadiene product. Although verticillene is proposed as an intermediate in the cyclization (Scheme 3), no direct evidence for this, or any other free intermediate in the reaction, has thus far been obtained (see below).

Geranyl Diphosphate: Taxadiene Cyclase

Of a range of tissues examined, young green shoots of *T. brevifolia* provided the highest levels of extractable cyclase, and the activity was localized in preparations from peeled bark containing adhering cambium cells; little activity was associated with the woody internal tissue. The enzyme was operationally soluble following rigorous tissue disruption, and negligible cyclase activity was associated with membranous fractions.

The enzyme has been purified greater than 300-fold by combination of ion-exchange, hydrophobic interaction, hydroxylapatite and gel permeation chromatography, and it appears to be a monomeric protein of about 75 kDa. The enzyme requires Mg^{2+} as the only cofactor and resembles in general properties the abietadiene synthase from grand fir, the only other diterpene cyclase from a gymnosperm to be examined in any detail (*26*). The partially purified cyclase yields taxadiene as the only detectable product. Therefore, the proposed 1*S*-verticillene intermediate must be very short-lived and/or tightly bound to the enzyme surface.

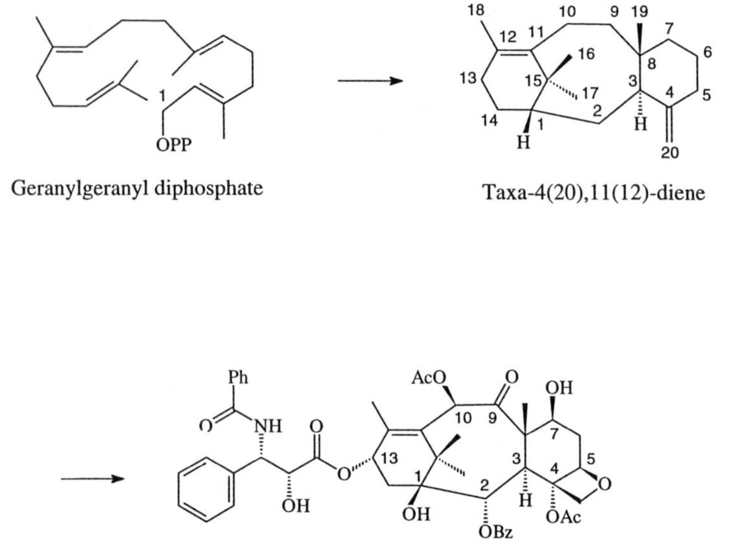

Geranylgeranyl diphosphate → Taxa-4(20),11(12)-diene → Taxol

Scheme 2.

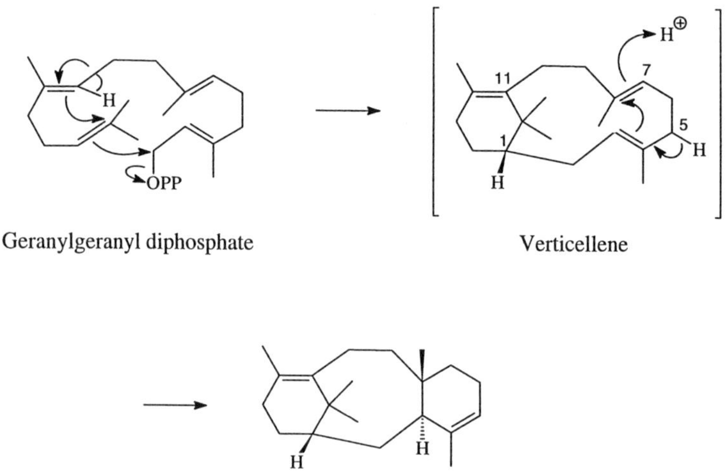

Geranylgeranyl diphosphate → Verticellene → Taxa-4(5),11(12)-diene

Scheme 3.

Precedent for such enzyme-bound olefinic intermediates exists in the case of several sesquiterpene cyclases (*27*). Floss and Mocek (*19*) have suggested that a cyclization of this type could involve intramolecular proton transfer from C11 of the verticillyl intermediate (to generate the 11,12-double bond) to C7 (to promote B/C-ring closure of the taxane skeleton), but this possibility has not yet been examined experimentally.

Taxadiene Oxygenase and Subsequent Reactions

The next step of the pathway on route to taxol is presumed to be oxygenation of taxadiene (*10,13,19*). Since most reactions of this type are catalyzed by membranous P450 cytochromes (*21,28,29*), microsomal preparations from *T. brevifolia* stem extracts were examined for their ability to transform biosynthetically prepared [2-^{3}H]taxadiene to more polar products. Such microsomal preparations were shown to catalyze the O_2- and NADPH-dependent conversion of taxadiene to a product with chromatographic properties indicative of a diterpene monool. The enzymatic reaction is strongly inhibited by CO, consistent with the involvement of a cytochrome P450 hydroxylase, but blue light reversal of inhibition (*28,29*) has been difficult to demonstrate, thus far, due to photochemical interference by pigments present in the crude microsomal preparation. Radiochemically-assisted metabolite isolation is under way to provide a sufficient amount of the diterpenoid product for structural identification. Little guidance to the possible identity of the initial oxidation product is provided by defined *Taxus* metabolites, since the least functionalized naturally occurring taxoid thus far described is at the level of a tetraol (e.g., taxusin). However, the observations that no oxygenated taxoids bearing the 4(5)-double bond have yet been reported, whereas taxoids with the exo-methylene at the 4(20)-position and that also bear an oxygen function at C5 are exceedingly common (*20*), suggest that hydroxylation at C5 of taxa-4(5),11(12)-diene, with migration of the double bond, must occur as an early, if not the first, oxygenation step of the pathway (Scheme 4).

The transformation of taxa-4(5),11(12)-diene to taxa-4(20),11(12)-dien-5-ol would also set the stage for the elaboration of the unusual oxetane moiety, appended to the C-ring of taxol and related taxoids, which is generally considered (19) to arise by conversion of the α-hydroxy-4(20)-methylene to the corresponding epoxide function followed by ring expansion (Scheme 4). This biogenetic supposition is supported by the occurrence of the three major structural classes of taxoids (Scheme 4; **a**,**b**,**c**) in decreasing order of abundance (*20*). Formation of the oxetane is thought to be a relatively late-stage transformation and, while several mechanisms for the expansion of the 4(20)-epoxide to the oxetane have been proposed (*11-13*), all remain untested biochemically.

Following conversion of taxadiene to the rearranged, allylic taxadien-5-ol, or other taxadienol, a succession of subsequent oxygenations and acylations is presumed to occur. The order of oxygenation of the taxane nucleus is at present quite speculative, since no taxoids bearing only one, two or three oxygen functional groups have yet been identified, and the tissue concentrations of such metabolites are apparently low. Based on comparison of the structures of over 100 naturally-occurring taxanes (*20*), it can be surmised from relative abundances (*19*) that oxygens at C5 and C10 are introduced first, followed by oxygenation at C2 and C9, then at C13. Thus, the tetraol corresponding to (+)-taxusin (taxa-4(20),11(12)-dien-5,9,10,13-tetraol) is unlikely to be on the pathway to taxol, and this compound was not detectably incorporated into taxol in *T. brevifolia* bark pieces (*19*). Oxygenations at C7 and C1 of the taxane nucleus are considered to be very late introductions, possibly occurring after oxetane ring formation; however, epoxidation and oxetane formation seemingly must precede acylation at C13 and oxidation of the C9 hydroxyl to a carbonyl (*19*).

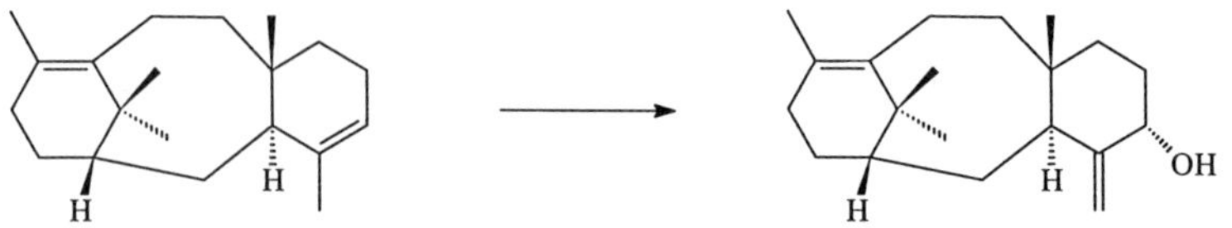

Taxa-4(5),11(12)-diene Taxa-4(20),11(12)-dien-5-ol

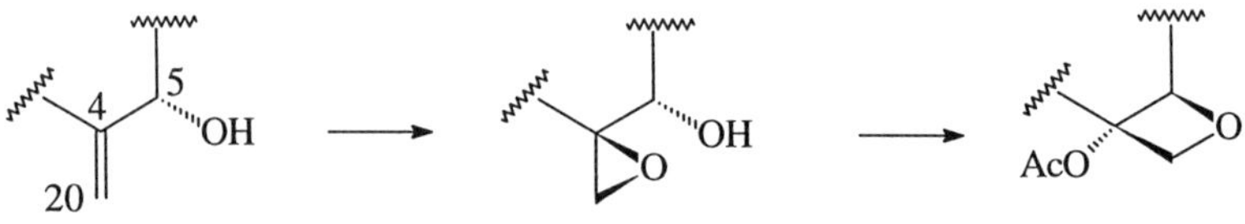

a b c

Scheme 4.

With the exception of the C13 side chain esterification, which occurs very late in the sequence (Scheme 1), the order and possible function of the acylation steps in the taxol pathway are also uncertain. It is possible that acylation of the various hydroxyl functions plays a role in promoting substrate recognition for the following enzymatic reaction, or in modifying solubility for trafficking of metabolites between organelles or cell cytosol. In either case, repetitive acylation/deacylation of a given hydroxy group may be involved as intermediates progress through the pathway and thereby confuse any attempt to deduce a biogenetic sequence based on metabolite co-occurrence. The challenge of proposing such a pathway is further magnified by the fact that over 100 highly functionalized taxoids have now been described (*20*) with little basis for distinguishing those which may be on the main pathway to taxol from those which may represent metabolic side products, dead ends, or even taxol catabolites (*24*). The number of chemically plausible biogenetic sequences leading from taxadiene to taxol is daunting. Thus, the most viable approach to deciphering the direct order of oxygenation and acylation steps in taxol formation is likely the systematic evaluation of the progression from simple to more complex metabolites in cell-free biosynthetic systems in which precursor and cofactor levels can be controlled, compartmentation barriers can be eliminated, and both intermediates and competing reactions can be most easily monitored.

Prospect

In contrast to the considerable experimental attention devoted to the biological effects of taxol and to the total and semi-synthesis of this compound, the biosynthesis of taxol has, until recently, received relatively little attention. In spite of the structural complexity of the target metabolite, and the experimental difficulties in working with gymnosperms, this seeming neglect is nevertheless surprising since the supply of taxol will rely on biological methods of production for some time to come. It will be difficult to improve yield in a directed fashion without first defining the slow steps of the reaction sequence and the controls on flux through the pathway. The successful demonstration that operationally soluble and membranous enzyme systems catalyzing the early steps can be isolated from *Taxus*, partially purified and characterized holds promise that full pathway definition will yield to a stepwise, systematic approach. While the demonstration of the key cyclization reaction and of a cytochrome P450 taxadiene oxygenase is encouraging, very much remains to be done to determine the subsequent order of oxygenation steps, the sequence and role of acylation processes, and the precise timing and mechanisms of oxirane and oxetane ring formation. Continued study of the enzymology of taxol biosynthesis may be expected to lead to the isolation of antibodies directed against, and cDNAs coding for, important catalysts of the pathway. These experimental tools should open additional avenues of research on regulatory mechanisms and can provide the strategies and the means for manipulating pathway flux.

Acknowledgments

Research by the authors was supported in part by National Institutes of Health grant CA-55254 and by McIntire-Stennis Project 0967 from the Washington State University Agricultural Research Center, Pullman, WA 99164.

Literature Cited

1. Nicolaou, K. C.; Yang, Z.; Liu, J. J.; Ueno, H.; Nantermet, P. G.; Guy, R. K.; Claiborne, C. F.; Renaud, J.; Couladouros, E. A.; Paulvannin, K.; Sorensen, E. J. *Nature* **1994,** *367*, 630-634.
2. Holton, R. A.; Somoza, C.; Kim, H. B.; Liang, F.; Biediger, R. J.; Boatman, P. D.; Shindo, M.; Smith, C. C.; Kim, S.; Nadizadeh, H.; Suzuki,

Y.; Tao, C.; Vu, P.; Tang, S.; Zhang, P.; Murthi, K. K.; Gentile, L. N.; Liu, J. H. *J. Am. Chem. Soc.* **1994,** *116*, 1597-1598.
3. Holton, R. A.; Kim, H. B.; Somoza, C.; Liang, F.; Biediger, R. J.; Boatman, P. D.; Shindo, M.; Smith, C. C.; Kim, S.; Nadizadeh, H.; Suzuki, Y.; Tao, C.; Vu, P.; Tang, S.; Zhang, P.; Murthi, K. K.; Gentile, L. N.; Liu, J. H. *J. Am. Chem. Soc.* **1994,** *116*, 1559-1600.
4. Borman, S. *Chem. Eng. News* **1994,** *72(7)*, 32-34.
5. Cragg, G. M.; Schepartz, S. A.; Suffness, M.; Grever, M. R. *J. Nat. Prod.* **1993,** *56*, 1657-1668.
6. Suffness, M.; Wall, M. E. In *Taxol: Science and Applications;* Suffness, M., Ed.; CRC Press: Boca Raton, FL, 1994; in press.
7. Suffness, M. In this volume.
8. Wani, M. C.; Taylor, H. L.; Wall, M. E.; Coggon, P.; McPhail, A. T. *J. Am. Chem. Soc.* **1971,** *93*, 2325-2327.
9. West, C. A. In *Biosynthesis of Isoprenoid Compounds;* Porter, J. W.; Spurgeon, S. L., Eds.; John Wiley and Sons: New York, NY, 1981, Vol 1; pp 375-411.
10. Harrison, J. W.; Scrowston, R. M.; Lythgoe, B. *J. Chem. Soc. (C)* **1966,** 1933-1945.
11. Della Cassa De Marcano, D. P.; Halsall, T. G.; Castellano, E.; Hodder, O. J. R. *J. Chem. Soc. Chem. Comm.* **1970,** 1382-1383.
12. Swindell, C. S.; Britcher, S. F. *J. Org. Chem.* **1986,** *51*, 793-797.
13. Guéritte-Voegelein, F.; Guénard, D.; Potier, P. *J. Nat. Prod.* **1987,** *50*, 9-18.
14. Begley, M. J.; Jackson, C. B.; Pattenden, G. *Tetrahedron* **1990,** *46*, 4907-4924.
15. Zamir, L. O.; Nedea, M. E.; Garneau, F. X. *Tetrahedron Letts.* **1992,** *33*, 5235-5236.
16. Strobel, G. A.; Stierle, A.; van Kuijk, F. J. G. M. *Plant Sci.* **1992,** *84*, 65-74.
17. Fleming, P. E.; Mocek, U.; Floss, H. G. *J. Am. Chem. Soc.* **1993,** *115*, 805-807.
18. Fleming, P. E.; Knaggs, A. R.; He, X.-G.; Mocek, U.; Floss, H. G. *J. Am. Chem. Soc.* **1994,** *116*, 4137-4138.
19. Floss, H. G.; Mocek, U. In *Taxol: Science and Applications;* Suffness, M., Ed.; CRC Press: Boca Raton, FL, 1994; in press.
20. Kingston, D. G. I.; Molinero, A. A.; Rimoldi, J. M. *Prog. Chem. Org. Nat. Prod.* **1993,** *61*, 1-206.
21. Gershenzon, J.; Croteau, R. In *Lipid Metabolism in Plants;* Moore, Jr., T. S., Ed.; CRC Press: Boca Raton, FL, 1993; pp 333-382.
22. Vidensek, N.; Lim, P.; Campbell, A.; Carlson, C. *J. Nat. Prod.* **1990,** *53*, 1609-1610.
23. Ellis, D.; Zeldin, E.; Russin, W.; Brodhagen, M.; Evert, R.; McCown, B. *Proc. Internat. Yew Res. Conf.;* Berkeley, CA, 1993.
24. Wheeler, N. C.; Jech, K.; Masters, S.; Brobst, S. W.; Alvarado, A. B.; Hoover, A. J.; Snader, K. M. *J. Nat. Prod.* **1992,** *55*, 432-440.
25. Kelsey, R. G.; Vance, N. C. *J. Nat. Prod.* **1992,** *55*, 912-917.
26. LaFever, R. E.; Stofer Vogel, B.; Croteau, R. *Arch. Biochem. Biophys.* **1994,** *313*, 139-149.
27. Cane, D. E. *Chem. Rev.* **1990,** *90*, 1089-1103.
28. Mihaliak, C. A.; Karp, F. *Methods Plant Biochem.* **1993,** *9*, 261-279.
29. West, C. A. In *The Biochemistry of Plants: A Comprehensive Treatise;* Stumpf, P. K.; Conn, E. E., Eds.; Academic Press: New York, NY, 1980, Vol. 2; pp 317-364.

RECEIVED September 7, 1994

Chapter 6

Bioactive Metabolites of the Endophytic Fungi of Pacific Yew, *Taxus brevifolia*

Paclitaxel, Taxanes, and Other Bioactive Compounds

Andrea Stierle[1,2], Donald Stierle[2], Gary Strobel[1], Gary Bignami[3], and Paul Grothaus[3]

[1]Department of Plant Pathology, Montana State University, Bozeman, MT 59717
[2]Department of Chemistry, Montana College of Mineral Science and Technology, Butte, MT 59701
[3]Hawaii Biotechnology Group Inc., Aiea, HI 96701

The endophytic microbes associated with the Pacific yew tree, *Taxus brevifolia*, were examined as potential alternative sources of the anticancer drug taxol, a secondary metabolite of the host organism. A novel fungus, *Taxomyces andreanae*, isolated from the inner bark of a yew tree growing in northwestern Montana, appears to produce taxol and other taxanes, in *de novo* fashion when grown in semi-synthetic liquid media. The presence of taxol in the fungal extract was confirmed by mass spectrometry, comparative chromatographic behavior with yew taxol, reactivity with taxol-specific monoclonal antibodies, and 9KB cytotoxicity studies. Both acetate-1-^{14}C and phenylalanine UL-^{14}C served as precursors of taxol-^{14}C in fungal culture labeling studies, confirming the *de novo* synthesis of taxol by the fungus. Immunoassay techniques are currently being used to screen extracts of *Taxomyces andreanae* for new taxanes, and to determine if other endophytic fungi are taxol producers. Fungal endophytes used in this study are further screened for additional biological activity following taxol/taxane analysis.

Background

Cancer is the second leading cause of death in the United States, and the incidence of cancer continues to climb annually (*1*). Chemotherapeutic agents are instrumental in the fight against this dreaded disease, and effective anticancer agents, particularly those that affect refractory tumors, are critical objectives in western medicine. Taxol (**1**) is a new anticancer drug with particular efficacy against refractory breast and ovarian cancers (*2, 3*). Although initially isolated and characterized in 1971, taxol did not achieve notoriety until 1977, when its strong activity against human tumor xenograph

NOTE: Paclitaxel is the generic name for Taxol, which is now a registered trademark.

0097–6156/95/0583–0081$08.00/0

systems and murine B16 melanoma cells prompted its development towards clinical trials (*4*). Interest in taxol intensified when its unique mode of antimicrotubule cytotoxicity was published in 1979 (*5, 6*). Microtubules are among the most strategic of the subcellular chemotherapeutic targets. Antimicrotubule agents are extremely potent, requiring only a few molecules to disrupt the microtubular structure of cancer cells (*2*). These compounds, which include the vinca alkaloids, are among the most important anticancer drugs currently employed (*2*).

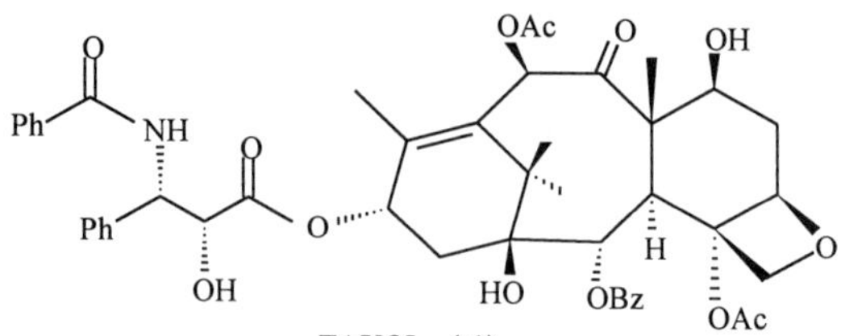

TAXOL (1)

Taxol: the Supply Dilemma. But there is a problem with taxol. This highly functionalized diterpene is isolated primarily from the inner bark of the relatively rare and slow growing Pacific yew tree, *Taxus brevifolia*, and a few related species, in extremely small yields (< .02% dry weight) (*7*). The emergence of taxol as an effective anticancer agent created a dilemma: how to insure an adequate supply of a compound of non-microbial origin. Athough the pharmaceutical potential of taxol elevated the status of the yew tree from a nuisance weed to a precious commodity and natural resource, it did not alter the underlying dilemma - there were simply not enough yew trees to supply the growing demands for taxol (*4*). Advanced preclinical and phase I clinical development of taxol required several collections ranging in size from 5,000 to 15,000 pounds of dry bark. A mature Pacific yew (100 years old) yields approximately 10 lb. of dry bark, so each collection required the sacrifice of 500 to 1500 trees (*4*). As the efficacy of the compound became more apparent, the demand for additional taxol increased (*4*).

In 1987-1988 a 60,000 lb bark collection was undertaken with little controversy. The need for a second 60,000 lb bark collection in 1989, however, sparked concern about the impact such collection sizes may have on the continued existence of the yew tree. Although no accurate inventories of the tree have ever been undertaken, the Fish and Wildlife Service of the U.S. Department of the Interior states that "Much of the range of the yew has not been subject to statistical inventories, especially the northern portion (i.e., Alaska and British Columbia). Nonetheless, based on stand information, together with satellite imagery, the U.S. Forest Service estimates that 130 million yew trees occur on 1,778,000 acres of National Forest in the Washington and Oregon Cascades, and Oregon Coast Range" (*4*).

Even with this estimate, however, it is clear that additional sources of taxol must be found. A single course of clinical treatment is 125-300 mg of taxol, and typical treatments may extend for 10 or more courses. Treatment of the 12,000 women who die annually of ovarian cancer alone would consume as much as 36 kg of the drug (*4*). With current isolation methodologies, 1 kg of taxol is isolated from 25,000 lb of dried bark, or the bark of 2500 yew trees. Therefore, simply treating ovarian cancer over a one year period would consume 90,000 mature yew trees (*4*). The recent approval of taxol for treatment of breast cancer will triple the projected demand. Should taxol prove as effective against other refractory cancers, including head and neck and non-small cell lung cancers as clinical trials have indicated (*8, 9*,), it is not unreasonable to assume that the demand for taxol may exceed 300 kg, or 750,000 trees per year (*4*). This represents an enormous demand on a limited resource. There are simply not enough yew trees growing in North America to satisfy projected needs of this drug over the next twenty years (*4*).

Several research groups have labored to alleviate the supply problem using a variety of strategies. Total syntheses from simple precurors were published virtually simultaneously by Robert Holton (*10*) and K.C. Nicolaou (*11*), and Paul Wender's elegant approach to total synthesis will probably be realized before this symposium is published (*12*). Although these synthetic efforts are chemical masterpieces, they will not answer the supply question. Semisynthetic methods using taxoid starting materials have proven successful, however, and will facilitate taxol availability. Most of these syntheses begin with either baccatin III or 10-deacetylbaccatin III which can be isolated from the needles of the European yew *Taxus baccata*. Yew needles are a renewable resource and should provide an adequate supply of the necessary starting materials. Several semisynthetic methods have been proposed (*13, 14*), but the most promising utilize the strategies devised by Holton (*15*), Georg (16) and Ojima (*16*). Plant tissue culture also shows promise, and some research groups are reporting yields commensurate with commercialization (*17*).

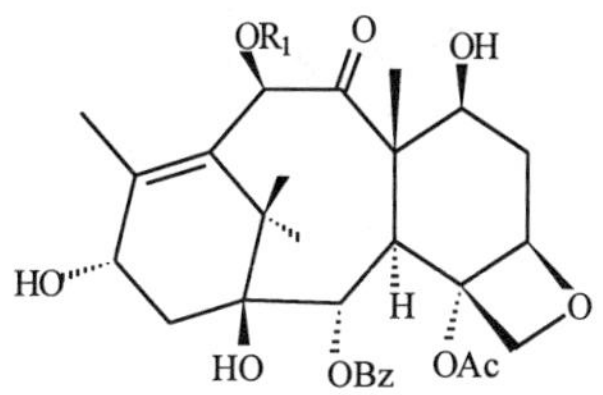

BACCATIN III R_1= Ac
10-DEACETYLBACCATIN III R_1 = H

Microbial Source for Taxol

Our own attempt at easing the supply dilemma focused on the discovery of a new biological source of the drug: an endophytic microbe colonizing the yew tree. Over the last two years we have isolated over 300 fungi from the bark and needles of yew trees in Montana, Washington, Idaho, and Oregon. Promising taxol producers have been studied using a variety of different techniques, including chromatography, mass spectrometry and antibody based immunoassays. Immunoassay is proving an effective tool not only in assessing the presence of taxol and taxanes in crude extracts, but also in providing an efficient fractionation guide.

We were painfully aware from project inception that our chances of success were pretty minimal. In an effort to justify the tremendous time expenditure of this venture we broadened our research goals. Microbial extracts were evaluated not only for evidence of taxoids but also for other bioactive components. Particular attention was paid to compounds with either antifungal or anticancer potential. Fungi already provide a number of important antibiotics, including the penicillins and cephalosporins

(*18*). Endophytic fungi, however, particularly those isolated from conifers, are an untapped reservoir of compounds with pharmaceutical potential. Fungi associated with medicinal plants might prove a good source of novel bioactive compounds, and a taxol-producing fungus would be a noteworthy beginning.

Advantages of Microbial Source. From a practical viewpoint, microbial fermentation as a means of producing bioactive substances has several advantages (*19*).

1. Industrial production of a bioactive substance like taxol requires reproducible, dependable productivity. If a microbe is the source organism, it can be grown in tank fermentors as needed, producing a virtually inexhaustible supply of taxol (*19*).

2. Microorganisms typically respond favorably to routine culture techniques. Cultivation of macroorganisms (tissue culture) is considerably more challenging, requiring either specialized techniques or months of growth before harvesting is feasible (*19*).

3. Productivity amplification is relatively easy in microorganisms. In the case of penicillin, improved culture conditions and genetic manipulation of producing strains of *Penicillium* increased drug yield from a few micrograms per milliliter to thousands of micrograms per milliliter (*19, 20*). With macroorganisms, larger collection sizes are the most reasonable option for improved productivity. In the case of taxol, larger collection sizes will lead to the eradication of the source organism within a few years if all of the demands for it are to be met.

4. Different bioactive compounds can be produced by altering culture conditions. The antibiotic aplasmomycins were produced by *Streptomyces griseus SS-20* only after NaCl was added to the medium (*21*). Directed changes in culture conditions can be explored indefinitely as a method of optimizing various biosynthetic pathways, that may lead to even more effective derivatives of taxol (*19*).

What all of this means is that a **microbial source of taxol could provide an inexhaustible supply of taxol and novel taxanes**.

The Gibberellins: Precedence for Taxol-Producing Microbe? The search for a taxol producing fungus was prompted by the advantages inherent in a microbial drug source. The real motivation for this search, however, was a discovery made forty years earlier by Yabuta, in his study of "foolish rice seedling disease". He determined that the gibberellins, also highly functionalized diterpenes, were responsible for the disease symptoms induced by the phytopathogenic fungus *Gibberella fujikuroi* (*22*). It has since been established that the gibberellins are ubiquitous phytohormones produced by most higher plants. The pathways of gibberellin biosynthesis in the fungus and the higher plant are identical up to gibberellic acid-12 (*23*). This suggests the possibility of intergeneric-genetic exchange between higher plant and fungus. This type of exchange would probably require an intimate association between the cells of the tree and its microbial associates. Therefore, a search for a taxol producing microorganism should (and did) commence in the tissues of *Taxus spp.*, particularly in the portions of the tree in which taxol is isolated (*7*).

Search for a Taxol-Producing Microorganism

Yew bark, needle, and root samples were collected in several national forests throughout Washington, Oregon, Idaho and Montana. Samples were taken from both healthy and diseased specimens. Each sample was placed on water agar, and developing microbial colonies were transferred to mycological agar as they appeared. Microbes were established in pure culture using standard methodology. Each microbe was grown in liquid medium (100 mL) following purification. Fungi were grown in mycological broth to which 2% yew needle broth was added. Bacteria were grown in tryptic soy broth with the same amendment. Endophytic microbes often cease production of secondary metabolites when removed from the host organism. Yew needle broth may serve either as a critical precursor reservoir or as a genetic promoter for the biosynthetic mechanism of taxol production. It must be noted, however, that **yew needle broth is added only in the first fermentation**. Subsequent fermentations of promising microbes use strictly defined synthetic or semi-synthetic media: all yew products are excluded.

Chemical Analysis of Fungal Extract. All microbial extracts were subjected to first-order examination, which, in the initial phase of this study, consisted of chemical extraction, thin layer chromatography, and nuclear magnetic resonance spectroscopy. Monoclonal antibody immunoassay analysis with taxol and taxane specific antibodies is now included in first-order examination. Following this initial examination, promising microbial extracts were grown in 1L cultures **without the addition of yew broth**. These cultures were processed as before, and then subjected to second order examination, which consists of a chromatographic step followed by mass spectrometry. Third order examination involves purification of the potential taxol fraction followed by both intramural and extramural mass spectral analysis, further monoclonal antibody immunoassay, and 9KB cytotoxicity determinations.

Twenty-one day cultures were filtered through cheesecloth. The residue (mycelia) was macerated and extracted thoroughly with methylene chloride-methanol (1:1). The filtrate was extracted with methylene chloride. The two organic extracts were examined by thin-layer chromatography on Whatman silica gel plates (0.5 mm, 5x10cm) with yew taxol as the standard, using four different solvent systems (*24*). The plates were visualized both by ultraviolet light and by sulfuric acid-vanillin spray reagent. Extracts with TLC spots reminiscent of taxol were prepared for second order analysis by flash silica gel chromatography using acetonitrile as the solvent, followed by high performance liquid chromatography in 3:1 hexane-isopropanol (Rainin Dynamax-60A 8μm Cyano, 4.6x250mm). The appropriate fraction was analyzed by electron impact mass spectrometry. EIMS does not give a prominent molecular ion but it does yield the important fragment peaks at m/z 509 and 569 amu.

Taxomyces andreanae, a Fungal Source of Taxol

Throughout our investigation we have examined microorganisms isolated from more than twenty-five trees from more than twenty locations. Of the three hundred microbes screened to date, one fungus has progressed to third-order analysis with consistent evidence of taxol synthesis. This previously undescribed fungus, which we named *Taxomyces andreanae*, has demonstrated the ability to produce taxol (*25*). It was isolated from the bark of a single yew tree in a unique location in an old growth cedar forest in northern Montana. Despite extensive searches, *T. andreanae* has not been found in any other yew tree examined to date.

Confirmation of Taxol Production by *T. andreanae*. The complete third order analysis of *T. andreanae* combined several different protocols, each of which was critical to confirmation. We will outline the analytical process to provide evidence of the ability of *T. andreanae* to produce taxol, although a detailed description of these processes has been previously published (*26, 27*).

Fermentation Procedure. In the initial investigation, *T. andreanae* was established in pure culture via hyphal tip transfer from water agar, on which the bark pieces had been placed, to mycological agar (DIFCO). The growing mycelium was then serially transferred three to six times, to fresh mycological agar. This eliminated the possibility that fungal hyphae carried a taxol or taxane "contaminant" from the source yew tree. Since the conidia of *T. andreanae* do not germinate in our hands, pieces of agar block (5x5mm) impregnated with mycelia were used as an inoculum source for broth cultures. Liquid cultures were grown primarily in modified mycological broth (*28*), 21 day still culture, 25°C, with a surface: volume ratio of 1.3:1 (cm^2:mL).

Fungal Taxol Isolation Protocol. After 21 days, the culture was filtered through 8 layers of cheesecloth. The filtrate was extracted thoroughly with methylene chloride. The aqueous phase was lyophilized and extracted with methylene chloride-methanol (1:1). The mycelium was macerated and thoroughly extracted with methylene chloride-methanol (1:1). The solvent was removed from the organic extracts by rotary evaporation at 30 to 35°C. Thin layer chromatography of all three organic extracts in four solvent systems (*24*) indicated that the taxol-like metabolite was concentrated in the methylene chloride extract of the filtrate.

This organic extract was dissolved in 2 mL of chloroform and chromatographed on a silica gel column (1x5 cm, 60-200 mesh) prewashed with chloroform, and eluted with 20 mL of acetonitrile. The resulting fraction was dried and the residual oil was chromatographed by preparative TLC (Merck silica gel plate, 0.5 mm) and developed in chloroform-acetonitrile, 7:3 v/v. The silica gel at the R_F of taxol-baccatin at ca. 0.17 - 0.30 was removed from the plate by scraping, and extracted with acetonitrile.

The acetonitrile soluble extract was analyzed by HPLC (silica gel 1.5 x 25 cm column) using chloroform-acetonitrile, 7:3 v/v in an isocratic mode. The peak eluting with the same retention time as taxol (5 min.) was collected, dried and subjected to the final preparative TLC on a prewashed Merck silica gel plate (0.25 mm) in ethyl acetate-isopropanol 95:5 v/v. The area with the identical R_F to taxol was eluted with acetonitrile and dried. Various modifications of this extraction and purification method were also successful in yielding fungal taxol.

The compound isolated from *T. andreanae* had identical R_F values as yew-derived taxol in four different TLC solvent systems (*24*). It reacted positively with vanillin-sulfuric acid spray reagent, yielding a blue spot which turned brown after 12-24 hr (*29*). Fungal taxol has the same retention time (5 min.) on HPLC as yew taxol on a 1.5 x 2.5 cm silica column using chloroform-acetonitrile, 7:3 v/v as the solvent system. It has the same retention time using analytical cyanopropyl bonded-phase HPLC with different solvent systems (*26, 27*). In addition, the UV spectrum of fungal taxol is superimposable on that of authentic taxol, with maxima at 273 nm and 235 nm (*7*).

Mass Spectral Analysis of Fungal Taxol. Fungal taxol was examined by several mass spectral techniques. Electron impact mass spectrometry exhibited strong fragment peaks at m/z 509 and 569, but the molecular ion at m/z 854 was not apparent (*26, 27*). EIMS of authentic taxol exhibited a virtually identical fragmentation pattern. Electrospray mass spectrometry of the fungal taxol fraction primarily yielded peaks at m/z 854 and 876.5 (*26, 27*). These masses represent the M^+ + H of taxol and the M^+ + Na of its sodiated adduct, respectively. Authentic taxol yielded the identical spectrum as fungal taxol following sodiation. In addition, the fast atom bombardment (FAB) spectrum of fungal taxol yielded the M^+ + H of taxol (854.3) with peaks characteristic of taxol at 509 and 569 when compared to authentic taxol (*26, 30*). These data have been corroborated at least 5 times on different fungal preparations. Evidence was also obtained by liquid chromatography-mass spectrometry for the presence of baccatin III in *T. andreanae*: the parent peak at m/z 604 is consistent with $M^+ + NH_4$ of this compound.

Immunoassay Techniques as a Screening Tool. The selection and monitoring of fungal strains like *Taxomyces andreanae* for secondary metabolite production can be expedited if sensitive, specific, rapid screening methods are available for compounds of interest. Antibody based immunoassays can either complement or provide an alternative to chromatographic or spectroscopic techniques, and often provide the advantages of greater sensitivity, simpler sample preparation and high sample throughput (*31*). Because the sensitivity and specificity of an immunoassay reflects the binding properties of the antibodies utilized, production of high affinity antibodies to the target analyte is a critical step in assay development. Most organic compounds smaller than 2500 amu, which includes taxol and its congeners, will not stimulate the antibody response in animals. It is necessary to covalently link such small molecules (haptens) to an immunogenic macromolecule, usually a protein. Serum taken from an animal immunized with hapten-conjugates will contain polyclonal antibodies with a wide range of specificity and affinity for the different components of the immunogen, including the hapten, the carrier protein, and the hapten-carrier complex (*32*). Immortal cell lines producing homogeneous monoclonal antibodies (mAbs) can be derived from individual antibody-producing lymphocytes by the hybridoma technique. Both monoclonal and polyclonal antibodies can be used for developing immunoassays as long as high affinity antibodies with reactivity to unconjugated hapten are present.

Development of Monoclonal Antibodies Specific to Taxol and its Congeners. A hybridoma cell line derived from a mouse immunized with keyhole limpet hemocyanin-7-succinyltaxol conjugate produces a high affinity mAb to taxol and its C-7 derivatives (*27*, *33*). This mAb, 3C6, is twenty-fold less reactive with cephalomannine and is virtually unreactive with baccatin III. MAb 8A10 was derived from a mouse immunized with 7-succinylbaccatin III: it cross reacts with taxol, cephalomannine, baccatin III, and 10-deacetylbaccatin III. It did not react, however, with an analog lacking the C-20 oxetane ring, 20-acetoxy-4-deacetyl-5-epi-20,O-secotaxol (*34*). Thus mAb 8A10 appears to bind a determinant common to the intact tetracyclic diterpenoid ring structure common to many natural taxanes.

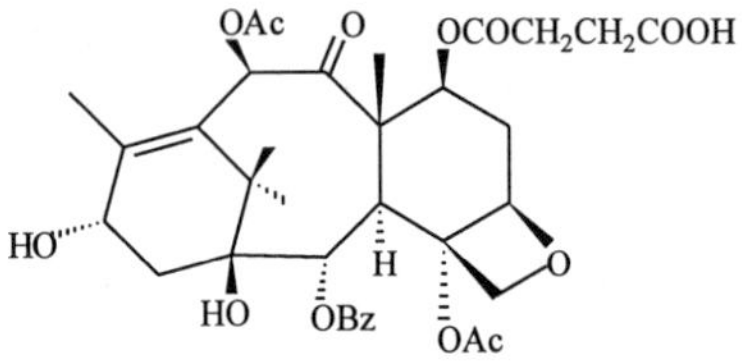

7-SUCCINYLBACCATIN III

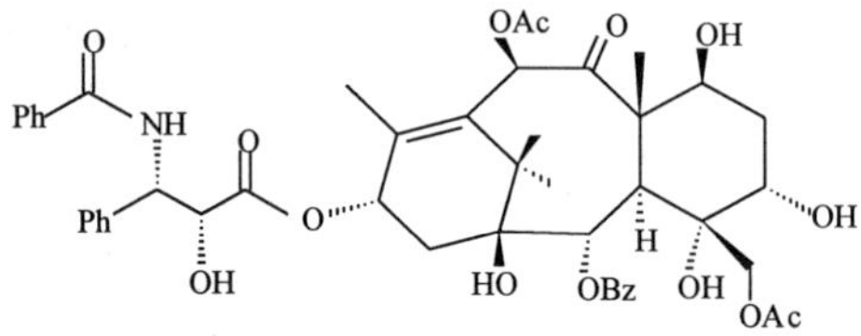

20-ACETOXY-4-DEACETYL-5-epi-20,O-SECOTAXOL

Indirect Competitive Inhibition Enzyme Immunoassay (CIEIA). Competitive inhibition enzyme immunoassays have been developed utilizing mAb 3C6 and mAb 8A10 to quantitate the amount of taxol vs. total taxanes in crude sample extracts according to the method shown in Figure 1. The assay is performed in 10% methanol to facilitate dissolution and processing. The CIEIA is conducted in 96-well microtiter plates coated with 100 µL of a bovine serum albumin (BSA) conjugate of the hapten, either 7-succinyltaxol for mAb 3C6 or 7-succinylbaccatin III for mAb 8A10. Additional BSA (200 µL) is then added to prevent non-specific antibody binding to the solid phase. To generate standard curves, analytical standards are serially diluted between 0.5 and 300 nM in phosphate buffered saline containing 0.25% BSA, 0.05% Tween-20 and 20% methanol: 50 µL is combined with an equal volume of optimally diluted antibody in phosphate buffered saline containing 0.25% BSA and 0.05% Tween-20. Dried extracts of fungal cultures are suspended in 0.2 mL MeOH, and diluted 1:4 with phosphate buffered saline containing 0.25% BSA and 0.05% Tween-20. The suspended extract is serially diluted in phosphate buffered saline containing 0.25% BSA, 0.05% Tween-20, and 20% MeOH, and combined with antibody as described for analytical standards. Analyte present in standard or test wells competitively inhibits antibody binding to the solid-phase BSA-hapten conjugate. The bound antibody is detected indirectly using an anti-mouse immunoglobulin enzyme-conjugate and appropriate chromogenic substrate, to generate a colorimetric assay endpoint which is inversely proportional to the analyte concentration (Figure 2).

CIEIA methods using these antibodies confirmed the presence of taxol in partially and totally purified preparations of fungal taxol. Extracts from twenty different fermentations of *T. andreanae* have been tested by CIEIA with consistent levels of taxol and taxane in each test. Furthermore, a quantitative comparison between the CIEIA's using monoclonal antibodies specific to taxol with a CIEIA using monoclonal antibodies class specific to taxanes in general revealed that taxol comprises

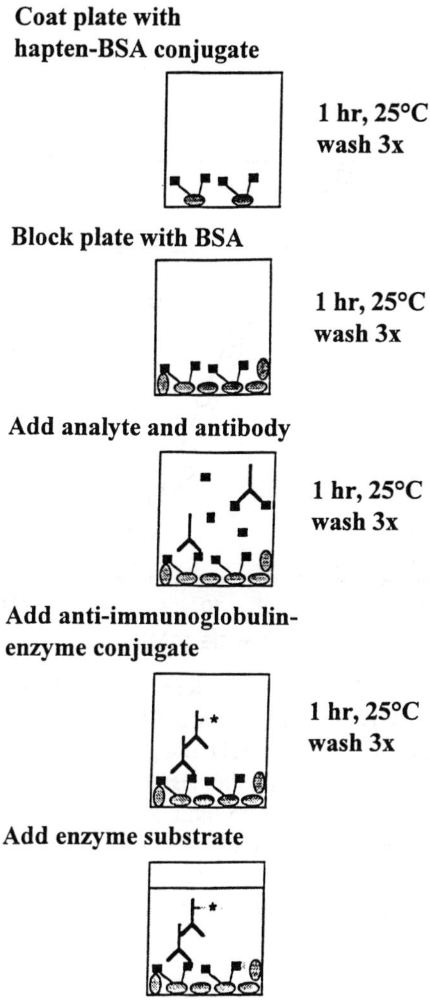

Figure 1. Schematic representation of the CIEIA method.

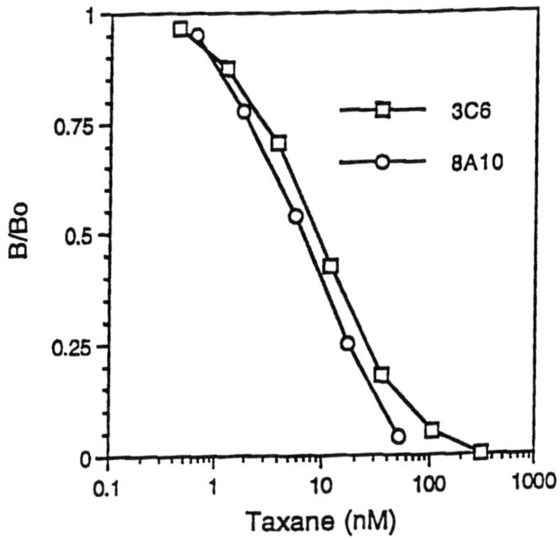

Figure 2. Standard curves for taxol (3C6) and taxane CIEIA's. B/B_0 values for each dilution were calculated by dividing the mean OD of a given set of replicates containing taxane inhibitor (B) by the mean OD of all the wells containing no inhibitor. Unknown taxane concentrations in test samples were calculated from the B/B_0 of the sample dilutions that fell within the log-linear portion of the standard curve.

only 15-20% of the total taxanes present in the semi-purified fungal extract (1st TLC step in purification) (*27*).

Fungal taxol was isolated from 3 week old culture fluids of this fungus and identified by mass spectrometry, immunochemistry and chromatographic methods. Although *T. andreanae* also makes baccatin III and other taxanes, cephalomannine has never been detected in the fungal extracts. In all yew tree bark extracts, however, taxol and cephalomannine are isolated together, and are difficult to separate using silica gel HPLC. The fungus and the tree may exploit biosynthetic pathways that differ to some degree, which may account for the absence of cephalomannine. An important corollary to this premise is the potential of the fungus to produce unique taxanes. Taxol is not the perfect drug. Its poor water solubility poses delivery problems that have not been adequately resolved (*35*). *T. andreanae* may produce related, more polar compounds with activity approaching that of taxol. We have already discovered taxanes with unique mass spectral profiles. These findings open the possibility for an unlimited source of taxol via fermentation technology.

Radioisotopic Labeling Studies. To demonstrate that the taxol found in *T. andreanae* is truly produced *de novo* by this fungus, we performed radiolabeling experiments with sodium acetate-1-^{14}C, phenylalanine-UL-^{14}C, sodium benzoate-7-^{14}C, and leucine-UL-^{14}C. Each precursor was added to a 20 day old *T. andreanae* liquid culture and then incubated for 4 days at 25°C. Taxol and baccatin III were isolated by repeated preparative TLC. Confirmation of the identity of ^{14}C-taxol in the fungal preparations was done by 2 dimensional thin layer co-chromatography with yew-derived taxol (*27*). The data are all normalized on the basis of 100 μCi administered per 2 g dry weight of fungal mycelium. Phenylalanine-UL-^{14}C was the best precursor for both ^{14}C-taxol (1241 ± 40 dpm) and ^{14}C-baccatin III (268 ± 25 dpm). Sodium acetate-1-^{14}C also yielded ^{14}C-taxol (261 ± 32 dpm) and ^{14}C-baccatin III (128 ± 20 dpm). Neither sodium benzoate-7-^{14}C, nor leucine-UL-^{14}C yielded any ^{14}C-taxol or baccatin III in this experiment, although leucine is a very effective taxol precursor in *Taxus brevifolia* (*36*). The size, shape, and location of the vanillin/sulfuric acid and UV absorbing spot on the TLC plate was identical to the exposed single-spot on the x-ray film. As a control, ethanol (60%) killed mycelium was incubated with acetate-1-^{14}C and the culture medium processed in the identical manner (*27*). No radioactivity appeared in the area coincident with taxol.

Controlled Experiments. Several control tests were run to eliminate the possibility that the detected taxol was either a carry over from the tree or an accidental contaminant of our fungal cultures. A 5L liquid culture inoculated with agar blocks of fungal mycelium yielded no detectable taxol at time zero. Likewise, it was impossible to detect taxol in the agar blocks of mycelium used as inoculum. The presence of 1 mg/liter of chlorocholinechloride in the medium completely abolished taxol production. This compound is also an effective inhibitor of gibberellin production in *G. fujikuroi* (*22, 23*), although it stimulates petasol (sesquiterpenoid) production in *Drechslera gigantea* (*37*). We can reasonably conclude that the taxol isolated from cultures of *T. andreanae* is actually a product of the metabolism of this organism.

Taxol Yield as a Result of Media Amendments. The amounts of taxoids produced by *T. andreanae* are disappointingly low. Estimates made by 2 different methods, electrospray mass spectrometry and the quantitative monoclonal antibody technique (CIEIA), indicate that 24-50 ng of taxol are produced per liter. We have observed, however, that many plant associated fungi require one or more plant metabolites to activate pathways critical to secondary product formation (*38*). Plant pathogenic fungi often produce mycotoxins at higher yields if the aqueous extract of the host plant is included in the medium (*38*). This appears to be true of fungal synthesis of taxol. Preliminary experiments showed an increase in taxol production if the water soluble components of yew needles and shoot tips were added to broth cultures of *T. andreanae* (*27*). Of course, because the initial purpose of this investigation was to prove *de novo* taxol synthesis by *T. andreanae*, no plant extracts were included in any growth media used to this end. However, such water extracts may indeed augment taxol production and may be an important component of yield enhancement.

Using CIEIA to monitor taxol titers, we have attempted to enhance taxol production. The complete experimental results will be presented at a later date, but we will outline briefly in this section several of our enhancement experiments. Initial experiments showed an increase in taxol titer if yew needle extract were added to the medium. In repeated experiments, the aqueous extract of yew needles, prepared by steeping 5 gm of needles in 1L boiling water for 5 minutes, was added to mycological broth to make a 1% solution. This amendment results in a hundred-fold increase in taxol production. Taxol monoclonal antibody analysis of this 1% yew medium (sterile) shows a taxol titer of between 95 and 200 ng/L. *T. andreanae* grown in this medium for 21 days consistently yields 2000-3000 ng/L taxol, with as much as 80% of the antibody active material sequestered in the mycelia. When the fungus is grown in unadulterated mycological broth it consistently yields 40-90 ng/L taxol.

Other amendments added to mycological broth have also augmented taxol titers. These include different sugars added at different points in the fermentation cycle, and compounds associated with the acetate biosynthetic pathway. We have compared still cultures to shaker cultures. Rapid shaking (>200rpm) results in poor growth and poor yield, but moderate shaking (100 rpm) results in a tenfold increase in mycelial mat. The filtrate of these cultures yields similar taxol titers to control (still) cultures of *T. andreanae*, but the mycelial mat, after grinding and extracting, yields several hundred additional ng/L taxol. Mycelial extracts of still cultures tend to be quite "wispy" and yield little taxol.

Several other experiments have been run, including a repetition of the original chlorocholinechloride experiment. CCC, a known blocking agent of gibberellin biosynthesis in *Gibberella fujikuroi* (*22, 23*) was added to *T. andreanae* in varying amounts. As CCC content increased, taxol titer decreased, to the point of total suppression. The addition of N(dimethylamino)succinic acid (alar), another known plant growth regulator, increased taxol production at certain concentrations. Other plant growth regulators, particularly those used to increase gibberellin production in *G. fujikuroi,* are currently being studied for their effect on taxol production in *T. andreanae*. Such improvements would not be unlike the course taken for virtually all microbes that have been fermented for industrial microbiological purposes (*20*).

Improved culturing techniques, the addition of "activators", and the application of genetic engineering methods may ultimately permit the commercialization of *T.andreanae* for taxane production.

Potential Microbial Sources of Taxol

Several other fungi have shown consistent cross-reactivity with the monoclonal antibodies for taxol and/or taxane. These organisms are currently being subjected to second and third order examinations to determine if the observed antibody reactivity is actually attributable to the presence of taxoids. The following tables indicate typical taxol and taxane titers observed for fungi grown in synthetic, yew-free media to examine "uninduced" potential of microbes to produce taxol/taxanes (Table I) and in media with 2% yew broth added (Table II). Titers are determined as taxoid/Liter.

Table I. Taxol and taxane titers of endophytic fungi

	CH_2Cl_2 sample	**total wt.**	**taxol/ sample**	**taxol/ Liter**	**taxane/ sample ug**	**taxane/ Liter ug/L**
CC45BD	0.0045g	0.0266g	0.0031ug	**0.018ug**	0.4112ug	**2.43**
CC50NA1	0.0071	0.0137	0.173	**0.669**	0.279	**1.077**
CC53NC	0.0046	0.0254	0.0046	**0.025**	0.0184	**0.102**
CC54BE	0.0051	0.0655	0.0032	**0.041**	0.0074	**0.095**
CC57BC2	0.0057	0.1658	0.0064	**0.186**	0.0078	**0.227**
CC64BB	0.0046	0.0699	0.0035	**0.053**	0.0089	**0.135**
H10BA2	0.0171	0.0366	0.0135	**0.058**	0.276	**1.189**
CC50NA22	0.0046	0.0146	0.0592	**0.376**	0.282	**1.790**
CC53NA1	0.0093	0.0193	0.0327	**0.135**	0.241	**1.003**
4BA	0.0083	0.0198	0.0576	**0.274**	0.377	**1.798**
Media	0.0085	0.0085	0		0	

Table II. Taxol and taxane titers of endophytic fungi grown in yew broth media

	CH_2Cl_2 ext.	taxol/ fraction	taxol/ liter (ug/L)	taxane/ fraction(ug)	taxane/ liter (ug/L)
CC45BD	0.0027	0.408	**4.08**	1.67	**10.67**
CC50NA1	0.0045	0.462	**4.62**	5.22	**52.2**
CC50NA2	0.0022	0.880	**8.80**	8.96	**89.6**
CC52NC	0.0024	0.758	**7.58**	7.96	**79.6**
CC53NA	0.0012	0.622	**6.22**	7.27	**72.7**
CC53NC	0.0020	0.614	**6.14**	5.65	**56.5**
CC54BA	0.0074 (1/2)	0.304	**6.08**	4.20	**84.0**
CC54BE	0.0056 (1/2)	0.174	**3.48**	1.82	**36.4**
CC57BC2	0.0065 (1/2)	0.192	**3.84**	3.59	**71.8**
CC64BB	0.0060	0.264	**2.64**	1.87	**37.4**
blank(1L)	0.0057	0.95	**0.95**	6.61	**6.6**

The Search for Unrelated Bioactive Compounds from this Fungal Collection.

The microorganisms isolated in this study were examined for evidence of a broad spectrum of biological activities unrelated to the presence or absence of taxol. Standard disk bioassays were utilized to determine activity against both fungi and Gram(+) and Gram(-) bacteria, with special attention paid to microbes whose organic extracts showed promising antifungal activity. Compounds with potential anticancer or antitumor activity were also desirable, although intramural screening of large numbers of extracts for these activities can be problematic. Simple in-house assays have been devised and perfected by a number of scientists to facilitate such activity assessments. Galsky and Ferrigni have promoted two assays, the crown gall tumor assay on potato discs and brine shrimp assay that can model the traditional 9KB and P388 in vivo mouse leukemic systems (*39-41*) and easily facilitate bioassay guided fractionation. All pure compounds with activity in our in-house screens will be sent to the National Cancer Institute for rigorous testing.

Our preliminary investigations show promise. We have found the endophytic fungi isolated from *Taxus brevifolia* to be a rich source of biologically active components. Fungi with significant antifungal activity or cytotoxicity are grown in large volume cultures and the bioactive components are currently being isolated and characterized.

Fungal Culture and Chemical Analysis. Bioactive fungi were grown and extracted in the manner described for *T. andreanae*. Both aqueous and organic extracts were tested to confirm desired activity, and subjected to bioassay guided fractionation. Bioassays include antimicrobial standard disc assays, brine shrimp toxicity assay, an effective method for predicting cytotoxicity (*41*), and crown gall potato disc assay (*39, 40*). Extracts will also be sent to our pharmaceutical collaborators for additional testing. Preliminary work on the endophytic fungi associated with the yew has yielded over 300 isolated fungi. Of the fungi tested, 21% are active against *Bacillus subtilis* and *Staphylococcus aureus*, 9% are active against *Escherichia coli,* and 4% are active against *Candida albicans*. We have isolated some of the compounds responsible for the biological activity observed in the various fungal extracts (Table III).

Table III. Compounds isolated from endophytic fungi of *Taxus brevifolia*

Fungal#	Isolated compounds	In-house	Lit.
CC50NB	gliovictin	af	af (*42*)
CC44BC-1	griseofulvin	*B. subtilis*, af	af (*43, 44*)
	dechlorogriseofulvin	*B. subtilis*	af (*45*)
	kojic acid		af, ab (*46*)
	compactin		
CC57BC-2	mycophenolic acid	*C.albicans*, *S.aureus*	af, ab (*47*)
EM1BD	mycophenolic acid	*C.albicans, S.aureus*	antiviral (*48*)
	pebrolides		
	ergosterol peroxides	*C.albicans*	
G29NE	6-pentyl-α-pyrone	*C.albicans, B.subtilis*	

Table III, cont'd. Compounds isolated from endophytic fungi of *Taxus brevifolia*

Fungal#	Isolated compounds	In-house	Lit.
H10BA2	Penitrem A		
	Penitrem B		
	dihydroisocoumarin		
	dihydroisocoumarin		
H1RE	phomopsolide B	*B.subtilis*	
	ergosterol peroxide		
	dimethylhydroxy-phthalide	*C.albicans, E.coli*	
H27RD	propenyl furanone		
	ergosterol peroxides		
	phomopsolide B		
	phomopsolide A		
BC4BA	ramulosin		
	6-hydroxyramulosin		
MSH1BA	fusaric acid	*C.albicans, E.coli, B.subtilis*	
WIC65NC	Penicillic acid	*C.albicans, B.subtilis,*	av, at, ab (*49-50*)

af- antifungal, ab- antibacterial, av- antiviral, at- antitumor

Significance

These research efforts are significant for both practical and philosophical reasons. First, they could have a profound effect on the supply issues concerning the important anticancer compound taxol. Taxol is currently isolated from the bark of the yew tree which represents a finite drug source at best. A fungus or bacterium capable of producing taxol at a rate of 50 mg/L would represent an inexhaustible source of the drug. From both an ecological and an economic viewpoint, a microbial source would supplant reliance on the yew tree. We would no longer be confronted with the choice of saving lives or saving yew trees. If any of the microbial sources isolated can provide reasonable, reliable quantities of taxol, more drug would be available for both studies and treatment regimen, at a lower cost to patients, and no cost to the environment.

Our results to date have raised some fundamental issues concerning the relationship between endophytes and their hosts. The discovery of two closely associated but taxonomically diverse organisms producing the same complex natural product is the most compelling. Although not the first discovery of this kind (the production of gibberellins by both higher plants and *Gibberella fujikuroi* predated our discovery by 40 years) this coincident taxol production suggests that some mechanism for genetic exchange between tree and fungus might exist. If taxol production is indeed coded by a transposable element, then it might be possible to amplify this element and insert it into a fast-growing procaryote, resulting in improved yields and shorter production time. Microbial production will also facilitate examination of the biosynthetic pathway of taxol and taxoids via feeding studies.

Our results also suggest that the endophytes of medicinal plants are a good source of compounds with biological activity. Endophytes represent a virtually untapped reservoir of potentially novel, effective drugs. They may operate in a mutualistic sense, protecting their hosts from infectious diseases, while being afforded a stable environment in return. Microbial defenses are probably chemical in nature, and could be exploited for our own purposes. These early results suggest that this area of endeavor will prove a worthy arena for study for years to come.

References

1. Munro, M.H.G.; Luibrand, R.T.; Blunt, J.W. In *Bioorganic Marine Chemistry*, Springer: Berlin, **1987,** Vol. 1; pp 94-95.
2. Rowinsky, E.K.; Cazenave, L.A; Donehower, R.C. *J. Nat. Cancer Inst.* **1990**, *82*, 1247-1259.
3. Holmes, F.A.; Walters, R.S.; Theriault, R.L.; Forman, A.D.; Newton, L.K.; Raber, M.N.; Buzdar, A.U.; Frye, D.K.; Hortobagyi, G.N. *J. Nat. Cancer Inst.* **1991**, *83*, 1797-1805.
4. Cragg, G.M.; Snader, K.M. *Cancer Cells* **1990**, *3*, 233-235.
5. Schiff, P.B.; Fant, J.; Horowitz, S.B. *Nature* **1979**, *277*, 665-667.
6. Schiff, P.B.; Horowitz, S.B. *Proc. Natl. Acad. Sci.* **1980**, *77*, 1561-1565.
7. Wani, M.C.; Taylor, H.C.; Wall, M.E.; Coggon, P.; McPhail, A.T. *J. Am. Chem. Soc.* **1971,** *93*, 2325-2327.
8. Forastriere, A.A. *Semin. Oncol. Suppl. 3* **1993**, *20*, 56-60.
9 Ettinger, D.S. *Semin. Oncol. Suppl. 3* **1993**, *20*, 46-49.
10. Holton, R.A.; Somoza, C.; Kim, H.; Liang, F.; Biediger, R.J.; Boatman, P.D.; Shindo, M.; Smith, C.C.; Kim, S.C.; Nadizadeh, H.; Suzuki, Y.; Tao, C.; Vu, P.; Tang, S.H.; Zhang, P.; Murthi, K.K.; Gentile, L.N.; Liu, J.H. *J. Am. Chem. Soc.* **1994,** *116,* 1597-1598.
11. Nicolaou, K.C.; Yang, Z.; Liu, J.J.; Ueno, H.; Nantermet, P.G.; Guy, R.K.; Claiborne, C.F.; Renaud, J.; Couladouros, E.A.; Paulvannan, K.; Sorensen, E.J. *Nature* **1994**, *367*, 630-634.
12. Wender, P.A.; Mucciaro, T.P. *J. Am. Chem. Soc.*, **1992,** *1149*, 5878-5879.
13. Georg, G.I.; Ali, S.M.; Zymunt, J.; Jayasinghe, L.R. *Exp. Opin. Ther. Pat.* **1994**, *4*, 109-120.
14. Nicolaou, K.C.; Dai, W.M.; Guy, R.K. *Angew. Chem., Int. Ed. Engl.* **1994**, *33*, 15-44.
15. Holton, R.A. *Eur.Pat. Appl.* **1990**, *EP O 400 971 A2.*
16. Ojima, I.; Habus, I.; Zhao, M.; Georg, G.I.; Jayasinghe, L.R. *J. Org. Chem***.** **1991**, *56*, 1681-1683.
17. Fett-Neto, A.G.; DiCosmo, F.; Reynolds, W.F.; Sakata, K. *BioTechnology* **1992**, *10*, 1572-1575.
18. *Antibiotics, Chemotherapeutics, and Antibacterial Agents for Disease Control*, Grayson, M., Ed.; John Wiley & Sons: New York, 1982; p 92.
19. Okami, Y. *Microbial Ecology* **1986,** *12,* 67-78.
20. Demain, A.L. *Science* **1981**, *214,* 987-995.
21. Nakamura, H.; Iitaka, Y.; Kitahara, T.; Okazaki, T.; Okami, Y. *J. Antibiot.* **1977**, *30*, 714-719.
22. Stowe, B.B.; Yamaki, T. *Ann. Rev. Plant. Physiol.* **1957**, *8*, 181-216.

23. Lang, A. *Ann. Rev. Plant Physiol.* **1970,** *21*, 537-570.
24. TLC's were run on Merck silica gel plates (5x10 cm, 0.25 mm) in the following solvent systems (v/v): chloroform-acetonitrile, 7:3; chloroform-methanol, 7:1; dichloromethane-tetrahydrofuran, 6:2; and ethyl acetate-isopropanol, 95:5.
25. Strobel, G.A.; Stierle, A.; Stierle, D.; Hess, W.M. *Mycotaxon* **1993**, *47*, 71-80.
26. Stierle, A.; Stierle, D.; Strobel, G.; Bignami, G.; Grothaus, P. In *Bioregulators for Crop Protection and Pest Control*; Hedin, P.A., Ed.; ACS Symposium Series 557; American Chemical Society: Washington, DC, **1994**, pp 64-77.
27. Stierle, A.; Strobel, G.; Stierle, D. *Science* **1993**, *260*, 214-216.
28. Modified mycological agar consists of 10 g bacto-soytone, 40 g glucose, 15 g bacto-agar, 1 g sodium acetate, and 50 mg sodium benzoate per liter.
29. Cardellina, J.H. *J. Liquid Chromatography* **1991**, *14*, 659-665.
30. McClure, T.D.; Schram, K.H.; Reimer, M.L.J. *J. Am. Soc. Mass Spec.* **1992**, *3*, 672-675.
31. Robins, R.J. In *Modern Methods of Plant Analysis. New Series, Vol.4: Immunology in Plant Sciences* Linskens, H.F.; Jackson, J.F., Eds.; Springer: Berlin, **1986**, pp 86-141.
32. Vanderlaan, M.; Stanker, L.; Watkins, B. In *Immunoassays for Trace Chemical Analysis*; Vanderlaan, M.; Stanker, L.; Watkins, B.E.; Roberts, D.W., Eds.; ACS Symposium Series 451; American Chemical Society: Washington, D.C. **1990**, pp 2-13.
33. Bignami, G.S.; Grothaus, P.G.; O'Malley, S.; Harada, K.E.; Byrnes, J.B.; Waller, D.F.; Rowinsky, E.K.; McGuire, M.T.; Alvarado, A.B.; Raybould, T.J.G., manuscript in preparation.
34. Kingston, D.G.I.; Samaranayake, G; Ivey, C.A. *J. Nat. Prod.* **1990**, *53*, 1-12.
35. Vyas, D.M.; Wong, H.; Crosswell, A.R.; Casazza, A.M.; Knipe, J.O.; Mamber, S.W.; Doyle, T.W. *Bioorg. Med. Chem. Lett.* **1993**, *3*, 1357-1360.
36. Strobel, G.A.; Stierle, A.; Van Kuijk, T.J.G.M. *Plant Science*, **1992**, *84,* 65-74.
37. Bunkers, G.J.; Kenfield, D.; Strobel, G.A. *Mycol. Res.* **1991**, *95*, 347-351.
38. Pinkerton, F.; Strobel, G.A. *Proc. Natl. Acad. Sci. USA* **1976**, *73*, 4007-4011.
39. Galsky, A.G.; Wilsey, J.P. *Plant Physiol.* **1980**, *65,* 184-185.
40. Galsky, A.G.; Kozimor, R.; Piotrowski, D.; Powell, R.G. *J. Natl. Cancer Inst.* **1981**, *67*, 689-692.
41. Ferrigni, N.R.; McLaughlin, J.L.; Powell, R.G.; Smith., C.R. *J.Nat.Prod.* **1984**, *47*, 347-352.
42. Cole, R.J.; Cox, R.H. In *Handbook of Toxic Fungal Metabolites*; Academic: New York, **1981**, p 571.
43. Cole, R.J.; Cox, R.H. *ibid.,* p. 857.
44. Klein, M.F.; Beall, J.R. *Science* **1972**, *175*, 1483.
45. Cole, R.J.; Cox, R.H. *op. cit.*, p 862.
46. Cole , R,J.; Cox, R.H. *op. cit.*, p 759.
47. Cole , R,J.; Cox, R.H. *op. cit.,* p 866.
48. Planterose, D.N. *J. Gen. Virol.* **1969**, *4*, 629-632.
49. Cole , R,J.; Cox, R.H. *op. cit.*, p 521.
50. Johnson, J.R.; Buchanan, J.B. *J. Am. Chem. Soc.* **1953**, *75*, 2103-2105.

RECEIVED August 23, 1994

Chapter 7

Metabolism of Taxoid Drugs

M. Vuilhorgne[1], C. Gaillard[2], G. J. Sanderink[2], I. Royer[3], B. Monsarrat[3], J. Dubois[4], and M. Wright[3]

[1]Rhône-Poulenc Rorer, Centre de Recherche de Vitry-Alfortville, 13 Quai Jules Guesde, F-94400 Vitry sur Seine, France
[2]Rhône-Poulenc Rorer, Institute de BioPharmacie, 20 Avenue Raymond Aron, F-92165 Antony, France
[3]Centre National de la Recherche Scientifique, Laboratoire de Pharmacologie et de Toxicologie Fondamentales, 205 route de Narbonne, F-31400 Toulouse, France
[4]Centre National de la Recherche Scientifique, Institut de Chimie des Substances Naturelles, F-91190 Gif-sur-Yvette, France

Abstract : The metabolism of the two most promising anti-cancer taxoid drugs - docetaxel (Taxotere) and paclitaxel (Taxol) - is fully reported. Complete *in-vitro* and *in-vivo* results in animals and humans are described while structures of the observed metabolites are identified. The chemical synthesis of the major metabolites of docetaxel as well as their biological activities are reported. Furthermore, in accordance with the chemical reactivity observed for some of the synthetic intermediates, a metabolic pathway is proposed.

Despite extensive drug research efforts over the past 20 years, cancer remains the second cause of death in the industrialized countries (*1*).
A new series of drugs, the taxoids paclitaxel (Taxol), originally extracted from the bark of the Pacific yew *(2)*, and docetaxel (Taxotere), the first semisynthetic compound, have generated wide interest due to their unique mechanism of action on the microtubule-tubulin system in eukaryotic cells (*3,4*).

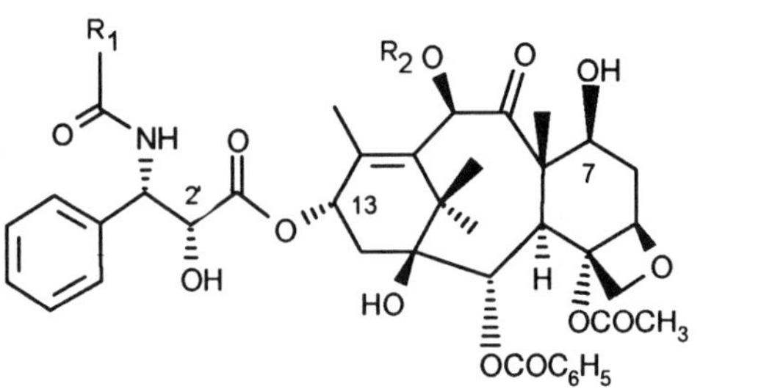

P : paclitaxel (Taxol)
$R_1 = C_6H_5$
$R_2 = COCH_3$

D : docetaxel (Taxotere)
R_1 = O-t-Bu
R_2 = H

Both compounds have demonstrated impressive *in vivo* activities (*5,6*) and are now considered among the most promising new agents in cancer therapy. Research into

0097–6156/95/0583–0098$08.00/0

this compound class has given rise to a large number of publications ranging from clinical studies (*7*) to total syntheses (8-9).

Pharmacokinetic results

Concerning the metabolism and pharmacokinetics of paclitaxel and docetaxel two recent reviews have been published (*10,11*). Pharmacokinetics have been studied in several phase I studies with various administration schedules. With sensitive analytical methods a triphasic elimination process can be observed for both compounds at sufficiently high doses (*12,13,14,15,16*).
Paclitaxel has been administered to patients at a dose of 135 or 175 mg/m^2 by 24 hour infusion or, more recently, by 3 hour infusion. Distribution of paclitaxel has been shown to be very rapid, with a first half-life of about ten minutes. In one study (135 mg/m^2, 3 h) elimination half-life was about 14 hours and systemic clearance 18 l/h/m^2 (*13*). At the shorter infusion time, paclitaxel pharmacokinetics are dose-dependent, with relatively higher AUC values at higher dose. One possible explanation is saturation of hepatic extraction and metabolism at higher plasma concentrations, which may reach 12.9 μM at 300 mg/m^2 *(12)*. Another interesting observation was a decrease in plasma clearance and a higher toxicity when paclitaxel was administered after cisplatin treatment in a combination study. No modification occurred with the inverse sequence. An effect of cisplatin on paclitaxel metabolism was considered as a possible explanation for this finding. The importance of hepato-biliary extraction in paclitaxel elimination was further demonstrated by lower clearance in patients with abnormal liver function, while renal function had little effect (*11*).
Docetaxel has generally been administered in phase II studies as a one-hour intravenous infusion at a 100 mg/m^2 dose. Phase I studies showed that docetaxel pharmacokinetics were best described by a three-compartment model with a very rapid initial phase ($t_{1/2}$ 5 min) and an elimination phase half-life of 12 hours. Systemic clearance of docetaxel was 21 l/h/m^2. The AUC of docetaxel increased in a linear manner with the dose from 20 to 115 mg/m^2, with no evidence for saturation of elimination. Maximum plasma levels of docetaxel were about 5 μM (*15,16*). Distribution studies in tumor-bearing mice confirmed very rapid tissue uptake. Tumor exposure was higher than plasma exposure (respectively 84 μg/g. hr and 17 μg/ml.hr). Elimination from tumor tissue ($t_{1/2}$ 22 h) was much slower than from plasma or from other tissues ($t_{1/2}$ 1-4 h) (*17*). No studies on the effect of individual factors on docetaxel pharmacokinetics have been published. However, an important population pharmacokinetic study, including over 600 patients, is being conducted in phase II studies. This should lead to valuable information on the effects of, for instance, pathophysiological status and co-medication on docetaxel elimination (*10*).

Preclinical studies : *in vitro* results

Studying drug metabolism in humans *in-vivo* is technically difficult, particularly the recovery of biliary excretion, and also because of generally low plasma levels of metabolites. However the use of microsomes is known to lead to reliable biotransformation results. Therefore both docetaxel and paclitaxel have been screened in these *in vitro* systems.

Recently [^{3}H] paclitaxel metabolism assays were conducted in freshly isolated rat hepatocytes and metabolite identification was carried out using HPLC and Tandem Mass Spectrometry (LSIMS) (*18*). Two major metabolites, P_2 and P_3, identical to those previously observed in the intact rat were formed. Inhibition studies suggested that the hydroxylation of paclitaxel was mediated by cytochrome P450 3A enzymes. The same authors (*19*) expanded these studies to human liver microsomes showing the formation of one major human metabolite. They provided evidence for the involvement of CYP 3A isoenzymes. The structures of the paclitaxel metabolites are reported in figure 1 while those of docetaxel are reported in schemes 1 and 2.
Monsarrat *et al.* (*11,20*) reported that when using human microsomes, the cytochromes P450 of the 2C family were responsible for the formation of P_5 while the cytochromes P450 of the 3A family led to P_2. The formation of P_5 was confirmed using hepatic microsomes and liver slices (*21*). Futhermore, two minor metabolites (unidentified) resulted from biotransformation by cytochrome P450 3A4 since antibodies raised against this enzyme inhibited their formation.
Docetaxel metabolism has also been extensively studied using these *in vitro* systems. Assays in mouse, rat, dog and human liver microsomes led to the conclusion that metabolic profiles were similar in these species and to those obtained *in vivo* (Sanderink G. ; unpublished results). Three to five metabolites were detected, while the inter-individual variability in terms of biotransformation rates in human liver microsomes (29 subjects) was reported to be high (*22*).
Inhibition studies have shown that the main enzymes responsible for biotransformation are from the cytochrome P450 3A family. However, other enzymes might also be involved (*23,24*).

Preclinical studies : *in vivo* metabolism in animals

The first studies on paclitaxel metabolism *in vivo* were conducted in the rat by Monsarrat *et al.* with unlabelled compound (*25*). From the bile of 20 treated animals, the formation of 9 metabolites was observed. After HPLC purification, metabolites P_2 and P_3 were identified using Mass Spectrometry (FAB and D/CI) and 400 MHz NMR experiments. The structures of other minor compounds showing an excess mass of + 16 and + 32 a.m.u., with respect to the molecular weight of paclitaxel, remain unknown.
In these earlier observations, the authors provided evidence for a significant hepatic metabolism of paclitaxel. Indeed, more than 40 % of the administered dose (I.V. 10 mg/kg) was recovered in the bile. On the other hand, less than 10 % of the dose was found in urine over a period of 24 hours. This preliminary work was followed by other evaluations confirming the prior results (*26,27,28*). Finally, the distribution of the main metabolites in bile was found to be : unchanged paclitaxel P : 12 % ; major metabolites : P_2 : 13 % and P_3 : 5 %, while minor compounds, including baccatin III and P_4, accounted for less than 10 %.
Very recently [^{14}C] labelled paclitaxel was used in rat studies (*29*). More than 95 % of the radioactivity was recovered in feces during the 6 days following administration. This result confirmed the essentially complete elimination of paclitaxel, with biliary excretion as the main clearance route.
Docetaxel metabolism has largely been studied in different animal species - mouse, rat, rabbit and dog - using [^{14}C] labelled drug (*30,31,32*). Fecal extracts have been

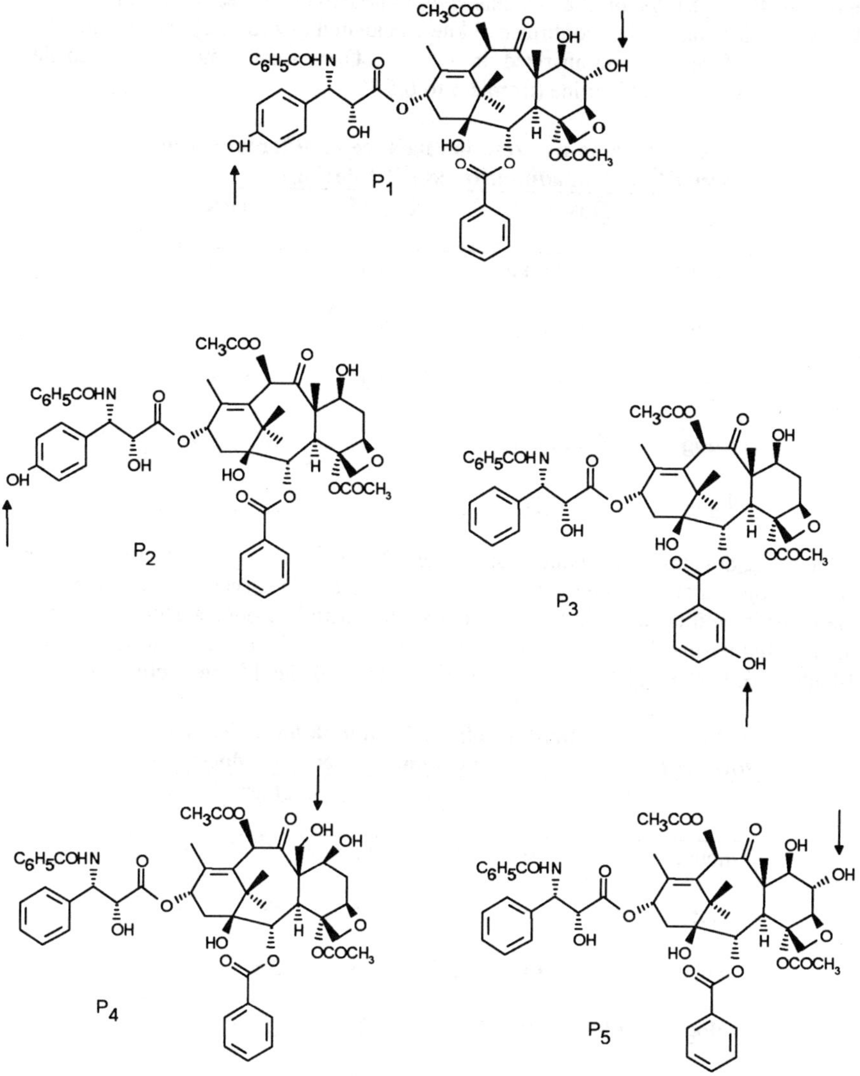

Figure 1 : Structures of paclitaxel metabolites
(Structural modifications are indicated by arrows)

analyzed for all the species. The results demonstrate that docetaxel excretion, as with paclitaxel, occurs mainly via the biliary route. The excretion balance shows that between 48 to 81 % of the administered radioactivity is recovered in the first 48 hours. This was further confirmed in bile-cannulated rats (*31,32*). The results thus obtained in feces are summarized in table I. Only small amounts of labelled compounds were found in urine (from 2.5 to 6.5 %).

Table I : 0-48 hours excretion balance in different species expressed as % of administered [^{14}C] docetaxel.

	Dose (I.V.)	Urine %	Feces %
mouse	37 mg/kg	6.5	81
rat	5 mg/kg	4.0	69
rabbit	0.5 mg/kg	4.3	48
dog	0.72 mg/kg	2.5	68
human	100 mg/m^2	2.5	61

Concerning docetaxel metabolite distribution, D_V , D_{VI} , D_{VII} , D_{XVI} and D as parent drug are the major identified compounds. Minor metabolites account for 3 to 5 % of the administered dose (table II). Recently studies with bile obtained *in-vivo* or with the isolated perfused rat liver were conducted. Docetaxel : D was then found to be 10 and 9 %, D_V : 3 and 4 %, D_{VI} : 15 and 7 %, D_{VII} : 9 and 4 % respectively (*32*).

Table II : Metabolites isolated from 0-48 hours feces of different species as % of the administered [^{14}C] docetaxel

	D_V	D_{VI}	D_{VII}	D_{XVI}	other	D
mouse	15	10	22.5	11.5	5	1.5
rat	5	17	6.5	5	3	9.5
rabbit	1.5	13	2.5	1	3.5	9.3
dog	2.5	29	4.5	2.6	5	13.2
human	7.5	6.5	7.7	23.3	n.r.	3

n.r. : not reported

Clinical trials : phase I and phase II human metabolites

In a one-patient single-experiment study approximately 20 % of the administered dose of paclitaxel was recovered in human bile while renal excretion accounted for 5 to 10 % (*11*). Five to seven metabolites were detected (*11, 25, 28*) with the following distribution : parent compound, P : 3 %, ; major metabolites : P_5 : 12 % ; P_2 : 2 %, while minor metabolites - including P_1 - accounted for less than 3 % (table III).

Table III : Paclitaxel metabolite distribution in human and rat bile as % of the total administered dose

	P_1	P_2	P_3	P_4	P_5	P	baccatin III
human	minor	2	n.r.	n.r.	12	3	n.r.
rat	n.r.	13	5	minor	n.r.	12	minor

n.r. : not reported

In this experiment no metabolite was found in urine or in plasma. However this last observation has to be tempered by the fact that Beijnen reported the possible presence of metabolites in plasma of patients at the end of infusion (*13*). Furthermore an additional HPLC peak (of unknown composition) eluting before paclitaxel has previously been observed in human plasma (*33*). This leads to the conclusion that some paclitaxel derivatives or metabolites could be present even at low quantities in these biological fluids (*11*).

[^{14}C] docetaxel excretion and metabolism have been studied in cancer patients after a one-hour infusion of 100 mg/m^2 (*34)*. Excretion occured almost exclusively in the feces (61 %) via the biliary route during the first 48 hours after drug administration. Urinary excretion was very low. Four main metabolites were detected in feces (*31,35*) with the following distribution : parent drug D : 3 % ; major metabolites : D_V : 7.5 % ; D_{VI} : 6.5 % ; D_{VII} : 7.7 % and D_{XVI} : 23.3 % (Table II). These results confirm those reported for studies in rats (*32*). Furthermore, a small amount (less than 2 %) of 7-epi docetaxel (isomer at C-7 of the diterpene moiety) was identified. This result is in good agreement with previous observations. It was reported that paclitaxel is also converted into its 7-epi isomer either in cell culture medium or in normal saline solution (*36,37*). This facile epimerization - presumably via a retroaldol/aldol mechanism (*38*) - generally occurs under basic conditions. As a consequence, a small amount of 7-epi docetaxel is also found in plasma extracts (pH=7.4) (Vuilhorgne M.,Gaillard C.; unpublished results).

Structural determinations, synthesis, metabolic pathway

Given the small amount of purified material usually available (often sub pico-molar quantities) structural identification of taxoid metabolites were mainly conducted using Mass Spectrometry (*18,25,28,31,32*).

Paclitaxel and docetaxel show D/CI, FAB (LSIMS) and electrospray pseudo molecular ions at respectively : MH^+, MNa^+ and MNH_4^+. Furthermore the main fragmentation pattern - i.e. the loss of the phenyl isoserine side chain - generally allows one to determine which part of the molecule is biotransformed (side chain, diterpene, or both moieties) (*39,40,41*). MS-MS experiments either in parent or daughter ion modes (*42*) and direct HPLC/MS-MS coupling experiment have been performed in order to obtain a better selectivity (*18,35)*.

As a consequence of HPLC/MS-MS results, the structure of an unstable human metabolite of docetaxel was postulated to be D_{IV}.
Despite careful interpretation of mass spectra, the structures of some minor metabolites of paclitaxel found in the rat bile remain unresolved (*25*). It has been reported that these compounds possess respectively a pseudo molecular ion showing : - 6, + 16, + 32 and + 38 a.m.u. in respect to the parent drug molecular weight.
When the amount of isolated metabolites exceeded a few µg, high field NMR spectroscopy greatly assisted the structural determinations (*25,35,43*). For instance 600 MHz NMR experiments conducted on 3 minor docetaxel metabolites found in mouse, rabbit, rat and dog showed that they were 7-epi isomers of D_V , D_{VI} and D_{VII}. Furthermore, in rabbit feces, the recovery of a very small amount of the side chain moiety, identified as a carboxylic acid derivative, confirmed that the loss of the side chain at C-13 had occured (Gaillard C., Vuilhorgne M. ; unpublished results). This cleavage has been previously reported in paclitaxel studies (*25*).
Synthetic efforts were undertaken to provide material for assessing their biological and toxicological effects. The four major human metabolites of docetaxel (D_V , D_{VI} , D_{VII} and D_{XVI}) were prepared (Commerçon A. *et al. ; Tetrahedron,* in press) from the previously described C-3' free amino C-10 deacetyl baccatin III derivative 1 (*44*) using the mixed carbonates 2 , 3 , 4 , as acylation reagents (scheme 1). The semisynthetic compounds obtained through synthesis have the same physicochemical properties (NMR, IR, MS) as the docetaxel metabolites, thus confirming the prior postulated structures.
The results demonstrate that, despite similarity in their chemical structures, docetaxel and paclitaxel show large differences in their metabolism. The metabolites of paclitaxel, i.e. P_2 ; P_3 and P_4 in rat ; P_1 ; P_2 and P_5 in man, are mono and dihydroxylated species. These biomodifications take place either on the phenyl group of the side chain or on the diterpene taxoid ring itself and are caused by different P450 enzymes. The "replacement" at the C-3' amido function of a phenyl (in paclitaxel) by a O-t-Butyl group (in docetaxel) leads to a different metabolic pathway as docetaxel biomodifications only occur on this side chain.
From chemical reactivity observations obtained during semisynthetic work, i.e. the very rapid conversion of the aldehyde and carboxylic acid derivatives into the expected metabolites, the following metabolic pathway can be proposed. First the parent compound docetaxel D is oxidized into the primary alcohol D_{VI} leading then to a putative aldehyde derivative which gives the two hydroxy oxazolidinone D_V and D_{VII} after prompt cyclisation. A further oxidative step leads to the unstable carboxylic derivative D_{IV} giving the cyclic oxazolidinedione D_{XVI} (scheme 2). The three oxidative steps : $D \rightarrow D_{VI} \rightarrow$ [aldehyde] $\rightarrow D_{IV}$ result from the involvement of cytochrome P450 enzymes (see the *in-vitro* results section) and may be applicable to all studied species (*31,32*). As expected, some significant interspecies differences are observed in the proportions of metabolites for a given parent drug. It should be noted that there is also a large interspecies variability concerning the chemical structures of paclitaxel metabolites.
Apparently the only metabolic pathway shared by paclitaxel and docetaxel leads to D_{VIII} (minor in rat bile) and P_2 (major in rat bile). The two resulting metabolites present the same chemical transformation, a para hydroxylation of the C-13 side chain C-3' phenyl group. Finally, almost no glucurono-conjugated or sulphated metabolites have been reported in any species (*18,25,28*).

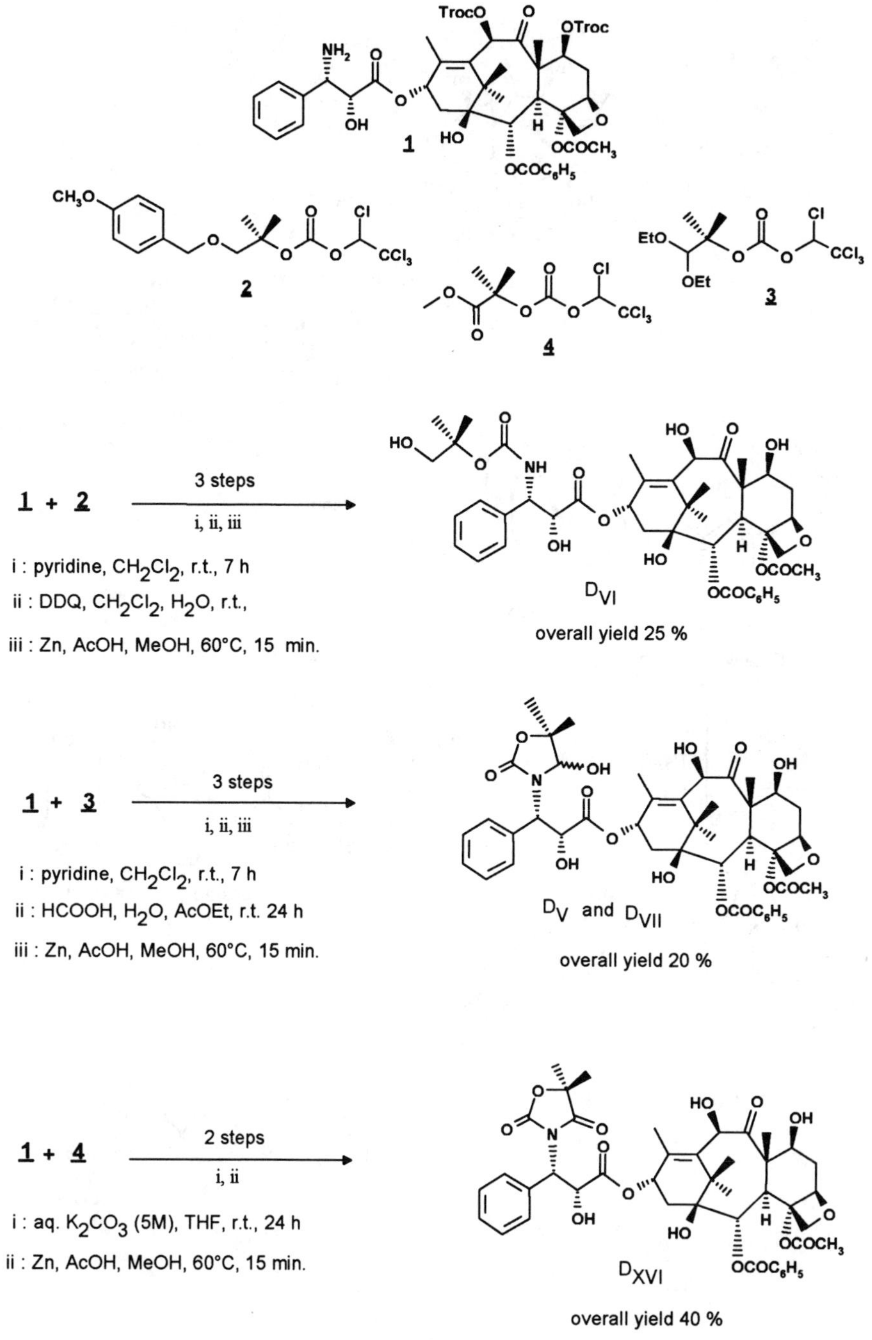

Scheme 1 : Synthesis of docetaxel metabolites

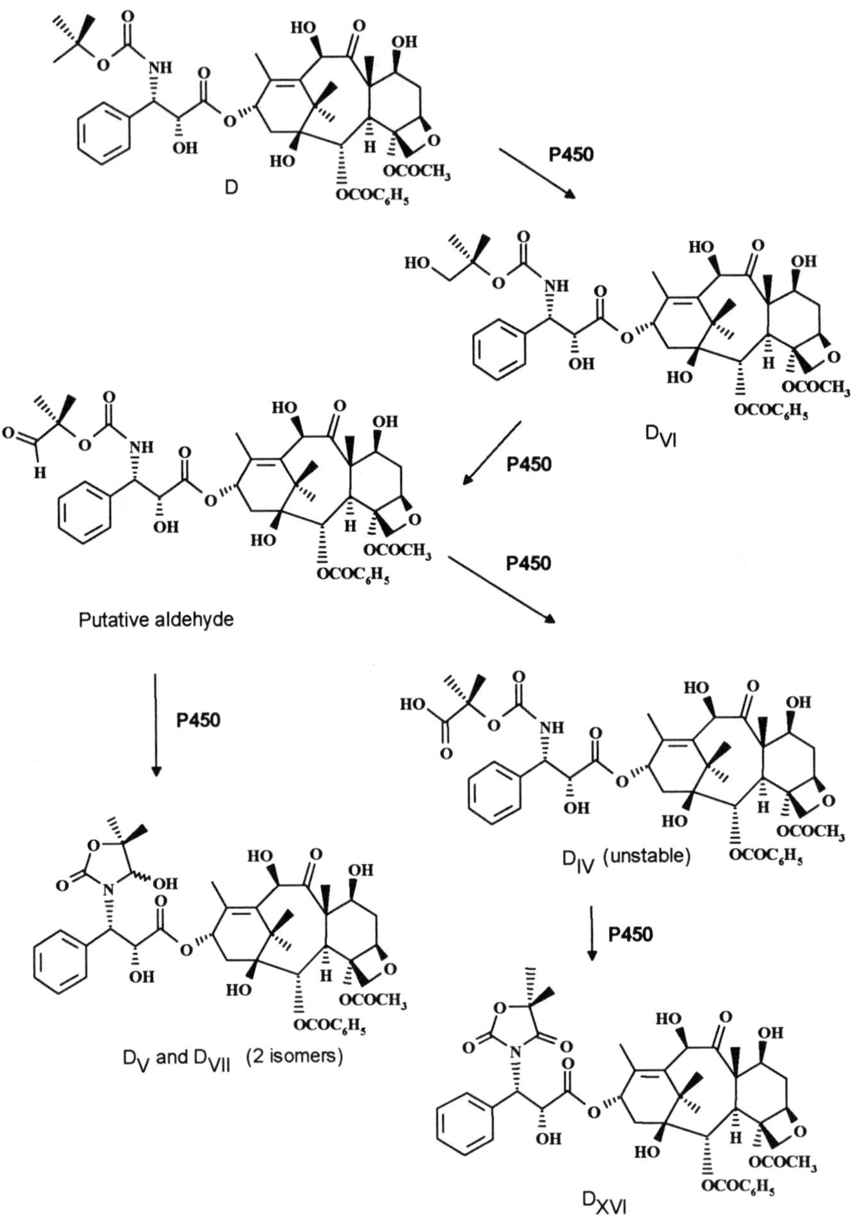

Scheme 2 : Postulated metabolic pathway for docetaxel

Biological activities of taxoid drug metabolites

Isolated metabolites of paclitaxel and docetaxel have been biologically studied using an *in-vitro* tubulin disassembly test (*6,45,46*) and L1210 or P388 leukemia cell growth assays (*6,47*). The metabolites P_2 and P_3 of paclitaxel found in rat bile, were as active as the parent drug in stabilizing microtubules against disassembly in cell - free tubulin assays (*25*). However, they were much less active than paclitaxel (respectively from 10 to 40-fold) against L1210 leukemia cells. Baccatin III isolated from rat bile is known to be inactive (*48*). The human bile metabolite P_5 was reported to be 30-fold less active in MOLT-4 and U-937cell line growth assays (*43*). Docetaxel metabolites : D_V , D_{VI} , D_{VII} and D_{XVI} were also evaluated *in vivo*. Indeed, and due to semisynthetic work, the amount of available material was sufficient for experiments with sub-cutaneous B16 advanced tumor-bearing in mice (*6,49*). These metabolites showed poor *in vitro* cytotoxicity (except D_{VI}) and were inactive against *in-vivo* B16 melanoma (*31*, Bissery M.C. *et al.* ; manuscript in preparation). Table IV summarizes these biological results.

Table IV : Biological activities of docetaxel metabolites

Compound	Tubulin test	P388 IC_{50} µg/ml	B16 Melanoma mg/kg/day	B16 Melanoma % T/C
P paclitaxel	T	0.075	31	27^a
D docetaxel	0.7 T	0.015	20	0^b
D_V	2.2 T	> 10	17	>> 42
D_{VI}	0.65 T	0.34	7.7	>> 42
D_{VII}	45 T	> 10	32	>> 42
D_{XVI}	100 T	> 10	32	>> 42

Log cell kill : a = 1.5 ; b = 3.4

Conclusion

Despite minor structural differences between paclitaxel and docetaxel they display major differences in their metabolic pathways. For both drugs clearance occurs via the biliary route while biotransformations are apparently effected by cytochrome P450 enzymes. Five major hydroxylated metabolites have been isolated in paclitaxel studies. Depending on the species, structural modifications take place either on the side chain and/or on the diterpene skeleton. In the case of docetaxel the "replacement" of the side chain C-3' phenyl by the tert-butyloxy group leads to different metabolic pathways. In this case, oxidized side chain products were the major metabolites observed while small amounts of epimerized compounds (7-epi derivatives) were also characterized. Since the same metabolites occur in mouse, rat, dog, rabbit and man, this leads to the conclusion that there is a principal and probably identical metabolic pathway for all studied species. All metabolites described herein are significantly less active *in vivo* than parent compounds.

Essentially no circulating taxoid derivatives (except the parent drug) were reported in plasma.

Finally the search for a second generation of potent taxoid drugs with improved solubility characteristics and enhanced ability to escape or overcome resistance phenomena will actively continue (*50,51,52*).

Moreover, studies of synergistic effects of taxoids with other drugs are required to refine multidrug treatment during cancer therapy. These efforts will most likely also require further extensive work on metabolism and pharmacokinetics.

Acknowledgments

We thank F. Lavelle and M.C. Bissery for supplying biological results ; A. Commerçon, D. Bourzat for providing semisynthetic compounds ; J. Nichele, J.P. Kettler for isolating metabolites ; B. Monegier for Mass determinations. We are indebted to F. Guéritte-Voegelein, D. Guénard for stimulating collaboration and to E. Guittet and J.Y. Lallemand for the use of their 600 MHz NMR spectrometers (BioAvenir and Organibio facilities). C. Burns is warmly acknowledged for critical reading of the manuscript.

Literature cited

1 EIU Marketing in Europe , *Trade Reviews* , *337* , December **1990**

2 Wani , M.C. ; Taylor , H.L. ; Wall , M.E. ; Coggon , P. ; McPhail , A.T. ; *J. Am. Chem. Soc.*, **1971** , *93* , 2325

3 Schiff, P.B. ; Fant , J. ; Horwitz , S.B. ; *Nature* , **1979** , *277* , 665

4 Ringel , I. ; Horwitz , S.B. ; *J. Natl. Cancer Inst.* , **1991** , *83* , 288

5 Mc Guire , W.P. ; Rowinsky , E.K. ; Rosenshein , N.B. ; Grumbine , F.C. ; Ettinger , D.S. ; Armstrong , D.K. ; Donehower , R.C. ; *Ann. Int. Medicine* , **1989** , *111* , 273

6 Bissery , M.C. ; Guénard , D. ; Guéritte-Voegelein , F. ; Lavelle , F. ; *Cancer Res* , **1991** , *51* , 4845

7 Pazdur , R. ; Kudelka , A.P. ; Kavanagh , J.J. ; Cohen , P.R. ; Raber , M.N. ; *Cancer Treatment Reviews* , **1993** , *19* , 351

8 Nicolaou , K.C. ; Yang , Z. ; Liu , J.J. ; Ueno , H. ; Nantermet , P.G. ; Guy , R.K. ; Claiborne , C.F. ; Renaud , J. ; Couladouros , E.A. ; Paulvannan , K. ; Sorensen , E.J. ; *Nature* , **1994** , *367* , 630

9 Holton , R.A. ; Somoza , C. ; Kim , H.B. ; Liang , F. ; Biediger , R.J. ; Boatman , P.D. ; Shindo , M. ; Smith , C.C. ; Kim , S. ; Nadizadeh , H. ; Suzuki , Y. ; Tao , C. ; Vu , P. ; Tang , S. ; Zhang , P. ; Murthi , K.K. ; Gentile , L.N. ; Liu , J.H. ; *J. Am. Chem. Soc.* , **1994** , *116* , 1597 and 1599

10 Bruno , R. ; Sanderink , G.J. ; *Cancer Surveys* , **1993**, *17* , 305

11 Rowinsky , E.K. ; Wright , M. ; Monsarrat , B. ; Lesser , G.J. ; Donehower , R.C. ; *Cancer Surveys* , **1993** , *17* , 283

12 Kearns , C. ; Gianni , L. ; Vigano, L . ; Giani, A. ; Spriggs , D. ; Tutsch , K. ; Schiller , J. ; Sridhara , R. ; Egorin , M. ; *Proc. Amer. Soc. Clin. Oncology,* **1993**, *13* , 135

13 Huizing , M.T. ; Keung , A.C.F. ; Rosing , H. ; Van der Kuij , V. ; ten Bokkel Huinink , W.W. ; Mandjes, I.M. ; Dubbelman , A.C. ; Pinedo , H.L. ; Beijnen ; J.H. ; *J. Clin. Oncol.* , **1993** , *11* , 2127

14 Vergniol , J.C. ; Bruno , R. ; Montay , G. ; Frydman , A. ; *J. Chromatography* , **1992** , *582* ,273

15 Bruno , R. ; Vergniol , J.C. ; Montay , G. ; Le Bail , N. ; Frydman , A. ; Clavel , M. ; Marty , M. ; *Proc. Amer. Assoc. for Cancer Research,* **1992** , *33* , 261

16 Bruno , R. ; Cosson , V. ; Montay , G. ; Frydman , A. ; *1st Meeting of NONMEM European Users* , June **1992** , Basel , Switzerland

17 Bissery , M.C. ; Renard , A. ; Montay , G. ; Bayssas , M. ; Lavelle , F. ; *Proc. Amer. Assoc. for Cancer Research* , **1991** , *32* , 401

18 Walle , T. ; Kumar , G.N. ; Mc Millan , J.M. ; Thornburg , K.R. ; Walle , U.K. ; *Biochem. Pharm.*, **1993**, *46* , 1661

19 Kumar , G.N. ; Walle , U.K. ; Walle , T. ; *J. Pharmacol. Exp. Ther.*, **1994** , *268* , 1160

20 Cresteil , T. ; Monsarrat , B. ; Alvinerie , P. ; Tréluyer , J.M. ; Vieira , I. ; Wright, M. ; *Cancer Res.* ; **1994** , *54* , 386

21 Harris , J.W. ; Rahman , A. ; Kim , B.R. ; Guengerich , F.P. ; Collins , J. ; *5th European ISSX Meeting*, Tours (France) , September 1993 , *ISSX Proceedings* , **1993** , *3* , 24

22 Zhou Pan , X.R. ; Marre , F. ; Zhou , X.J. ; Gauthier , T. ; M. Placidi , R. Rahmani ; *Second Interface of Clinical and Laboratory Response to Anticancer Drugs* ; *Drugs and Microtubules* , Marseille (France) , April **1992** , S23

23 Sanderink , G. ; Martinet , M. ; Touzet , A. ; Chapelle , P. ; Frydman , A. ; *5th European ISSX Meeting*, Tours (France), September 1993 , *ISSX Proceedings* , **1993** , *3* , 35

24 Marre , F. ; de Sousa , G. ; Placidi , M. ; Rahmani , R. ; *5th European ISSX Meeting* , Tours (France), September 1993, *ISSX Proceedings* , **1993** , *3* , 36

25 Monsarrat , B. ; Mariel , E. ; Cros , S. ; Garès , M. ; Guénard , D. ; Guéritte-Voegelein , F. ; Wright , M. ; *Drug Metab. Dispos.* , **1990** , *18* , 895

26 Wright , M. ; Monsarrat , B. ; Alvinerie , P. ; Dubois , J. ; Guéritte-Voegelein , F. ; Guénard , D. ; Donehower , R. ; Rowinsky , E. ; *2nd National Cancer Institute Workshop on Taxol and Taxus* , Alexandria (USA), **1992**

27 Monsarrat , B. ; Alvinerie , P. ; Gares , M. ; Wright , M. ; Dubois , J. ; Guéritte-Voegelein ,F. ; Guénard , D. ; Donehower , R. ; Rowinsky , E. ; *7th NCI-EORTC Symposium on New Drug Development* , Amsterdam (Holland) , September **1993** , 123

28 Monsarrat , B. ; Alvinerie , D. ; Gares , M. ; Wright , M. ; Dubois , J. ; Guéritte-Voegelein , F. ; Guénard , D. ; Donehower , R. ; Rowinsky , E. ; *Cell. Pharm.* , **1993** , *1* , (1) , 577

29 Gaver , R.C. ; Deeb , G. ; Willey , T. ; Dandekar , K.A. ; *Proc. Amer. Assoc. for Cancer Research* , **1993** , *34* , 390

30 Marlard , M. ; Gaillard , C. ; Sanderink , G. ; Roberts , S. ; Joannou , P. ; Facchini V. , Chapelle , P. ; Frydman , A. ; *Proc. Amer. Assoc. for Cancer Research* , **1993** , *34* , 373

31 Bissery , M.C. ; Bourzat , .D ; Commerçon , A. ; Frydman , A. ; Gaillard , C. ; Lavelle , F. ; Monegier , B. ; Sablé , S. ; Vuilhorgne , M. ; *207th Amer. Chem. Soc. Meeting* , San Diego (USA) , March **1994**

32 Gaillard , C. ; Monsarrat, B. ; Vuilhorgne , M. ; Royer , I. ; Monegier , B. ; Sablé , S. ; Guénard , D. ; Gires , P. ; Archimbaud , Y. ; Wright , M. ; Sanderink , G. ; *Proc. Amer. Assoc. for Cancer Research* , **1994** , *35* , 428 .

33 Rizzo , J. ; Riley , C.M. ; von Hoff , D. ; Kuhn , J. ; Phillips , J. ; Brown , T. ; *J. Pharm. Biomed. Analysts,* **1990,** *8,* 159

34 De Valeriola , D. ; Brassinne , C. ; Gaillard , C. ; Ketler , J.P. ; Tomiak , E. ; Van Vreckem , A. ; Frühling , J. ; Frydman , A. ; Kerger , J. ; Piccart , M. ; Chapelle , P. ; Blanc , C. ; *Proc. Amer. Assoc. for Cancer Research,* **1993,** *34,* 373

35 Monegier , B. ; Gaillard , C. ; Sablé , S. ; Vuilhorgne , M. ; *Tetrahedron Letters* , **1994** , *35* , 3715

36 Ringel , I. ; Horwitz , S.B. ; *J. Pharm. Exp. Ther.,* **1987** , *242* , 692

37 Brown , T. ; Havlin , K. ; Weiss , G. ; *J. Clinical Oncology* , **1991** , *9* , 1261

38 Kingston , D.G.I. ; Molinero , A.A. ; Rimoldi , J.M. ; *Prog. Chem. Org. Nat. Prod.* , **1993** , *61* , 68

39 McClure , T.D. ; Schram , K.H. ; Reimer ; M.L.J. ; *J. Am. Soc. Mass Spectrom.,* **1992,** *3,* 672

40 Bitsch , F. ; Kim , J. ; Shackleton , C.H.L. ; *41st Amer. Soc. of Mass Spectrometry* , San Francisco (USA), June **1993**

41 Kerms , E.H. ; Volk , K.J. ; Hill , S.E. ; Lee , M.S. ; *41st Amer. Soc. of Mass Spectrometry* , San Francisco (USA), June **1993**

42 Mc Lafferty , F.W. ; *Tandem Mass Spectrometry* , John Willey and sons , **1983**

43 Harris , J.W. ; Katki , A. ; Anderson , L.W. ; Chmurny , G.N. ; Paukstelis , J.V. ; Collins ,J.M. ; *J. Med. Chem.* , **1994** , *37* , 706

44 Commerçon , A. ; Bézard , D. ; Bernard , F. ; Bourzat , J.D. ; *Tetrahedron Letters* , **1992** , *33* , 5185

45 Shelansky , M.L. ; Gaskin , F. ; Cantor , C.R. ; *Proc. Natl. Acad. Sci ., U.S.A.* , **1973** , *70* , 765

46 Gaskin ,F. ; Cantor , C.R. ; Shelansky , M.S. ; *J. Mol. Biol.* , **1974** , *89* , 737

47 Dixon , G.L. ; Dulmadge , E.A. ; Schabel , F.M. ; *Cancer Chem. Rep.* , **1966,** *50* , 247

48 Lataste , H. ; Senilh , V. ; Wright , M. ; Guénard , D. ; Potier , P. ; *Proc. Natl. Acad. Sci. U.S.A.* , **1984,** *81* , 4090

49 Lavelle , F. ; Guéritte-Voegelein , F. , Guénard , D. ; *Bull. Cancer* , **1993** , *80* , 326

50 Nicolaou , K.C. ; Dai , W.M. ; Guy , K. ; *Angew. Chem. Int. Ed.* , **1994** , *33* , 15

51 Georg , G.I. ; Ali , S.M. ; Zygmunt, J. ; Jayashinge , L.R. ; *Exp. Opin. Ther. Patents* , **1994** , *4* , 109

52 Lavelle, F. ; *Curr. Opin. Invest. Drugs* , **1993** , *2* , 627

RECEIVED September 19, 1994

Chapter 8

Pharmacology and Antitumor Effect of Novel Paclitaxel Formulations

R. M. Straubinger, A. Sharma, U. S. Sharma, and S. V. Balasubramanian

Department of Pharmaceutics, State University of New York–Buffalo, Amherst, NY 14260–1200

Taxol (paclitaxel) is a highly-promising but poorly water-soluble anticancer agent for which the development of safe and effective formulations has been a challenge. Clinical experience with the drug has shown vehicle-dependent toxicity, and experiments in animal models have shown a relationship between formulation and antitumor effect. Therefore, formulation of taxol in better-tolerated vehicles was undertaken. Taxol was encapsulated in liposomes (phospholipid vesicles) composed of phosphatidylglycerol and phosphatidylcholine and tested for activity and toxicity against (*a*) subcutaneous C-26, a taxol-resistant murine model for colon carcinoma, and (*b*) intraperitoneal P388 leukemia. Antitumor potency of taxol was retained or slightly enhanced in liposome-based formulations, while adverse effects were reduced considerably. Physical studies were initiated to identify aspects of taxol-taxol and taxol-lipid interaction that determine stability and efficacy of these liposome-based formulations.

Taxol is a novel antineoplastic agent isolated from the Western Yew *Taxus brevifolia* (*1*) that is active clinically against advanced ovarian and breast cancer (*2,3*), and which is undergoing clinical trial for efficacy against a variety of other cancers. Taxol has been approved for use in the U.S. for treatment of advanced and refractory ovarian cancer, a point that is noteworthy given the few treatment alternatives that exist for that disease.

A number of problems have been encountered in the pharmaceutical development of taxol (*4,5*), including scarcity of the drug (owing to low abundance in Yew tissue), lack of alternative sources (owing to the lack of economically-feasible routes of complete or partial synthesis), and extremely low aqueous solubility. Problems in drug supply largely have been alleviated, not only as a result of more efficient collection and extraction of plant material (*6*), but also because of the progress in complete and semi-synthesis that has yielded both taxol and a wealth of new taxane derivatives (reviewed elsewhere in this volume). Efforts to solve the problem of taxol aqueous solubility include synthesis of taxol analogs and taxol prodrugs, as well as intensive efforts to devise safe and bio-compatible formulations. To date, no prodrugs have shown the stability, solubility, or activity necessary for clinical development (*7-12*),

NOTE: Paclitaxel is the generic name for Taxol, which is now a registered trademark.

0097–6156/95/0583–0111$08.00/0

although more recently-described prodrugs appear to offer some promise for further development (*13*). A semisynthetic taxane, Taxotère (docetaxel), has somewhat greater solubility and potency than taxol (*14,15*), and has entered human trial (*16-18*). In addition, new taxanes (reviewed elsewhere in this volume) offer the hope of both activity and acceptable solubility. Efforts to develop safe and efficacious formulations of taxol and other poorly-soluble taxanes will be discussed here.

Taxol as a Formulation Problem

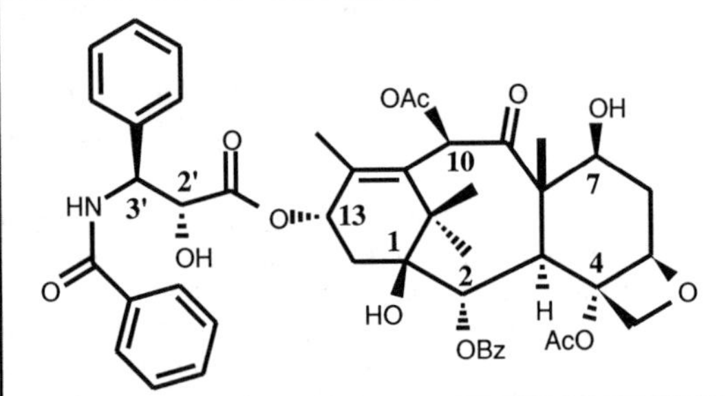

Figure 1: Structure of taxol. Figure shows the taxane backbone, side chain, and principal functional groups, as well as the numbering scheme used in the text.

Taxol is a complex diterpenoid natural product (Figure 1) consisting of a bulky, fused ring system and an extended side chain (at C13) that is required for activity. Although the molecule has relatively hydrophilic domains (in the vicinity of C7-C10 and C1'-C2'), hydrophobic domains of the taxane backbone and side chain (*19,20*) contribute to the overall poor aqueous solubility of the drug. In order to administer human doses in a reasonable volume, taxol currently is formulated for clinical use in an organic co-solvent system (described below) at a concentration of 6 mg/mL (7 mM). The aqueous solubility of taxol is considerably lower, underscoring the need for vehicles or carriers as an aid to administration. Estimates of taxol aqueous solubility vary widely, depending in part on whether the measurement is made under equilibrium or non-equilibrium conditions (*5*). Estimates range from ~35 μM (~30 μg/mL) (*14,21*) and ~7 μM (~6 μg/mL) (*22*) to ≤ 0.77 μM (~0.7 μg/mL). (*9*) A solubility of approximately 0.4 μM was determined in this laboratory (*23*). In the approach to equilibrium conditions, Taxol appears to undergo a time-dependent 10-100-fold decrease in solubility (*5,23,24*). Because taxol precipitates from aqueous media in a time-dependent manner, extended intravenous infusions must be performed using an in-line filter to protect patients from the potentially life-threatening infusion of solids (*25*).

Taxol Solubility in Co-Solvents and Non-aqueous Media. Because of insufficient aqueous solubility, an intensive effort was devoted to the development of vehicles for parenteral administration of taxol (*4,5*). Figure 2 shows data on taxol solubility in a number of solvent systems. Taxol can be prepared at millimolar concentrations in a variety of waxes, oils, polymers, and alcohols (*4,5,26*), as well as in dimethylsulfoxide (DMSO). Although some of these materials are sufficiently low in toxicity to permit parenteral use, and high concentrations of taxol can be prepared in them, a general problem with cosolvent systems is precipitation upon dilution. Figure 2 also illustrates a general pharmaceutical problem that is observed commonly, which arises from the fact that drug *solubility* may decrease exponentially upon dilution, but drug *concentration* decreases in a linear fashion. During the dilution that occurs upon *i.v.* infusion, intermediate concentrations of drug and vehicle may be achieved in which the drug concentration exceeds the solubility afforded by the

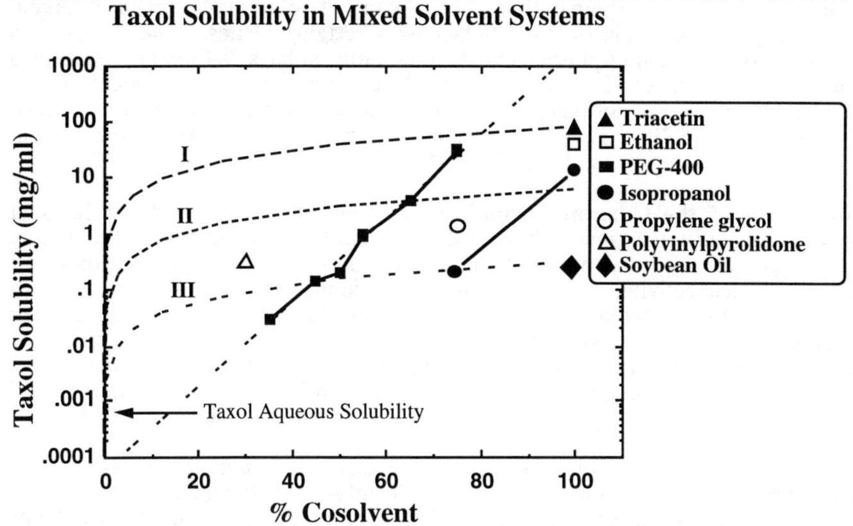

Figure 2: Taxol solubility in cosolvents. Figure shows estimates of taxol solubility in a variety of vehicles, as indicated in the inset. Data is taken from (*5*). A best-fit line shows the relationship between taxol solubility and concentration of polyethylene glycol (PEG-400) (filled squares), emphasizing the exponential change in solubility with cosolvent concentration. Also shown (dotted and dashed lines labeled **I**, **II**, and **III**) is the concentration of drug resulting from dilution of three different starting concentrations of drug to the concentration of cosolvent indicated. Over the range indicated, drug concentration changes linearly as the cosolvent mixture is diluted. In contrast, solubility may decrease exponentially. In regions of the graph where the drug concentration lies above and to the left of a data point for maximal taxol solubility in a given cosolvent, the solution would be supersaturated and prone to precipitation *Adapted from* (*27*).

diluted cosolvent, and precipitation from the super-saturated solution may occur (*27,28*).

The Current Clinical Formulation of Taxol. A search for a safe and efficacious parenteral form of taxol was undertaken to devise a formulation that would allow administration at > 5 mg/mL (5.85 mM) by the intravenous route and have stability for > 24 hours when diluted in common intravenous solutions such as 5% dextrose (D5W) or 0.9% saline (*4*). The formulation actually developed, which is used currently in the clinic, consists of taxol solubilized at a concentration of 6 mg/mL (7 mM) in a 1:1 (vol:vol) mixture of anhydrous ethanol and polyethoxylated castor oil (Cremophor EL), a clear, oily, viscous, yellow surfactant. Shelf-life studies (reviewed in (*5*)) suggest that the formulation is stable in unopened vials for 5 years at 4°C. The Cremophor solution is diluted before use with saline or D5W to a taxol concentration of 0.3-1.2 mg/mL (0.35-1.4 mM). Published and unpublished studies suggest that the diluted material is physically and chemically stable for ≥27 h (*5*). However, dilution to certain concentrations (*cf.* Figure 2) may produce a supersaturated solution (*5*) that could be prone to precipitation if used outside of established guidelines (*25*). An in-line filter is required during administration as a safeguard against the infusion of particulates, and it is recommended that diluted taxol solutions be used within 24 h of preparation (*25*). Hazing of solutions has been observed, particularly in taxol solutions in contact with certain plastics. The hazing was attributed to extraction of plasticizers from the infusion bags and tubing, rather than from taxol precipitation (*29*). The potential dose of plasticizer to the patient may be considerable if care is not taken in the selection of materials contacted by diluted taxol solutions (*5,29*). The development process and properties of the current formulation are reviewed in greater detail elsewhere (*5*).

Limitations of the Current Formulation. Beyond the potential problems of physical instability mentioned above, the most significant problem with the current clinical taxol formulation is that the Cremophor EL vehicle possesses pharmacological activity. A variety of drugs are administered in Cremophor EL, such as cyclosporine (*30*) and teniposide. (*31*) However, the dose of Cremophor EL that accompanies a dose of taxol is the highest for any marketed drug (*32*). Cremophor has been observed to cause serious or fatal hypersensitivity episodes (*33*), and vehicle toxicity may be largely responsible for fatal or life-threatening anaphylactoid reactions observed upon rapid infusion of taxol into animals or humans (*2,34,35*).

Mechanisms of Toxicity. Histaminergic mechanisms were implicated as the basis of the physiological reaction to Cremophor EL infusion (*34*), and adverse reactions appeared to be associated with rapid infusion rates (*35,36*). Therefore, patients receiving taxol are supported with prophylactic administration of antihistamines and corticosteroids, and infusion times are extended. Three, 6-24, and 96 h infusions are common in the literature (*32,36,37*). The rationale for the prophylaxis regimen (*5,35*) apparently derives from earlier experience with hypersensitivity reactions to radiocontrast media (*38*). Premedication and prolonged infusion has reduced the incidence of serious hypersensitivity reactions (*39*), although milder reactions appear in about 30% of patients (*35,40*).

In addition to the well-documented hypersensitivity effects of the Cremophor EL vehicle, the picture of taxol activity *in vivo* is complicated further by reports that Cremophor EL can enhance the cytostatic effect of taxol, putatively through inhibition of the Multidrug Resistance (MDR) transporter (*41-43*). MDR is a mechanism, mediated by cell-membrane glycoproteins, by which tumor cells can acquire resistance to taxol (*44,45*) and a variety of structurally-unrelated compounds (*46,47*). The importance of Cremophor-mediated alteration of MDR is unknown with respect to the clinical activity of taxol.

Although premedication and prolonged infusion has made it possible to institute clinical testing of taxol in a wide range of clinical trials, the current formulation cannot be considered optimal. In a general sense, multi-drug pharmacological intervention is less desirable than a safer, better-tolerated formulation. With the administration of multiple agents, there is the potential for pharmacological, pharmacokinetic, or metabolic interactions that may alter taxol antitumor effect or toxicity. Such interactions are under investigation (*48-52*), and no clear problem has yet emerged. However, considerably more investigation is required before concluding that no interactions exist that have clinical impact on taxol therapy.

Formulation Alternatives

Efforts to develop safe, convenient, and efficacious taxol formulations continue to the present, perhaps somewhat remarkable given the advanced clinical status of the drug. In addition to co-solvent systems (mentioned above), other approaches for taxol formulation include emulsions, micellar systems, liposomes, and implants. More detailed reviews are given elsewhere (*5,24*); the present chapter will focus on results with a liposome-based formulation developed recently (*53-56*).

Liposome-based formulations. Liposomes are microscopic particulate carriers consisting of an internal aqueous space enclosed within one or more delimiting lipid bilayer membranes. Liposomes are prepared most commonly from synthetic or naturally-occurring phospholipids and neutral lipids such as cholesterol. Encapsulation of a drug in liposomes often results in distinct changes in pharmacokinetic and pharmacodynamic properties of the agent, in some cases causing a marked decrease in toxicity or an increase in potency (*57*). A more detailed explanation of the rationale for choosing liposomes as a first-line approach to taxol formulation is given elsewhere (*53*). Compared to other experimental drug delivery systems, liposomes are a relatively mature technology. Several comprehensive works examine both the applications and the methodology for production of liposomes (*58,59*). Given the novelty of the approaches in the drug carrier field in general, and with liposomes specifically, there have been many basic and applied pharmaceutical concerns to overcome in the clinical development of liposome carriers (*57,60-63*). Nonetheless, liposomes have advanced to human testing in a number of trials (*57*).

Liposomes may be produced from a wide variety of phospholipids, and other lipids and amphipathic molecules may be included; the variety of possible constituents provides the opportunity to control a broad range of liposome physical properties, such as diameter, membrane fluidity, electrostatic charge, and surface properties (*64*). Such liposome physical properties can exert major effects on liposome behavior *in vivo* and *in vitro* (*eg.* during preparation, storage, and use). Although simple liposome-based formulations of many drugs have been prepared, liposome constituents in some cases must be chosen carefully, and optimization of formulation properties can be somewhat complex owing to the many parameters that may be varied in order to achieve the goals set. Such was the experience during the formulation of taxol in liposomes, in which over 300 sets of formulations were examined, representing a systematic variation of liposome properties including diameter, charge, membrane fluidity, length of lipid acyl chain, surface hydration, and inclusion of specific dopants (*53,55*).

***In Vitro* Activity of Prototype Liposomes.** A family of related taxol-liposome formulations were developed that had properties suitable for further evaluation (*54*). One example of a promising formulation consisted of taxol and phospholipid in a 1:33 mole ratio, and was prepared from phosphatidylglycerol (PG) and phosphatidylcholine (PC) in a 1:9 mole ratio. Formulations were prepared and stored as a lyophilized powder of drug and lipid, which simply requires reconstitution in saline before use (*54,55*). In order to determine whether the cytostatic potency of taxol was retained when encapsulated in liposomes, formulations were tested for activity against a panel of cultured cell lines, including several lines used as tumor models in mice. Different cell lines varied considerably in their sensitivity to taxol (Figure 3); Colon-26, the murine colon tumor line selected for subsequent therapeutic experiments *in vivo*, was the most taxol-resistant. On several cell lines, taxol liposomes appeared to be significantly less potent than free taxol. However, it was found that 0.5% dimethylsulfoxide (DMSO), used to solubilize unencapsulated taxol prior to addition to cells, enhanced the potency of taxol. If taxol was dissolved directly in serum-containing medium and added to cells without DMSO, free taxol potency was nearly identical to that of liposome-encapsulated taxol, in most cases.

***In Vivo* Activity of Taxol Liposomes.** The toxicity and antitumor activity of taxol-containing liposomes has been reported for several formulations and tumor model systems (*53,54,65-67*). Although not all of the reported experiments are comparable, some generalizations may be possible regarding the efficacy of liposome-encapsulated taxol. Such formulations appear to show consistently lower toxicity than the conventional Cremophor-based formulation, not only upon rapid bolus administration, but also in delayed (non-anaphylactoid) toxicity. Stability under the conditions of use is not described for all formulations in the literature, and precipitation of taxol after injection could, in some cases, reduce toxicity by reducing bioavailability. Selected findings on the efficacy of liposome-associated taxol are described below.

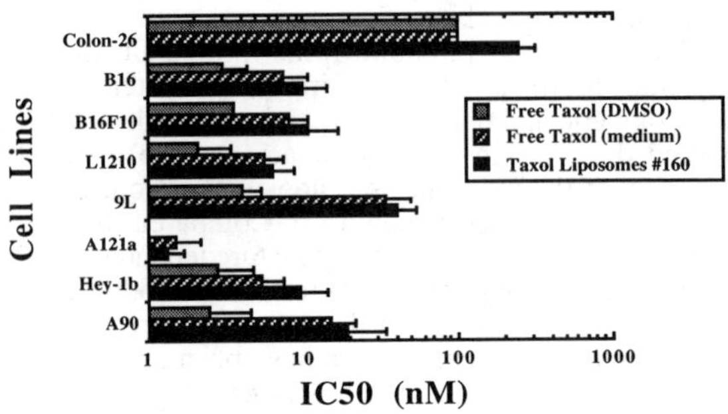

Figure 3: Activity of taxol formulations *in vitro*: Cells were plated at a density of 2 x 10^4/mL in multiwell plates and allowed to adhere overnight. Triplicate wells were exposed to various concentrations of taxol, either added as PG:PC liposomes (see text for details) (*solid bar*), as a 200x concentrated stock in DMSO (*stippled bar*), or absorbed to serum proteins (*hatched bar*) in the absence of organic solvent. Cells were enumerated after 72 h, and the IC_{50} (50% growth inhibition) value for each concentration-effect curve was calculated graphically. Cell lines: *Colon-26*: murine colon carcinoma; *B16, B16F10*: murine melanoma; wild type and highly-metastatic variants, respectively; *L1210*: murine leukemia; *9L*: rat gliosarcoma; *A121a, HEY-1b, A90*: human ovarian tumor lines. *Adapted from* (*24*)

Activity Against Intraperitoneal P388 Leukemia. Liposomes containing approximately 2% taxol in liposomes composed of soybean PC were administered by *i.p.* injection to animals inoculated with intraperitoneal P388 (*65*), a taxol-sensitive leukemia. For comparison, free taxol was given in 5% DMSO/5% Cremophor in saline. Treatment consisted of 4 consecutive daily taxol treatments with 12.5 mg/kg/day, initiated 24 h after tumor inoculation. Lifespan was increased 66% and 68% for the free- and liposome-encapsulated taxol groups, respectively, compared to untreated controls. Toxicity appeared to be similar for the two forms of taxol, judging from animal body weight changes during treatment. Compared to controls which received the DMSO/Cremophor vehicle without drug, the mean body weight of both free- and liposome-taxol -treated animals was decreased 10.5% at the nadir. (*65*)

Recently this laboratory has initiated similar testing of taxol-liposome activity against P388 leukemia (Figure 4). The primary rationale is that taxol displays unusual pharmacokinetic behavior following *i.p.* infusion into human cancer patients, showing a prolonged retention in peritoneal fluids. Thus *i.p.* administration offers the potential for sustained exposure of *i.p.* malignancies to taxol, with reduced side effects to critical normal tissues (*50,68*). An important limitation to *i.p.* taxol is severe pain (*69*), and it is our goal to investigate whether this side effect can be ameliorated by encapsulation of taxol in liposomes. As a prelude to such experiments, we investigated the activity of *i.p.* taxol against *i.p.* P388 leukemia, a fast-growing tumor that we have used as a probe for retention of formulation activity in the schedule of administration chosen (*70*). Preliminary results from these experiments are presented in Figure 4. It appears that both free- (conventional) and liposome-encapsulated taxol have essentially equal antitumor potency. However, the liposome-based formulation shows lower dose-dependent toxicity. In the experiment shown, the optimal dose appears to be lower than the lowest cumulative dose tested,

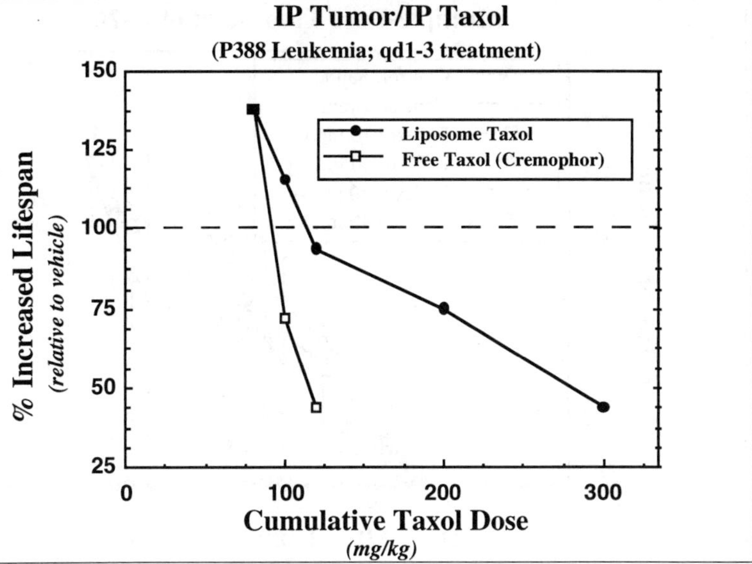

Figure 4: Effect of Taxol and Taxol Liposomes on Survival of Murine P388 Leukemia (*70*). Groups of 10 mice were treated with taxol in Cremophor:ethanol (squares) or in multilamellar liposomes (PG:PC:taxol 1:9:0.3) (circles). Treatment was started on Day 1 after initiation of the *i.p.* tumor and continued for 3 days. The figure shows the median extension of lifespan for treated animals, compared to vehicle-treated animals that received no taxol. Vehicle control animals had a median survival of 22 days, defined as 100% in this figure; the horizontal dashed line is drawn to aid the comparison of survival in treated animals.

perhaps consistent with previous results using a different formulation and treatment schedule (*65*).

Activity Against Subcutaneous C-26 Murine Colon Tumor. Liposomes of diverse physical characteristics were tested for activity against Colon-26 (C-26) murine colon tumor by intravenous administration (Figure 5) (*53,54*). Colon-26 is highly taxol-resistant *in vitro* (*cf.* Figure 3), and has been used in previous investigations of taxane activity (*15,26*). Colon-26 was chosen for several reasons: first, a lethal end-stage complication of cancer chemotherapy is the emergence of drug-resistant tumor; activity of formulations against highly taxol-sensitive tumor targets (*eg.* the A121a human ovarian carcinoma, Figure 3) may be of interest, but may not reflect the apparent trend of dosing at or near the maximum tolerated dose (MTD) for "salvage" therapy in treatment-resistant disease (*cf.* (*71*)). Second, previous experiments investigating the relationship between formulation and activity (*26*) have underscored the importance of investigating activity of formulations against distal sites (*eg.* intravenous administration against subcutaneous tumors), given the observation that some formulations have only local activity when applied directly to the tumor-containing compartment (*eg.* intraperitoneal taxol against intraperitoneal tumor).

Single and multiple dosing schemes were tested for activity against C-26, and liposome properties were varied to investigate the role of liposome size and lipid composition on antitumor effect (*54*). It was observed in general that liposomes were

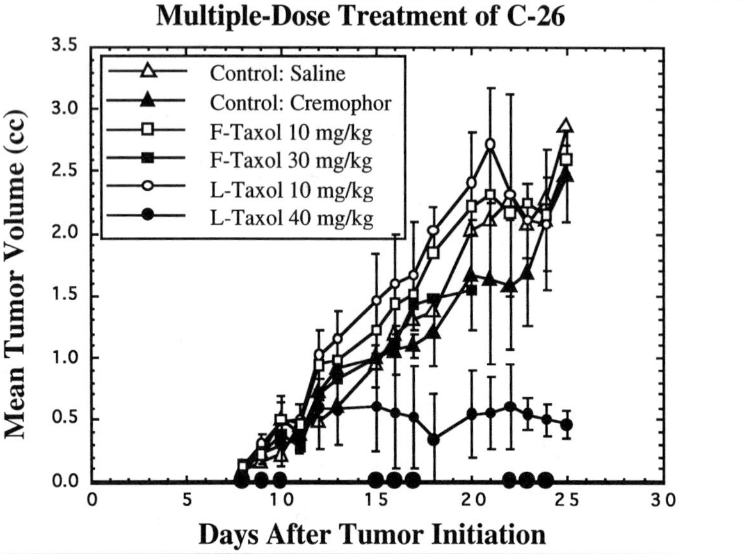

Figure 5: Activity of Free- and Liposome-Encapsulated Taxol Against Taxol-Resistant C-26 Tumors. Subcutaneous Colon-26 tumors were initiated in Balb/C mice by inoculation with 10^6 tumor cells. Eight days after initiation, the tumors were measurable, and treatment was carried out with the indicated doses of free- or liposome-encapsulated taxol. Liposomes were composed of PG:PC 1:9. Free taxol was given in Cremophor EL/ethanol, and liposomes were administered in buffered saline. The amount of drug given in each injection is indicated in the figure inset. Control animals were given saline or the volume of Cremophor EL/ethanol equivalent to that required to administer free taxol. Treatment was given thrice weekly, as indicated by filled circles along the abscissa, for 3 weeks. Data shows mean tumor volume for groups of 10 mice, and vertical bars indicate the standard deviation for the group. Error bars are representative, and some are omitted for clarity. *Adapted from* (*54*).

equipotent to or slightly more potent than free taxol given in the clinically used vehicle of Cremophor EL and ethanol. However, the Maximum Tolerated Dose was 2- to 7-fold greater for the liposome-based formulations, compared to free taxol. The acute, vehicle-mediated toxicity was abolished by encapsulation of taxol in liposomes, and the delayed, taxol-mediated toxicity likewise was reduced by encapsulation. No dose of free taxol exerted a significant effect on the progression of C-26 tumors (Figure 5). Drug was injected up to the maximum acutely-tolerated dose; by using repetitive dosing schemes, it was possible to reach high cumulative doses of taxol. Even at a dose of free taxol that showed delayed uniform lethality (30 mg/kg/injection x 6 injections) no effect on tumor progression was observed. In contrast, taxol liposomes maintained tumor at a reduced volume throughout the period of dosing, and the antitumor effect was observed at a dose level that was approximately 30% higher than the dose that would have been uniformly lethal if given as the free drug.

Pharmaceutical Properties of Taxol-Containing Liposomes. Initial success in small-scale formulation of taxol in liposomes was followed by difficulties in scale-up to the quantities required for animal antitumor experiments (*53,55*). As a result, a systematic approach was initiated both to develop formulations and production processes resulting in suitable material for antitumor testing (*55*), and to understand taxol-taxol (*19*) and taxol-lipid (*56*) interactions in sufficient molecular detail to guide further formulation development.

Empirical Development of Taxol Liposomes. Early studies suggested that the phospholipid composition of liposomes modulated the amount of taxol that could be accommodated; because PC membranes appeared to incorporate the greatest mole fraction of taxol (*72*), that lipid was chosen as the primary constituent. Subsequent detailed work revealed extensive aggregation of taxol:PC liposomes, and that the conferral of electrostatic charge on the liposomes, in the form of PG, phosphatidylinositol (PI), or synthetic anionic amphipaths, abolished aggregation (*54,55*). However, the increase in mole fraction of non-PC lipid was observed to decrease stability, and 10-30 mole% negatively-charged lipid was observed to be a reasonable tradeoff between reduction in aggregation and physical stability (*55*).

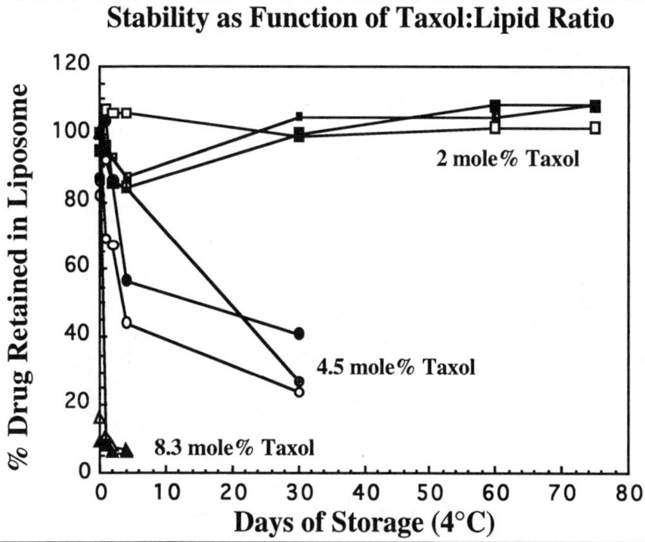

Figure 6: Physical Stability of Taxol-Containing Liposomes. Small (sonicated) liposomes were prepared from PG:PC (3:7) and contained the indicated mole% taxol. Formulations were stored at 4°C in buffered saline. At various intervals indicated along the abscissa, differential centrifugation was used to separate liposomes from precipitated taxol, and the taxol and lipid concentrations were analyzed by (*73*) and (*74*), respectively. Symbols represent preparations stored at phospholipid concentrations of 50 mM (open symbols), 100 mM (crossed symbols), and 150 mM (filled symbols). *Adapted from* (*55*).

The taxol content of liposomes also modulated physical stability (*55*). Figure 6 shows that liposomes of PG:PC (3:7) and containing ~2 mole% taxol were stable for months in solution at 4°C. Increasing the taxol content decreased stability to days or hours, depending on the taxol content and lipid composition. It must be emphasized that the stability for some of the "unstable" formulations actually may exceed the stability of the taxol-Cremophor EL formulation currently used clinically. Thus although efforts to increase the taxol content of liposomes are underway, it is possible to increase the taxol content of liposomes already developed, if stability criteria are selected that resemble those accepted for the current clinical taxol formulation.

Molecular Basis of Formulation Stability. Because the liposome membrane provides a hydrophobic environment for the accommodation of taxol, and because one objective of our current developmental work is to increase the taxol:lipid ratio of formulations, we investigated environment- and concentration-dependent interactions of taxol (*19*). Although a number of previous studies have investigated taxol conformation in different solvents and concentrations, we focused on conditions that would be anticipated to occur upon interaction of taxol with membrane bilayers. NMR, Circular Dichroism, and fluorescence spectroscopy were used to build a model of taxol monomer conformation consistent with previous studies (*75-79*). Further studies yielded support for a concentration-dependent aggregation of taxol (Figure 7), in which hydrogen bonding between adjacent taxol molecules, promoted in non-aqueous environments such as the membrane bilayer interior, stabilizes taxol into a stacked structure that may be propagated along both faces (*19*).

Such concentration-dependent stacking of taxol occurs in the mM concentration range. Physical studies with model membranes suggests that as the taxol:lipid ratio increases, taxol submerges into the phospholipid bilayer and undergoes concentration-dependent aggregation into structures that may be similar to those observed to occur in solution (*56*). Such aggregation may represent the initiation of crystal nucleation, and our operant hypothesis is that this event precedes and foretells formulation instability. Broader studies on taxol-lipid interaction are underway, and suggest that modulation of liposome composition may be one approach to enhancing the stability of taxol formulations by providing the appropriate molecular environment for taxol.

Concentration- and Environment-Dependent Aggregation of Taxol

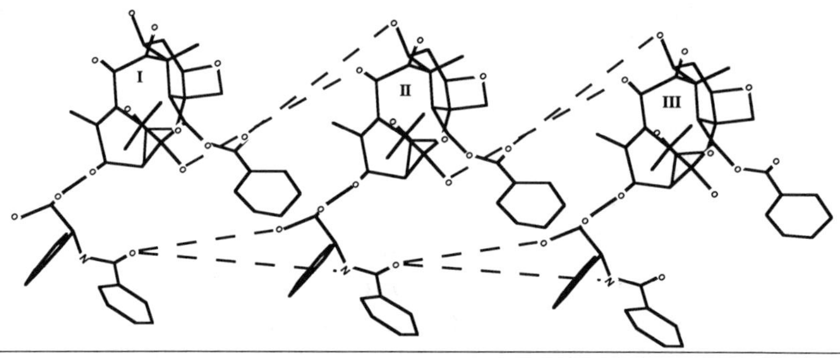

Figure 7: Proposed Model of Taxol-Taxol Interactions. Dashed lines indicate intermolecular hydrogen bonds inferred from NMR spectroscopy. The stacked structure may be propagated from both faces and thus grow indefinitely. Taxol aggregates form in a concentration-dependent manner in solvents which do not compete with the intermolecular hydrogen bonds. *Adapted from* (*19*).

Conclusions

Formulation of taxol has posed difficulties that have had impact on drug development from the initial stages of testing and continuing to the present clinical use. Safe, efficacious, and pharmaceutically acceptable formulations may improve not only the therapeutic benefits of taxol, but also would aid in the development of taxol and taxanes, by providing a means for accurate testing of biological or antitumor activity. Liposomal formulations of taxol show promise, given the observed reduction in both acute- and delayed taxol toxicity and the maintenance of antitumor potency.

Other taxanes may be formulated in liposomes as well, suggesting the additional beneficial role as a simple, reliable vehicle for testing and administration of new compounds.

Literature Cited

1. Wani, M.C., Taylor, H.L., Wall, M.E., Coggon, P., McPhail, A.T. *J. Am. Chem. Soc.* **1971**, *93*, 2325-2327.
2. Rowinsky, E.K., Cazenave, L.A., Donehower, R.C. *J. Natl. Cancer Inst.* **1990**, *82*, 1247-1259.
3. McGuire, W.P., Rowinsky, E.K., Rosenshein, N.B., Grumbine, F.C., Ettinger, D.S., Armstrong, D.K., Donehower, R.C. *Ann. Intern. Med.* **1989**, *111*, 273-279.
4. Suffness, M. *Ann. Rep. Med. Chem.* **1993**, *28*, 305-314.
5. Adams, J.D., Flora, K.P., Goldspiel, B.R., Wilson, J.W., Finley, R., Arbuck, S.G., Finley, R. *J. Natl. Cancer Inst. Monographs* **1993**, *15*, 141-147.
6. DeFuria, D., Horovitz, Z. *J. Natl. Cancer Inst. Monographs* **1993**, *15*, 195-198.
7. Deutsch, H.M., Glinski, J.A., Hernandez, M., Haugwitz, R.D., Narayanan, V.L., M., S., Zalkow, L.H. *J. Med. Chem.* **1989**, *32*, 788-792.
8. Zhao, Z., Kingston, D.G.I., Crosswell, A.R. *J. Nat. Prod.* **1991**, *54*, 1607-1611.
9. Mathew, A.E., Mejillano, M.R., Nath, J.P., Himes, R.H., Stella, V.J. *J. Med. Chem.* **1992**, *35*, 145-151.
10. Nicolaou, K.C., Riemer, C., Kerr, M.A., Rideout, D., Wrasidlo, W. *Nature* **1993**, *364*, 464-466.
11. Vyas, D.M., Wong, H., Crosswell, A.R., Casazza, A.M., Knipe, J.O., Mamber, S.W., Doyle, T.W. *Bioorganic Med. Chem. Lett.* **1993**, *3*, 1357-1360.
12. Georg, G. In *Taxol;* Suffness, M. Ed.; CRC Press: Boca Raton, 1994, In Press.
13. Ueda, Y., Mikkilineni, A.B., Knipe, J.O., Rose, W.C., Casazza, A.M., Vyas, D.M. *Bioorganic Med. Chem. Lett.* **1993**, *3*, 1761-1766.
14. Ringel, I., Horwitz, S.B. *J. Natl. Cancer Inst.* **1991**, *83*, 288-291.
15. Bissery, M., Guénard, D., Guéritte-Voegelein, F., Lavelle, F. *Cancer Res.* **1991**, *51*, 4845-4852.
16. Bisset, D., Setanoians, A., Cassidy, J., Graham, M.A., Chadwick, G.A., Wilson, P., Auzannet, V., Le Bail, N., Kaye, S.B., Kerr, D.J. *Cancer Res.* **1993**, *53*, 523-7.
17. Extra, J.-M., Rousseau, F., Bruno, R., Clavel, M., Le Bail, N., Marty, M. *Cancer Res.* **1993**, *53*, 1037-42.
18. Aapro, M., Pujade-Lauraine, E., Lhomme, C., Lentz, M.-A., Le Bail, N., Fumoleau, P., Chevallier, B. *Proc. Amer. Soc. Clin. Oncol.* **1993**, *12*, 256a.
19. Balasubramanian, S.V., Alderfer, J.L., Straubinger, R.M. *J. Pharm. Sci.* **1994**, In Press.
20. Guénard, D., Guéritte-Voegelein, F., Dubois, J., Potier, P. *J. Natl. Cancer Inst. Monographs* **1993**, *15*, 79-82.
21. Swindell, C.S., Krauss, N.E. *J. Med. Chem.* **1991**, *34*, 1176-1184.
22. Tarr, B.D., Yalkowski, S.H. *J. Parenteral Sci. & Technology* **1987**, *41*, 31-33.
23. Sharma, U., Straubinger, R.M. **1994**, Submitted.
24. Straubinger, R.M. In *Taxol;* Suffness, M. Ed.; CRC Press: Boca Raton, 1994, In Press.

25. USPHS. In *NCI Investigational Drugs, Pharmaceutical Data;* National Cancer Institute, National Institutes of Health, Department of Health and Human Services, U.S Public Health Service: Bethesda, MD, 1990, pp. 151-153.
26. Rose, W.C. *Anti-Cancer Drugs* **1992**, *3*, 311-321.
27. Yalkowski, S.H., Valvani, S.C. *Drug Intell. and Clin. Pharmacy* **1977**, *11*, 417-419.
28. Flynn, G.L. *J. Parenteral Science & Technology* **1984**, *38*, 202-9.
29. Waugh, W., Trissel, L., Stella, V. *Am. J. Hosp. Pharm.* **1991**, *48*, 1520-1524.
30. Howrie, D.L., Ptachcinski, R.J., Griffith, B.P., Hardesty, R., Rosenthal, J., Burckart, G., Venkataramanan, R. *Drug Intell. Clin. Pharmacy* **1985**, *19*, 425-7.
31. O'Dwyer, P.J., King, S.A., Fortner, C.L., Leyland-Jones, B. *J. Surg. Oncol.* **1986**, *4*, 1262-9.
32. Rowinsky, E.K., Onetto, N., Canetta, R.M., Arbuck, S.G. *Seminar. Oncol.,* **1992**, *19*, 646-662.
33. Dye, D., Watkins, J. *Br. Med. J.* **1980**, *280*, 1353.
34. Lorenz, W., Riemann, H.J., Schmal, A., Schult, H., Lang, S., Ohmann, C., Weber, D., Kapp, B., Luben, L., Doenicke, A. *Agents Actions* **1977**, *7*, 63-67.
35. Weiss, R.B., Donehower, R.C., Wiernik, P.H., Ohnuma, T., Gralla, R.J., Trump, D.L., Baker, J.R., VanEcho, D.A., VonHoff, D.D., Leyland-Jones, B. *J. Clin. Oncol.* **1990**, *8*, 1263-1268.
36. Rowinsky, E.K., Eisenhauer, E.A., Chaudhry, V., Arbuck, S.G., Donehower, R.C. *Seminar. Oncol.,* **1993**, *20*, 1-15.
37. Eisenhauer, E., ten Bokkel-Huinink, W., Swenerton, K., Mangioni, C., van der Burg, M., Kerr, I., Gianni, L., Buser, K., Vermorken, J., Onetto, N., Winograd, B., Canetta, R. In *Second National Cancer Institute Workshop on Taxol and Taxus;* National Cancer Institute: Alexandria, VA, 1992, pp.
38. Greenberger, P.A., Halwig, J.M., Patterson, R., Wallemark, C.B. *J. Allergy Clin. Immunol.* **1986**, *77*, 630-4.
39. Arbuck, S.G., Canetta, R., Onetto, N., Christian, M.C. *Semin. Oncol.* **1993**, *20*, 31-9.
40. Runowicz, C.D., Wiernik, P.H., Einzig, A.I., Goldberg, G.L., Horwitz, S.B. *Cancer* **1993**, *71*, 1591-1596.
41. Woodcock, D.M., Jefferson, S., Linsenmeyer, M.E., Crowther, P.J., Chojnowski, G.M., Williams, B., Bertoncello, I. *Cancer Res.* **1990**, *50*, 4199-03.
42. Webster, L., Linsenmeyer, M., Millward, M., Morton, C., Bishop, J., Woodcock, D. *J. Natl. Cancer Inst.* **1993**, *85*, 1685-1690.
43. Fjällskog, M.-L., Frii, L., Bergh, J. *Lancet* **1993**, *342*, 873.
44. Roy, S.N., Horwitz, S.B. *Cancer Res.* **1985**, *45*, 3856-63.
45. Waud, W.R., Gilbert, K.S., Harrison, S.D.J., Griswold, D.P.J. *Cancer Chemother. Pharmacol.* **1992**, *31*, 255-7.
46. Pastan, I.H., Gottesmann, M.M. *Imp. Adv. Oncol.* **1988**, *1*, 3-16.
47. Georges, E., Sharom, F.J., Ling, V. *Adv. Pharmacol.* **1990**, *21*, 185-220.
48. Monsarrat, B., Mariel, E., Cros, S., Garés, M., Guénard, D., Guéritte-Voegelein, F., Wright, M. *Drug Metab. Dispos.* **1990**, *18*, 895-901.
49. Jamis-Dow, C.A., Klecker, R.W., Katki, A.G., Collins, J.M. *Proc. Ann. Meet. Am. Assoc. Cancer Res.* **1993**, *34*, A2198.
50. Rowinsky, E.K., Donehower, R.C. *Seminars in Oncol.* **1993**, *20*, 16-25.
51. Cresteil, T., Monsarrat, B., Alvinerie, P., Tréluyer, J.M., Viera, I., Wright, M. *Cancer Res.* **1994**, *54*, 386-392.

52. Kumar, G.N., Walle, U.K., Bhalla, K.N., Walle, T. *Res. Comm. Chem. Pathol. Pharmacol.* **1993**, *80*, 337-44.
53. Straubinger, R.M., Sharma, A., Murray, M., Mayhew, E. *J. Natl. Cancer Inst. Monographs* **1993**, *15*, 69-78.
54. Sharma, A., Mayhew, E., Straubinger, R.M. *Cancer Res.* **1993**, *54*, 5877-5881.
55. Sharma, A., Straubinger, R.M. *Pharm. Res.* **1994**, *11*, 889-896.
56. Balasubramanian, S.V., Straubinger, R.M. *Biochemistry* **1994**, *33*, In Press.
57. Szoka Jr., F.C. In *Membrane Fusion;* Wilschut, J. and Hoekstra, D. Ed.; Marcel Dekker: New York, 1991, pp. 845-890.
58. *Liposome Technology;* Gregoriadis, G., Ed; CRC Press: Boca Raton, 1988.
59. *Liposomes: From Biophysics to Therapeutics;* Ostro, M.J., Ed; Marcel Dekker: New York, 1987.
60. Mayhew, E., Papahadjopoulos, D. In *Liposomes;* Ostro, M. Ed.; Marcel Dekker: New York, 1983, pp. 289-341.
61. Popescu, M.C., Swenson, C.E., Ginsberg, R.S. In *Liposomes: From Biophysics to Therapeutics;* Ostro, M. Ed.; Marcel Dekker: New York, 1987, pp. 219-251.
62. Martin, F.J. In *Specialized Drug Delivery Systems;* Tyle, P. Ed.; Marcel Dekker: New York, 1990, pp. 267-316.
63. Fielding, R.M. *Clin. Pharmacokinet.* **1991**, *21*, 155-64.
64. Szoka Jr., F.C., Papahadjopoulos, D. *Ann. Rev. Biophysics Bioeng.* **1980**, *9*, 467-508.
65. Bartoli, M.-H., Boitard, M., Fessi, H., Beriel, H., Devissaguet J-H., Picot, F., Puisieux, F. *J. Microencapsulation* **1990**, *7*, 191-7.
66. Riondel, J., Jacrot, M., Fessi, H., Puisieux, F., Poiter, P. *In Vivo* **1992**, *6*, 23-28.
67. Rafaeloff, R., Husain, S.R., Rahman, A. *Proc. Ann. Meet. Am. Assoc. Cancer Res.* **1992**, *33*, A2883.
68. Markman, M., Rowinsky, E., Hakes, T., Reichman, B., Jones, W., Lewis, J.L., Jr.,, Rubin, S., Curtin, J., Barakat, R., Phillips, M., Hurowitz, L., Almadrones, L., Hoskins, W. *J. Clin. Oncol.* **1992**, *10*, 1485-1491.
69. Markman, M., Rowinsky, E., Hakes, T., Reichman, B., Jones, W., Lewis Jr., J., Rubin, S., Curtin, J., Barakat, R., Almadrones, L., Hoskins, W. *J. Natl. Cancer Inst. Monographs* **1993**, *15*, 103-6.
70. Sharma, A., Straubinger, R.M. *Unpublished Observations.* 1994.
71. Seidman, A., Reichman, B., Crown, J., Yao, T., Heelan, R., Hakes, T., Lebwohl, D., Gilewski, T., Surbone, A., Currie, V., Hudis, C., Klecker, R., Jamis-Dow, C., Collins, J., Quinlivan, S., Berkery, R., Toomasi, F., Canetta, R., Norton, L. *J. Natl. Cancer Inst. Monographs* **1993**, *15*, 171-176.
72. Straubinger, R.M. *Unpublished Observations.* 1992.
73. Sharma, A., Conway, W., Straubinger, R. *J. Chromatog. B* **1994**, *655*, 315-319.
74. Bartlett, G.R. *J. Biol. Chem.* **1959**, *234*, 466-468.
75. Kingston, D.G.I., Hawkins, D.R., Ovington, L.J. *J. Natl. Prod.* **1982**, *45*, 466-470.
76. Williams, H.J., Scott, A.I., Dieden, R.A., Swindell, C.S., Chirlian, L.E., Francl, M.M., Heerding, J.M., Krauss, N.E. *Tetrahedron* **1993**, *49*, 6545-6560.
77. Vander Velde, D.G., Georg, G.I., Gruewald, G.L., Gunn, C.W., Mitscher, L.A. *J. Am. Chem. Soc.* **1993**, *115*, 11650-11651.
78. Falzone, C.J., Benesi, A.J., Lecomte, J.T.J. *Tetrahedron. Lett.* **1992**, *33*, 1169-1172.
79. Chmurny, G.N., Hilton, B.D., Brobst, S., Look, S.A., Witherup, K.M., Beutler, J.A. *J. Nat. Prod.* **1992**, *55*, 414-423.

RECEIVED August 31, 1994

Chapter 9

Phosphatase-Activated Prodrugs of Paclitaxel

D. M. Vyas[1], Y. Ueda[1], H. Wong[1], J. D. Matiskella[1], S. Hauck[1], A. B. Mikkilineni[1], Vittorio Farina[1], W. C. Rose[2], and A. M. Casazza[1]

[1]Bristol-Myers Squibb Pharmaceutical Research Institute, 5 Research Parkway, P.O. Box 5100, Wallingford, CT 06492–7660
[2]Bristol-Myers Squibb Pharmaceutical Research Institute, P.O. Box 4000, Princeton, NJ 08540

In an effort to supplant the Cremophore EL® based intravenous formulation of paclitaxel, a program on synthesis and evaluation of water-soluble, phosphatase activated prodrugs was initiated. Prototype, water-soluble prodrugs of paclitaxel namely, C-2' and C-7 phosphate derivatives were found to be unsuitable as prodrugs of paclitaxel in-vivo. This was attributed to the steric congestion about the C-2' and C-7 phosphate moieties of these derivatives resulting in them being poor substrates for phosphatase enzymes in-vivo. To surmount this steric hurdle, synthesis of novel phosphates as pro-prodrugs of paclitaxel was undertaken. The successful pro-prodrugs incorporated a self-immolative linker carrying a phosphate group which served as a solubilizing group and a 'phosphatase trigger' required to unravel paclitaxel after activation by phosphatase enzymes in-vivo.

The natural diterpene Taxol® (**1**, paclitaxel) (*1*) and its close semisynthetic analog Taxotere® (**2**, docetaxel) (*2*) have emerged as promising anticancer agents

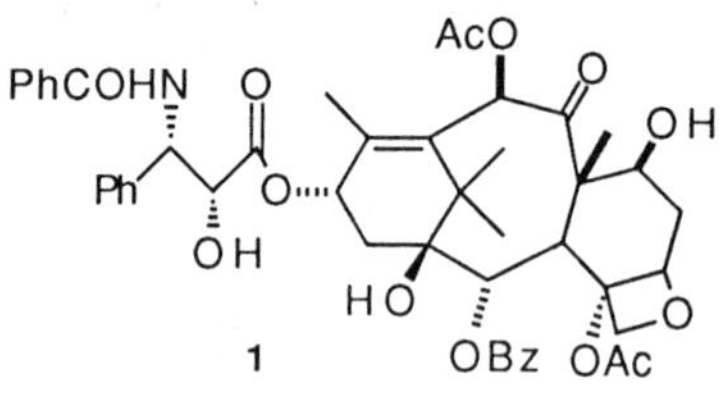

1

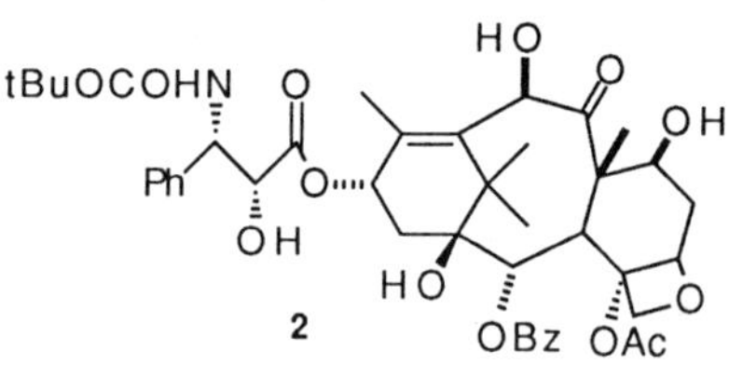

2

efficaceous against a variety of human solid tumors such as ovary, breast, head & neck and lung cancers (*3*). Paclitaxel is currently an FDA approved agent in the USA for the treatment of cis-platinum refractory ovarian cancer and refractory metastatic breast cancer. Several expanded clinical trials are currently in progress (*4*) to fully exploit the utility of this novel tubulin interacting agent. Taxotere®, is

0097–6156/95/0583–0124$08.00/0

also on the verge of being registered in Europe and Japan for the treatment of metastatic breast cancer.

Inspite of their great promise in treatment of refractory and untreatable human neoplasms, both agents are afflicted with formulation and systemic administration problems. These problems stem from their extreme low solubility in water; with paclitaxel's aqueous solubility at 0.25 ug/ml (*5*) and docetaxel's at 6-7 ug/ml (Agharkar, S., BMS, personal communication, 1993). Consequently, special formulations requiring excipients such as Cremophor EL® for paclitaxel (*6*) and Tween 80 for docetaxel (*7*) have been necessitated for intravenous (iv) administration. In the case of paclitaxel the amount of Cremophor EL® required to administer the therapeutic dose (135-200 mg/m^2) represents the highest amount ever to be used with any drug. Exposure to this large amounts of Cremophor EL® has produced major hypersensitivity reactions (HRs) in patients (*8*). It is perceived that such adverse effects are vehicle related, since it is well documented that Cremophor EL® alone causes hypotension and histamine release in dogs (*9*). The high incidences and severity of HRs to paclitaxel almost led to termination of some earlier Phase I clinical trials. However, prolonged infusions(*10*) and prophylactic medications(*11*) with antihistamines and corticosteroids have avoided the adverse episodes and allowed continuation of clinical use of Cremophore EL® formulation.

Water-Soluble Paclitaxel Prodrugs.

In face of the pharmaceutical liabilities of Cremophore EL® in the current intravenous formulation of paclitaxel, various alternatives to replace this excipient have been evaluated; one being delivery of liposome encapsulated (*12*) paclitaxel. However, the majority of approaches involve synthesis and evaluation of water-soluble prodrugs of paclitaxel. In this vein a number of C-2' and C-7 ester derivatives carrying water-solubilizing functionalities have been synthesized and evaluated for their antitumor activity in-vivo (*13, 14, 15*).

Paclitaxel C-2' Esters. The C-2' hydroxyl of paclitaxel provides an ideal chemical handle for the synthesis of esterase cleavable prodrugs. It has been well established from earlier structure activity studies that for paclitxel to impart its tubulin polymerization activity and cytotoxicity, the C-2' hydroxyl has to be free. Masking of the C-2' hydroxyl with stable functionality or its replacement has led to inactive analogs (*16*). The plethora of C-2' derivatives synthesized and evaluated for their suitability as water-soluble prodrugs of paclitaxel include succinate and glutarate derivatives (e.g. **3**, **4**); sulfonic acid derivatives (e.g. **5**, **6**) and amino acid derivatives (e.g. **7**, **8**).

Though, several of these derivatives (e.g. **3** and **4**) possess adequate solubility (up to 1%) for an iv formulation, they were found less suitable as prodrugs of paclitaxel. The main reason for this inadequacy was their instability in aqueous solution at neutral pH; a property deemed unacceptable for intravenous administration of a highly insoluble agent such as paclitaxel. To date, amongst this class of prodrugs, derivative **8** appears to be best suited for prodrug delivery of paclitaxel since it has acceptable solution stability at physiological pH and is active in-vivo against murine tumor models.

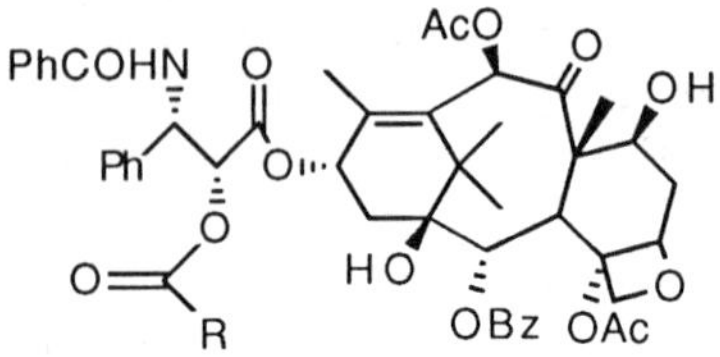

3 R = $COCH_2CH_2COO(HOCH_2CH_2)_3NH^+$
4 R = $COCH_2CH_2CH_2COO^-Na^+$
5 R = $COCH_2CH_2SO_3^-Na^+$
6 R = $COCH_2CH_2CONHCH_2CH_2SO_3^-Na^+$
7 R = $COCH_2CH_2NH_2.HCOOH$
8 R = $COCH_2CH_2NEt_2.CH_3SO_3H$

Paclitaxel C-7 Esters. In the context of prodrugs it is noteworthy that, unlike C-2' derivatives several of the C-7 derivatives have been shown to promote microtubule assembly and consequently possess bioactivity (*17*). Thus, C-7 derivatives, especially esters have the potential to be either prodrugs or analogs of paclitaxel depending upon the stability of these derivative in-vivo. In general, the C-7 esters have found little utility as prodrugs of paclitaxel. This has been attributed to the stability of these derivatives in-vivo towards esterase cleavage. Intrestingly, the cationic water soluble C-7 ester conterpart of **8**(*15*) reported by Stella and coworkers was shown to promote microtubule assembly as effectively as paclitaxel, but was poorly bioactive in whole cell assays. This was attributed to its poor cell permeability properties.

Protaxols.

Very recently, K. C. Nicolaou's group (*18*) disclosed the synthesis and in-vitro biological evaluation of a novel class of prodrugs classified as 'protaxols'. These are claimed to possess greater aqueous solubility than paclitaxel and a novel mechanism of bioconversion to active drug in-vivo. Protaxols, chemically fall into two classes namely, C-2' carbonates (e.g. **9**, **10**) and C-2' monoesters (e.g. **11-14**). The authors postulate, that C-2' carbonates

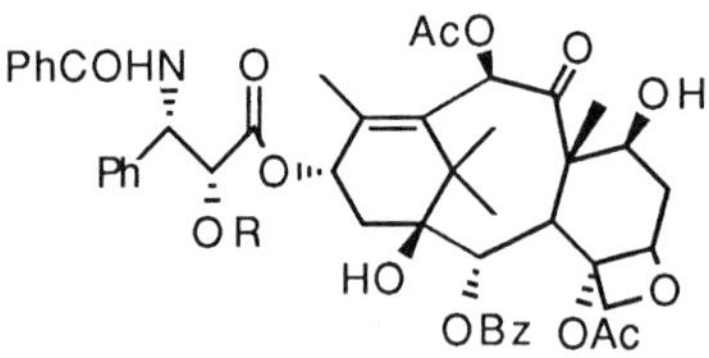

9 R = $COOCH_2CH_2SO_2Ph$
10 R = $COOCH_2CH_2SO_2pNO_2Ph$
11 R = $COCH_2OCH_2COOH$
12 R = $COCH_2SCH_2COOH$
13 R = $COCH_2SOCH_2COOH$
14 R = $COCH_2SO_2CH_2COOH$

in-vivo could generate paclitaxel through a base-induced beta-elimination in the basic microenvironment of certain tumors. In contrast, the monoesters **11-14** are concieved to release paclitaxel in-vivo via an intramolecular hydrolysis mechanism. There are no in-vivo antitumor results on these class of prodrugs to support their hypothesis. In the design of above protaxols, the main reason for the introduction of hetero atom in esters **11-14** was to impart water solubility; ~1mg/ml solubility is reported for derivatives **11**, **12** and **14**. It should be noted that for this class of prodrugs the toxicity of the promoieties is unknown.

Phosphatase Cleavable Prodrugs of Paclitaxel

In face of the above prior art, we have elected, for reasons of novelty, to focus on synthesis and evaluation of water-soluble phosphatase cleavable prodrugs of paclitaxel. The rationale behind this strategy is the ubiquitous nature of phosphatase enzymes in the mammalian systems (*19*). Also, there are several documented reports claiming that certain tumors express high levels of alkaline phosphatases (*20*). More importantly, currently, there are several examples of successful delivery of clinically useful drugs such as etoposide (*21*) and dexamethasone (*22*) via their respective phosphates as phosphatase cleavable prodrugs.

C-2' and C-7 Paclitaxel Phosphates. In our prodrug program, the initial prototype targets synthesized were the C-2' phosphate (**15**) and the C-7 phosphate (**16**). Their synthesis from paclitaxel has been previously reported by us (*23*).

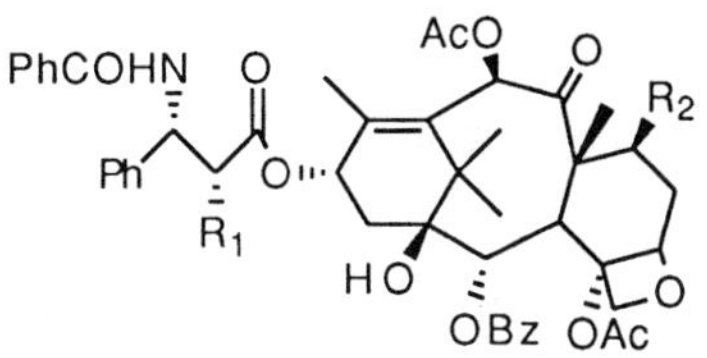

15 R_1 = $OPO(OH)_2$, R_2 = OH

16 R_1 = OH, R_2 = $OPO(OH)_2$

Although, the sodium salts of **15** and **16** were endowed with adequate water solubility (~10 mg/ml), their in-vitro and in-vivo performance indicated that they were poor prodrugs of paclitaxel. Neither of them upon treatment with isolated bovine intestinal alkaline phosphatase in-vitro generated paclitaxel; this was further corroborated by their extreme stability and failure to generate paclitaxel in rat plasma. Also both were inactive in promoting microtubule assembly in-vitro. In-vivo evaluation of **15** and **16** against the intraperitoneal (ip) Madison (M) 109 murine lung tumor model (*24*) and in a head to head comparison with paclitaxel demonstrated that they possessed marginal antitumor activity at best.

The lack of in-vitro enzymatic cleavage and inferior in-vivo activity of **15** and **16** was rationalized as due to them being poor substrates for the phosphatase enzymes in-vivo. Sterically, both the C-2' phosphate and the C-7 phosphate moieties in **15** and **16** are too close to the congested taxane core. Consequently, there is good possibility the enzymes are not able to process these derivatives efficiently in-vivo to generate paclitaxel.

Pro-Prodrug Strategies.

To surmount the steric hurdle encountered in **15** and **16** towards phosphatase cleavage in-vivo, a decision was made to pursue the synthesis of pro-prodrugs of paclitaxel that can be activated by phosphatase enzymes in-vivo. The two most widely employed strategies in medicinal chemistry to design pro-prodrugs include the exploitation of the fragmentation cascade approach (*25*) and the masked lactone approach (*26*).

Fragmentation Cascade Approach. Several examples of fragmentation cascade approaches are documented in the classical antibiotics and non-steroidal antiinflammatory drug discovery literature (*27*). In this vein, most recently Stella and his group reported (*28*) the synthesis of phosphoryloxymethyl carbamate **17** and carbonate **18** derivatives as water-soluble pro-prodrugs of hindered amines and alcohols respectively.

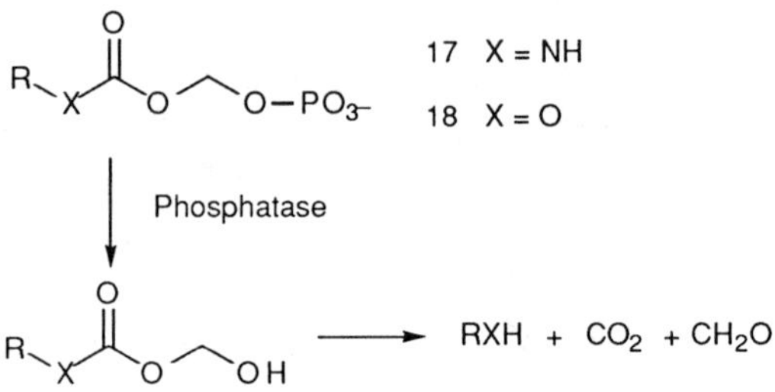

In the pro-prodrugs **17** and **18** the phosphate moiety provides both the 'phosphatase trigger' and the water-solubilizing functionality. The release of active drug in-vivo will be initiated by a fragmentation cascade only after the dephosphorylation step by phosphatases is accomplished. The authors caution towards the use of **18** as a viable pro-prodrug for hindered alcohols because of inherent instability problems.

Masked Lactone Approach. Masked lactone approach has been well explored in medicinal chemistry, since the factors that influence the rates of lactonization reactions are generally well understood (*29*) and have already been successfully used in prodrug design. The success of the masked lactone approach will depend on the appropriate choice of linkers. An effective example to illustrate this is that of the 'trimethyl lock' linker employed by Borchardt and co-workers (*30*) in the design of an esterase cleavable pro-prodrug of an amine. It was elegantly demonstrated by in-vitro experiments with isolated enzyme preparation and plasma studies that **19** was an efficient pro- prodrug of para-methoxy aniline. The success of this approach is ascribed to the phenomenal lactonization rate enhancement (of the order of $>10^5$) achieved by the presence of methyl substituents on the alkyl and the aromatic ring (*31*). The half life of lactonization for the desacetyl analog of **19** was reported to be of the order of 65 seconds.

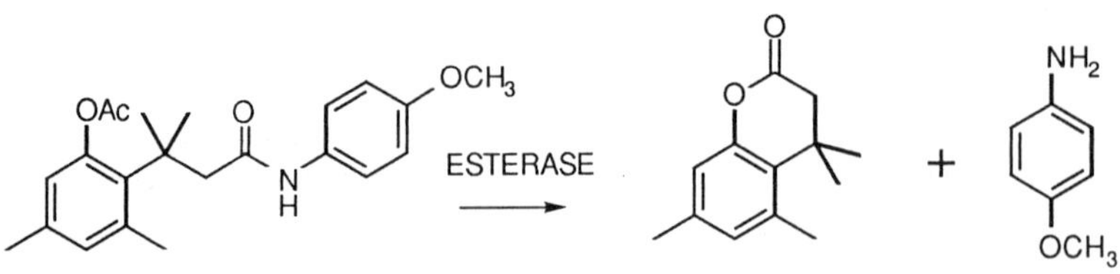

19 20 21

Synthesis of Pro-prodrug of Paclitaxel - Masked Lactone Approach.

Our initial attempt in this program was to exploit the masked lactone strategy. This approach entailed in exploration of several self-immolating linkers for their utility in successful design of pro-prodrugs of paclitaxel. The factors which may prove to be crucial to succeed are substrate specificity of the designed phosphates to enzymes in-vivo and the the efficiency of self-immolation of linker via lactonization to generate paclitaxel. To this end several linkers were studied. Our initial approach involving the use of the 'trimethyl lock' linker has been published (*32*) and briefly summarized here.

'Trimethyl lock' Linker. This approach mandated synthesis of esters either at the C-2' or C-7 hydroxyl functionality in paclitaxel; designed such to incorporate a self immolative linker with a 'phosphatase trigger' and a solubilizing group. To this end initially linker **24** was synthesized and utilized in the synthesis of the desired target **26** (Scheme I).

In synthesis of the C-7 phosphate counterpart of **26** it was neccessiated to employ the C-2' blocked paclitaxel derivative **27** in the coupling reaction with linker **24** (Scheme II).

SCHEME I

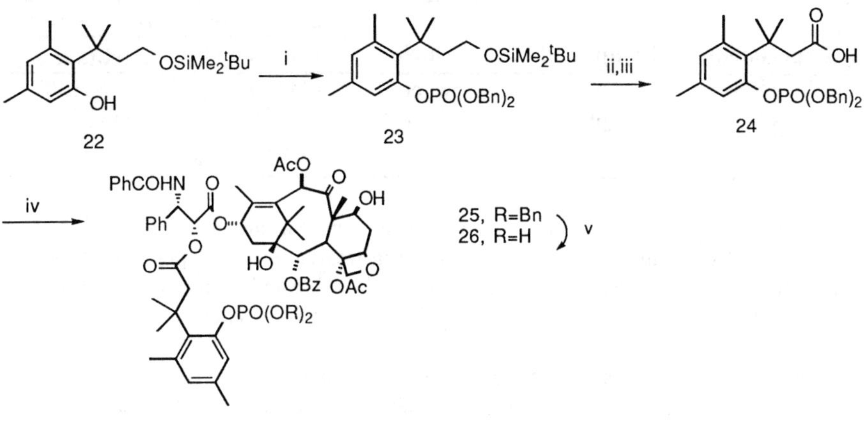

Conditions: (i) BuLi, THF,0°C, then $[(BnO)_2PO]_2O$ (95%); (ii) HCl, iPrOH (100%); (iii) H_2CrO_4, rt (77%); (iv) DCC, DMAP, CH_2Cl_2, paclitaxel, rt (65%); (v) H_2/Pd(C) (quant.);

The sodium salts of **26** and **29** were found to possess adequate water solubility (~10mg/ml) and solution stability at physiological pH (see later table II). In the tubulin polymerization assay both failed to promote microtubule assembly. As expected based on the design of these prodrugs, both derivatives upon treatment with isolated purified alkaline phosphatase in-vitro afforded paclitaxel. However, on 24 hr incubation with rat or dog plasma no sign of paclitaxel liberation from

SCHEME II

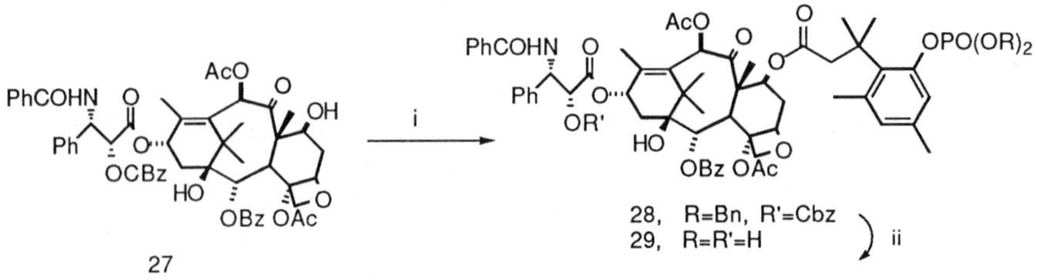

Conditions: (i) DCC, DMAP, **24**, CH_2Cl_2, rt (72%); (ii) H_2/Pd(C) (quant.).

either of the derivatives was discernible by hplc analysis. We attributed this behaviour to their tight binding to plasma proteins; this is well precedented for paclitaxel(*33*). Subsequently, they were evaluated ip for their in-vivo antitumor against an ip M109 tumor model (see Table I). In this tumor model a %T/C of 125 or greater is considered to be an active result. Accordingly, the C-2' ester **26** was considered to be marginally active; whereas the C-7 ester **29** was found to be as active as paclitaxel.

Table I: In-vivo Antitumor Activity of Compounds 26 and 29 in M109 Tumor Model[a]

			% T/C[b] (mg/Kg/Injection)[c,d]	
Experiment	Compound	Vehicle	Compound	Paclitaxel[e]
1[c]	**26**	water	144 % (100)	275 % (30)
2[d]	**29**	water	156 % (140)	144 % (40)
			153 % (70)	

a, Murine lung carcinoma, i.p. (intraperitoneal) implant model. **b**, T/C refers to the percentage of the median survival time of drug-treated mice (six per dose) to saline-treated controls. **c**, Dose administered i.p. on days 1,5 and 9. **d**, Dose administered i.p. on days 5 and 8. **e**, Administered in 10 % Tween 80 in saline.

'Gem -dimethyl' Gamma-Hydroxybutyric Acid Linker. It is well known that rate of lactonization is enhanced for gamma-hydroxybutyric acid with 'gem dimethyl' groups over their corresponding acid derivative devoid of these substituents (*29*). Accordingly, we synthesized (D. Vyas et. al. BMS - unpublished results) paclitaxel derivatives **35** and **37** (Scheme III) for their full biological evaluation. Their synthesis is based on similar lines to the one described above.

Ortho-hydroxyphenylacetic acid Linker. Our choice of this linker was based on the published work of Cohen and Hillery (*34*) on the rate and equilibrium enhancement in lactonization of ortho-substituted phenyl acetic acids. Although the reported hydronium ion catalysed rate of lactonization for ortho-hydroxyphenyl acetic acid was not as efficient in comparison to that of the 'trimethyl lock' linker

SCHEME III

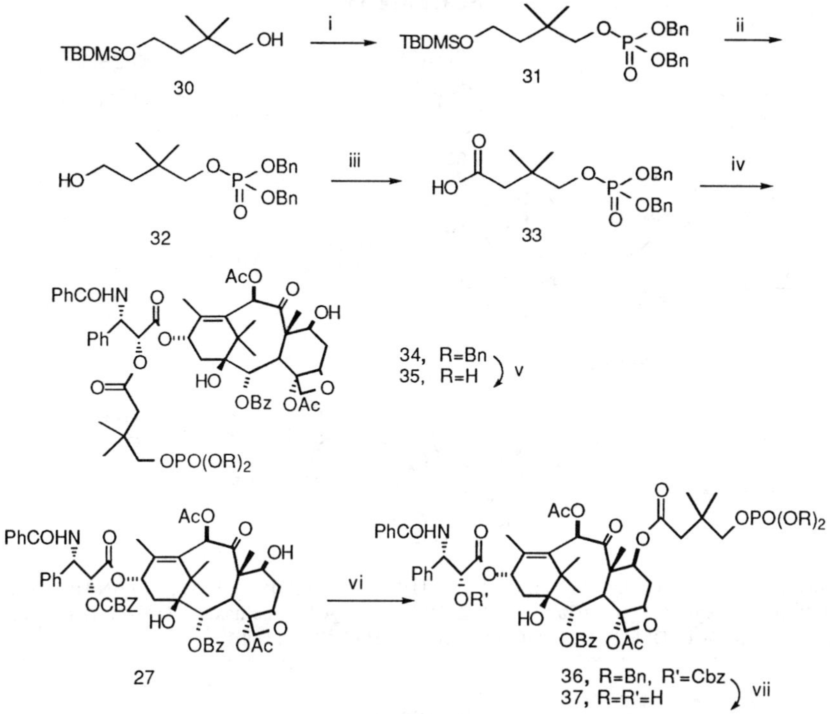

Conditions: (i) iPr_2N-$P(OBn)_2$, tetrazole, CH_2Cl_2, rt, then MCPBA, -40°C to 0°C (89%); (ii) TBAF, THF (89%); (iii) H_2CrO_4, rt (89%); (iv) paclitaxel, DCC, DMAP, CH_2Cl_2 (59%); (v) H_2/Pd(C) (quant); (vi) DCC, DMAP, **33**, CH_2Cl_2 (69%); (vii) H_2/Pd(C) (quant).

($k_{H^+} = 26.2\ M^{-1}\ sec^{-1}$) present in **26** and **29**, we thought this linker ought to give us a measure for the rate neccessary to translate into in-vivo activity. Accordingly synthesis (Y. Ueda et al., BMS-unpublished results) of target derivatives **42** and **44** was accomplished via the route outlined in Scheme IV.

Ortho-hydroxymethyl benzoic acid Linker. According to the published work of Fife and Benjamin (*35*), the hydronium ion catalysed rate of lactonization for this linker appears to be intermediate between that of the 'trimethyl lock' linker and of the ortho-phenylacetic acid linker mentioned above. Consequently, we believed this would yield vital SAR information on in-vivo activity in relation to their rates of lactonization. Also, this linker will facilitate in establishing the utility of a benzylic phosphate versus phenolic phosphate (as in **42** and **44**) in prodrug design.

Accordingly, synthesis of target derivatives **51** (Scheme V) and **54** (Scheme VI) were accomplished (Y. Ueda et al., BMS-unpublished results). C-7 phosphate derivative **54** is a C-2' carbonate derivative. Usefulness of C-2' carbonates as

SCHEME IV

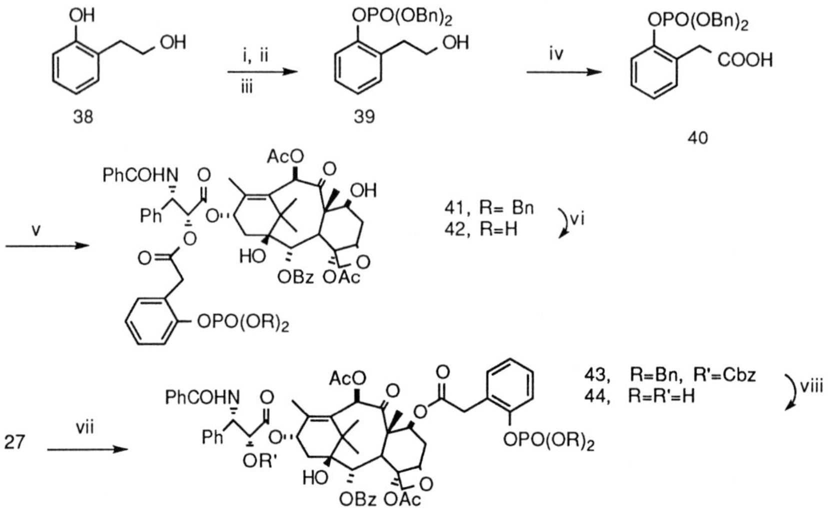

Conditions: (i) TBSCl, Im (100%); (ii) BuLi, THF, $[(BnO)_2PO]_2O$ (95%); (iii) 6N HCl, iPrOH (100%); (iv) H_2CrO_4, rt (60%); (v) DCC, DMAP, CH_2Cl_2, paclitaxel, (75%); (vi) H_2/Pd(C) (quant.); (vii) DCC, DMAP, **40**, CH_2Cl_2 (35%); (viii) H_2/Pd(C) (quant.).

SCHEME V

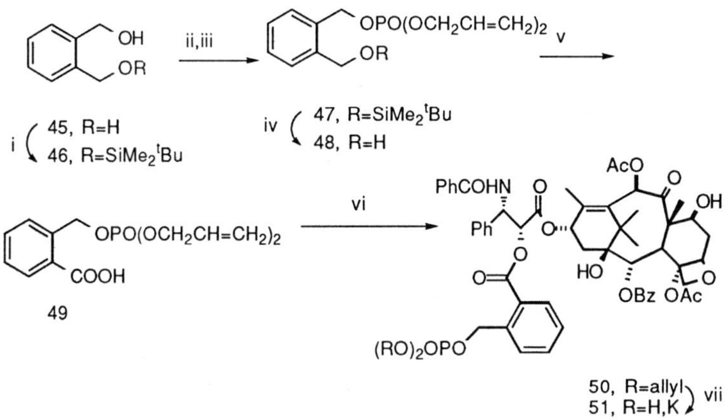

Conditions: (i) TBSCl, Im; (ii) $iPr_2N\text{-}P(OCH_2\text{-}CH{=}CH_2)_2$, tetrazole, rt; (iii) MCPBA (65% overall); (iv) HCl, iPrOH (78%); (v) H_2CrO_4, rt (88%); (vi) DCC, DMAP, paclitaxel, CH_2Cl_2 (87%); (vii) $Pd(PPh_3)_4$, Bu_3SnH, AcOH, K-ethylhexanoate; then C-18 chrom. (38%).

prodrugs of paclitaxel has been demonstrated in our labratory (Y. Ueda et al., Bioorg. Med. Chem. Lett; in press). We envision **54** to be first metabolized by either a phosphatase or an esterase to an inactive intermediate which subsequently can be bioconverted to active paclitaxel by one of the two enzyme systems. Thus for a meaningful comparison of in-vivo biological results, a C-7 phosphate derivative **55** was also synthesized.

SCHEME VI

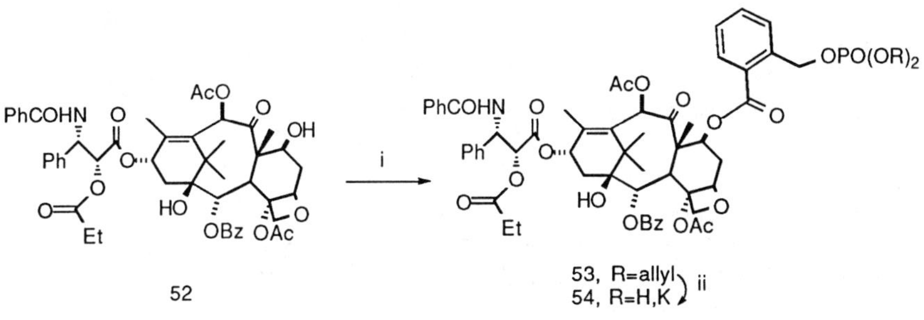

Conditions: (i) DCC, DMAP (ii) $Pd(PPh_3)_4$, Bu_3SnH, AcOH, K-ethylhexanoate; then C-18 chrom.

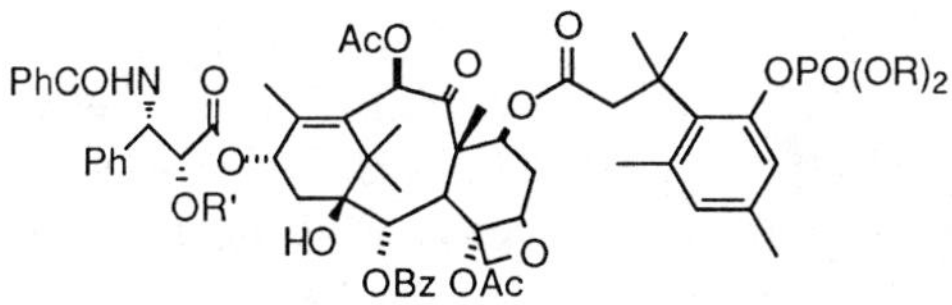

55, R' = -COOEt , R = H

IN-VITRO Evaluation of Pro-produgs.

In-vitro evaluation of these prodrugs involved: (a) measurements of their water solubility in mg/ml; (b) solution stability measured in 50mM tris buffer as time required (t_{90} at 37° C and pH =7.4) for 10% decomposition of the prodrug: and (c) linker self-cleavage half life in mins. or hrs. as time ($t_{1/2}$) required to generate paclitaxel after dephosphorylation (in 50 mM tris buffer at pH 7.4 and 37°C) by an isolated commercial phosphatase is accomplished. For solution stability, our acceptable criteria for a phosphate derivative to be a suitable pro-prodrug was a t_{90} value much greater than 24 hours (preferably >100 hrs for drugs like paclitaxel) at pH 7.4 and at 37°C; this should ensure minimum drug precipitation occuring during infusion time course. It should be noted that in earlier studies (*15*) involving amine based prodrugs, the optimal t_{90} values were at pH range of 3.5 to 4.5. Thus pH range for optimal t_{90} values may vary depending upon classes of prodrugs. Table II lists the properties of prodrugs evaluated.

Table II. Solubility, Stability and Linker Self-Cleavage Rates for Prodrugs

Compound	Water-solubility (mg/ml)[a]	Stability (t_{90} at 37° C, pH 7.4)[b]	Linker Self Cleavage ($t_{1/2}$)[c]
26	>10	20h	<5min
29	>10	>100h	<5min
35	>10	8.3h	<5min
37	>10	23h	~80h
42	>10	41h	<5min
44	5.5	15.2h	>200h
51	5.0[d]	20.5h	<5min
54	2.5[d]	60h	ND
55	4.0	>150h	ND

a. Determined on the sodium salts by HPLC of saturated solutions after filteration through 0.2 µm Nylon filters. **b**. determined in 50mM tris buffer. **c**. Bovine alkaline phosphate (Sigma) in 50mM tris. buffer containing 1% DMSO, pH=7.4 at 37° C. **d**. mono-potassium salt was used.

IN-VIVO Evaluation of Pro-prodrugs.

The in-vivo evaluation assay involved M109 murine lung tumor implanted subcutaneously (sc) in CDF_1 mice; it was initiated by implantation of 0.1 ml of 2% tumor brei. Mice received their drug treatments via intravenous (iv) injections on a five day consecutive daily treatment schedule beginning on either Day 4 or Day 5 post-tumor implant. There were eight mice per treatment group in the experiments, and eight mice in all parallel untreated control groups. Each derivative in an experiment was evaluated at a minimum of three dose levels. Drug-treated mice dying prior to the first death occurring in the parallel control group were considered to be toxic deaths. Also, any mouse dying prior to its tumor reaching 1 gm in size was considered to have died of drug toxicity. If more than one toxic death occurred in a drug-treated group, the therapeutic results from that treatment group were not utilized.

Antitumor activity was judged by the relative median time for tumors to reach 1 gm in drug-treated as compared to control groups (i.e. T- C values, in days). Tumors were measured by calipers once or twice weekly and tumor weights determined according to published methods (*36*). A drug was considered active by this parameter if it produced a T- C value of ≥ 4 days. The results of phosphate containing paclitaxel pro-prodrugs evaluated in-vivo are tabulated in Table III.

Results and Discussion.

Perusal of Table II reflects that all of the pro-prodrugs synthesized in this program are endowed with acceptable water-solubility to allow their iv administration in water. However, from a stability perspective only derivatives **29**, **42**, **54** and **55**

Table III. IN-VIVO antitumor activity of paclitaxel phosphates against Sc-implanted Madison 109 Lung Carcinoma

Compound	Schedule[a]	Optimal dose[b]	Delay in Tumor Growth-(T-C) days[c] Compound	Paclitaxel
29	qd 4-8	36	34.8	28.5
35	qd 4-8	61	9.0	14.5
37	qd 4-8	31	8.3	14.0
42	qd 4-8	34	16.5	15.8
44	qd 4-8	61	8.0	15.8
51	qd 4-8	23	6.0	14.0
54	qd 4-8	40	17.0	10.8
55	qd 4-8	40	18.0	12.3

a. Consecutive daily treatment schedule begining on day 4 post-tumor implant. **b**. Optimal dose given intravenously in mg/kg/injection. **c**. Relative median time for tumors to reach 1 gm in drug treated as compared to control groups (8 mice per dose).

appear to meet our criteria of t_{90} >24 hrs. Interestingly, these same four derivative demonstrated antitumor activity (Table III) comparable to that of concomitantly evaluated paclitaxel. All the other derivatives tested (**35**, **37**, **44** and **51**) produced delays in tumor growth that were inferior to those caused by paclitaxel assayed parallel, and inferior to the smallest T-C value associated with paclitaxel in any of these experiments. Compounds showing comparable activity demonstrated that self-immolating linkers associated with them were functional in generating paclitaxel efficiently after activation by phosphatase enzymes (see self cleavage rate in Table II). It is noteworthy that ortho-hydroxymethyl benzoic acid linker in **54** is a viable linker for pro-prodrug design even though its rate of lactonization is slower than that of the 'trimethy lock' linker in **29**. In the case of **42** it is likely that activation by an esterase and/or a phosphatase is responsible in generation of paclitaxel in-vivo. The activity seen in **54** and **55**, however, is certainly due to activation by both enzymes; esterase for cleavage of the C-2' carbonate and phosphatase for the cleavage of the C-7 phosphate group. In contrast, the in-vivo efficacy of **29** is achieved presumably, solely through phosphatase activation. Detailed in-vitro enzyme kinetic studies and in-vivo pharmacokinetic studies on selected derivatives reported herein is in progress and will be reported in full elsewhere. These studies should reveal detailed mechanistic insight into the function of these prodrugs.

In conclusion, of the four self-immolative linkers studied in this program only the gamma-hydroxybutyric acid linker incorporated in **35** and **37** proved to be of little utility in the design of pro-prodrugs of paclitaxel. The others led to derivatives (either at C-2' or C-7) which behaved as true pro-prodrugs of paclitaxel endowed with acceptable water-solubility and aqueous stability.

Literature cited.

1. Wani, M. C.; Taylor, H. L.; Wall, M. E.; Coggon, P.; McPhail, A. T. *J. Am. Chem. Soc.* **1971**, *93*, 2325.
2. Bissery, M. C.; Guénard, D.; Guéritte-Voegelein, F.; Lavelle, F. *Cancer Res.* **1991**, *51*, 4845.
3. Donehower, R. C.; Rowinsky, E. K. *Cancer Treatment Rev.* **1993**, *19*, 63-78.
4. Arbuck, S. G. *J. of Clin. Oncol.* **1994**, *12*, 233.
5. Adams, J. D.; Flora, K. P.; Goldspiel, B. R.; Wilson, J. W.; Finley, R. *J. Nat'l Cancer Inst. Monographs*. **1993**, *15*, 141.
6. Rowinsky, E. K.; Donehower, R. C. *Pharmacol. Ther.* **1991**, *52*, 35.
7. Pazdur, R.; Newman, R. A.; Newman, B. M.; Fuentes, A.; Benvenuto, J.; Bready, B.; Moore, D.; Jaiyesimi, I.; Vreeland, F.; Bayssas, M. M. G.; Raber, M. N. *J. Nat'l Cancer Inst.* **1992**, *84*, 1781.
8. Rowinsky, E. K.; Cazenave, L. A.; Donehower, R. C. *J. Nat'l Cancer Inst.* **1990**, *82*, 1247.
9. Lorenz, W.; Riemann, H.J.; Schmal, A. *Agents Actions* **1977**, *7*, 63.
10. Rowinsky, E. K.; Onetto, N.; Canetta, R. M.; Arbuck, S. G.; *Semin. Oncol.* **1992**, *19*, 646.
11. Lubejko, B. G.; Sartorius, S. E. *Semin. Oncol.* **1993**, *20*, 26.
12. Sharma, A.; Mayhew, E.; Straubinger, R. M. *Cancer Res.* **1993**, *53*, 5877.
13. Deutsch, H. M.; Glinski, J. A.; Hernandez, M.; Haugwitz, R. D.; Narayanan, V. L.; Suffness, M.; Zalkow, L. H. *J. Med. Chem.* **1989**, *32*, 788.
14. Zhao, Z.; Kingston, D. G. I.; Croswell, A. R. *J. Nat. Prod.* **1991**, *54*, 1607.
15. Mathew, A. E.; Mejillano, M. R.; Nath, J. P.; Himes, R. H.; Stella, V. *J. Med. Chem.* **1992**, *35*, 145.
16. Kant, J.; Stella, H.; Wong, H.; Fairchild, C.; Vyas, D. M.; Farina, V. *Bioorg. Med. Chem. Lett.* **1993**, *3*, 2471.
17. Lataste, H.; Sénilh, V.; Wright, M.; Guenard, D.; Potier, P. *Proc. Nat'l Acad. Sci. U.S.A.* **1984**, *81*, 4090.
18. Nicolaou, K. C.; Riemer, C.; Kerr, M. A.; Rideout, D.; Wrasidlo, W. *Nature*. **1993**, *364*, 464.
19. Harris, H. *Clin. Chim. Acta.*, **1989**, *186*, 133.
20. Fishman, W. H.; Inglis, N. R.; Stolbach, L. L.; Krant, M. J. *Cancer Res.* **1968**, *28*, 150.
21. Rose, W. C.; Basler, G. A.; Trail, P. A.; Saulnier, M. G.; Crosswell, A. R.; Casazza, A. M. *Invest. New Drugs,*. **1990**, *8*, S25.
22. Hare, L. E.; Yeh, K. C.; Ditzler, C. A.; McMahon, F. G.; Duggan, D. E. *Clin. Pharmacol. Ther.* **1975**, *18*, 330.
23. Vyas, D; Wong, H.; Crosswell, A. R.; Casazza, A. M.; Knipe, J. O.; Mamber, S. W.; Doyle, T. W. *Bioorg. Med. Chem. Lett.* **1993**, *3*, 1357.
24. Rose, W. C., *Anti-Cancer Drugs.* **1992**, *3*, 311.
25. Ferres, H. *Med. Actual/Drugs Today*. **1983**, *19*, 499.
26. Johnson, C. D.; Lane, S.; Edwards, P. N.; Taylor, P. J. *J. Org. Chem.*, **1988**, *53*, 5130.
27. Bundgaard, H. *Drugs of the Future*. **1991**, *16*, 443.
28. Safadi, M.; Oliyai, R.; Stella, V. J. *Pharma. Res.* **1993**, *10*, 1350.
29. Kirby, A. J. *Adv. Phys. Org. Chem.* **1980**, *17*, 183.

30. Amsberry, K. L.; Gerstenberger, A. E.; Borchardt, R. T. *Pharma. Res.* **1991**, *8*, 455.
31. Milstien, S.; Cohen, L. A. *J. Am. Chem. Soc.* **1972**, *94*, 9158.
32. Ueda, Y.; Mikkilineni, A. B.; Knipe, J. O.; Rose, W. C.; Casazza, A. M.; Vyas, D. M. *Bioorg. Med. Chem. Lett.* **1993**, *3*, 1761.
33. Kumar, G. N.; Walle, U. K.; Bhalla, K. N.; Walle, T. *Research Communications in Chemical Pathology and Pharmacology*, **1993**, *80*, 337.
34. Hillery, P. S.; Cohen, L. A. *J. Org. Chem.* **1983**, *48*, 3465.
35. Fife, T. H.; Benjamin, B. M. *J. Am. Chem. Soc.* **1973**, *95*, 2059.
36. Rose, W. C. *Cancer Treat. Repts.* **1981**, *65*, 299.

RECEIVED July 21, 1994

Chapter 10

Microtubule Polymerization Dynamics, Mitotic Block, and Cell Death by Paclitaxel at Low Concentrations

Mary Ann Jordon and Leslie Wilson

Division of Molecular, Cellular, and Developmental Biology, Department of Molecular and Cellular Biology, University of California–Santa Barbara, Santa Barbara, CA 93106

Microtubules are dynamic protein polymers that are important in cell reproduction and are the target of antitumor drugs including taxol. Microtubule dynamics are suppressed by low concentrations of taxol or another class of drugs, the vinca alkaloids. In a human tumor cell line we found that low concentrations of taxol (10 nM) block cell division at a critical point called the metaphase/anaphase transition and kill cells without inducing microtubule bundles often associated with taxol's antitumor action. The blockage resembles that induced by low concentrations of the vinca alkaloids. Kinetic stabilization of microtubule dynamics by taxol and other antimitotic drugs appears to be the most potent mechanism by which these drugs block the cell cycle, inhibit cell proliferation, and induce cell death.

Taxol, a complex diterpene isolated from the bark of the western yew *Taxus brevifolia,* is a potent new antitumor drug that has been approved by the US Food and Drug Administration for use in refractory ovarian cancer. It is a potent inhibitor of cell proliferation that arrests cells in mitosis and induces multinucleation of cells in interphase *(1-5)*. At high concentrations the therapeutic and cytotoxic actions of taxol are associated with stabilization of microtubules in cells and large increases (as high as five-fold) in the mass of cellular microtubules that result in multiple asters of microtubules in mitosis and large bundles of microtubules in interphase *(2-7)*. Despite the considerable knowledge gained in the past several years, the precise molecular cytotoxic mechanism of taxol is not fully elucidated.

Recently we examined in detail the antimitotic and antitumor action of taxol to elucidate the most sensitive and potent antiproliferative mechanism of the drug and the actions of taxol on the polymerization dynamics of microtubules in vitro. We found that in HeLa cells, a human tumor cell line, very low concentrations of taxol can block cell division and kill cells *without* inducing the formation of large bundles of stable microtubules. Many of the cells look normal, except that they

NOTE: Paclitaxel is the generic name for Taxol, which is now a registered trademark.

0097–6156/95/0583–0138$08.00/0

cannot progress beyond a critical point in cell division called the metaphase/anaphase transition. The cell cycle block induced by low concentrations of taxol appears to result from suppression of microtubule dynamics by taxol rather than by an effect on the mass of cellular microtubules *(5)*.

Our results suggest that taxol shares important properties with another group of antitumor drugs, the vinca alkaloids, vinblastine and vincristine. Although there are important differences in their molecular interactions with tubulin and microtubules, they have a common ability to suppress microtubule dynamics by binding to microtubules. Kinetic stabilization of microtubule dynamics by taxol and other antimitotic drugs appears to be the most potent mechanism by which these drugs block the cell cycle at mitosis, inhibit cell proliferation, and induce cell death.

Microtubule Structure and Polymerization Dynamics

Microtubules are one of the major components of cells. They form a constantly changing, dynamic framework, the cytoskeleton, that undergoes continual reorganization depending on the cell's activities. Microtubules are long, tubular polymers composed of tubulin heterodimers (mw 100,000) of two related tubulin polypeptides, α and β (Figure 1). The tubulin dimers are arranged head-to-tail in protofilaments and in a helical array in the walls of the microtubule. As a result of this arrangement, each microtubule has a plus (+) end that is structurally and kinetically distinct from the minus (-) end. This polarity is important in determining the arrangement and behavior of microtubules in cells.

Figure 2 is an illustration of the main characteristics of microtubule dynamics. Microtubules are in equilibrium with a pool of soluble tubulin dimers at a concentration that is known as the critical concentration. Interestingly, microtubules are not simple equilibrium polymers (reviewed by Gelfand and Bershadsky, *8*). The energy provided by hydrolysis of tubulin-liganded GTP during addition of tubulin at the microtubule ends creates transitions at the microtubule ends between more stable and less stable states that give rise to two remarkable dynamic behaviors. One such behavior, called *dynamic instability (9,10) is* a stochastic switching between shortening and growing phases at both ends of individual microtubules. Growing and shortening behavior may be due to a stochastic gain and loss of a stabilizing "cap" at both microtubule ends. The cap is thought to consist of a short region of GTP- or GDP-P_i-liganded tubulin at the ends of the microtubules *(11-14)*. However, the chemical nature of the stabilizing cap and the mechanisms governing the gain and loss of the cap are not yet known.

A second dynamic behavior called *treadmilling* or *flux (15,16)*, is the net growing of microtubules at one end and net shortening at the opposite end. Treadmilling appears to be a consequence of the inequality of the balance of the association and dissociation rate constants for tubulin at the two microtubule ends *(17,18)*. Although treadmilling must occur simultaneously with dynamic instability, treadmilling can be best observed under conditions where microtubule length changes associated with dynamic instability are minimal. Tubulin exchange at microtubule ends due to treadmilling is readily distinguished from tubulin exchange due to dynamic instability in experiments using radiolabeled GTP to monitor tubulin exchange in microtubules at steady state *(e.g., 19-21)*.

Both dynamic instability and treadmilling or flux have been observed in vivo as well as in purified microtubule preparations *(17,22-29)*. In cells, the rates of tubulin addition and loss at the two ends of a microtubule vary in a predictable

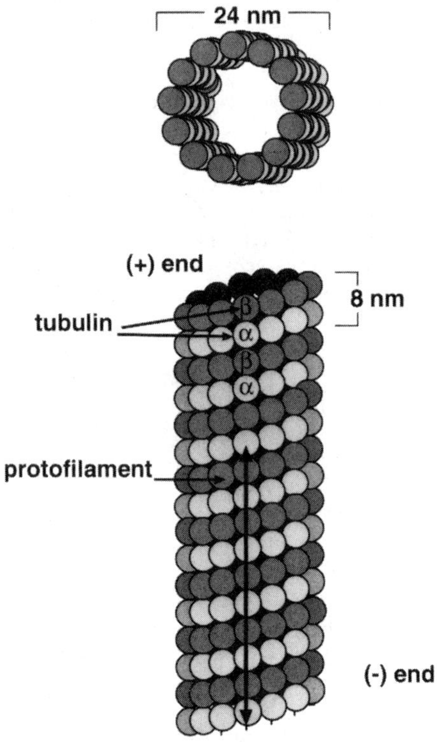

Fig. 1. Diagrammatic representation of the structure of a microtubule in cross-section (top) and longitudinal view (bottom). A typical microtubule is composed of heterodimers of α and β tubulin arranged head-to-tail in 13 protofilaments that make up the wall of the microtubule.

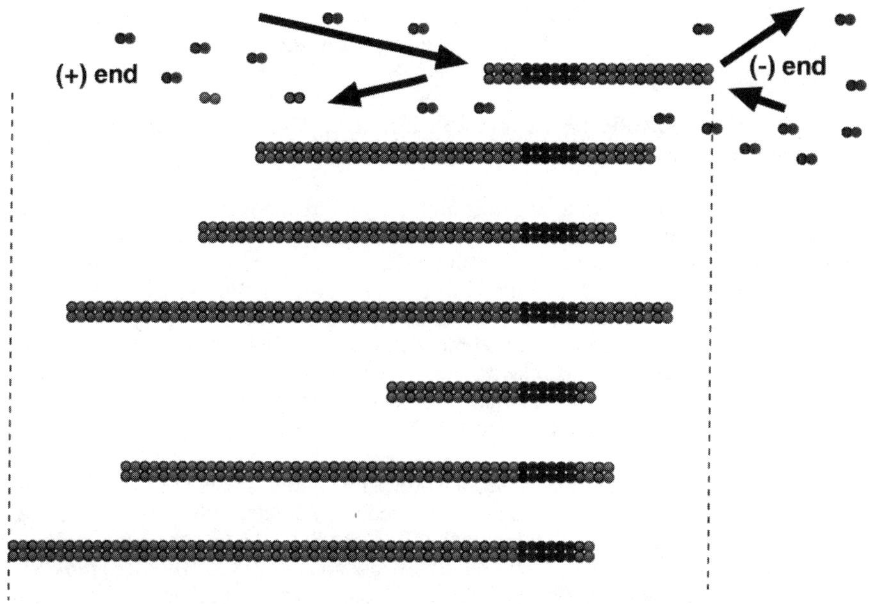

Fig. 2. Illustration of the main characteristics of microtubule dynamics. Tubulin dimers add to and are lost from both ends of a microtubule. The rate constants (indicated by arrows) for addition and loss are often greater at the plus end than at the minus end. Upon addition of a tubulin dimer to a microtubule, GTP bound to the dimer is hydrolyzed to GDP. For some time following addition of dimers, the microtubule end may be "capped" by unhydrolyzed GTP (not shown). The sequential microtubule images (top to bottom) illustrate the two kinds of microtubule dynamics known as *dynamic instability* and *treadmilling*. *Dynamic instability* is a stochastic switching between shortening and growing phases at both ends of the microtubule indicated by the changing lengths of the two microtubule ends. The extents of growing and shortening at the plus microtubule end are greater than the extents of growing and shortening at the minus end. *Treadmilling* is the net growing of microtubules at one end and net shortening at the opposite ends. Treadmilling appears to be a consequence of the inequality of the balance of the association and dissociation rate constants for tubulin at the two microtubule ends. Thus the marked (dark) segment of microtubule represents tubulin dimers that have just added to the plus end of the microtubule in the top image. The marked dimers traverse the microtubule with time, and are about to be lost from the minus end of the microtubule in the last image.

manner in different populations of microtubules and through the cell cycle. During interphase of the cell cycle, microtubules exchange their tubulin with tubulin in the cytoplasmic pool with half-times on the order of minutes to hours. When a cell enters mitosis, its microtubules become much more dynamic, turning over with half-times of 11-14 sec *(30)*. Microtubule dynamics can be regulated both by naturally occurring cell molecules and by drugs, and this regulation has profound effects on cellular function.

Mechanism of Mitotic Block and Inhibition of Microtubule Dynamics by Antimitotic Drugs Such as Vinblastine

The target for many antimitotic drugs including taxol appears to be microtubules. However, the action of taxol on microtubules generally has been considered to be unique. At high concentrations, both in vitro and in cells, other antimitotic compounds such as vinblastine inhibit microtubule polymerization *(31-33)*, whereas at high concentrations, taxol enhances microtubule polymerization *(34,35)*. Depolymerization of microtubules has been thought to be responsible for the antitumor activity of vinblastine and most of the other previously known antimitotic compounds. In contrast, the antitumor activity of taxol has been thought to be due to enhanced microtubule polymerization and the formation of large bundles of microtubules in cells.

However, the interaction of vinblastine with microtubules is more complex and interesting than has often been recognized. Vinblastine binds directly to the ends of microtubules with relatively high affinity (5.3×10^5 M^{-1}) and binds along the surface of the microtubule with relatively low affinity 3-4 $\times 10^3$ M^{-1}) *(36-38)*. The low affinity binding of vinblastine, which occurs at high concentrations, induces microtubule depolymerization by the peeling of protofilaments at the microtubule ends *(37)*. We have found that the binding of 1.2 ± 0.3 molecules of vinblastine to high affinity sites at the microtubule ends is sufficient to inhibit tubulin exchange at the ends by 50% *(39)*. In addition, *low* concentrations of vinblastine inhibit both dynamic instability and treadmilling of microtubules *in vitro* without appreciably affecting the equilibrium between the microtubule polymer mass and the pool of soluble tubulin dimers *(21)*.

In HeLa cells, we found that at its lowest effective concentrations (0.4-2 nM), vinblastine inhibits mitosis and cell proliferation without depolymerizing microtubules and with only subtle changes in the organization of the mitotic microtubules (the mitotic *spindle) (33,40,41)*. With low concentrations of vinblastine, mitosis was blocked at a critical stage, the *metaphase/anaphase transition,* in the presence of a normal complement of microtubules. A 14-times higher concentration of vinblastine was required to depolymerize the microtubules than was required to block cells in mitosis.

More recently we have found that vinblastine inhibits dynamic instability of interphase microtubules in cells in the absence of microtubule depolymerization (R. Iyengar, M. A. Jordan, D. Thrower, E. Sheldon, L. Wilson, and P. Wadsworth, unpublished results). Thus, the results we obtained *in vivo* and *in vitro* strongly suggest that mitotic block induced by low concentrations of vinblastine occurs by suppression of tubulin exchange at the ends of mitotic spindle microtubules.

As a consequence of finding that vinblastine, as well as other antimitotic drugs, can suppress microtubule dynamics and block mitosis in the absence of changes in the mass of polymerized microtubules, we examined in similar fashion the antitumor and antimitotic mechanism of taxol at low drug concentrations. The results, summarized below, suggest that vinblastine and taxol, as well as other antimitotic drugs, share a common mechanism for inducing mitotic block and cell death.

Mechanism of Mitotic Block and Inhibition of Microtubule Dynamics by Low Concentrations of Taxol

Binding of Taxol to Microtubules. Taxol does not bind to free tubulin subunits *(34,42,43)*. Taxol binds reversibly to microtubules in vitro with an affinity of approximately 1 μM and a stoichiometry of one mol of taxol per mol of tubulin *(34,44)* and is assumed to bind to sites on tubulin that are exposed along the surface of the microtubule. Taxol binds specifically to microtubules in cells *(45)*. The use of photoactivatable analogs of taxol has localized at least one binding site of taxol to the amino terminus of β tubulin with indications of additional involvement of α tubulin *(46,47)*.

Stabilization of Microtubules by Taxol. Taxol profoundly affects the polymerization characteristics and stability of microtubules. It stimulates microtubule polymerization *in vitro*, promoting both the nucleation and elongation phases of the polymerization reaction, and it can reduce the critical tubulin subunit concentration (*i.e.*, soluble tubulin concentration at steady state) to near zero *(35,48,49)*. Microtubules polymerized in the presence of taxol are extremely stable. Howard and Timasheff *(49)* found that the linkage free energy provided by taxol binding to tubulin in the microtubule is -3 kcal/mol which results in a stable polymer under conditions that normally do not support polymerization. Microtubules in the presence of taxol resist depolymerization by cold, calcium ions, dilution, and other antimitotic drugs *(35,48-50)*. Taxol induces the self-assembly of tubulin into microtubules under conditions that normally do not support microtubule assembly, i.e., at 0°C *(51)*, from GDP-tubulin in the absence of exogenous GTP *(44,52,53)*, without microtubule-associated proteins (MAPs) *(48)*, and at alkaline pH *(54)*. Taxol and its analog taxotere are the only ligands known to fully dispense with the requirement for microtubule assembly for a γ phosphate at the exchangeable GTP-binding site on tubulin *(44)*.

Although the critical concentration and the rate of microtubule depolymerization can be significantly reduced by taxol, taxol-induced microtubule assembly is intrinsically reversible. Depolymerization of taxol-stabilized microtubules can be effected with millimolar Ca^{++} *(55)* or at mildly alkaline pH *(54)*.

Inhibition of Treadmilling and Dynamic Instability by Taxol at Low Concentrations. Although the molecular mechanism is not fully elucidated, it is known that taxol inhibits the dynamic reactions at microtubule ends which in turn results in inhibition of treadmilling. A primary effect of taxol at concentrations that are stoichiometric or near-stoichiometric with respect to tubulin is to reduce the magnitudes of the dissociation rate constants equally at both microtubule ends, without exerting detectable effects on the apparent association rate constants *(56,57)*.

We have examined the effects of low concentrations of taxol on the dynamics of individual microtubules by polymerizing microtubules from purified bovine brain tubulin to a steady state in the absence of microtubule-associated proteins and observing them by video-enhanced differential interference contrast

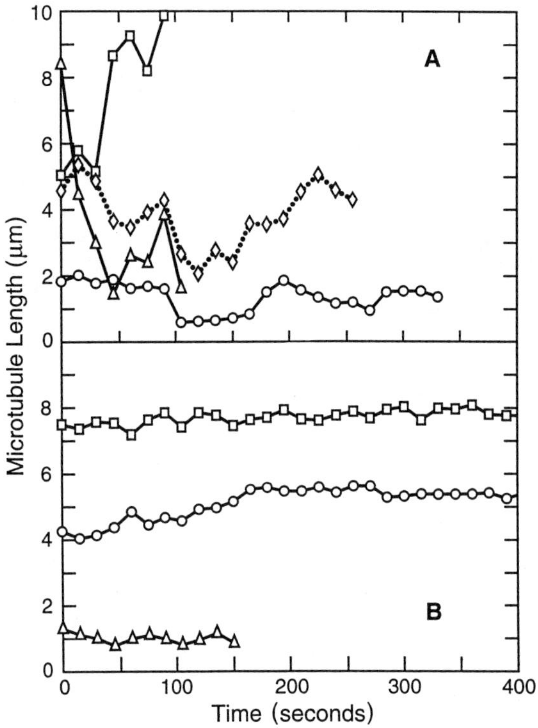

Fig. 3. Changes in microtubule lengths in the absence (A) and presence (B) of 0.5 μM taxol. Each tracing shows the growing and shortening of an individual microtubule. Microtubules were polymerized at the plus ends of axonemal seeds in the absence or presence of taxol.

microscopy. The change in length with time of several typical control microtubules is shown in Figure 3A. The microtubules alternate between phases of growing, shortening, and *pause* or *attenuation* (a state in which microtubule length changes are undetectable by this method, i.e., < 0.2 μm, equivalent to 330 tubulin dimers).

At low concentrations of taxol the dynamics of microtubules become strikingly suppressed (Figure 3B). Measurements of more than 14 microtubules each observed over 25 min indicate that with taxol concentrations as low as 100 nM, the rate of shortening is reduced by 47%. The overall dynamic behavior of the microtubules, termed *dynamicity* (total detectable tubulin exchange per unit time) is reduced by 63%. The fraction of time that microtubules remain in the pause state increases from 19% to 45%. With 100 nM taxol, we found a binding stoichiometry of 1 taxol molecule per 72 tubulin dimers in microtubules and a concomitant shift in the dimer-polymer equilibrium that resulted in a 0 or 30% reduction in the critical concentration of tubulin dimers depending upon the buffer conditions (W. B. Derry, L. Wilson, and M. A. Jordan, unpublished data). In other words, binding of low numbers of taxol molecules to the microtubules suppressed excursions of dynamic instability and resulted in microtubules that were static, rather than dynamic. One model that may explain this phenomenon is that following GTP-cap loss, an ensuing shortening excursion at the microtubule end occurs until depolymerization reaches a bound taxol molecule. The taxol then sufficiently slows the tubulin off-rate so that recapping of the microtubule end by tubulin-GTP is favored and further shortening is suppressed.

Mitotic Block Induced by Low Concentrations of Taxol. As diagrammed in Figure 4, after completion of DNA synthesis and passage through critical checkpoints in *interphase* of the cell cycle, the cell enters *mitosis* during which the equal partitioning of genetic material to two new cells occurs. In mitosis, chromosomes condense and microtubules of the interphase cytoskeletal array depolymerize and then new microtubules repolymerize into a structure called the *mitotic spindle*. Microtubule dynamics speed up 10 - 100 fold, and their dynamics appear to be crucial for several phases of mitosis, i.e., attachment of microtubules to chromosomes and congression of the chromosomes to the metaphase plate *(58)*, maintenance of the spindle shape, chromosome separation and spindle elongation during anaphase, and perhaps even for the signals involved in passage from metaphase to anaphase *(40)*.

To examine the effects of low taxol concentrations on cell proliferation, mitotic block, bundling of microtubules, the mass of microtubules in cells, and subsequent cell killing, we incubated HeLa cells with a range of taxol concentrations for 20 h (the duration of one cell cycle) followed by extensive washing of the cells and then continued incubation for several days in the absence of added taxol. The mass of cellular microtubules was measured by isolating cytoskeletons in a microtubule-stabilizing buffer and measuring the tubulin that was in polymeric form by means of an enzyme-linked immunoadsorbent assay *(59,60)*.

The effects of taxol at a range of concentrations on proliferation of HeLa cells, mitosis, and the mass of microtubules in the cells are shown in Figure 5. Taxol inhibits cell proliferation in parallel with a block in mitosis at the metaphase/anaphase transition. Cell proliferation is inhibited half-maximally at a taxol concentration of 8 nM, and inhibition is complete at taxol concentrations >33 nM. Similarly, cells accumulate in mitotic metaphase half-maximally at a

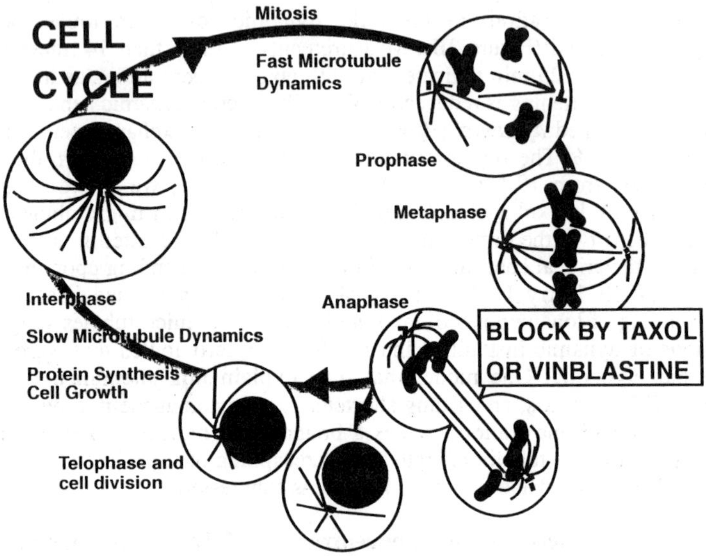

Fig. 4. Diagram of the cell cycle showing the arrangements of microtubules and chromosomal material in *interphase* and *mitosis*. During *interphase,* the cell grows in size, synthesizes new proteins and replicates its DNA-containing chromosomal material (in the spherical nucleus). During interphase microtubules undergo relatively slow exchange of tubulin subunits with the soluble tubulin in the cell. They function in maintenance of cell shape and in intracellular transport. After passage through critical checkpoints, the cell enters *mitosis* during which the equal partitioning of genetic material to two new cells occurs. In mitosis, microtubule dynamics speed up 10 - 100 fold. During *prophase* of mitosis the plus ends of dynamic microtubules appear to probe the cytoplasm until they attach to a chromosome. Following attachment of dynamic microtubules from both poles to the chromosomes, chromosomes oscillate toward and away from each pole, and their attached microtubules undergo many cycles of lengthening and shortening until a balanced *metaphase* configuration is attained. During metaphase, microtubules have also been shown to undergo treadmilling behavior. Following passage of another cell cycle checkpoint, cells enter *anaphase* during which chromosomes separate and segregate to the two forming daughter cells. Low concentrations of taxol or vinblastine or of several other antimitotic drugs block mitosis of HeLa and other cells at the critical transition between metaphase and anaphase.

concentration of ~8 nM, and maximal mitotic accumulation (80-95%) occurs at taxol concentrations > 33 nM. No cells enter anaphase during an extended 48-h incubation with 10 nM taxol. Mitotic block occurs in the absence of any increase in the amount of microtubule polymer at taxol concentrations < 10 nM. As the taxol concentration is raised above 10 nM, the mass of microtubules increases, attaining 50% of the maximal achievable level with 80 nM taxol and a maximal level of microtubule polymer that is 5 times the normal level, at 330 nM taxol (5).

As shown in Figure 6, spindles blocked in mitosis by low concentrations of taxol are often slightly abnormal in organization and strongly resemble spindles blocked by low concentrations of other antimitotic drugs including vinblastine, vincristine, nocodazole, colchicine, and podophyllotoxin *(5,33,40,61)*. For example, at 10 nM taxol, 28% of the spindles appear normal in that they are bipolar with a compact equatorial metaphase plate of condensed chromosomes; however, they are abnormal in that they contain some chromosomes near the spindle poles. These chromosomes apparently are unable to congress normally to the metaphase plate, a process which has been shown to require dynamic microtubules *(58,62)*. In addition, these spindles are, on the average, 46% shorter than control spindles (pole to pole) and their astral microtubules are more prominent than in control spindles. In the absence of any detectable change in the equilibrium between soluble tubulin dimers and microtubule polymers in these cells, it appears as though there is a shift in favor of astral microtubules at the expense of microtubules in the central spindle suggesting that the balance of microtubule dynamics in spindles is altered. Many cells exhibit a more extreme spindle reorganization at 10 nM taxol; 64% of the spindles are severely aberrant; they are essentially monopolar or multipolar asters of microtubules surrounded by disorganized ball-shaped aggregations of condensed chromosomes resembling classical colchicine-blocked mitoses.

It is difficult at this time to dissect the role of microtubule dynamics in metaphase spindle function mechanistically. However, it is clear that microtubules undergo distinct changes in their dynamics during mitosis. In addition to undergoing dynamic instability behavior in *prophase* during chromosome congression *(58,62)*, Hamaguchi *(23)* and Mitchison *(27)* have shown that microtubules of the central spindle in *metaphase* undergo treadmilling. When *anaphase* begins, the dynamics of spindle microtubules change once again as the microtubules attached to the chromosomes shorten while another subset of spindle microtubules, the interpolar microtubules, simultaneously lengthens. The evidence we have obtained indicates that five chemically diverse classes of drugs (taxol, vinca alkaloids, nocodazole, podophyllotoxin, and colchicine) that bind to at least three distinct regions on the tubulin molecule (reviewed by Correia, *63*) *all* suppress microtubule dynamics *(5,21,57,61,64-66)* and *all* induce mitotic block with a common rearrangement of spindle microtubules *(5,40,61)* suggests 1) that subtle regulation of spindle microtubule dynamics is important for the organization of the mitotic spindle and 2) that dynamic microtubules are crucial for the cell cycle transition from metaphase to anaphase.

Mitotic block and inhibition of cell proliferation by taxol at low concentrations do not appear to involve induction of massive microtubule bundles. Some loosely packed parallel arrays of microtubules occasionally occur at 33 nM taxol. However, microtubule bundles become prominent only at taxol concentrations of 100 nM and above, at which as much as a five-fold increase in microtubule polymer is observed as compared with controls (5). Since one-third of the total tubulin in control HeLa cells is in the form of microtubule polymer *(33)*,

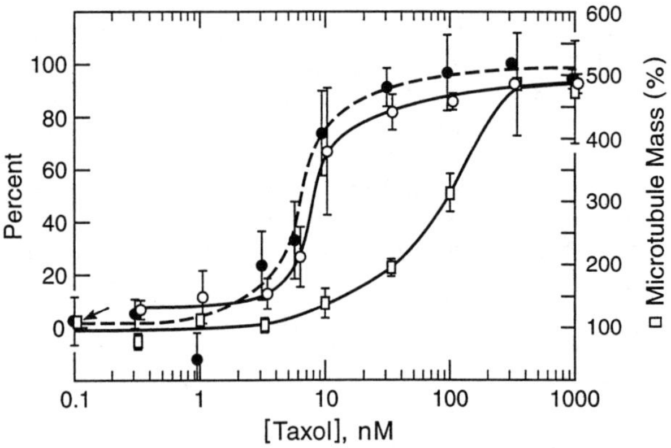

Fig. 5. Effect of taxol on cell proliferation (solid circles, left axis), the percentage of cells in mitotic metaphase (open circles, left axis; arrow denotes control value), and the mass of cellular microtubules (open squares, right axis) in HeLa cells following incubation with taxol for one cell cycle (20 h). Proliferation was measured by counting cells using a hemocytometer, metaphase arrest was assayed by immunofluorescence microscopy of fixed cells, and microtubule polymer mass was measured by an enzyme-linked immunoadsorbent assay for tubulin in stabilized cytoskeletons. Reproduced from (5) with permission of the publisher.

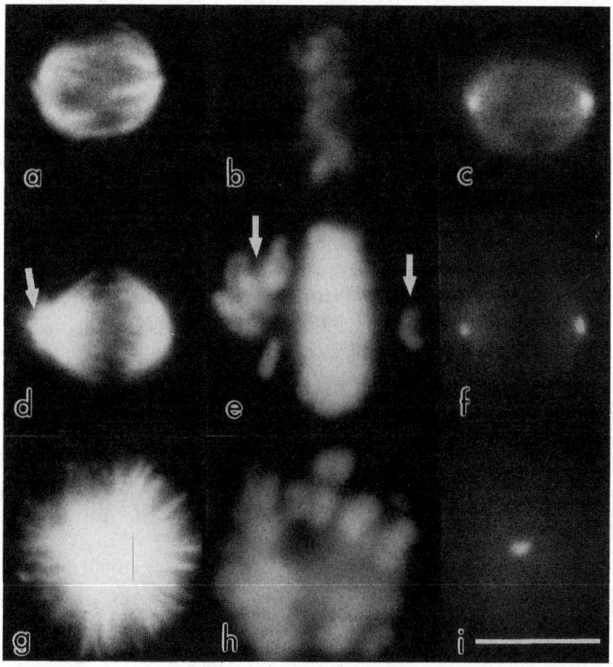

Fig. 6. Effect of taxol on microtubules, chromosomes, and centrosomes of HeLa cells. Cells were incubated with taxol for one cell cycle (18-20 h). Each row shows different components of a single cell. a,d,g - spindle microtubules; b,e,h - chromosomes of the same spindles; c,f,i - centrosomes (poles) of the same spindles. a,b,c - spindle of a control cell, d,e,f - abnormal bipolar spindle from a cell incubated with 6 nM taxol (arrows in d indicate a prominent cluster of astral microtubules that is characteristic of blocked spindles; arrows in e indicate chromosomes that have failed to congress to the brightly-stained compact equatorial metaphase plate of chromosomes); g,h,i - monopolar spindle from a cell incubated with 1 μM taxol. Bar = 10 μm. Reproduced from (5) with permission of the publisher.

it is clear that a large excess of tubulin must have been synthesized in HeLa cells in response to the 20-h incubation with taxol concentrations of 100 nM and above. Thus the massive microtubule bundles that form in cells at high taxol concentrations appear to result in part from a reduction in the critical subunit concentration for assembly and in part from an upregulation of tubulin synthesis *(67,68)*. While the microtubule bundles that occur at high concentrations of taxol undoubtedly contribute to the antiproliferative action of taxol in HeLa cells, formation of microtubule bundles appears to be a secondary effect that is not essential to the antiproliferative mechanism.

Induction of Apoptotic Cell Death as a Result of Mitotic Block by Low Concentrations of Taxol

To determine whether suppression of microtubule dynamics and mitotic block at low concentrations of taxol is related to cell killing and the therapeutic antitumor mechanism of the drug, we followed the fate of HeLa cells that were blocked in mitosis by incubation with 10 nM taxol for 20 h (one cell cycle). Following the initial 20-h incubation, we washed the cells three times with culture medium free of taxol over a period of two h and then continued incubation of the washed cells in taxol-free medium. Over the next 3 days we found that the cells were unable to resume proliferation and in fact, 50% of them died, undergoing morphological changes and DNA degradation characteristic of programmed cell death, or apoptosis (M. A. Jordan, W. B. Derry, K. Wendell, and L. Wilson, unpublished results). Similar results have been obtained following mitotic block with low concentrations of vinblastine (M. A. Jordan, E. Tsuchiya, and L. Wilson, unpublished results). The results suggest that mitotic block at the metaphase/anaphase transition for a sufficient period of time induces a genetic program whereby the cell commits suicide.

Although the molecular mechanisms have not been elucidated, it is noteworthy that programmed cell death or apoptosis is hypothesized to result from conflicting growth regulatory signals which ultimately lead to an unsuccessful attempt to traverse the cell cycle *(69-71)*. We suggest that extended mitotic block by antimitotic drugs such as taxol may lead to just such a significant perturbation of cell cycle regulation resulting in cell death.

Conclusions

The most potent antiproliferative mechanism of action of taxol in HeLa cells appears to be a cell cycle block at the metaphase/anaphase transition of mitosis. The mitotic block appears to result from kinetic stabilization of spindle-microtubule dynamics and it occurs without increased polymerization of microtubules. Five chemically diverse classes of drugs (taxol, vinca alkaloids, nocodazole, podophyllotoxin, and colchicine) *all* induce mitotic block in HeLa cells with a similar abnormal rearrangement of spindle microtubules, not by increasing or decreasing the spindle microtubule polymer mass, but apparently by suppressing spindle microtubule dynamics. The results suggest that spindle-microtubule dynamics are critical for the organization and function of the mitotic spindle and in particular for the transition from metaphase to anaphase. The mitotic block in HeLa cells with as little as 10 nM taxol induces extensive apoptotic cell death within a few days following the block. The results suggest that the critical mechanism underlying successful chemotherapeutic use of antimitotic agents in humans may be a prolonged kinetic stabilization of spindle microtubules at the metaphase/anaphase transition.

Acknowledgments

We thank W. B. Derry for critically reading the manuscript. This work was supported by Grant CA57291-02 from the National Cancer Institute.

References

1. Fuchs, D. A.; Johnson, R. K. *Cancer Treatment Reports* **1978**, *62*, 1219-1222.
2. Schiff, P. B.; Horwitz, S. B. *Proc. Natl. Acad. Sci. USA* **1980**, *77*, 1561-1565.
3. Rowinsky, E. K.; Donehower, R. C.; Jones, R. J.; Tucker, R. W. *Cancer Research* **1988**, *49*, 4093-4100.
4. Roberts, J. R.; Rowinsky, E. K.; Donehower, R. C.; Robertson, J.; Allison, D. C. *J. Histochem. Cytochem.* **1989**, *37*, 1659-1665.
5. Jordan, M. A.; Toso, R. J.; Thrower, D.; Wilson, L. *Proceedings of the National Academy of Science, USA* **1993**, *90*, 9552-9556.
6. De Brabander, M.; Geuens, G.; Nuydens, R.; Willebrords, R.; De Mey, J. *Cold Spring Harbor Symposia on Quantitative Biology* **1982**, *XLVI*, 227-240.
7. Turner, P. F.; Margolis, R. L. *The Journal of Cell Biology* **1984**, *99*, 940-946.
8. Gelfand, V.; Bershadsky, A. D. *Annu. Rev. Cell Biol.* **1991**, *7*, 93-116.
9. Mitchison, T. J.; Kirschner, M. *Nature* **1984a**, *312*, 232-237.
10. Mitchison, T. J.; Kirschner, M. *Nature* **1984b**, *312*, 237-242.
11. O'Brien, E. T.; Voter, W. A.; Erickson, H. P. *Biochemistry* **1987**, *26*, 4148-4156.
12. Carlier, M.-F. *Int. Rev. Cytol.* **1989**, *115*, 139-170.
13. Stewart, R. J.; Farrell, K. W.; Wilson, L. *Biochemistry* **1990**, *29*, 6489-6498.
14. Melki, R.; Carlier, M.-F.; Pantaloni, D. *Biochemistry* **1990**, *28*, 8921-8932.
15. Margolis, R. L.; Wilson, L.; Kiefer, B. *Nature* **1978**, *272*, 450-452.
16. Margolis, R. L.; Wilson, L. *Nature (London)* **1981**, *293*, 705-711.
17. Farrell, K. W.; Jordan, M. A.; Miller, H. P.; Wilson, L. *J. Cell Biol.* **1987**,*184*, 1035-1046.
18. Walker, R. A.; O'Brien, E. T.; Pryer, N. K.; Soboeiro, M. F.; Voter, W. A.; Erickson, H.; Salmon, E. D. *J. Cell Biol.* **1988**, *107*, 1437-1448.
19. Jordan, M. A.; Farrell, K. W. *Anal Biochem* **1983**, *130*, 41-53.
20. Wilson, L.; Farrell, K. W. *Ann. NY Acad. Sci.* **1986**, *466*, 690-708.
21. Jordan, M. A.; Wilson, L. *Biochemistry* **1990**, *29*, 2730-2739.
22. Mitchison, T. J.; Evans, L.; Schulze, E.; Kirschner, M. *Cell* **1986**, *45*, 515-527.
23. Hamaguchi, Y.; Toriyama, M.; Sakai, H.; Hiramoto, Y. *Cell Struct. Funct.* **1987**, *12*, 43-52.
24. Sammak, P. J.; Borisy, G. G. *Cell Motility and the Cytoskeleton* **1988**, *10*, 237-245.
25. Cassimeris, L. S. I.; Salmon, E. D. *Cell Motility and the Cytoskeleton* **1988**, *10*, 185-196.
26. Cassimeris, L.; Pryer, N. K.; Salmon, E. D. *Journal of Cell Biology* **1988**, *107*, 2223-2231.
27. Mitchison, T. J. *J. Cell Biol.* **1989**, *109*, 637-652.

28. Horio, T.; Hotani, H. *Nature* **1986**, *321*, 605-607.
29. Sawin, K. E.; Mitchison, T. J. *Mol. Biol. of the Cell* **1994**, *5*, 217-226.
30. Saxton, W. M.; Stemple, D. L.; Leslie, R. J.; Salmon, E. D.; Zavortnik, M.; McIntosh, J. R. *Journal of Cell biology* **1984**, *99*, 2175-2186.
31. George, P.; Journey, L. J.; Goldstein, M. N. *J. Natl. Cancer Institute* **1965**, *35*, 355-375.
32. Jordan, M. A.; Himes, R. H.; Wilson, L. *Cancer Research* **1985**, *45*, 2741-2747.
33. Jordan, M. A.; Thrower, D.; Wilson, L. *Cancer Res* **1991**, *51*, 2212-2222.
34. Parness, J.; Horwitz, S. B. *The Journal of Cell Biology* **1981**, *91*, 479-487.
35. Schiff, P. B.; Fant, J.; Horwitz, S. B. *Nature* **1979**, *277*, 665-667.
36. Wilson, L.; Jordan, M. A.; Morse, A.; Margolis, R. L. *J. Mol. Biol.* **1982**, *159*, 129-149.
37. Jordan, M. A.; Margolis, R. L.; Himes, R. H.; Wilson, L. *J. Mol. Biol.* **1986**, *187*, 61-73.
38. Singer, W. D.; Jordan, M. A.; Wilson, L.; Himes, R. H. *Mol Pharmacol* **1989**, *36*, 366-70.
39. Wilson, L.; Morse, A.; Jordan, M. A.; Margolis, R. L. In *Biological Functions of Microtubules and Related Structures;* H. Sakai, M. Kurokawa and G. Borisy, Eds.; Academic Press: Japan, 1982; pp 61-72.
40. Jordan, M. A.; Thrower, D.; Wilson, L. *J Cell Sci* **1992**, 401-416.
41. Wendell, K. L.; Wilson, L.; Jordan, M. A. *J Cell Sci* **1993**, 261-274.
42. Takoudju, M.; Wright, M.; Chenu, J.; Gueritte-Voegelein, F.; Guenard, D. *FEBS Lett.* **1988**, *227*, 96-98.
43. Diaz, J. F.; Menendez, M.; Andreu, J. M. *Biochemistry* **1993**, *32*, 10067-10077.
44. Diaz, J. F.; Andreu, J. M. *Biochemistry* **1993**, *32*, 2747-2755.
45. Manfredi, J. J.; Parness, J.; Horwitz, S. *B. The Journal of Cell Biology* **1982**, *94*, 688-696.
46. Rao, S.; Krauss, N. E.; Heerding, J. M.; Swindell, C. S.; Ringel, I.; Orr, G. A.; Horwitz, S. B. *J. Biol. Chem.* **1994**, *269*, 3132-3134.
47. Combeau, C.; Commercon, A.; Mioskowski, C.; Rousseau, B.; Aubert, F.; Goeldner, M. *Biochemistry* **1994**, *33*, 6676-6683.
48. Kumar, N. *The Journal of Biological Chemistry* **1981**, *256*, 10435-10441.
49. Howard, W. D.; Timasheff, S. *N. The Journal of Biological Chemistry* **1988**, *263*, 1342-1346.
50. Hamel, E. In *Microtubule Proteins;* J. Avila, Ed.; CRC Press: Boca Raton, 1990; pp 89-191.
51. Thompson, W. C.; Wilson, L.; Purich, D. L. *Cell Motility* **1981**, *1*, 445-454.
52. Schiff, P. B.; Horwitz, S. B. *Biochemistry* **1981**, *20*, 3247-3252.
53. Hamel, E.; del Campos, A. A.; Lowe, M. C.; Lin, C. M. *The Journal of Biological Chemistry* **1981**, *256*, 11887-11894.
54. Ringel, I.; Horwitz, S. B. *J. Pharmacol. Exp. Ther.* **1991a**, *259*, 855-860.
55. Collins, C. A.; Vallee, R. B. *J. Cell Biol.* **1987**, *105*, 2847-2854.
56. Caplow, M.; Zeeberg, B. *Eur. J. Biochem.* **1982**, *127*, 319-324.
57. Wilson, L.; Miller, H. P.; Farrell, K. W.; Snyder, K. B.; Thompson, W. C.; Purich, D. L. *Biochemistry* **1985a**, *24*, 5254-5262.
58. Hayden, J. J.; Bowser, S. S.; Rieder, C. *J. Cell Biol.* **1990**, *111*, 1039-1045.
59. Thrower, D.; Jordan, M. A.; Wilson, L. *J Immunol Methods* **1991**, *136*, 45-51.
60. Thrower, D.; Jordan, M. A.; Wilson, L. In *Methods in Cell Biology;* D. J. Asai, Ed.; Academic Press: San Diego, 1993; Vol. 37; pp 129-145.

61. Wilson, L.; Toso, R. J.; Jordan, M. A. *J Cellular Pharmacology* **1993**, *1 (Suppl. 1), 35-40.*
62. Rieder, C. L.; Alexander, S. P. *Electron Microsc. Rev.* **1990**, *3*, 269-300.
63. Correia, J. J. *Pharmacol. Therapeut.* **1991**, *52*, 127-147.
64. Schilstra, M. J.; Martin, S. R.; Bayley, P. M. *J. Biol. Chem.* **1989**, *264*, 8827-8834.
65. Toso, R. J.; Jordan, M. A.; Farrell, K. W.; Matsumoto, B.; Wilson, L. *Biochemistry* **1993**, *32*, 1285-1293.
66. Vandecandelaere, A.; Martin, S. R.; Schilstra, M. J.; Bayley, P. M. *Biochemistry* **1994**, *33*, 2792-2801.
67. Ben-Ze'ev, A.; Farmer, S. R.; Penman, S. *Cell* **1979**,*17*, 319-325.
68. Cleveland, D. W.; Lopata, M. A.; Sherline, P.; Kirschner, M. W. *Cell* **1981**, *25*, 537-546.
69. Ucker, D. S. *New Biol.* **1991**, *3*, 103-109.
70. Colombel, M.; Olsson, C. A.; Ng, P. Y.; Buttyan, R. *Cancer Research* **1992**, *52*, 4313-4319.
71. Lee, S.; Christakos, S.; Small, M. B. *Curr. Opin. Cell Biol.* **1993**, *5*, 286-291.

RECEIVED September 21, 1994

Chapter 11

Photoincorporation of a Paclitaxel Photoaffinity Analogue into the N-Terminal 31 Amino Acids of β-Tubulin

Susan Band Horwitz[1], Srinivasa Rao[1], Nancy E. Krauss[2], Julie M. Heerding[2], Charles S. Swindell[2], Israel Ringel[3], and George A. Orr[1]

[1]Department of Molecular Pharmacology, Albert Einstein College of Medicine, Bronx, NY 10461
[2]Department of Chemistry, Bryn Mawr College, Bryn Mawr, PA 19010
[3]Department of Pharmacology, The Hebrew University, Jerusalem 91120, Israel

3′-(p-azidobenzamido)taxol, a photoaffinity analogue of taxol, covalently binds to the N-terminal domain of β-tubulin after irradiation of the microtubule-drug complex. The azido analogue has biological properties that are similar to those of taxol, such as the ability to polymerize tubulin into stable microtubules in the absence of GTP. Taxol competes with [^{3}H]3′-(p-azidobenzamido)taxol binding, indicating that both compounds interact with the same site on the microtubule. Cleavage of [^{3}H]3′-(p-azidobenzamido)taxol-photolabeled β-tubulin by formic acid and subsequent mass analysis and protein sequencing have identified the N-terminal 31 amino acids as the major site for [^{3}H]3′-(p-azidobenzamido)taxol photoincorporation.

Taxol is a compound of considerable interest due to its clinical activity as an antitumor drug in humans, its unusual chemical structure, and its unique mechanism of action. The drug was isolated originally from the bark of the tree Taxus brevifolia and the elucidation of its structure was accomplished in 1971 (*1*). Taxol displayed cytotoxicity in a number of rodent tumor model systems and in tumor cells growing in tissue culture. Studies reported in the late seventies and early eighties clearly indicated that taxol blocked cells in the late G_2/M phase of the cell cycle, thereby acting as an antimitotic agent (*2*). Taxol is unusual in its ability to enhance the assembly of microtubules in vitro, even in the absence of GTP that is normally required to achieve assembly. The drug also stabilizes microtubules against depolymerization by cold and Ca^{++}, treatments that depolymerize normal microtubules to their α-and β- tubulin dimers (*3,4*). Taxol binds to the microtubule with a stoichiometry, relative to the constituent tubulin heterodimers, approaching

NOTE: Paclitaxel is the generic name for Taxol, which is now a registered trademark.

0097–6156/95/0583–0154$08.00/0

1 (*5,6*). When cells are treated with taxol, the microtubule cytoskeleton is reorganized and stable bundles of microtubules that are diagnostic of taxol treatment are present in the cell (*2*).

Although it has been known for the past fifteen years that taxol interacts with the tubulin/microtubule system, there has been no definitive information available on the taxol binding site on the microtubule. An excellent method for deciphering the binding site is photoaffinity labeling. In the absence of effective radiolabeled-photoaffinity analogues of taxol, [^{3}H]taxol was used directly to photolabel tubulin. When a complex of microtubule protein and [^{3}H]taxol was irradiated at 254nm and analyzed by denaturing gel electrophoresis, it was observed that the drug bound covalently to the β-subunit of tubulin (*7*). However, the extent of photoincorporation was too low to pursue these studies further.

The ability to prepare 3′-(p-azidobenzamido)taxol (Figure 1) and to radiolabel it with tritium to approximately 2.0 Ci/mmol has allowed the further delineation of the taxol binding site (*8*). A comparison of the effects of GTP, taxol and the azido analogue on microtubule assembly in vitro is seen in Figure 2. Both taxol and the azido analogue induced the assembly of Ca^{++} stable microtubules in the absence of GTP, a unique characteristic of taxol activity. In a series of related experiments, it was observed that the binding of ^{3}H-taxol to microtubules was specifically inhibited by both unlabeled taxol and the azido analogue, indicating that both are occupying the same binding site (Figure 3). Although the azido analogue was slightly less active than taxol in vitro, its effects in cells were similar to those of taxol (*9*). Like taxol, it induced the formation of normal microtubules plus hoops and ribbons as determined by electron microscopy, and immunofluorescence studies revealed the presence of stable microtubule bundles in Chinese hamster ovary cells (CHO). In terms of cytotoxicity, the azido analogue was 3-fold less active than taxol when tested in the murine cell line, J774.2 (*9*).

β-Tubulin was specifically labeled with ^{3}H-(p-azidobenzamido)taxol (Figure 4), and after 30 min of irradiation the efficiency of photoincorporation of the azido analogue ranged from 2 to 6%. A ten-fold excess of taxol or unlabeled 3′-(p-azidobenzamido)taxol reduced photoincorporation. In contrast, baccatin III, that is known not to have taxol-like activity (*10*), had no effect on photoincorporation. Such experiments indicated that 3′-(p-azidobenzamido)taxol specifically labeled β-tubulin under the conditions of our experiments.

It is known that formic acid cleaves Asp-Pro bonds (*11-13*) and since there are two such bonds in β-tubulin, as indicated below, formic acid digestion was carried out with purified β-tubulin to determine into which fragment the azido analogue had been incorporated (*8*). Since the molecular weight of each fragment differs, they can be readily resolved on denaturing gels.

NH2 A1 ↓(31 32) A2 ↓(304 305) A3 COOH

A time course of formic acid treatment demonstrated that after complete digestion of β-tubulin all of the radiolabel was associated with the lowest molecular weight fragment. High performance electrophoresis chromatography (HPEC) was used to analyze further the photoaffinity labeled β-tubulin and its formic acid digestion products (Figure 5). As can be seen, the radiolabeled β-tubulin migrated as a

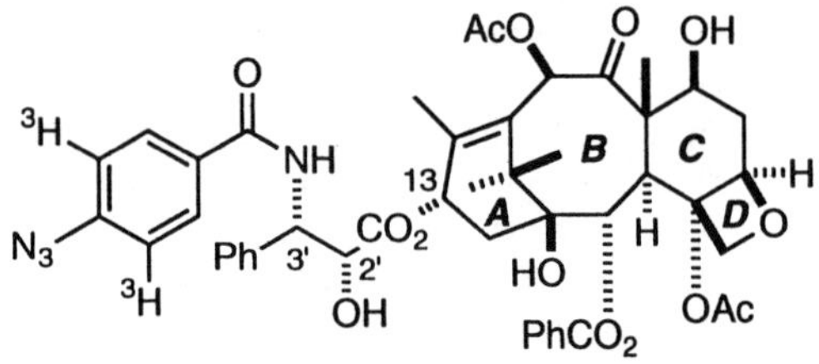

^{3}H-3'-(*p*-Azidobenzamido)taxol

Figure 1. Molecular structure of [^{3}H]3′-(p-azidobenzamido)taxol.

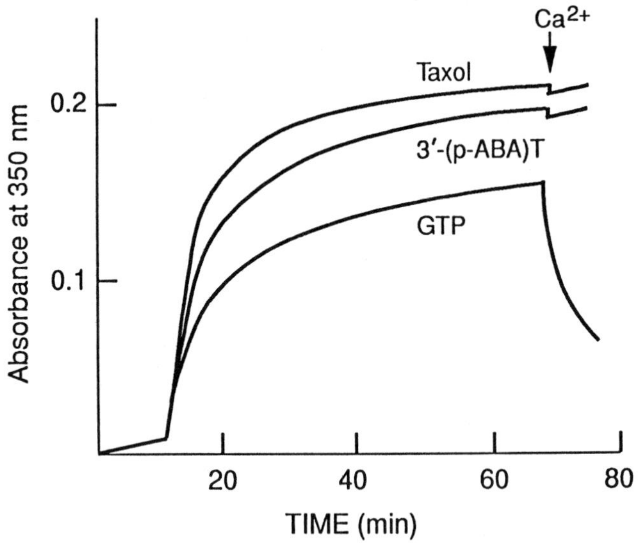

Figure 2. Assembly of microtubule protein (MTP) in the presence of GTP. MTP (1.5 mg/ml) was incubated at 35°C with either 1mM GTP or 20μM taxol or 20μM 3′-(p-azidobenzamido)taxol [3′-(P-ABA)T]. The assembly reaction was followed turbidimetrically. At the time designated, 4mM $CaCl_2$ was added to each sample. (Reproduced with permission from ref. #9).

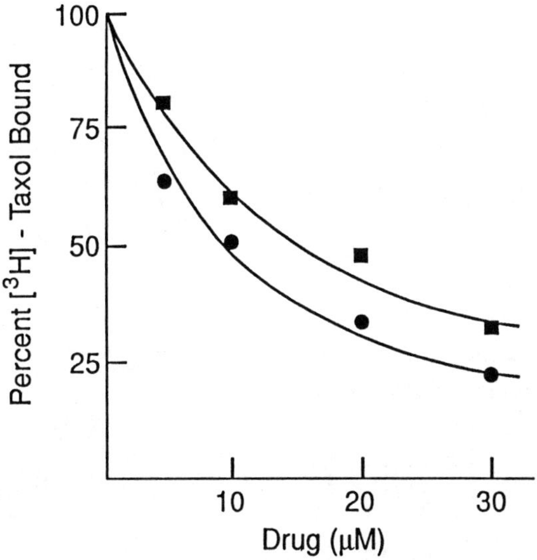

Figure 3. Competition between taxol and 3′-(p-azidobenzamido)taxol for the binding site(s) on microtubules. MTP (1.5 mg/ml) was assembled in the presence of 10μM ^{3}H-taxol and various concentrations of unlabeled taxol (●), or 3′-(p-azidobenzamido)taxol (■). ^{3}H-taxol and its competitor were added simultaneously. The microtubule polymer was centrifuged through a 50% sucrose layer and the pellet was analyzed for ^{3}H-taxol. Binding obtained in the presence of ^{3}H-taxol alone represents the 100% value (counts/min/μg protein). (Reproduced with permission from ref. #9)

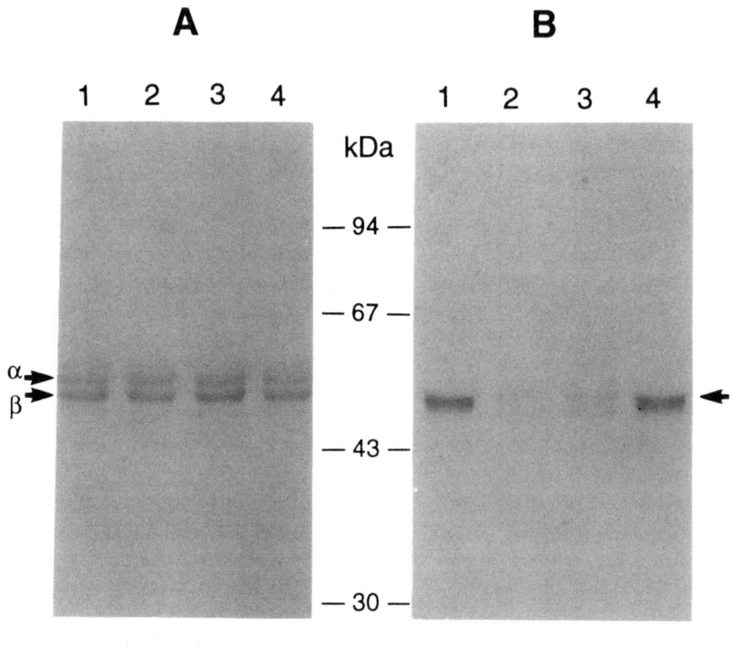

Figure 4. [^{3}H]3′-(p-Azidobenzamido)taxol specifically photoincorporates into β-tubulin. [^{3}H]3′-(p-Azidobenzamido)taxol (5μM, 1.7 Ci/mmol) was added to MTP (5 μM tubulin) and incubated at 37°C for 30 min. After the incubation, the sample was irradiated at 254nm for 30 min and analyzed by SDS-PAGE and by fluorography. A, SDS-PAGE followed by Coomassie staining. B, fluorography of A. *Lane* 1, no additions; *lane* 2, addition of 50 μM taxol; *lane* 3, 50 μM unlabeled 3′-(p-azidobenzamido)taxol; *lane* 4, 50 μM baccatin III. The gel was exposed for 10 days. (Reproduced with permission from ref. #8)

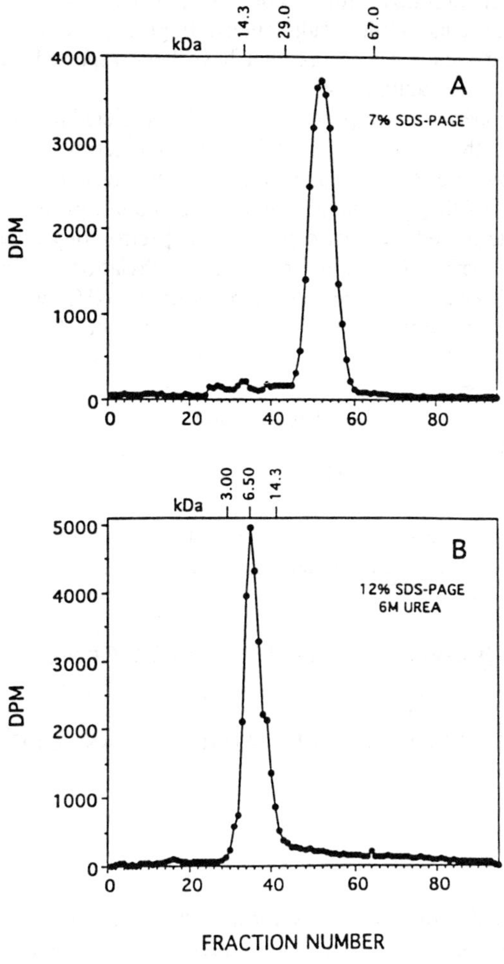

Figure 5. High performance electrophoresis chromatographic analysis of [^{3}H]3′-(p-azidobenzamido)taxol-photolabeled β-tubulin (A) and its formic acid digestion products (B). Intact β-tubulin was resolved on a 7% polyacrylamide gel (A) and digested β-tubulin on a 12% polyacrylamide gel containing 6 M urea (B). In each case, 75,000 dpm in a total volume of 25 μl was loaded. Electrophoresis was conducted either at 0.3 mA for 30 min followed by 1.5 mA for 5 h (A) or at 0.4 mA for 30 min followed by 1.3 mA for 5 h (B). Fractions were counted by liquid scintillation spectrometry. Insulin (3 kDa, A/B chains), aprotinin (6.5 kDa), lysozyme (14.3 kDa), carbonic anhydrase (29 kDa), and bovine serum albumin (67 kDa) were used as markers. (Reproduced with permission from ref. #8)

single peak of the appropriate size. The major radiolabeled peptide after a 120 hr formic acid digest migrated with an apparent molecular weight of 6500 daltons. Amino acid sequencing of this radiolabeled fragment revealed that the first 20 amino acids were identical to those that have been reported for the N-terminal sequence of human β_2-tubulin (*14*).

Due to the apparent discrepancy between the predicted molecular weight of the A_1 fragment and the apparent molecular weight determined by HPEC, it was necessary to prove rigorously that formic acid was cleaving β-tubulin at Asp^{31}-Pro^{32}. The mass of the N-terminal peptide of β-tubulin generated by formic acid cleavage was determined by electrospray mass spectroscopy after modification of Cys^{12} by either pyridylethylation or carboxymethylation. The calculated and experimentally determined monoisotopic mass of the NH_2-terminal domain of β-tubulin corresponded very closely, thereby proving that formic acid cleaves at Asp^{31}-Pro^{32} in β-tubulin.

Modification	Experimental Mass	Theoretical Mass
pyridylethylation	3584.97	3584.73
carboxymethylation	3537.85	3537.68

The sequence of the N-terminal 32 amino acids of β-tubulin, with an arrow indicating the site of formic acid cleavage, is shown below.

10
MET-ARG-GLU-ILE-VAL-HIS-ILE-GLN-ALA-GLY-GLN-CYS-GLY-ASN-GLN-

20 30
ILE-GLY-ALA-LYS-PHE-TRP-GLU-VAL-ILE-SER-ASP-GLU-HIS-GLY-ILE-

ASP-PRO
↑

The N-terminal region of β-tubulin isoforms is highly conserved throughout nature (*15*). Since taxol is known to interact with microtubules from a variety of cell lines and sources, it is reasonable that its binding site be in a conserved domain of β-tubulin. It is interesting to note that recent experiments have implicated Cys^{12} in β-tubulin as being part of the binding site for the exchangeable GTP that is hydrolyzed during microtubule polymerization (*16*). The only compound that can substitute for GTP in the in vitro polymerization of tubulin is taxol, a drug that is now known to bind to β-tubulin in the same region as the exchangeable GTP. A recent study that examined the binding site in tubulin for the antimitotic agent colchicine, indicated that there are two distinct domains of β-tubulin that form part of the binding site for colchicine (*17*). One of these regions maps to the N-terminal region of β-tubulin. Although both taxol and colchicine are antimitotic agents, colchicine, in contrast to taxol, interacts with the tubulin dimer and inhibits the polymerization of tubulin in vitro. All of these studies implicate the N-terminal domain of β-tubulin as an important regulatory site for microtubule assembly. The interaction of taxol or colchicine with this domain of β-tubulin could result in distinct conformational changes that would regulate tubulin polymerization in different ways.

To understand the interaction of taxol with its cellular target, the microtubule, which is crucial to comprehending the basis of the antitumor activity of this drug, it is necessary to determine the binding site for the drug. This information, in addition to structure-activity data for taxol, will provide knowledge that could lead to the design and synthesis of new and improved taxol analogues.

Literature Cited

1. Wani, M.C.; Taylor, H.L.; Wall, M.E.; Coggon, P.; McPhail, A.T. *J. Am. Chem. Soc.* **1971**, *93*, 2325-2327.
2. Schiff, P.B.; Horwitz, S.B. *Proc. Natl. Acad. Sci. U.S.A.* **1980**, *77*, 1561-1565.
3. Schiff, P.B.; Fant, J.; Horwitz, S.B. *Nature* **1979**, *277*, 665-667.
4. Schiff, P.B.; Horwitz, S.B. *Biochemistry*, **1981**, *20*, 3247-3252.
5. Parness, J.; Horwitz, S.B. *J. Cell Biol.*, **1981**, *91*, 479-487.
6. Diaz, J.F.; Andreu, J.M. *Biochemistry*, **1993**, *32*, 2747-2755.
7. Rao, S.; Horwitz, S.B.; Ringel, I. *J. Natl. Cancer Inst.*, **1992**, *84*, 785-788.
8. Rao, S; Krauss, N.E.; Heerding, J.M.; Swindell, C.S.; Ringel, I.; Orr, G.A.; Horwitz, S.B. *J. Biol. Chem.*, **1994**, *269*, 3132-3134.
9. Swindell, C.S.; Heerding, J.M.; Krauss, N.E.; Horwitz, S.B.; Rao, S.; Ringel, I. *J. Med. Chem.*, **1994**, *38*, 1446-1449.
10. Parness, J.; Kingston, D.G.I.; Powell, R.G.;, Harracksingh, C.; Horwitz, S.B. *Biochem. Biophys. Res. Commun.*, **1982**, *105*, 1082-1089.
11. Sonderegger, P.; Jaussi, R.; Gehring, H.; Brunschweiler, K.; Christen, P. *Anal Biochem.*, **1982**, *122*, 298-301.
12. Hall, J.L.; Dudley, L.; Dobner, P.R.; Lewis, S.A.; Cowan, N.J. *Mol. Cell. Biol.*, **1983**, *3*, 854-862.
13. Lee, M.G.-S.; Loomis, C.; Cowan, N.J. *Nucleic Acids Res.*, **1984**, *12*, 5823-5836.
14. Lewis, S.A.; Gilmartin, M.E.; Hall, J.L.; Cowan, N.J. *J. Mol. Biol.*, **1985**, *182*, 11-20.
15. Kimmel, B.E.; Samson, S.; Wu, J.; Hirschberg, R.; Yarbrough, L.R. *Gene (Amst).*, **1985**, *35*, 237-248.
16. Shivanna, B.D.; Majillano, M.R.; Williams, T.D.; Himes, R.H. *J. Biol. Chem.*, **1993**, *268*, 127-132.
17. Uppuluri, S.; Knipling, L.; Sackett, D.L.; Wolff, J. *Proc. Natl. Acad. Sci. U.S.A.* **1993**, *90*, 11598-11602.

RECEIVED September 20, 1994

Chapter 12

Paclitaxel, a Mimetic of Bacterial Lipopolysaccharide (LPS) in Murine Macrophages

S. N. Vogel[1], J. M. Carboni[2], and C. L. Manthey[1]

[1]**Department of Microbiology and Immunology, Uniformed Services University of the Health Sciences, Bethesda, MD 20814**
[2]**Molecular Drug Mechanisms, Bristol Myers-Squibb, Princeton, NJ 08543**

The mechanism by which paclitaxel (henceforth referred to as "taxol") induces its anti-tumor effects has been largely attributed to its ability to stabilize microtubules during mitosis, thus blocking cell division. In addition to this well characterized mode of action, taxol has been found in murine macrophages to elicit a series of intracellular signals that are indistinguishable from those elicited by endotoxin, the lipopolysaccharide (LPS) component of Gram negative bacteria. This chapter will review the current status of this relationship and its potential therapeutic and pharmacologic implications.

Approximately 100 years ago, William Coley observed that a patient who had contracted a bacterial infection responded with high fever and regression of his solid tumor. Although Coley was not the first to observe this phenomenon, he reasoned that the infection had been the cause of both the fever and the tumor regression and created a cocktail of killed bacterial filtrates which he administered to cancer patients, in the hope of causing their tumors to regress. Although Coley enjoyed a significant degree of success, the side effects of his extracts were too toxic for consideration as a cancer therapy. Shear and his colleagues analyzed Coley's extracts and in 1936 isolated the active component, which they called "bacterial polysaccharide." This substance, now known as lipopolysaccharide (LPS), was very effective at inducing tumor necrosis; however, as with Coley's original extracts, its toxicity in humans precluded its general use as a chemotherapeutic agent. Interest in LPS-induced tumor regression was rekindled by the observation that serum from mice injected with LPS contained a

NOTE: Paclitaxel is the generic name for Taxol, which is now a registered trademark.

0097–6156/95/0583–0162$08.00/0

soluble factor that induced tumor regression in tumor-bearing recipients. In contrast to LPS, the active factor in serum was heat labile, toxic for several transformed cell lines *in vitro*, and was subsequently shown to be derived from LPS-treated macrophages. It was appropriately called, "Tumor Necrosis Factor" (TNF) and since that time has been purified, cloned, and expressed in a number of different species (reviewed in *1-3*). In addition to tumor regression, TNFα induces a spectrum of biologic activities that are both beneficial (i.e., TNFα is also essential for the elimination of certain infectious agents) and harmful (i.e., TNFα is a major mediator in systemic inflammatory response syndrome associated with sepsis) to the host (reviewed in *3, 4*).

It is now clear that TNFα is only one of many gene products that are induced in macrophages stimulated by LPS (reviewed in *5, 6*). The exact nature and sequence of biochemical signals transmitted to the nucleus following LPS interaction with cell surface receptor(s) is an area of intense investigation. Early signaling events include the translocation of NFκB to the nucleus, as well as activation of serine/threonine and tyrosine kinase cascades. Specifically, LPS activates in human macrophages several src-family kinases (e.g., $p56^{lyn}$, $p58^{hck}$, and $p59^{c\text{-}fgr}$; *7*) and induces in both murine and human macrophages the phosphorylation on tyrosine of three proteins (~40-45 kDa species). These include two mitogen activated protein (MAP) kinase isoforms, p44 erk1 and p42 erk2, and a third yet to be identified tyrosine phosphoprotein (*8*). In mice, sensitivity to LPS is controlled by a single autosomal gene, *Lps*, that maps to Chromosome 4 near the brown coat color locus (*9*). Several mouse strains have been described that have a mutated allelic form of this gene. Not only do these mice fail to respond to LPS *in vivo*, but also their macrophages fail to exhibit any of the signaling events or inducible gene products that are stimulated by LPS in normal (Lps^{n}) macrophages (rev. in *5*).

Discovery that Taxol Induces TNFα in Macrophages

Ding and her colleagues made the initial observation that taxol, like LPS, induced TNFα secretion in murine macrophages (*10-12*). This seminal observation developed out of early studies designed to determine if the anti-inflammatory action of colchicine might be due to its ability to down-regulate expression of receptors for TNFα on human macrophages and endothelial cells. These investigators found that colchicine and four other microtubule depolymerizing agents resulted in the rapid involution of TNFα receptors in the absence of TNFα release. They suggested that microtubule disassembly resulting in a loss of cytokine receptor expression might underlie the anti-inflammatory actions of these drugs. In previous studies, LPS had also been demonstrated to result in a very rapid loss of TNFα receptor expression that preceded detection of TNFα and they extended their findings by testing taxol, a microtubule stabilizing agent, for its ability to modulate both TNFα receptor expression and TNFα production in murine macrophages. The authors were struck by the capacity of taxol to mimic these two effects of LPS. Perhaps the single most important observation made by this group was that macrophage responsiveness to taxol was observed only in macrophages derived from mice that are genetically responsive to LPS. Specifically, macrophages derived from mice that carry the *Lps*^{n} allele are fully responsive to both taxol and LPS, whereas mice that are genetically defective for LPS responsiveness (i.e., they express the *Lps*^{d} allele), failed to respond to taxol or LPS to secrete TNFα or to down-regulate TNFα receptor expression. Using a series of recombinant inbred mouse strains, complete concordance between LPS and taxol sensitivity or lack thereof confirmed the genetic linkage of the two responses. This, in turn, led to the intruiging speculation that LPS signaling in macrophages was dependent upon a microtubule-associated protein.

Taxol interacts preferentially with β-tubulin in the context of polymerized microtubules and very recent studies have demonstrated that the N-terminal 31 amino acids of β-tubulin contains a high affinity binding site(s) (*13-15*). Ding and her colleagues sought to test the hypothesis that LPS physically interacts with microtubules or their associated proteins. LPS was found to bind to β-tubulin and to a high molecular weight microtubule-associated protein (MAP-2) present in the tubulin preparation, but bound poorly to α-tubulin (*16*). The preferential binding to β-tubulin is shared by taxol and several distinct MAPs and, thus, the authors hypothesized that signaling induced by LPS may be the result of LPS binding to the microtubule network, causing the displacement of a yet to be identified MAP.

Gene Induction and Tyrosine Phosphorylation Induced in Murine Macrophages by LPS and Taxol Are Indistinguishable.

Manthey *et al.* (*17*) extended these findings by demonstrating that a panel of six LPS-inducible genes (*viz.*, TNFα, IL-1β, TNFR-2, IP-10, D3, and D8) were induced by taxol and LPS in both primary macrophages and a macrophage cell line, RAW 264.7, derived from genetically responsive (*Lps*^{n}) mouse strains. Taxol was optimally active over a dose range of 10 - 35 μM, and induced gene expression with amplitude and kinetics that were indistinguishable from LPS-induced gene expression. In contrast, C3H/HeJ (*Lps*^{d}) macrophages failed to express any of these genes in response to taxol (*17*) or LPS (*18*), but responded normally to other non-LPS-containing macrophage stimuli (reviewed in *6*). The ability of taxol to induce tyrosine phosphorylation of proteins was also examined (*17*). In fully LPS-responsive murine macrophages, taxol induced the appearance of the same three newly phosphorylated proteins, erk1, erk2, and the additional ~41 kDa tyrosine phosphoprotein that are induced by LPS (*8*). Taxol failed to induce tyrosine phosphorylation in *Lps^d^* macrophages (*17*). The ability of taxol to induce protein tyrosine phosphorylation and cytokine gene expression in murine macrophages was confirmed in other laboratories (*19-21*). In contrast to the activation of this signaling pathway by taxol, other anti-tumor compounds such as colchicine, vinblastine, vincristine, adriamycin, and mitomycin C (10 μM each) failed to induce either TNFα secretion or tyrosine phosphorylation.

Since LPS is a ubiquitous bacterial contaminant, it was very important to insure that the observed actions of taxol were not due to contaminating LPS in the taxol preparations. Taxol stock solutions were tested in the *Limulus* amoebocyte lysate assay, a standard measure of LPS activity, and were found to be inactive (*12, 17*). Moreover, neither the LPS-binding antibiotic, polymyxin B (*17*), nor a recombinant LPS-binding peptide corresponding to the terminal 23 amino acids of bactericidal permeability-increasing protein, $rBPI_{23}$ (*6*), inhibited taxol-induced TNFα secretion, gene expression, or tyrosine phosphorylation under conditions where these LPS-inducible events were inhibited in a dose-dependent fashion.

Examination of Taxol Analogs as LPS Mimetics: Dissociation of LPS-Mimetic Activity from Microtubule Binding Activity

The recent success of TAXOL® in the clinical setting has prompted much interest in the development of taxol-like structures with improved therapeutic efficacy (reviewed in *22*). As a result, taxol, related taxoids, and partial structures (reviewed in *23, 24*) have been purified and some of these have been tested for their ability to induce TNFα secretion from murine macrophages. Table I provides a comparison of the TNFα-inducing activities of a variety of analogs and partial structures and provides a semi-quantitative measure of microtubule stabilizing activity, based on the concentration of each compound required to induce mitotic arrest in a proliferating murine macrophage cell line, RAW 264.7 (*25*). The data illustrate that there is no direct correlation between the effects of these compounds on microtubule binding (as measured indirectly by the effect on mitosis) and their ability to induce TNFα in murine macrophages. Perhaps the most extreme example is illustrated by comparing the activities of taxol *versus* taxotere (docetaxel). The structural differences between taxol and taxotere are subtle (*23*) and both taxol and taxotere are known to compete for binding to the same site on β-tubulin (*15*). However, taxotere was found to exhibit anti-mitotic activity at concentrations that were ~threefold less than taxol, yet failed to induce TNFα or gene expression, even at the highest concentration tested (100 μM). Another taxoid, 2-Debenzoyl-2(m-azidobenzoyl) taxol (*24*), was approximately one-tenth as active as taxol with respect to induction of TNFα, yet exhibited greater anti-mitotic activity than taxol. In addition to the compounds shown in Table I, baccatin III, the basic core

Table I. Comparison of Taxol and Related Structures for the Ability to Induce TNFα Secretion and Mitotic Arrest in Murine Macrophages

Compound	*TNFα-Inducing Index (TI)*[a]	*Mitotic Index (MI)*[b]	*TI/MI*
Taxol	3	0.1	30
2-Debenzoyl-2-(m-azidobenzoyl) taxol[c]	30	<0.1	>300
Cephalomannine	30	0.3	100
7-Epi-10-deacetyltaxol	33	1	33
Taxotere	>100	0.03	>3000

[a]TI represents the lowest concentration of compound (μM) that causes the secretion of TNFα in murine macrophage cultures.

[b]MI represents the concentration of compound (μM) that yielded a half maximal increase in the number of RAW 264.7 macrophages arrested in mitosis.

[c]This compound was kindly provided by the laboratory of Dr. David Kingston, Dept. of Chemistry, Virginia Tech, Blacksburg, VA (*24*). All other compounds were provided by the Drug Synthesis and Chemistry Branch, NCI, NIH, Bethesda, MD.

structure of taxol lacking an active side chain (*23*), was completely inactive in both assays, as were the taxol and taxotere side chain precursors, (2R,3S)-(-)-N-Benzoyl-3-phenylisoserine methyl ester (NSC #D651433-N) and (2R,3S)-(-)-N-(tert-Butoxycarbonyl)-3-phenylisoserine methyl ester (NSC #D651434-O). Cephalomannine, a taxol analog that differs only in the R_3 substitution within taxol's side chain (*23*), exhibits microtubule effects at ~3-fold higher concentrations than taxol, yet has TNFα-inducing effects at ~10-fold the concentration seen with taxol. Semi-synthetic taxol (Bristol Myers-Squibb) was comparable to natural taxol in both systems (data not shown).

In non-dividing macrophages, such as the primary murine macrophage cultures used extensively in these studies, taxol-induced microtubule polymerization is readily detected by the presence of arrays of crystalline "bundles" in cells fixed and immunofluorescently stained with anti-tubulin antibody (*17*). Indeed, a striking dissociation between the capacity of taxol to induce TNFα and microtubule stabilization is illustrated again using macrophages derived from the LPS-hyporesponsive (Lps^d) C3H/HeJ mouse strain. Fluorescent staining of taxol-treated C3H/HeJ macrophages shows extensive microtubule "bundling" in the complete absence of TNFα secretion, tyrosine phosphorylation, or gene expression. Conversely, LPS fails to induce obvious microtuble bundling, even in C3H/OuJ (Lps^n) macrophages (*17, 25*). This latter finding is completely consistent with the observation of Carboni *et al.* (*21*) that, in contrast to taxol, LPS failed to induce assembly of microtubules *in vitro*. Taken collectively, these data suggest the existence of a microtubule-independent pathway for taxol-induced LPS-mimetic effects.

LPS Analogs Block the LPS-Mimetic Effects of Taxol

Taxol and LPS bear no obvious structural relationship, yet all of the LPS-mimetic effects of taxol are blocked by two LPS structural antagonists, *Rhodobacter sphaeroides* diphosphoryl lipid A (*Rs*DPLA) and 3-aza-lipid X 4-phosphate (SDZ 880.431) (*25*). Thus, occupancy of putative LPS receptors by these structural antagonists block taxol-induced signaling. The failure of these antagonists to block taxol-induced microtubule bundling again suggests that the LPS-like signaling induced by taxol is somehow distinct from its microtubule binding activity. In the simplest model, LPS and taxol share the receptor that initiates LPS-like signaling. It is clear that CD14, a glycosylphosphatidylinositol-linked macrophage membrane protein, binds LPS in association with specific serum proteins, although recent evidence suggests that CD14 *per se* may not be a true signaling receptor (*26*). Rather, it appears that CD14 may serve to concentrate and/or focus LPS to the true signaling apparatus. Thus, it is possible that taxol engages a membrane-associated structure identical to or closely associated with the LPS signaling apparatus. Occupancy of the LPS signaling receptor by LPS structural antagonists may sequester or immobilize membrane proteins that are necessary for taxol-induced LPS-mimetic activity.

Perhaps the major disconcerting observation to date is that human macrophages respond very well to LPS, but have yet to be shown to respond to taxol as an LPS mimetic. However, human and murine macrophages also differ with respect to their ability to be stimulated by lipid IV_A, a precursor in the synthesis of LPS (*27*). Specifically, in human macrophages, lipid IV_A is an LPS antagonist (*28*). In contrast, in murine macrophages, lipid IV_A is a strong agonist, and exhibits activity comparable to intact LPS (*29*). Nonetheless, it is possible that identification of LPS-signaling molecules shared by taxol in the mouse will provide insights in the search for homologous or overlapping signaling proteins in human macrophages. In this respect, taxol may provide an extraordinary tool in the ultimate elucidation of LPS-signaling pathways.

Other Possible Anti-tumor Actions of Taxol

We started this chapter with a brief review of the long-recognized role of LPS in the elimination of solid tumors. In addition to its ability to induce cytokines such as TNFα, LPS has also been shown to participate in the differentiation of murine macrophages to a fully activated, tumoricidal state (*30-34*). In this "two signal" model of macrophage activation, murine macrophages are "primed" to a more differentiated state (yet still non-tumoricidal) by "Macrophage Activating Agents," the prototype being Interferon-γ (IFN-γ). "Primed" macrophages are exquisitely sensitive to "trigger" signals, such as LPS, that induce the fully tumoricidal state (reviewed in *3*). Again, macrophages derived from the LPS-hyporesponsive C3H/HeJ strain illustrate the specificity of this genetic defect for LPS; macrophages derived from *Lps*^{d} mice and "primed" with IFN-γ, fail to respond to LPS as a "trigger" signal to become tumoricidal, but respond normally to alternative "trigger" signals such as heat killed *Listeria monocytogenes* (*35*) or endotoxin associated proteins (*34*).

The capacity of taxol to serve as a "trigger" signal in this model of tumoricidal activity was recently assessed by Manthey *et al.* (*36*). IFN-γ-primed C3H/OuJ macrophages responded to taxol to become cytotoxic for the P815 mastocytoma cell line *in vitro*, and synergized with IFN-γ to invoke tumoricidal activity that was indistinguishable from LPS. As anticipated, taxol failed to deliver an appropriate "trigger" signal to C3H/HeJ macrophages. The cellular mechanism underlying the induction of tumoricidal activity by IFN-γ in combination with either LPS or taxol is related to the generation of nitric oxide by the macrophages. IFN-γ and LPS or taxol synergized to induce production of nitric oxide and expression of the inducible nitric oxide synthase gene by macrophages. Moreover, tumoricidal activity was blocked in the presence of the N^{G}-monomethyl-L-arginine, a competitive inhibitor of nitric oxide synthase. These findings illustrate a potential for taxol to activate macrophage-mediated anti-tumor mechanisms, in addition to its well characterized role as an anti-mitotic agent.

Since macrophage-derived nitric oxide has also been shown to be a potent anti-microbial mechanism, one could speculate that taxol may also synergize with IFN-γ to kill nitric oxide-sensitive intracellular pathogens. A model that outlines the known and postulated pathways by which taxol mimics LPS in murine macrophages is presented in Figure 1.

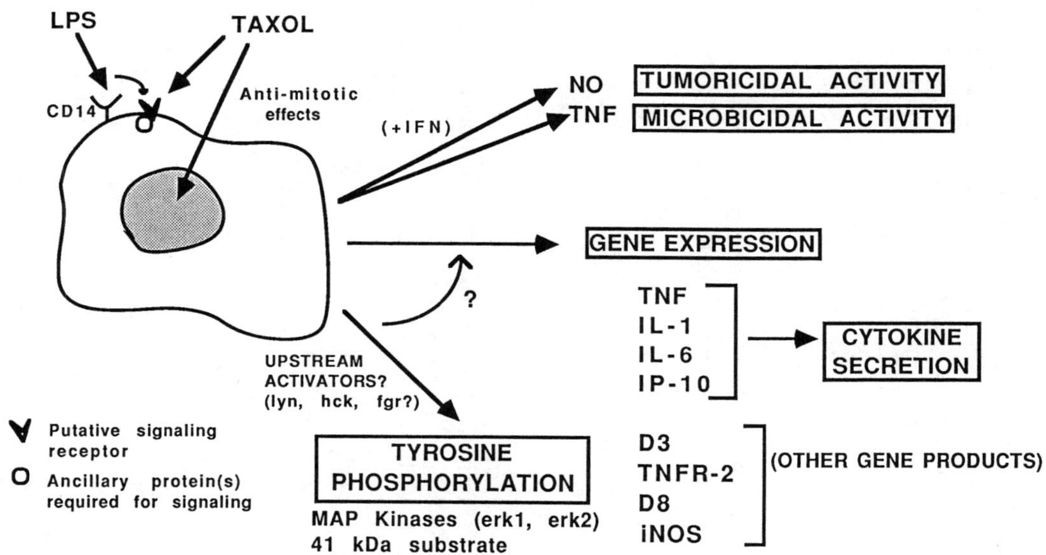

Figure 1. Hypothetical model for the action of taxol on murine macrophages.

Summary

TAXOL® represents one of the most promising anti-tumor drugs and novel analogs are being designed to improve delivery and therapeutic efficacy. Most of the approaches to date have focused on the well characterized action of taxol as a microtubule stabilizing, anti-mitotic agent. The finding that taxol, but not taxotere, serves as an LPS surrogate in murine macrophages suggests that microtubule-independent pathways of signaling may exist in addition to its better characterized mode of action. Using taxol and its derivatives to characterize novel associating proteins in murine macrophages may provide important insights into the signaling pathways utilized by Gram negative LPS.

Acknowledgments

This work was supported by NIH Grant #AI-18797 (S.N.V.). The authors wish to express their appreciation to Dr. David Kingston, Dept. of Chemistry, Virginia Tech, Blacksburg, VA and the Drug Synthesis and Chemistry Branch, Division of Cancer Treatment, National Cancer Institute, NIH for the taxol analogs presented in Table I.

Literature Cited

1. Beutler, B.; Cerami, A. In *Annual Review of Immunology:* Paul, W.E., Fathman, C. G., Metzger, H., Eds.; Annual Reviews Inc.: Palo Alto, CA, 1990, Vol. 7; pp 625-655.
2. Beutler, B.; Cerami, A. In *Tumor Necrosis Factors - The Molecules and Their Emerging Role in Medicine.* Beutler, B., Ed.; Raven Press: NY, NY, 1992, pp 1-10.
3. Vogel, S.N.; Hogan, M. M. In *Immunophysiology - The Role of Cells and Cytokines in Immunity and Inflammation.* Oppenheim, J. J.; Shevach, E. M., Eds.; Oxford University Press: NY, 1990, pp 238-258.
4. *Tumor Necrosis Factors - The Molecules and Their Emerging Role in Medicine.* Beutler, B., Ed.; Raven Press: NY, NY, 1992.
5. Vogel, S.N. In *Tumor Necrosis Factors - The Molecules and Their Emerging Role in Medicine;* Beutler, B., Ed.; Raven Press: NY, NY, 1992, pp 485-513.
6. Manthey, C.L.; Vogel, S.N. *J. Endotoxin Res.* **1994,** *(in press).*
7. Stefanova, I.; Corcoran, M.L.; Horak, E.M.; Wahl, L.M.; Bolen, J.B.; Horak, I.D. *J. Biol. Chem.* **1993,** *268,* 20725-20728.
8. Weinstein, S.L.; Sanghera, J.S.; Lemke, K.; DeFranco, A.L.; Pelech, S.L. *J. Biol. Chem.* **1992,** *149,* 2459-2465.

9. Watson, J.; Kelly, K.; Largen, M.; Taylor, B. A. *J. Immunol.* **1978**,*120*, 422-424.
10. Ding, A.H.; Sanchez, E.; Stimal, S.; Nathan, C.F. *J. Biol. Chem.* **1989**, *264*, 3924-3929.
11. Ding, A.H.; Porteu, F.; Sanchez, E.; Nathan, C.F. *J. Exp. Med.*, **1990**, *171*, 715-727.
12. Ding, A.H.; Porteu, F.; Sanchez, E.; Nathan, C.F. *Science*, **1990**, *248*, 370-372.
13. Rao, S.; Horwitz, S.B.; Ringer, I. *J. Natl. Can. Inst.*, **1992**, *84*, 785-788.
14. Rao, S.; Krauss, N.E.; Heerding, J. M.; Swindell, C.S.; Ringel, I.; Orr, G.A.; Horwitz, S.B. *J. Biol. Chem.*, **1994**, *269*, 3132-3134.
15. Diaz, J.F.; Andreu, J.M. *Biochemistry*, **1993**, *32*, 2747-2755.
16. Ding, A.H.; Sanchez, E.; Tancinco, M.; Nathan, C.F. *J. Immunol.*, **1992**, *148*, 2853-2858.
17. Manthey, C.L.; Brandes, M.E.; Perera, P.-Y.; Vogel, S.N. *J. Immunol.*, **1992**, *149*, 2459-2465.
18. Manthey, C.L.; Perera, P.-Y.; Henricson, B.E.; Hamilton, T.A.; Qureshi, N.; Vogel, S.N. *J. Immunol.* **1994** (*in press*).
19. Ding, A.H.; Sanchez, E.; Nathan, C.F. *J. Immunol.*, **1993**, *151*, 5596-5602.
20. Bogdan, C.; Ding, A.H. *J. Leuk. Biol.*, **1992**, *52*, 119-121.
21. Carboni, J.M.; Singh, C.; Tepper, M.A. *J. Natl. Can. Inst.*, **1993**, *15*, 95-101.
22. Rowinsky, E.K.; Onetto, N.; Canetta, R.M.; Arbuck, S.G. *Sem. in Oncology*, **1992**, *19*, 646-662.
23. Kingston, D.G.I. *Pharmac. Ther.*, **1991**, *52*, 1-34.
24. Chaudhary, A.G.; Ghapure, M.M.; Rimoldi, J.M.; Chordia, M.D.; Gunatilaka; Kingston, D.G.I. *J. Am. Chem. Soc.*, **1994**, *116*, 4097-4098.
25. Manthey, C.L.; Qureshi, N.; Stütz, P.L.; Vogel, S.N. *J. Exp. Med.*, **1993**, *178*, 695-702.
26. Lee, J.-D.; Kravchenko, V.; Kirkland, T.N.; Han, J.; Mackman, N.; Moriarty, A.; Leturcq, D.; Tobias, P.S.; Ulevitch, R.J. *Proc. Natl. Acad. Sci. USA*, **1993**, *90*, 9930-9934.
27. Rick, P.D.; Fung, L. W.-M.; Ho, C.; Osborn, M.J. *J. Biol. Chem.*, **1977**, *252*, 4904-4912.
28. Golenbock, D.T.; Hampton, R.Y.; Qureshi, N.; Takayama, K.; Raetz, C.R.H. *J. Biol. Chem.*, **1991**, *266*, 19490-19498.
29. Perera, P.-Y.; Manthey, C.L.; Stütz, P.L.; Hildebrandt, J.; Vogel, S.N. *Infect. Immun.*, **1993**, *61*, 2015-2023.
30. Russell, S.W.; Doe, W.F.; McIntosh, T. *J. Exp. Med.*, **1977**, *146*, 1511-1520.
31. Hibbs, J.B.; Taintor, R.R.; Chapman, H.A.; Weinberg, J.B. *Science*, **1977**, 279-285.

32. Meltzer, M.S. *J. Immunol.*, **1981**, *127*, 179-183.
33. Pace, J.L.; Russell, S.W.; Schreiber, R.D.; Altman, A.; Katz, D.H. *Proc. Natl. Acad. Sci. USA*, **1983**, *80*, 3782-3786.
34. Hogan, M.M.; Vogel, S.N. *J. Immunol.*, **1987**, 139, 3697-3702.
35. Pace, J.L.; Russell, S.W.; LeBlanc, P.A.; Murasko, D.M. *J. Immunol.*, **1985**, *134*, 977-981.
36. Manthey, C.L.; Perera, P.-Y.; Salkowski, C.A.; Vogel, S.N. *J. Immunol.*, **1994**, *152*, 825-831.

RECEIVED August 31, 1994

Chapter 13

The Roles of the Cytoskeleton in Steroidogenesis

The Actions of Paclitaxel and Other Cytoskeletal Agents

Pamela J. Massey[1,2], Michael R. Caudle[1], Jeffrey K. Keenan[1], and Thomas T. Chen[1,2]

[1]Department of Obstetrics and Gynecology, University of Tennessee, Medical Center, Knoxville, TN 37920
[2]Department of Zoology, University of Tennessee, Knoxville, TN 37996

The primary action of taxol is the cytoskeleton. All three cytoskeletal elements (microfilaments, intermediate filaments, and microtubules) have been implicated in steroidogenesis. This information was based on studies using a variety of cytoskeletal inhibitors including taxol. The rate-limiting step governing the transport of steroid precursor from lipid droplets to mitochondria appears to be controlled by all three cytoskeletal elements. The mechanism by which taxol affects steroid secretion is complex. Its action may be mediated by both microtubules and cytokines. Cremophor, the solvent used in the clinical formulation of Taxol, also has effects on steroidogenesis. This information may be useful for physicians when Taxol is used in treating cancer patients.

Taxol (Paclitaxel, Bristol-Myers Squibb), discovered in the early 1960's, was the first taxane found to have anticancer activity. In tests conducted by the National Cancer Institute, crude extracts from the Pacific yew tree were found to have activity against a number of murine tumors. In 1971 taxol was identified as the active substance in these extracts and its chemical structure was described (*1*). In the late 1970's, Horwitz's group discovered taxol's unique antimicrotubule mechanism of action (*2*)

Clinical trials in the 1980's demonstrated a 25% response rate in patients with refractory ovarian tumors, compared to a 6% rate with standard regimens. Despite the success of these early studies, a number of problems delayed widespread use in humans until recently. These included drug hypersensitivity, neutropenia, poor solubility of the drug, and scarcity. Changes in drug administration rates, premedication with corticosteroids and antihistamines, and the recent clinical use of granulocyte colony stimulation factors have permitted widespread clinical trials (*3*).

Trimble et al (*4*) reported on the first 1,000 patients with platinum-refractory ovarian cancer treated with taxol in 1993. They reported an objective response rate

NOTE: Paclitaxel is the generic name for Taxol, which is now a registered trademark.

0097–6156/95/0583–0173$08.00/0

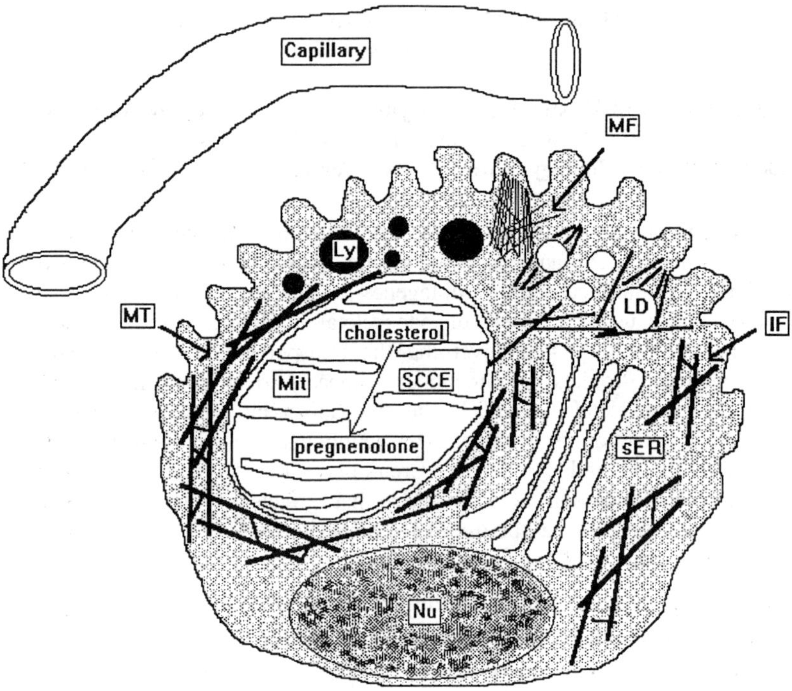

Figure 1. Diagram of a steroidogenic cell depicting architecture of microfilaments, intermediate filaments, and microtubules; and showing conversion of cholesterol to pregnenolone utilizing side-chain cleavage enzyme. LD = lipid droplet, sER = smooth endoplasmic reticulum, Mit = mitochondrion, Nu = nucleus, SCCE = side-chain cleavage enzyme, Ly = lysosome, MT = microtubule, IF = intermediate filament, MF = microfilament.

of 22%, consisting of a 4% complete response rate and an 18% partial response rate. They concluded that the drug could be safely administered using the proper precautions, reporting 15 treatment-related deaths. Similar studies have reported overall response rates of 25-30% (*5,6*).

A number of investigators have found taxol effective in the treatment of metastatic breast cancer. Holmes et al (*3*), for example, reported a 56% response rate to taxol in such patients. Similarly, taxol has been found useful in a number of other human malignancies including lung carcinoma, lymphoma, carcinoma of the head and neck, and malignant melanoma.

The anticancer activity is believed to be mediated by a combination of its primary action on microtubule assembly and secondarily by inhibiting DNA synthesis and promoting DNA fragmentation (*5*), inducing interleukin-1β (IL-1β) and tumor necrosis factor-α (TNF-α) gene expression (*7*), suppressing TNF-α receptor expression (*8*), and augmenting radio-sensitization (*9,10*) of the cell.

Since these primary and secondary targets of taxol action are not unique to cancer cells, taxol could potentially affect the physiology of virtually all cells. This is particularly relevant for steroid secreting cells (adrenal cortical cells and gonadal cells) because cytoskeletal elements (microfilaments, MF; intermediate filaments IMF; and microtubules, MT) are thought to play critical roles in steroid synthesis. Disruption of the steroidogenic capacity of these cells will have an enormous impact on how an organism deals with stress, disease states, and the propagation of species.

In this chapter, we will review our current knowledge about the involvement of all three cytoskeletal elements in steroidogenesis in the adrenal and gonad, focus on the effect of taxol and its vehicle on steroid secretion, and relate it to clinical implications.

General Mechanisms of Steroidogenesis

Steroidogenic cells acquire their steroid precursor cholesterol by binding and internalizing the plasma sterol-carrier LDL through receptor-mediated endocytosis (*11*). The internalized LDL, which is trapped in the endosome, is degraded by hydrolytic enzymes after fusing with primary lysosomes. This releases cholesterol ester, which is stored in lipid droplets. In response to trophic hormone stimulation, cholesterol ester is rapidly hydrolyzed to free cholesterol, then transported to mitochondria where a side chain cleavage enzyme complex converts it to pregnenolone, the immediate precursor for all steroids. This conversion constitutes the rate limiting step of steroidogenesis (*12*). Disruption of which invariably leads to impaired steroid secretion.

Involvement of Cytoskeleton in Steroidogenesis

All the above steps are highly dependent on an intact cytoskeletal system (Figure 1). A great deal of information is available about the importance of the cytoskeletal elements in steroid synthesis. Most experimental evidence has been derived from studies using cytoskeleton disrupting agents. However, there are drawbacks to using inhibitors in delineating the function of a particular structure, as inhibitors may affect processes other than the one under investigation (e.g., colchicine inhibits

TABLE I. THE ACTION OF MICROFILAMENT AGENTS ON STEROID SECRETION IN STEROIDOGENIC CELLS

INHIBITOR	RESPONSE	CELL TYPE	REFERENCE
Cytochalasin B	↓ basal	Y-1	53
	↓ ACTH	Y-1	54
	↑ basal	adr tumor	55
	↑ basal	rat ovary	23
	↓ basal	bovine luteal	56
	↓ hCG-P4		
	- basal	rat luteal	57
	↓ hCG-P4		
Cytochalasin D	↓ ACTH	Y-1	12
	↑ basal	rat adrenal	58
	↓ ACTH		
	↑ basal	bovine adrenal	59
	- ACTH		
	↓ basal	rat Leydig	12
Cytochalasins (B and D)	↑- basal	hamster GC	60
	↓ LH/hCG-P4		
	↓ LH-testosterone	rat Leydig	21
Anti-actin in liposome	↓ ACTH	adrenal tumor	19
DNase	↓ ACTH	adrenal	61
	↓ ACTH	rat adrenal	62
	↑ basal	Y-1	63
		adrenal	34

↑: stimulate; ↓: inhibit; --: no effect; ND: not determined.

hexose transport). Taxol is valuable in this regard because, in contrast to other microtubule disrupting agents, taxol promotes MT assembly and thus offers a unique probe for investigators in this area. An important thing to keep in mind is that all three types of cytoskeletal elements are structurally and functionally linked to each other. Interference of one element may lead to disruption of the entire interconnected network. Further, the action of a given inhibitor may be dependent on the dose used (*13,14*). Nevertheless, by combining specific cytoskeleton inhibitors with specific steroidogenic enzyme blockers coupled with ultrastructural approaches, useful information has been derived about the roles these three cytoskeletal elements play in steroidogenesis.

The general consensus about the role of the cytoskeleton in steroidogenesis is that the effect occurs primarily at the level of steroid synthesis rather than release, taking place at a step preceding pregnenolone formation. This process also involves cell rounding and the redistribution of structures such as lysosomes and mitochondria to a more compact configuration. There is evidence that all three cytoskeletal elements participate in steroid synthesis to various degrees. In the following sections, we will point out common scheme and unique ways by which cytoskeletal elements participate in the steroidogenic pathway.

Microfilaments. Microfilaments (MF) are 5-9nm diameter filamentous structure arranged as polymers of α-actin in a two-stranded helical configuration. By virtue of the contractile property of their functional unit actin, MF have been implicated in several key steps in the transport of steroid precursors. They have been shown to facilitate the transport of steroid precursors to their site of activation/metabolism (12). Microfilaments facilitate intracellular organelle movement and molecular interaction. They are involved in changing cell shape and inducing cell rounding (*15*), both of which are obligatory in the enhancement of steroid synthesis. The importance of cell shape is best illustrated by polyHEMA-induced cell rounding in adrenal tumor cells. The degree of cell rounding in vitro was found to correlate well with increased cholesterol transport to the inner mitochondrial membrane, enhanced pregnenolone synthesis by the mitochondria, and overall steroid output by adrenal tumor cells (Y-1) (*16*). Polylysine, which causes cell flattening, decreased cholesterol transport and the production of pregnenolone. Furthermore, trophic hormones and cAMP, which augment steroidogenesis, are invariably coupled to cell rounding (*17*). However, it should be noted that there is no concrete evidence that cell rounding is a prerequisite for the enhancement of steroid synthesis in vivo.

Adrenal Cells. Evidence for the involvement of MF in adrenal cells comes primarily from using different types of inhibitors (Table I). Although results vary somewhat in the way MF influence basal steroid secretion, there is a general consensus that disruption of MF attenuates hormone-stimulated steroid secretion.

Using analogues with different potencies toward MF assembly (cytochalasins B, D, E and reduced B), Hall et al (*18*) showed that despite wide differences in potencies, the concentrations required for a given cytochalasin to inhibit actin polymerization are highly correlated with its ability to cause cell rounding and inhibition of steroid synthesis in adrenal tumor cells. Additional evidence corroborating the inhibitor studies are provided by Hall (*12*), who used anti-actin antibody encapsulated in liposomes and DNase, another MF inhibitor, and found reduced steroid synthesis in Y-1 cells at the level of cholesterol transport (*19,20*). These findings show a close relationship exists between MF and steroidogenic responses.

Testicular Cells. When inhibitors were used to disrupt MF in Leydig cells, essentially an identical steroidogenic pattern as in adrenal cells was observed (Table I). Hall et al. (*19*) showed that anti-actin antibody entrapped in liposomes inhibited LH-stimulated testosterone production. The same treatment blocked the conversion of cholesterol to pregnenolone by isolated mitochondria in rat Leydig cells. (*21*). The importance of MF in steroid synthesis is further supported by the electron microscopic observation in Leydig cells that numerous MF arrange themselves in circular bundles, making direct contact with the organelles crucial in steroid synthesis, which are mitochondria, smooth endoplasmic reticulum and lipid droplets (*22*).

Ovarian Cells. The role of MF in ovarian steroidogenesis is less clear. Conflicting results were obtained using cytochalasins: ranging from inhibitory, no effect, or, to a delayed stimulatory response (*23*, Table I). However, the importance of a compact cellular configuration in steroid synthesis is clearly indicated. Granulosa cells cultured in collagen gel undergo cell rounding with redistribution of steroidogenic organelles to the perinuclear region. They secrete maximum amounts of progesterone (P_4) and cannot be further stimulated by hormones. A reorganization of lysosomes and mitochondria from a peripheral to a primarily perinuclear location and redistribution of the MF network with associated changes in cell shape were also noted in human granulosa cells stimulated with steroidogenic hormones (*24*).

Intermediate Filaments. Intermediate filaments (IMF) are fibrous protein with a diameter of around 10 nm with associated polypeptides that vary according to cell types. These polypeptides include keratin, vimentin, and desmin. Cholesterol and cholesterol ester form tight associations with IMF proteins such as vimentin. Vimentin rearranges around these molecules to form a cage that surrounds the forming lipid droplets during their biogenesis (*25*). It is speculated that this fiber network either promotes movement of lipid droplets or acts as a barrier to migration of droplets toward the mitochondrion, which is near or attached to MT and/or IMF. Disruption of MT or IMF leads to a redistribution of mitochondria and shortening of the distance between lipid droplets and mitochondrion (*26*). Thus, the combined action of the uncoating of lipid droplets and mitochondrial movement facilitates the entry of cholesterol to the side chain cleavage site in the mitochondrion.

Adrenal cells. Two reports have appeared in recent years implicating involvement of intermediate filaments in adrenal steroid synthesis. Almahbobi and Hall (*26*) found that after removal of MT and MF proteins, intermediate filaments were attached to lipid droplets and vimentin, and could be extracted from these droplets. Vimentin was further localized to lipid droplets by specific antibodies at both the light and EM levels. Subsequently, Shiver et al (*27*) showed that acrylamide, a specific intermediate filaments disrupting agent, induces cell rounding with concomitant stimulation of pregnenolone synthesis. The action of acrylamide occurs after cAMP formation and before pregnenolone conversion.

Testicular cells. No studies have been reported on the role of intermediate filaments in steroidogenesis in testicular cells.

Ovarian cells. Although no information is available about the direct role of intermediate filaments in steroidogenesis, there are indications that the intermediate filament protein vimentin is essential for the differentiation of ovarian cells to a steroidogenic competent cell type. In the absence of hormonal influences (human chorionic gonadotropin [hCG] or follicular stimulating hormone [FSH]), cells synthesize both intermediate filament proteins, vimentin and cytokeratins; however, in the presence of hCG, only vimentin is synthesized. This pattern of cytoskeletal appearance is consistent with immunohistochemical studies (*28*), which show coexpression of cytokeratins and vimentin in undifferentiated human granulosa cells (primordial and primary follicles), whereas a major decrease of cytokeratin, but not vimentin, occurs in mature follicles and corpora lutea (*29*).

Microtubules. Microtubules are polymers of α and β tubulins arranged in hollow tubes with a diameter of 25 nm. Although MT are important primary constituents of the mitotic spindle apparatus, they are also critical for the performance of many vital interphase functions such as maintenance of shape, cell motility, and intracellular transport. In steroid secreting cells, they also play a key role in modulating the availability of cholesterol to the steroid biosynthetic pool (*15,24*).

Most experimental evidence shows that basal steroid secretion is enhanced while hormone stimulated steroid secretion is inhibited when MT assembly is disrupted by colchicine or other congeners (Table II). These observations suggest that MT form a barrier preventing the access of cholesterol ester droplets to the mitochondrion.

Adrenal Cells. In MT-based organelles motility increases the proximity of mitochondria to cholesterol-containing lipid droplets, thereby facilitating the transfer of cholesterol to the mitochondrial outer membrane in adrenal tumor cells. Adrenocorticotrophic hormone increases opportunities for the transfer of cholesterol by promoting organelle motility and by increasing the number and accessibility of cholesterol-containing lipid droplets (*30,31*).

Testicular Cells. Clark and Shay (*32*) used anti-tubulin antibody to visualize the distribution of tubulin in Leydig cells. They found that under unstimulated conditions, tubulin was associated with discrete granules containing primarily cholesterol. They postulated that it is the monomeric tubulin, not MT, that acts as a sequestering barrier inhibiting the cholesterol transport.

Ovarian Cells. Colchicine and nocodazole reduce basal and FSH-stimulated P_4 by rat ovarian cells (*23*). The inhibitory action occurs in hormone synthesis rather than release since the cellular steroid content is increased (*33,34*).

TABLE II. THE ACTION OF MICROTUBULAR AGENTS ON STEROID SECRETION IN STEROIDOGENIC CELLS

INHIBITOR	RESPONSE	CELL TYPE	REFERENCE
Colchicine	↑ basal	adrenal cells	33
	↑ basal - ACTH	Y-1	31
	↑ basal	Y-1	27
	↓ plasma P4	ovary*	64
	↑ basal P4 ↓ FSH-P4	rat ovary	36
	↑ basal	porcine GC	37
	↓ hCG-P4	ov luteal	66
Vinblastin	biphasic ↓ ACTH	Y-1	31
Nocodazole	basal P4 FSH-P4	rat ovary	36
	↓ basal ↓ ACTH	Y-1	31
	biphasic FSH-P4	rat ovary	36
	↑ P4	porcine GC	37
	↑ basal	adrenal	34
	↓ ACTH		62
	ND	Leydig tumor	32

↑: stimulate; ↓: inhibit; --: no effect; ND: not determined; *: in vivo.

In vivo injection of colchicine in the ewe reduces plasma P_4 levels (*64*). In luteal cells, basal P_4 is unaffected, but hormone-stimulated P_4 secretion is reduced by colchicine treatment (*66*).

Gonadotropins facilitate reorganization of the cytoskeleton into a steroidogenic phenotype, resulting in the perinuclear clustering of organelles, and making the conversion of free cholesterol to P_4 more efficient (*29*). Human CG causes a decrease in vinculin (involved in cell-substrate adherence), non-muscle tropomyosin isoforms, actin binding protein, α-actinin, and actin, without changes in α- and β-tubulin synthesis in human granulosa cells.

The Action of Taxol on Steroid Synthesis

Because of the potential therapeutic benefits of taxol in the treatment of a wide variety of cancers, the action of taxol on steroid secreting cells have been studied extensively. Therapeutic doses of taxol inhibit basal and hormone-stimulated steroidogenesis in adrenal Y-1 cells (*30*) and granulosa cells (GC) (*13,36*), but only hormone-stimulated steroidogenesis in murine leydig tumor cells (MLTC-1) and rat GC (*35*). The inhibition does not take place at the level of cAMP production nor does it occur past the step of P_4 formation or steroid release (*31*). The inhibition appears to occur at the transport of cholesterol to mitochondria (*31,35*) where cholesterol side-chain cleavage enzyme complex is located (*38*). Alternatively, the action of taxol may be mediated by IL-1β and TNF-α, both of which have been shown to be present in ovarian cells and inhibit P_4 steroid secretion (*39*). Subclinical doses of taxol, on the other hand, stimulate granulosa P_4 production (Figure 2). Low concentrations of taxol may promote MT assembly in such a way that is conducive to the mobilization and transport of cholesterol to the mitochondria for steroid synthesis. Regardless of the action of taxol, the steroidogenic capacity is recoverable after Taxol removal (Figure 3).

The Action of Cremophor on Steroid Synthesis

The current injectable form of taxol is virtually insoluble in water and, thus, must be dissolved in cremophor, which is a polyoxyethylated castor oil. Cremophor is used as a vehicle for delivery of certain hydrophobic drugs, including taxol (*40*). Due to the extremely poor solubility of taxol, a very large amount of cremophor mixed with ethanol is required to dissolve a clinical dose (*41*). Cremophor, although long believed inert, is now thought to be responsible for the incidents of severe hypersensitivity reactions which almost resulted in the termination of phase I trials of taxol (*42*). Because taxol is used for the treatment of many types of cancer and is formulated in cremophor, it is important to know what effect cremophor has on steroidogenesis.

Cremophor exhibits biphasic effect on steroidogenesis in cultured porcine GC (Figure 2). Doses of cremophor equivalent to that present in a therapeutic dose of taxol suppress P_4 production while subclinical doses (10-100 fold less) stimulate P_4 production. However, the effect of cremophor is less pronounced than that seen with taxol treatment (Figure 2) (*14*).

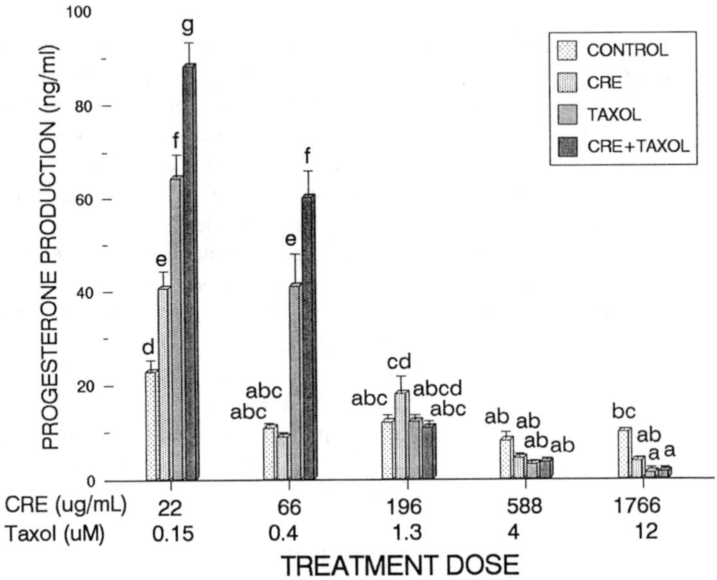

Figure 2. Dose-dependent, biphasic response of progesterone production by porcine granulosa cells treated (24 h) with cremophor (CRE), taxol or cremophor plus taxol. Bars: standard error. Statistical analysis performed was analysis of variance. Different letters of the alphabet denotes significant differences among means ($p<0.05$). (Adapted from ref. 14.)

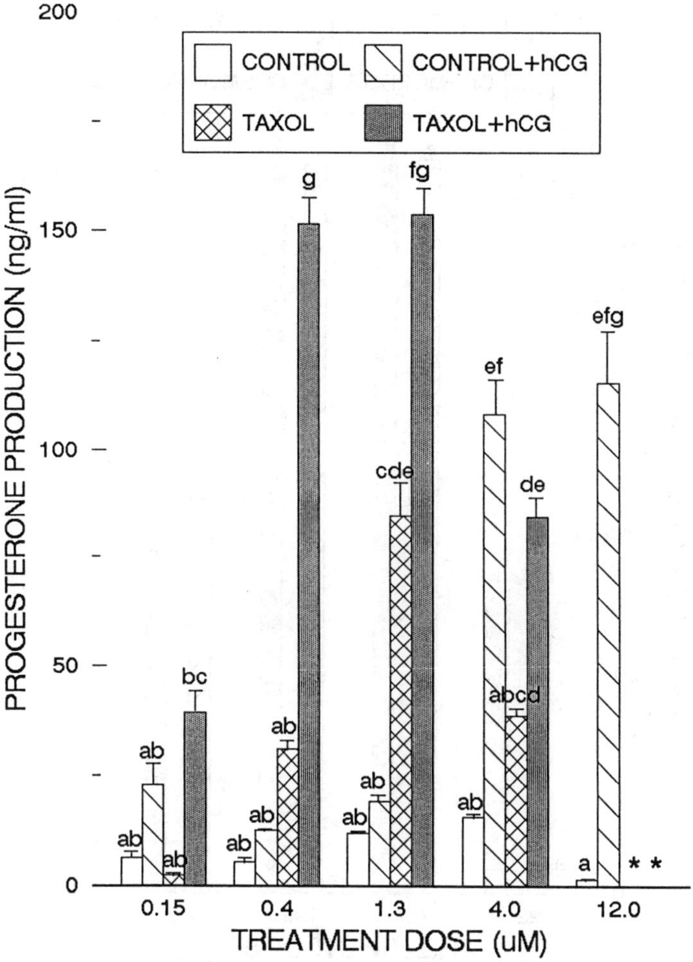

Figure 3. Basal (-hCG) and hCG-stimulated (+hCG) progesterone production by porcine granulosa cells pre-treated (24 h) with taxol and subsequently cultured for an additional 24 h in fresh culture medium. Different letters of the alphabet denotes significant differences among means ($p<0.05$). Bars: standard error. **Progesterone from these groups was undetectable. (Adapted from ref. 14.)

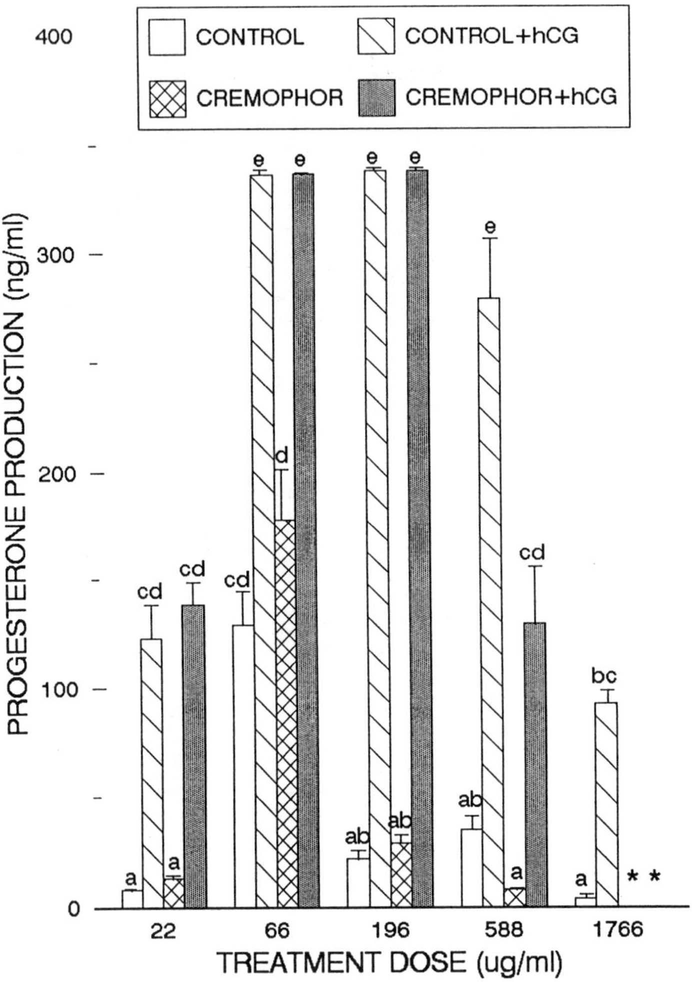

Figure 4. Basal (-hCG) and hCG-stimulated (+hCG) progesterone production by porcine granulosa cells pre-treated (24 h) with cremophor and subsequently cultured for an additional 24 h in fresh culture medium. Different letters of the alphabet denotes significant differences among means ($p<0.05$). Bars: standard error. **Progesterone from these groups was undetectable. (Adapted from ref. 14.)

The mechanism of cremophor action is as yet uncertain, but appears to be different from that of taxol. Several possibilities exist for the stimulatory effects of low levels of cremophor on P_4 production. Cremophor activates prostaglandins (PG) at physiologic concentrations (*43*). Certain PG's, especially PGE_2 and PGI, are potent stimulators of steroid secretion (*44,45*). Cremophor may cause membrane perturbations leading to PG synthesis by GC. Cremophor also inhibits protein kinase C (*46*). Protein kinase C activators has been shown to inhibit basal and gonadotropin-stimulated P_4 production by granulosa cells (*47,48*) and androstenedione production by theca cells (*49*). The effect of subclinical doses of cremophor is also transient. The cells are fully capable of recovering from the inhibitory action of cremophor (Figure 4).

Reversibility of Taxol and Cremophor Effects

In several steroidogenic cell types, ultrastructural alterations caused by therapeutic concentrations of taxol persist when the cells are allowed to recover from treatment (*50-52*). However, cells treated with therapeutic doses of taxol demonstrate acute steroidogenic responses to hormone stimulation when washed and recovered. These cells also reverse the severe inhibition of basal steroid production (*13,14,35,52*). Granulosa cells treated with cremophor are also responsive to subsequent gonadotropin stimulation (*14*).

Conclusion

Due to its unique mechanism of action, taxol has great potential as both a first and second line chemotherapeutic agent. Taxol's primary mechanism of action involves the induction of a large number of cytoplasmic microtubule bundles which bind to the endoplasmic reticulum, thereby, affecting cellular processes that require microtubule assembly and disassembly such as steroidogenesis. In vitro, taxol inhibits steroid production in adrenal, testicular, and ovarian cells. Recent clinical trials demonstrate that adverse reactions can be limited with the proper premedication. The clinical use of granulocyte colony stimulating factors has further enhanced this drug's potential as neutropenia occurs in over 75% of patients (*42*). Lack of solubility of taxol remains a clinical problem. Currently, taxol administration requires the use of large amounts of cremophor, which itself may be responsible for many reported hypersensitivity reactions. Furthermore, as we have noted above, cremophor itself inhibits steroidogenesis. A better solubilizing agent, or a modification of taxol's structure to enhance solubility as is currently being investigated by several labs, is needed to address these problems. The action of taxol is not specific to malignant cells. As clinical indications for taxol expand, knowledge of its effect on various steroid-secreting organ systems will become more critical in clinical decisions involving this drug's use.

Acknowledgments

The authors would like to thank Jane Gracia, Shelley Paden, and Kimberly Kim for editorial assistance.

Literature Cited

1. Suffness, M.; Wall, M.E. In *Taxol-Science and Application*; Suffness, M., Ed.; CRC Press: Boca Raton, FL, in press 1994.
2. Schiff, P.B.; Fant, J.; Horwitz, S.B. *Nature*, **1979**, 22, 665-667.
3. Holmes, F.A.; Walters, R.S.; Theriault, R.L.; Forman, A.D; Newton, L.K.; Raber, M.N.; Buzdar, A.U.; Frye, D.K.; Hortobagy, G.N. *J. Nat. Cancer Inst.* **1991**, *83,* 1797 -1805.
4. Trimble, E.L.; Adams, J.D.; Vena, D.; Hawkins, M.J.; Friedman, M.A.; Fisherman, J.S.; Christian, M.C.; Canetta, R.; Onetto, N.; Hayn, R.; Arbuck, S.G. *J. Clin. Oncol.* **1993**, *11*, 2405-2410.
5. Slichenmeyer, N.J.; Von Hoff, D.D. *J. Clin. Pharmacol.* **1990**, *30,* 770-788.
6. Seewaldt, V.L.; Greer, B.E.; Cain, J.M.; Figge, D.C.; Tamimi, H.K.; Brown, W.S.; Miller, S.A. *Am. J. Obstet. Gynecol.* **1994**, *170,* 1666-1671.
7. Manthey, C.L.; Brandes, M.E.; Perera, P.Y.; Vogel, S.N. *J. Immunol.* **1992**, *149*, 2459-2463.
8. Ding, A.H.; Porteu, F.; Sanchez, E.; Nathan, C.F. *Science* **1990**, *248*, 370-372.
9. Tishler, R.B.; Schiff, P.B.; Geard, C.R.; Hall, E.J. *Int. J. Radiat. Oncol. Biol. Physiol.* **1992**, *22,* 613-617.
10. Steren, A.; Sevin, B.U.; Perras, J.; Angioli, R.; Nguyen, H.; Guerra, L.; Koechli, O.; Averette, H.E. *Gynecol. Oncol.* **1993**, *48,* 252-258.
11. Goldstein, J.L.; Anderson, R.G.W.; Brown, M.S. *Nature* **1979**, *279,* 679-685.
12. Hall, P.F. *Rec. Prog. Hormone Res.* **1985**, *41*, 1-39.
13. Chen, T.T.; Massey, P.J.; Caudle, M.R. *Endocrinology* **1994**, *134*, 2178-2183.
14. Massey, P.J.; Chen, T.T.; Caudle, M.R. *Endocrine J.* 1994, in press.
15. Herman, B.; Langevin, M.A.; Albertini, D.F. *Eur. J. Cell. Biol.* **1983**, *31*, 34-45.
16. Betz, G.; Hall, P.F. *Endo. Soc.* **1987**, *120*, 2547-2554.
17. Kawaoi, A.; Uchida, T.; Okano, T. *Acta. Pathol. Jpn.* **1977**, *27*, 841-856.
18. Hall, P.F.; Nakamura, M.; Mrotek, J.J. *Biochem. Biophys. Acta* **1981**, *676,* 338-344.
19. Hall, P.F.; Charponnier, C.; Nakamura, M.; Gabbiani, G. *J. Biol. Chem.* **1979**, *254,* 9080-9084.
20. Nakamura, M.; Watanuki, M.; Tilley, B.; Hall, P.F. *J. Endocrinol.* **1980**, *84*, 179-188.

21. Murono, E.P.; Lin, T.; Osterman, J.; Nankin, H.R. *Biochim. Biophys. Acta* **1980**, *633*, 228-236.
22. Aguas, A.P. *J. Ultrastruc. Res.* **1981**, *74,*175-182.
23. Carnegie, J.A.; Tsang, B.K. *Biol. Reprod.* **1988**, *38,* 100-108.
24. Soto, E.A.; Kliman, H.J.; Strauss, J.F., III; Paavola, L.G. *Biol. Reprod.* **1986**, *34*, 559-569.
25. Franke, W.W.; Hergt, M.; Grund, C. *Cell* **1987**, *49,* 131-141.
26. Almahbobi, G.; Hall, P.F. *J. Cell Science* **1990**, *97*, 679-687.
27. Shiver, T.M.; Sackett, D.L.; Knipling, L.; Wolff, J. *Endocrinology* **1992**, *131*, 201-207.
28. Czernobilsky, B.; Moll, R.; Levy, R.; Franke, W.W. *Eur. J. Cell Biol.* **1985**, *37*, 175-190.
29. Ben-Ze'ev, A.; Amsterdam, A. *Endocrinology* **1989**, *124*, 1033-1041.
30. Benis, R.; Mattson, P. *Tissue Cell.* **1989**, *21*, 507-515.
31. Sackett, D.L.; Wolff, J. *Biochem. Biophys. Acta.* **1986**, *888,* 163-170.
32. Clark, M.A.; Shay, J.W. *Endocrinology.* **1981**, *109,* 2261-2263.
33. Temple, R.; Wolff, J. *J. Biol. Chem.* **1973**, *218*, 2691-2698.
34. Ray, F.; Strott, C.A. *Endocrinology* **1978**, *103*, 1281-1288.
35. Rainey, W.E.; Kramer, R.E.; Mason, J.I.; Shay, J.W. *J. Cell. Physiol.* **1985**, *123,*17-24.
36. Carnegie, J.A.; Dardick, I.; Tsang, B.K. *Endocrinology.* **1987**, *120*, 819-828.
37. Denkova, R.; Ivanov, I.; Dimitrova, M. *Endocrine Reg.***1992**, *26,* 195-199.
38. Keller, H.U.; Zimmerman, A. *Invasion Metastasis* **1986**, *6*, 33-43.
39. Adashi, E.Y. *Endocr. Rev.* **1990**, *11*, 454-464.
40. Takada, K.; Furuya, Y.; Yoshikawa, H.; Muranishi, S. *J. Pharmacobiodyn.* **1988**, *11*, 80-87.
41. Peereboom, D.M.; Donehower, R.C.; Eisnhauer, E.A.; McGuire, W.P.; Onetto, N.; Hubbard, J.L.; Piccart, M.; Gianni, L.; Rowinsky, E.K. *J. Clin. Oncol.* **1993**, *11*, 885-890.
42. Rowinsky, E.K.; Eisnhauer, E.A.; Chaudhry, V.; Arbuck, S.G.; Donehower, R.C. *Sem. Oncol.* **1993**, *20(suppl 3)*, 1-15.
43. Luderer, J.R.; Demwrs, L.M.; Nomides, C.T.; Hayes, A.H., Jr. *Adv. Prostaglandin Thromboxane Res.* **1980**, *8*, 1633-1635.
44. Weems, C.W.; Vincent, D.L.; Weems, Y.S. *Reprod. Fertil. Dev.* **1992**, *4,* 289-295.
45. Dennefors, B.; Hamberger, L.; Hillensjo, T.; Holmes, P.; Janson, P.O.; Magnunsson, C.; Nilsson, L. *Acta. Obstet.* Gynecol. Scand. Suppl. **1983**, *113*, 31-41.

46. Zhao, F.-K.; Chuang, L.F.; Israel, M.; Chuang, R.Y. *Biochem. Biophys. Res. Comm.* **1989**, *159*, 1359-1367.
47. Hylka, V.W.; Kaki, M.K.; diZerega, G.S. *Endocrinology* **1989**, *124*, 1204-1209.
48. Velduis, J.D.; Demers, L.M. *Biochem. J.* **1986**, *239,* 505-511.
49. Hofeditz, C.; Magoffin, D.A.; Erickson, G.F. *Biol. Reprod.* **1988**, *39*, 873-881.
50. Tokunaka, S.; Friedman, T.M.; Toyama, Y. Pacifici, M.; Holtzer, H. *Differentiation.* **1983**, *24,* 39-47.
51. Forry-Schaudies, S.; Murray, J.M.; Toyama, Y.; Holtzer, H. *Cell Mot. Cytoskel.* **1986**, *6,* 324-338.
52. Benis, R.; Mattson, P. *Tissue Cell.* **1989**, *21,* 687-698.
53. Mrotek, J.J.; Hall, P.F. *Biochem. Biophys. Res. Commun.* **1975**, *64*, 891-896.
54. Mrotek, J.J.; Hall, P.F. *Biochemistry* **1977**, *16*, 3177-3181.
55. Mattson, P.; Kowal, J. *Endocrinology* **1982**, *111*, 1632-1647.
56. Gwynne, A.; Condon, W.A. *J. Reprod. Fertil. 1982*, *65*, 151-156.
57. Azhar, S.; Menon, M.J. *Biochem. J.* **1981**, *194,* 19-27.
58. McPherson, M.A.;Ramachandran, J. *J Cell Biol.* **1980**, 86,129-134.
59. Rainey, W.E.; Shay, J.W.; Mason, J.I. *Mol. Cell Endocrinol.* **1984**, *35*, 189-197.
60. Silavin, S.L.; Javitt, N.B.; Strauss, J.F., III *Endocrinology* **1984**, *115*, 1511-1516.
61. Osawa, S.; Betz, G.; Hall, P.F. *J. Cell Biol.* **1984**, *99*, 1335-1342.
62. Crivello, J.F.; Jefcoate, C.R. *Biochem. Biophys. Acta* **1978,** *542*, 315-329.
63. Cortese, F.; Wolff, J. *J. Cell Biol.* **1978**, *77*, 507-516.
64. Gemmell, R.T.; Stacy, B.D. *J. Reprod. Fertil.* **1977**, *49,* 115-117.
65. Azhar, S.; Reaven, E. *Am. J. Physiol.* **1982**, *243,* E380-386.
66. Sawyer, H.R.; Abel, Jr. J.H.; McClelland, M.C.; Schmitz, M.; Niswender, G.D. *Endocrinology* **1979**, *104*, 476-486.

RECEIVED September 15, 1994

Chapter 14

Chemistry and Structure–Activity Relationships of Taxoids with Modified Skeletons

F. Guéritte-Voegelein, D. Guénard, J. Dubois, A. Wahl, R. Marder, R. Müller, M. Lund, L. Bricard, and P. Potier

Centre National de la Recherche Scientifique, Institut de Chimie des Substances Naturelles, F-91198 Gif-sur-Yvette, France

Taxoids bearing structural modifications of the taxane skeleton were prepared from 10-deacetylbaccatin III, a diterpene isolated from the leaves of *Taxus baccata*. Chemical transformations include reduction of the C11-C12 double bond, contraction of the A ring, rearrangements of the B and C rings as well as opening of the oxetane and other modifications at carbons 2, 4 and 5. Coupling of these 10-deacetylbaccatin III modified analogues with the acid side chain of docetaxel led to compounds which were tested in a microtubule disassembly assay.

Antitumor taxoids constitute one of the most important discovery in cancer chemotherapy in the last twenty years and have been the subject of intensive research since the first isolation of Taxol® (paclitaxel) **1** from the bark of the Pacific yew tree (*1*). Clinical studies led to the approval of Taxol® **1** ,in 1992 , for use in the treatment of ovarian cancer (*2*). Taxotere® (docetaxel) **2** (*3*), a semi-synthetic taxoid, was shown to possess a better bioavailability than paclitaxel and a broad spectrum of antitumor activity (*4*). It is currently in phase II and III clinical studies. These agents act by inhibiting the disassembly of microtubules into tubulin (*5*) in contrast to the other spindle poisons which are inhibitors of the assembly process.

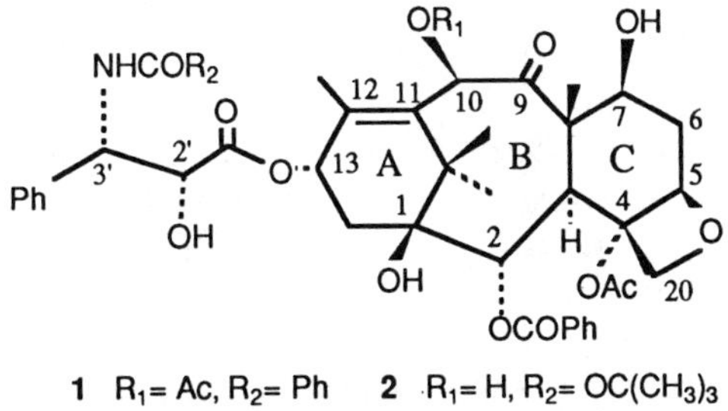

1 R_1= Ac, R_2= Ph **2** R_1= H, R_2= $OC(CH_3)_3$

Because of the limited availability of natural taxol from yew trees, different strategies including total synthesis, semisynthesis, cell and tissue culture of *Taxus spp.*,

0097–6156/95/0583–0189$08.00/0

as well as fungal cultures, have been studied with a view to produce paclitaxel in large amounts.

Although the total synthesis of paclitaxel **1** has been successfully performed (*6-8*), the most practical approach to a large scale production of paclitaxel and docetaxel is currently their semisynthesis from 10-deacetylbaccatin III **3a**, a natural taxoid, isolated from the leaves of the European yew tree, *Taxus baccata* L., in reasonable yield (*9-10*). Baccatin III **3b** is also found in the yew but in smaller amounts than **3a**. The first chemical conversion of 10-deacetylbaccatin III **3a** into paclitaxel **1** and 10-deacetylpaclitaxel was achieved through the hydroxyamination of the 13-cinnamoylbaccatin III derivatives **4a** and **4b** (*11-12*)) This strategy led also to the discovery of docetaxel **2** (*11, 13*).

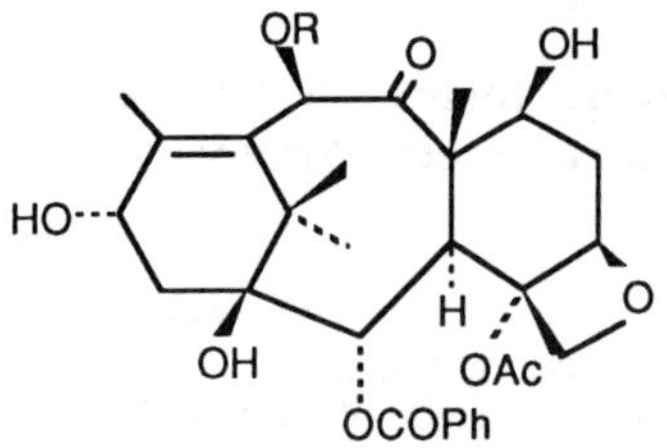

3a R= H
3b R= Ac

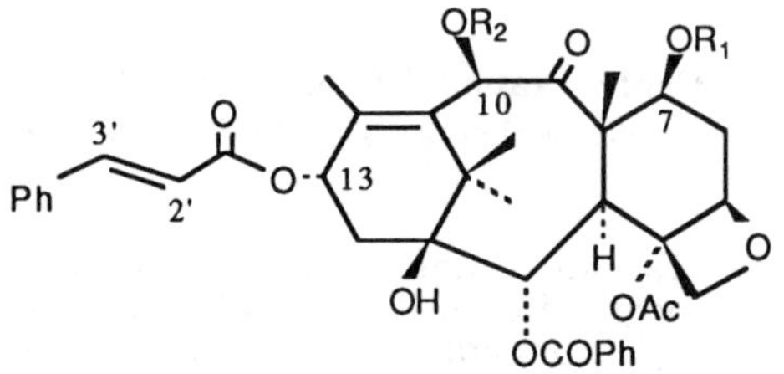

4a R_1= R_2= $CO_2CH_2CCl_3$
4b R_1= $CO_2CH_2CCl_3$, R_2= Ac

A more convenient procedure for a large scale production of paclitaxel **1** and docetaxel **2** was then developed (*10,14-15*). This approach, based on the direct coupling of *O*-protected baccatin III derivatives **5a** or **5b** with the acid side chain of paclitaxel **6a** or docetaxel **6b**, made paclitaxel **1** as well as docetaxel **2** available for clinical use. Other improved and practical methods using cyclic protection (**7a, 7b**) (*16*) as well as β-lactams **8a, 8b** (*17-19*) have been developed. Moreover, oxazinones such as **8c** can be used as acylating agents (*17b*). These procedures have been reviewed recently (*3,20-23*).

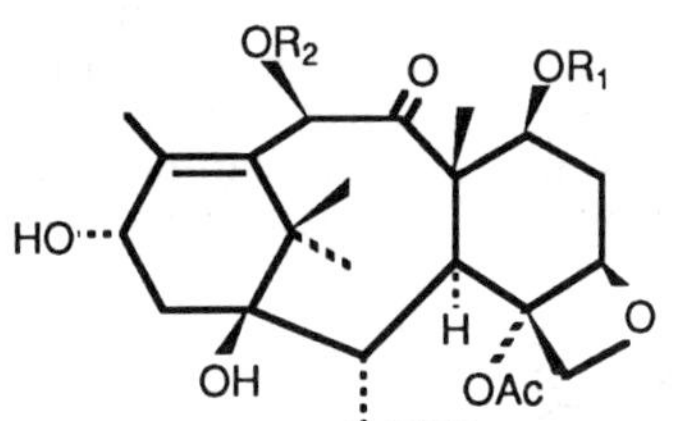

5a R_1= R_2= $CO_2CH_2CCl_3$
5b R_1=$SiEt_3$, R_2= Ac
5c R_1=$SiEt_3$, R_2= H

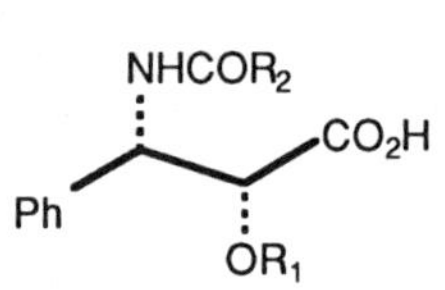

6a R_1= protective group R_2= Ph
6b R_1= protective group R_2= OtBu

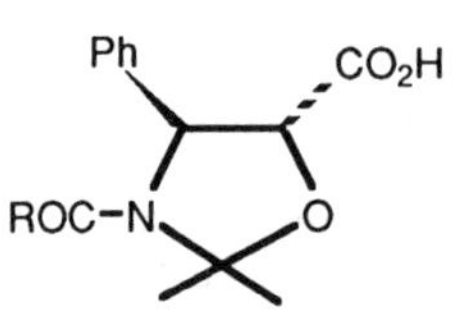

7a R= Ph
7b R= OtBu

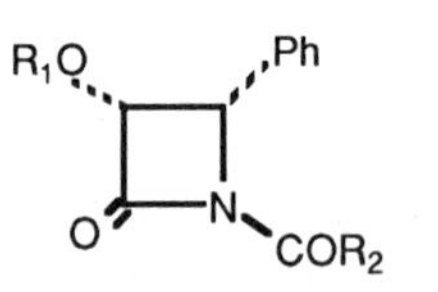

8a R_1= protective group, R_2= Ph
8b R_1= protective group, R_2= OtBu

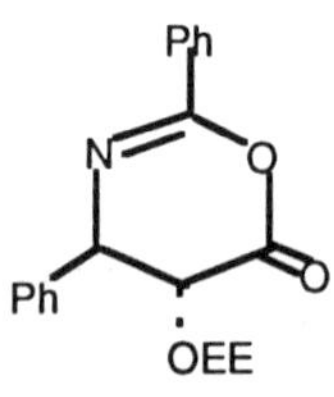

8c

Owing to the availability of 10-deacetylbaccatin III **3a** from the leaves of the yew, a number of new semi-synthetic taxoids with potential antitumor activity can be prepared. Thus, the coupling of protected baccatin III derivatives **5a** and **5b** with different acid side chains led to compounds with structural modifications at the 2' and 3' positions whereas modification of the taxane core of 10-deacetylbaccatin III and coupling with the acid side chain of paclitaxel and docetaxel give taxoids with modified skeletons. In addition to paclitaxel **1**, 10-deacetylbaccatin III **3a** and baccatin III **3b**, other naturally occurring substances such as brevifoliol (*24*), 10-deacetyl-19-hydroxybaccatin III (*25*), 13-acetyl-9-dihydrobaccatin III (*26*) and 14-hydroxy-10-deacetylbaccatin III (*27*) have been used to prepare modified taxoids. Recent reviews reported the effects of these analogues on the assembly of tubulin, on the disassembly of microtubules or on the growth of cancer cells *in vitro* and *in vivo* (*3, 21-22, 28-29*).

The purpose of this chapter is to report the chemistry of 10-deacetylbaccatin III **3a** which has been investigated in our laboratory for preparation of taxoids possessing selected modifications in the skeleton. These modifications involved reduction of the C_{11}-C_{12} double bond, contraction of the A ring, rearrangements of the B and C rings as well as opening of the oxetane and other modifications at carbons 2, 4 and 5. The inhibitory activity of most of the compounds was evaluated by measuring the disassembly of microtubules into tubulin and compared to paclitaxel **1**, in parallel tests at the same time and under the same conditions.

Reduction of the C_{11}-C_{12} Double Bond

Most of the natural taxoids possess a double bond at carbons 11 and 12 which induces a boat conformation of the A ring. An initial chemical study of taxoids showed that the C_{11}-C_{12} double bond was resistant to hydrogenation (*20*). To activate the double bound, 10-deacetylbaccatin III **3** was converted into its 13-oxo derivative **9** by treatment with chromic oxide (*30*). The zinc-promoted reduction of the α,β–unsaturated ketone was then investigated.

Treatment of **9** with zinc in a solution of aqueous acetic acid and methanol led to the dihydro derivative **10** (29%) and to the rearranged product **11** (35%) (Scheme 1).

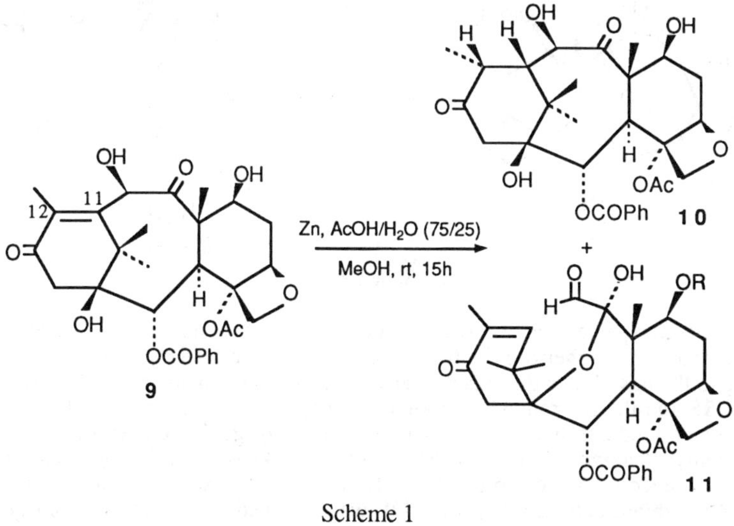

Scheme 1

The structure of compound **10** was determined by spectral methods. NOE experiments and the value of the coupling constant ($J_{11,12}$= 5Hz) indicated that the protons at C-11 and C-12 are oriented on the β-face of the molecule. Ring A adopts a chair conformation. The mass spectrum of compound **11** showed a molecular ion identical to that of 13-oxo-10-deacetylbaccatin III **9**. Its infrared band at 2560 cm^{-1} showed the presence of an aldehyde group. This was confirmed by the presence of a proton signal at δ 9.55 as well as a carbon signal at δ 195. Analysis of the ^{13}C- and ^{1}H-NMR spectra of **11** and comparison with **10** indicated the absence of both the keto group at C-9 and the secondary hydroxyl group at C-10, and the presence of an olefinic proton at C-11. In addition, the carbon signal at δ 95.3 suggested the presence of a quaternary carbon corresponding to an hemiacetal group. From these data, structure **11** was assigned to this rearranged product.

Reduction of the α,β-unsaturated ketone was then investigated under basic conditions using the 7- protected 10-deacetylbaccatin III derivative **12** as starting material to avoid epimerisation at C-7. Thus, a 95% yield of the dihydro derivative **13** was obtained when a solution of **12** in ethanol and ethyl acetate was treated under argon with zinc in a buffered solution of ammonia-ammonium chloride. When the reaction was carried out in air, compound **13** (70%) was isolated together with the hydroperoxide **14** (20%), which rearranged readily to give compounds **15** and **16** (Scheme 2).

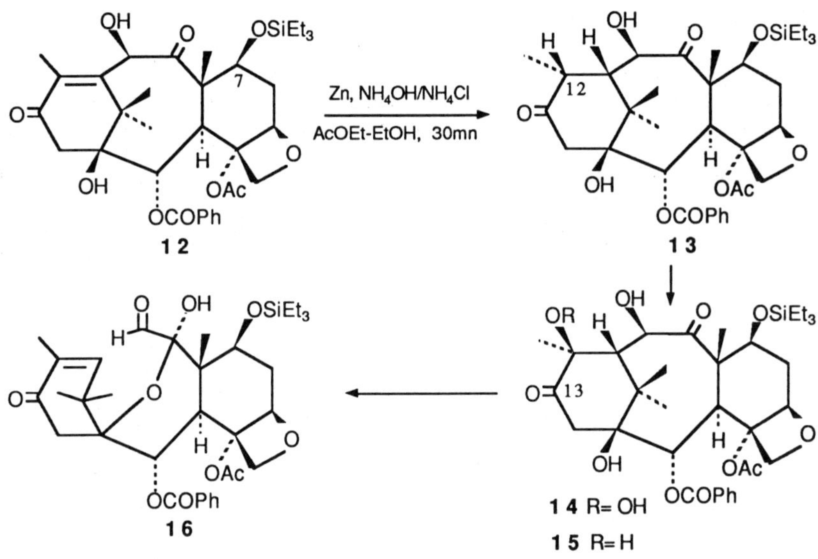

Scheme 2

The unexpected formation of an aldehydic compound such as **16** can now be rationalized : atmospheric oxidation occurs first at the tertiary C-12 carbon of **13** leading to the unstable hydroxyperoxide **14**. This hydroperoxide reacts further to give alcohol **15** and the rearranged compound **16** after cleavage of the C_{10}-C_{11} bond. It should be pointed out that different autoxidative cleavages have been found for the A ring of dihydrotaxinol (*31*). Reduction of the C-13 keto group in compound **13** was cleanly achieved with sodium borohydride, affording the alcohol **17a** in quantitative yield. After deprotection, compound **17b** was obtained. The 13α-OH configuration was assigned using NOESY experiments.

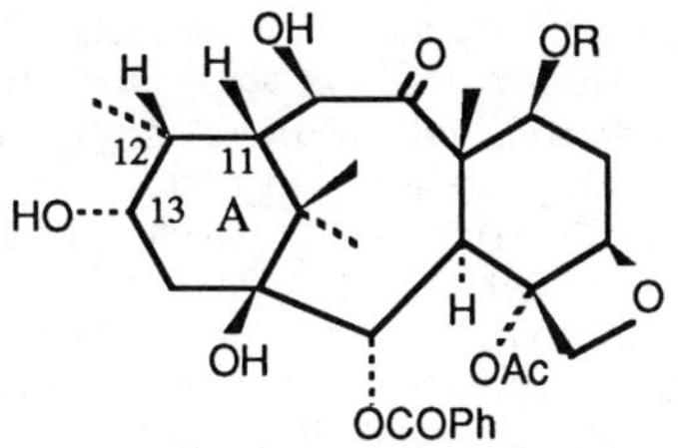

17a R= $Si(Et)_3$ **17b** R= H

To prepare 11,12-dihydrodocetaxel, esterification of **17a** with the acid side chain was studied. Unfortunately, the hydroxyl group at C-13 was resistant to acylation under any conditions. The lack of reactivity of this hydroxyl group can be explained by its hindered axial position due to the chair conformation of the A ring. To determine if the presence of the A ring double bond was biologically important, the antitubulin activity of compound **17b** was evaluated and compared to that of 10-deacetylbaccatin III **3a** (IC_{50} (**3a**) / IC_{50} (taxol) = 50) (*32*). A major reduction of antitubulin activity was noted for compound **17b** (IC_{50} (**3a**) / IC_{50} (taxol) = 200). It is therefore quite clear that the particular folding of antitumor taxoids, influenced by the A ring conformation is important for the interaction with microtubules.

A-Ring Contracted Taxoids from 10-Deacetylbaccatin III

As with other terpenoids, Wagner-Meerwein rearrangement occurs to the A-ring of taxoids bearing a tertiary hydroxyl group at C-1 (*33-35*). Kingston and his collaborators (*33*) found that treatment of the 7, 2'-*O*-diprotected paclitaxel derivative with mesyl chloride and subsequent deprotection led to the A-ring contracted derivative **18** which possesses an antitubulin activity similar to that of paclitaxel (*33*).

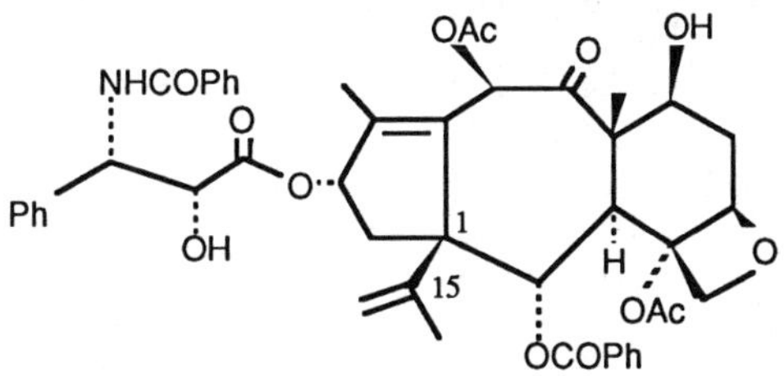

18

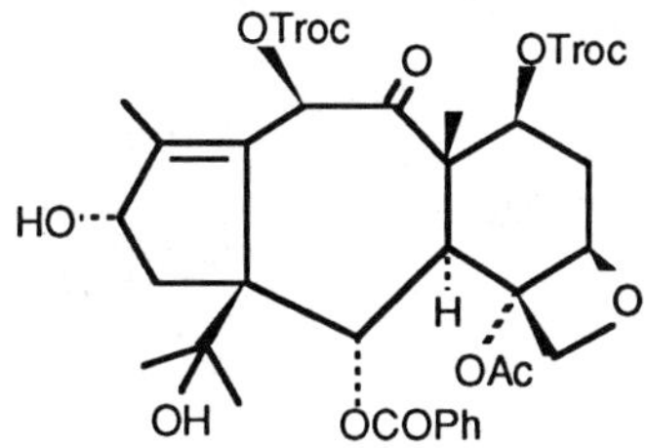

19

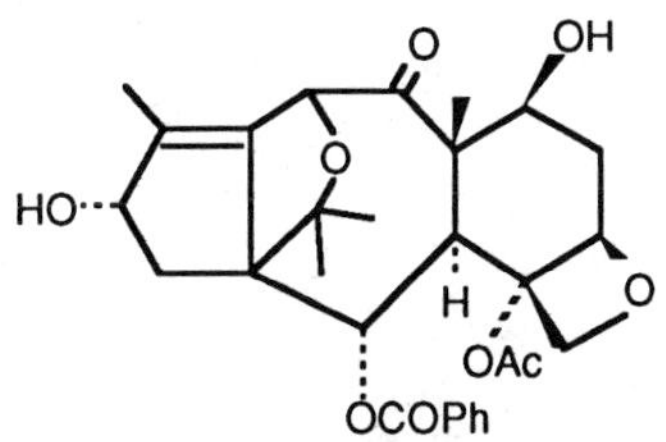

20

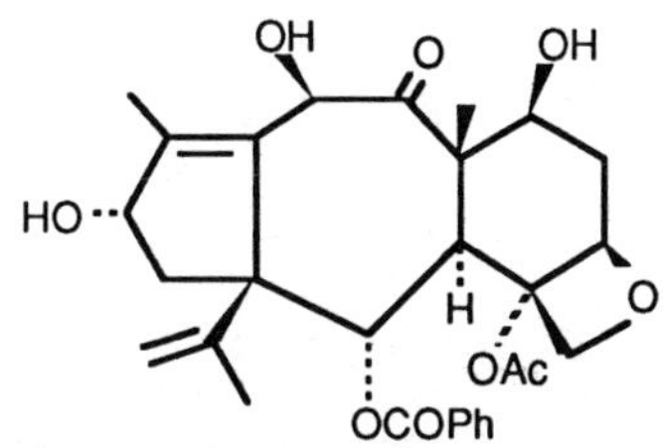

21

Under acidic conditions (aqueous trifluoroacetic acid), the diprotected 10-deacetyl baccatin III derivative **5a** led mainly to the hydroxyisopropyl derivative **19** (*34*).

On the other hand, compound **20** and **21** were obtained after treatment of 10-deacetylbaccatin III with *para*-toluene sulfonic acid in methanol (*35*).

The coupling of the diprotected hydroxyisopropyl derivative **19** with the acid side chain of docetaxel led mainly to ester **22** in which an intramolecular rearrangement has occurred (*36*) (Scheme 3). The five-membered carbonate ring in **22** is characterized by the presence of an IR band at 1820 cm^{-1}. Molecular modeling of compound **19** showed that the hydroxyl at carbon 15 is very close to the C-9 keto group, allowing the formation of an hemiacetal intermediate which reacts with the carbonate protecting group at C-10. Similar rearrangements have been previously noticed with the diprotected derivative of 10-deacetylbaccatin III (*34*). In order to avoid this intramolecular rearrangement, docetaxel was treated directly with trifluoroacetic acid giving the A-ring contracted docetaxel derivative **23** (Scheme 3).

Similar to **18**, the A-ring contracted analogue **23** was as active in the microtubule depolymerization assay as paclitaxel, but was devoid of cytotoxicity (*34*, *36*). Conformational study of compounds **18** and **23** shows that the structures of the taxane core and of the side chain are close to that of docetaxel **2** (*36*). This explains the unexpected interaction of the A-ring contracted derivative with microtubules.

On the other hand, the rearranged, A-ring contracted derivative **24** is forty times less active than paclitaxel in the microtubule disassembly assay. This compound was prepared from 7,10-diTroc-docetaxel after reaction with trifluoroacetic acid and deprotection (*36*) (Scheme 3).

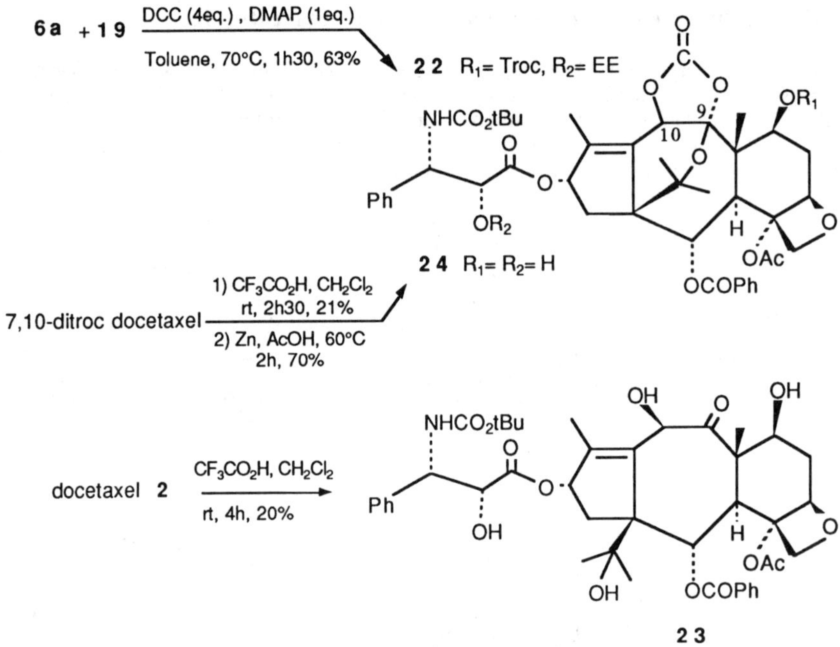

Scheme 3

It should be pointed out that natural A-ring contracted taxoids (*abeo*-taxoids) have recently been isolated from the genus *Taxus* (*35,37-39*).

Opened-Oxetane Analogues and Chemical Reactivity

Based on the inactivity of the secotaxol derivative **25**, it has been reported that the oxetane ring was essential for interaction with microtubules and cytotoxicity (*33*).

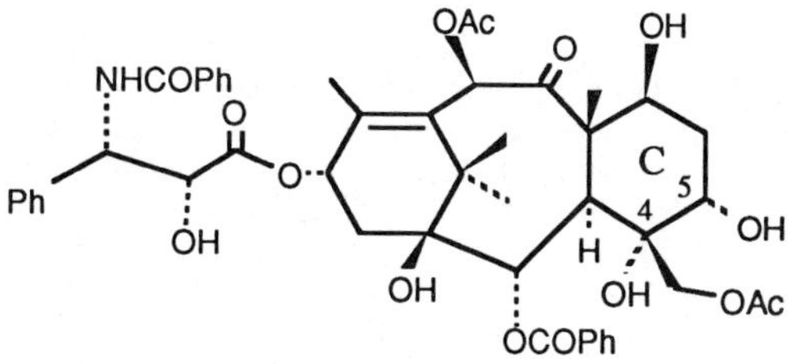

25

Although the presence of the oxetane ring induces a strained chair conformation for the C ring which may be an important feature for interaction with the receptor, it must be pointed out that, in comparison to the structure of paclitaxel, the secotaxol analogue **25** lacks the acetyl group at C-4 and possesses a C-5 hydroxyl group with the configuration reversed from taxol. Thus, preparation of opened oxetane analogue bearing a β-hydroxyl group at C-5 and an acetyl group at C-4 was needed to know the real influence of the oxetane ring on the antitubulin activity. Another interest with the seco analogues is that they can serve as precursors for new derivatives possessing different ring systems at carbons 4 and 5 or other structural groups.

The oxetane ring can be selectively opened using electrophilic reagents. Cleavage of the vicinal acetyl oxetane occurs by intramolecular attack of the acetyl group upon the methylene at carbon 5. The reaction was carried out with various electrophilic reagents from paclitaxel **1** (*33, 40*), baccatin III **3b** (*40*) and 10-deacetylbaccatin III **3a** (*34*). The use of tin tetrachloride gave the best yield of opened oxetane derivatives (*36, 40*). In our hands, treatment of 7,10-diTroc-10-deacetylbaccatin III **5a** with tin tetrachloride in methylene chloride (rt, 7 min.) afforded 5-acetyl **26** and 20-acetyl **27** analogues with a yield of 10% and 76%, respectively. Considering that compound **27** could be used as a precursor for analogues bearing various substituents at carbons 4 and 5, we studied its chemical reactivity.

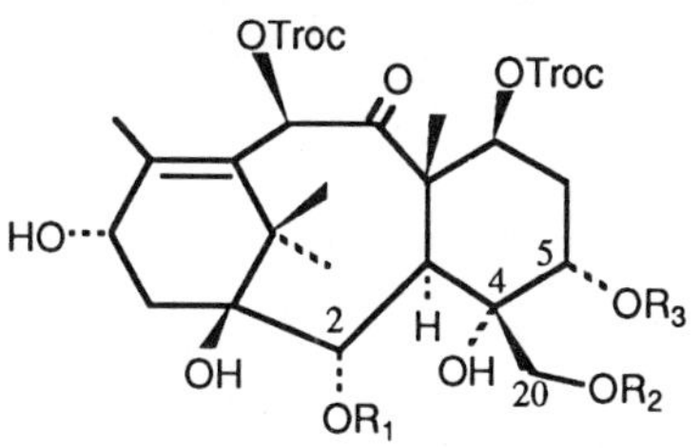

26 R_1= COPh, R_2= H, R_3= Ac

27 R_1= COPh, R_2= Ac, R_3= H

28 R_1= COPh, R_2= R_3= H

29 R_1= H, R_2= COPh, R_3= H

First attempts to produce the pentaol **28** from **27** were unsuccessful because transbenzoylation side reactions occurred. Indeed, hydrolysis of compound **27** with an aqueous THF solution of K_2CO_3 gave mainly the pentaol **29** with a benzoyl group at C-20. The migration of the C-2 benzoyl group to the C-20 alcohol was determined by analysis of the ^1H NMR spectra : the signal for H-2 is shifted upfield from 5.52 ppm in **27** to 3.86 ppm in **29** and the signals for 2H-20 are shifted downfield from 3.97 and 4.08 ppm in **27** to 4.59 and 5.22 ppm in **29**. The structure of compound **29** was confirmed by COSY and NOESY experiments.

Similar transbenzoylation has been reported recently by other teams (D.G.I. Kingston, Advances in the Chemistry of Taxane and Taxoid anticancer agents, 207th ACS National Meeting, San Diego, 1994). The formation of compound **29** under basic hydrolysis involves the cleavage of the primary C-20 acetyl group followed by the migration of the benzoyl group from the C-2 to the C-20 position which are close enough to allow the transformation.

Similar experiments have been carried out on the diprotected 13-oxo-10-deacetylbaccatin III derivative **30** which was obtained quantitatively by oxidation of 7,10-diTroc-10-deacetylbaccatin III **5a** with chromium trioxide (*30*). Compound **30** was treated with 2 equivalent of $SnCl_4$ in methylene chloride at room temperature for 10 min to give the opened oxetane derivatives **31** (27%) and **32** (66%). With prolonged reaction time, compound **33** was also obtained. Formation of **33** was unexpected under these conditions, but it was shown that the 5-acetyl derivative **31** is very unstable and forms the 20-acetyl (**32**) and 2-hydroxy-20-benzoyl (**33**) isomers during chromatographic purification. The difference in reactivity of the 13-hydroxy (**26**) and 13-oxo (**31**) 5-acetyl opened-oxetane derivatives can be explained by the presence in compound **26** of a hydrogen bond between the C-13 hydroxyl group and the acetyl group at C-5 which stabilizes the molecule and prevents the easy migration of the acyl groups as was found for compound **31**.

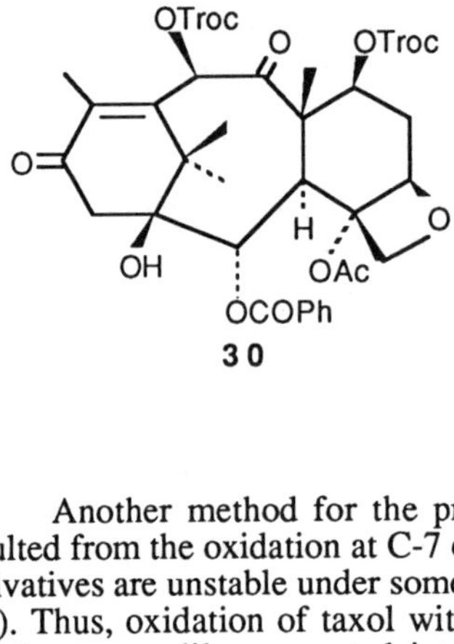

30

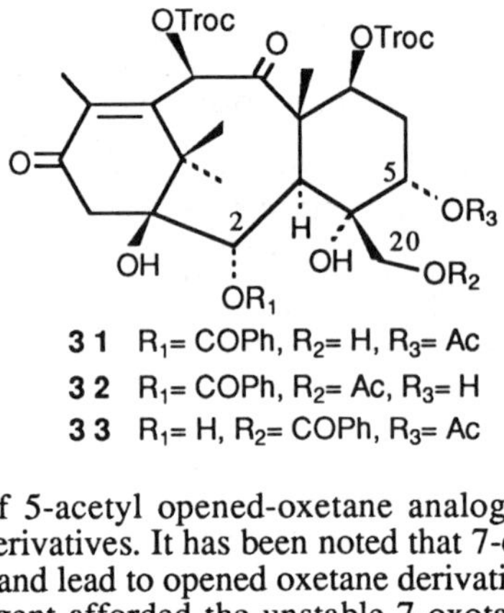

31 R_1= COPh, R_2= H, R_3= Ac
32 R_1= COPh, R_2= Ac, R_3= H
33 R_1= H, R_2= COPh, R_3= Ac

Another method for the preparation of 5-acetyl opened-oxetane analogues resulted from the oxidation at C-7 of oxetane derivatives. It has been noted that 7-oxo derivatives are unstable under some conditions and lead to opened oxetane derivatives (*41*). Thus, oxidation of taxol with Jones' reagent afforded the unstable 7-oxotaxol which was readily converted into the D-seco derivative **34** after purification by preparatrive TLC (*41*).

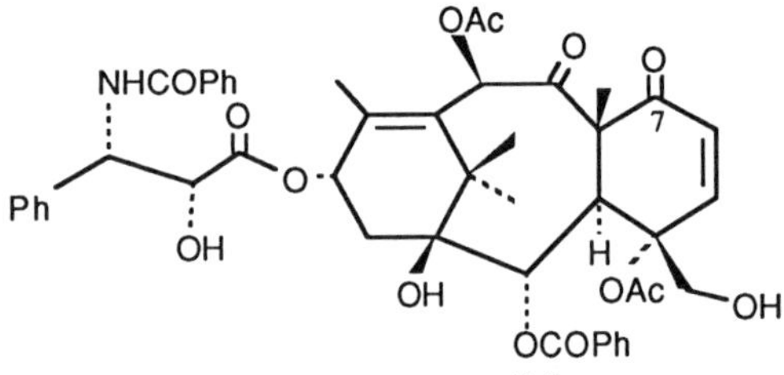

34

In our hands, oxidation of 13-oxobaccatin III **35** with PCC in pyridine at 80°C afforded the unsaturated ketone **36** directly (Scheme 4). Analysis of the NMR spectra showed two doublets at 5.97 (J=10 Hz) and 6.65 (J=10 Hz) ppm corresponding to the C-5 and C-6 protons. The resonance of the C-20 protons at 4.25 and 4.44 ppm (each a doublet, J=12Hz) is consistent with the presence of an acetyl group at C-20 (*33, 36, 40*).

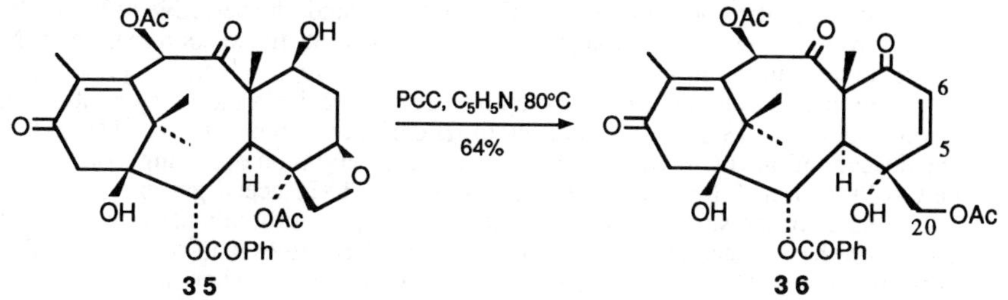

Scheme 4

Further chemical studies on these opened-oxetane derivatives are in progress with the goal of preparing taxoids bearing a modified ring system at carbons 4 and 5.

Taxoids with Modifications at Carbons 2 and 4

Preliminary studies on the reactivity of 7-triethylsilyl-10-deacetylbaccatin III **5c** (*34*) and 7-triethylsilyl-baccatin III **5b** (*42*) showed that the acyl groups at carbon 2 and 4 could be removed under basic or reductive conditions. During our first assays, we found that alkaline hydrolysis (NaOH, rt, 5h), methanolysis (NaOMe, rt, 3h) and metal hydride reduction ($LiAlH_4$, THF, rt) of 7-triethylsilyl-10-deacetylbaccatin III **5c** led to the 2-debenzoyl **37** and 2-debenzoyl-4-deacetyl **38** derivatives.

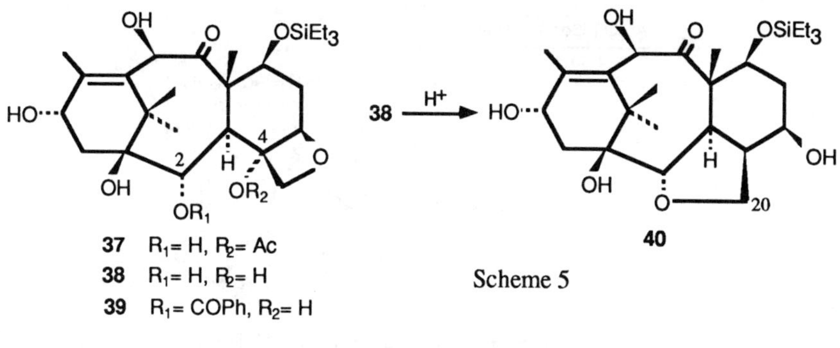

Scheme 5

The formation of the 4-deacetyl derivative **39** also occurred together with the 2-debenzoyl derivative **37**, under milder reduction conditions ($LiAlH_4$, THF at -30°C) (*34, 36*). Selective deacetylation at C-4 was also described by other teams (Kingston, D.G.I et al.; Georg, G.I. et al., Advances in the Chemistry of Taxane and Taxoid anticancer agents, 207th ACS National Meeting, San Diego, 1994).

Along with the deacylated derivatives **37** and **38** obtained during hydrolysis, the formation of rearranged compounds such as **40** occurred under longer reaction times or at higher temperatures (*34,36,42*). It should be noted that **38** isomerises readily to **40** under mildly acidic conditions (*34,42*) (Scheme 5).

Different metal-hydride systems have been investigated in order to find the appropriate conditions for selective reduction of the acetyl group. We found that treatment of 7-triethylsilyl-10-deacetylbaccatin III with super hydride in THF at -78 °C led quantitatively to the desired 4-deacetyl derivative **39**. In contrast, Chen and collaborators showed that the benzoyl group at C-2 could be selectively removed after

Red-Al reduction of 7,13-di(triethylsilyl)baccatin III (*43*). The selective cleavage of the tertiary acetyl group which takes place in the 7-protected derivative **5c** can be explained by the formation of an alkoxy-hydride complex at C-13 followed by intramolecular hydride reduction of the C-4 acetyl group. On the other hand, the selective reduction of the benzoyl group at C-2 in a 13-protected derivative could be due to coordination of the metal hydride reactant with the C-1 hydroxyl group (*43*). Such selective reductions of the C-4 acetyl or C-2 benzoyl groups have also been noted during the methanolysis of the 7-protected and 7, 13-diprotected baccatin III derivatives respectively (*42*). In that case, the deacetylation at C-4 was explained by a transacetylation occurring from the C-4 to the C-13 positions followed by the hydrolysis of the C-13 acetyl group.

The effect of the substituents at carbons 2 and 4 on antitubulin activity was then investigated on the 7-protected 10-deacetylbaccatin III derivatives, and compared to that of paclitaxel. As noted above, antitubulin interaction is retained when no side chain is present at C-13. Thus, 7-triethylsilyl-10-deacetylbaccatin III **5c** is 40 times less active than paclitaxel. The absence of the C-4 acetyl (**39**, IC_{50}/IC_{50} (paclitaxel)= 80) or C-2 benzoyl (**37**, IC_{50}/IC_{50} (paclitaxel)= 200) groups reduced activity but not to the same extent. Antitubulin activity was greatly diminished when the benzoyl group was lacking. Removal of the two acyl groups (**38**, IC_{50}/IC_{50} (paclitaxel) > 300) resulted in complete loss of activity (*36*). These observations lead to the suggestion that hydrophobic interaction of the benzoyl group plays a major role in the binding of these compounds to microtubules.

Acylation of compounds **37**, **38** and **40** was then studied. Treatment of 2-debenzoyl-7-triethylsilyl-10-deacetylbaccatin III **37** with acetic anhydride in pyridine at room temperature led to the 2-acetyl derivative **41**. On the other hand, the 2-debenzoyl, 4-deacetyl derivative **38** was easily acetylated at C-13 leading to compound **42** (*36, 42*) (Scheme 6). This confirms that the presence of an acetyl group at C-4 hinders the C-13 hydroxyl group (*30, 42, 44*), which is then resistant to mild acylation conditions.

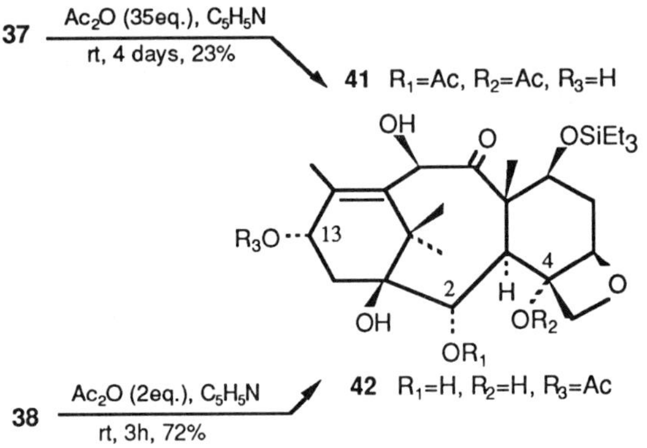

Scheme 6

Coupling of compound **38** with the acid side chain of docetaxel **6a** occurs at room temperature with a yield of 85% without epimerization at carbon 2' leading to the docetaxel derivative **43**. Concomitant deprotection at C-7 and C-2' under acidic conditions led to the tetrahydrofuran derivative **44** (*36*) (Scheme 7). The structure of the rearranged product **44** formed during the acidic deprotection was largely determined by NMR spectroscopy. The comparison of proton and carbon signals at positions 2, 5 and 20 with those of compounds **43** and **40** revealed that **44** possesses a tetrahydrofuran ring system. Indeed, the coupling constants of the C-20 protons change from J=8Hz, in compounds bearing an oxetane, to J=10Hz in tetrahydrofuran derivatives. Moreover, the carbon signals for C-2 and C-5 in tetrahydrofuran compounds are shifted downfield and upfield, respectively (*34, 36*).

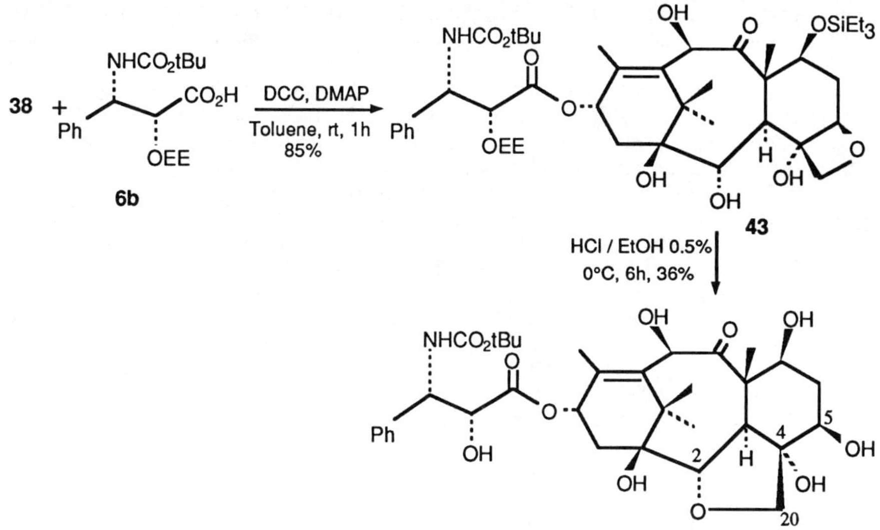

Scheme 7

Coupling of the tetrahydrofuran derivative **40** with the acid side chain of docetaxel and deprotection led to compound **45**, bearing the side chain at C-5 (*36*).

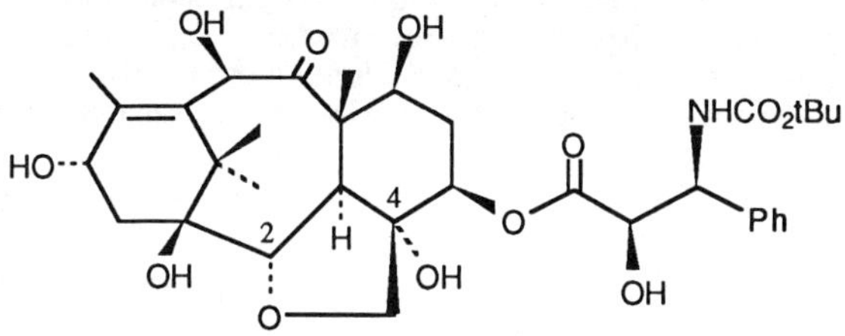

45

To avoid rearrangement of the oxetane ring, acetylation was carried out on compound **43**. Acetylation occurred first at C-10, then at C- 2 leading, respectively, to the mono- (**47**) and di- (**48**) acetates. Partial acidic deprotection of **48** gave compound **49** which was characterized by spectral data (*36*).

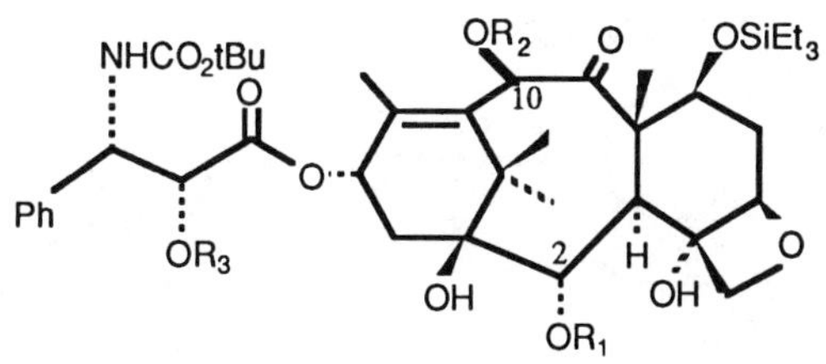

47 R_1= H, R_2= Ac, R_3= EE **48** R_1= R_2= Ac, R_3= EE

49 R_1= R_2= Ac, R_3= H

It is noteworthy that the order of reactivity of the hydroxyl groups at C-2 and C-10 towards acylation is related to the nature of the substituent at C-13. Indeed, on acetylation of 13-acetyl-10-deacetylbaccatin III **42**, the 2-OH group is less reactive than the 10-hydroxyl group (*42*). This difference in reactivity can be explained by the fact that the side chain at C-13 in **43** hinders the hydroxyl group at C-2, making it less reactive.

Both tetrahydrofuran derivatives **44** (IC_{50}/IC_{50} (taxol) = 100) and **45** (IC_{50}/IC_{50} (taxol) = 200) were substantially less active than paclitaxel **1** on microtubules. Compound **49** (IC_{50}/IC_{50} (taxol) > 300) was completely inactive. Molecular modeling studies of 2-debenzoyldocetaxel and 4-deacetyldocetaxel (*36*) were performed using the conditions described earlier for docetaxel (*45*). In docetaxel, the side chain adopts a particular conformation due to a network of intramolecular hydrogen bonding between the 1'C=O, 2'-OH and 3'-NH substituents of the side chain. In the case of 2-debenzoyldocetaxel, hydrogen bonding can occur between the C-4 acetyl group and the hydroxyl group at C-2' of the side chain. On the other hand, molecular modeling of 4-deacetyldocetaxel led to a structure in which hydrogen bonding occurs between the hydroxyl group at C-4 and the 3' NH group of the side chain. Consequently, the lack of acyl groups at C-2 or C-4 and the different positions of the side chain in 2-debenzoyldocetaxel and 4-deacetyldocetaxel, when compared to the "active" structure of docetaxel, must lead to a decrease in activity.

Reactivity of the C-1 hydroxyl group.

With the exception of the rearranged products formed under acidic or electrophilic conditions, few other experiments showed rearrangement involving the hydroxyl group at C-1. The conformation of docetaxel obtained by X-ray analysis (*46*) and molecular modeling showed that the tertiary hydroxyl group at C-1 is very close to the benzoyl group at C-2 (2.8Å). Transbenzoylation could thus occur under special conditions.

We noticed such a benzoyl migration when baccatin III derivatives were refluxed in pyridine in the presence of base. Under these conditions, 7-acetylbaccatin III **50** led to the A-ring modified analogue **51** which possesses the benzoyl group at C-1

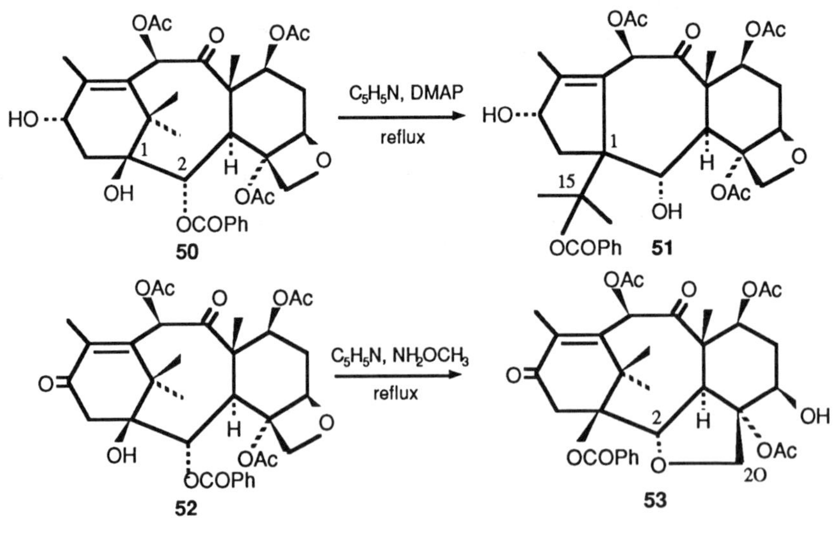

Scheme 8

(Scheme 8). One possible mechanism for the formation of **51** would be first the migration of the benzoate from C-2 to C-1. Then, migration of the benzoate from C-1 to C-15 and contraction of the A ring could take place through a concerted mechanism. On the other hand, treatment of 13-oxo-7-acetylbaccatin III **52** with methoxylamine in refluxing pyridine gave compound **53** (Scheme 8). This experiment was carried out in an attempt to prepare new analogues bearing a nitrogen atom at carbon 13 instead of an oxygen. In that case, migration of the benzoate led to the free hydroxyl group at C-2, which attacks to give the tetrahydrofuran derivative **53.** Such benzoyl migrations have recently been noticed in 14β-hydroxy-10-deacetylbaccatin III derivatives as well(*47*).

Acknowledgments

This work was supported by Rhône-Poulenc Rorer S.A. and the Ligue Nationale contre le Cancer.

Literature Cited

(1) Wani, M. C.; Taylor, H. L.; Wall, M. E.; Coggon, P.; McPhail, A. T. *J. Am. Chem. Soc.* **1971**, *93*, 2325

(2) Rowinski, E. K.; Donehower, R. C. *Pharm. Ther.* **1991**, *52*, 35-84.

(3) Guénard, D.; Guéritte-Voegelein, F.; Potier, P. *Acc.Chem.Res.*, **1993**, *26*, 160.

(4) Lavelle, F.; Guéritte-Voegelein, F.; Guénard, D. *Bull.Cancer*, **1993**, *80*, 326

(5) Schiff, P.B.; Fant, J.; Horwitz, S.B. *Nature*, **1979**, *277*, 665

(6) Nicolaou, K.C.; Zang, Z.; Liu, J.J.; Ueno, H.; Nantermet, P.G.; Guy, R.K.; Claiborne, C.F.; Renaud, J.; Couladouros, E.A.; Paulvannan, K.; Sorensen, E.J. *Nature*, **1994**, *367*, 630

(7) Holton, R.A.; Somoza, C.; Kim, H-B.; Liang, F.; Biedeger, R.J.; Boatman, P.D.; Shindo, M.; Smith, C.C.; Kim, S.; Nadizadeh, H.; Suzuki, Y.; Tao, C.; Vu, P.; Tang, S.; Zhang, P.; Murti, K. K.; Gentile, L. N.; Liu, J. H. *J.Am.Chem.Soc.*, **1994**, *116*, 1595

(8) Holton, R.A.; Kim, H-B.; Somoza, C.; Liang, F.; Biedeger, R.J.; Boatman, P.D.; Shindo, M.; Smith, C.C.; Kim, S.; Nadizadeh, H.; Suzuki, Y.; Tao, C.; Vu, P.; Tang, S.; Zhang, P.; Murti, K. K.; Gentile, L. N.; Liu, J. H. *J.Am.Chem.Soc.*, **1994**, *116*, 1599

(9) Chauvière, G.; Guénard, D.; Picot, F.; Sénilh, V.; Potier, P. *C. R. Acad. Sci. Paris.*, **1981**, *293*, 501

(10) Denis, J-N.; Greene, A.; Guénard, D.; Guéritte-Voegelein, F.; Mangatal, L.; Potier, P. *J.Am.Chem.Soc.*, **1988**, *110*, 5917

(11) Mangatal, L; Adeline, M-T.; Guénard, D.; Guéritte-Voegelein, F.; Potier, P. *Tetrahedron*, **1989**, *45*, 4177

(12) Colin, M.; Guénard, D.; Guéritte-Voegelein, F.; Potier, P. Eur. Pat Appl. EP 253,739, 1988, *Chem.Abstr.*, **1988**, *109*, 22763x.

(13) Colin, M.; Guénard, D.; Guéritte-Voegelein, F.; Potier, P.Eur. Pat Appl. EP 253,738, 1988. *Chem.Abstr.*, **1988**, *109*, 22762w.

(14) Denis, J-N.; Greene, A.; Guénard, D.; Guéritte-Voegelein, F. Eur. Pat Appl. EP 336, 840, 1989. *Chem.Abstr.* **1990**, *112*, 158670j

(15) Colin, M.; Guénard, D.; Guéritte-Voegelein, F.; Potier, P. Eur. Pat Appl. EP 336,841, 1989. *Chem.Abstr.* **1990**, *112*, 178487h

(16) Commerçon, A.; Bézard, D.; Bernard, F. and Bourzat, J.D., *Tetrahedron Letters*, **1992**, *33*, 5185

(17) a) Holton, R.A. Eur.Pat.Appl. EP 400,971, 1990. *Chem Abstr.* **1991**, *114*, 164568q b) Holton, R.A. Eur.Pat.Appl. EP 428,376, 1991. *Chem Abstr.* **1991**, *115*, 114817v

(18) Palomo , C.; Arrieta, A.; Cossio, P.; Aizpurua, J.M.; Mielgo, A.; Aurrekoetxea, N. *Tetrahedron Letters*. **1990**, *31*, 6429-6432

(19) Ojima, I.; Habus, I.; Zhao, M; Georg, G.; Jayasinghe, L.R. *J. Org. Chem.* **1991**, *56*, 1681

(20) Blechert, S.; Guénard, D. *Taxus* Alkaloids in *The Alkaloids. Chemistry and Pharmacology*; Brossi, A., Ed; Academic press: San Diego, 1990; Vol 39, pp 195-238.
(21) Kingston, D.G.I.; Molinero, A.A.; Rimoldi, J.M. The Taxane Diterpenoids in *Progress in the Chemistry of Organic Natural Products,* Hertz, W., Kirby, G.W., Moore, R.E., Steglish, W. and Tamm, Ch., Ed.; Springler-Verlag, Wien-New York, 1993, 1-206
(22) Georg, G.I.; Ali, S.M.; Zygmut, J.; Jayasinghe, L.R. *Exp.Opin.Ther.Patents* **1994**, *4*, 109
(23) Nicolaou, K.C.; Dai, W-M.; Guy, R.K. *Angew.Chem.Int.Ed.Engl.*, **1994**, *33*, 15
(24) Georg, G.I.; Cheruvallath, Z.S.; Vander Velde, D.; Ye, Q-M.; Mitscher, L.A.; Himes, R.H. *Bioorg.Med.Chem.Lett.*, **1993**, *3*, 1349
(25) Margraff, R.; Bézard, D.; Bourzat, J-D.; Commerçon, A. *Bioorg. Med. Chem. Lett.*, **1994**, *4*, 233
(26) Klein, L.L. *Tetrahedron Lett.*, **1993**, *34*, 2047
(27) Ojima, I.; Fenoglio, I.; Park, Y.H.; Sun, C-M.; Appendino, G.; Pera, P.; Bernacki, R.J. *J.Org.Chem.*, **1994**, *59*, 515
(28) Kingston, D. G. I. *Pharm. Ther.* **1991**, *52*, 1-34.
(29) Suffness, M. *Ann. Rep. Med. Chem.*, **1993**, *28*, 305
(30) Sénilh, V.; Guéritte, F.; Guénard, D.; Colin, M.; Potier, P. *C.R.Acad.Sci. Paris.*, **1984**, *T.299*, Série II, 1039
(31) Maki, Y.; Yamane, K. *Chem.Pharm.Bull.*, **1969**, *17*, 2071
(32) Guéritte-Voegelein, F.; Guénard, D.; Lavelle, F.; Le Goff, M-T.; Mangatal, L; Potier, P. *J. Med. Chem.* **1991**, *34*, 992
(33) Samaranayake, G.; Magri, N. F.; Jitrangsri, C.; Kingston, D. G. I.*J. Org. Chem.* **1991**, *56*, 5114
(34) Wahl, A.; Guéritte-Voegelein, F.; Guénard, D.; Le Goff M-T.; Potier, P. *Tetrahedron* **1992**, *48*, 6965
(35) Appendino, G.; Ozen, H.C.; Gariboldi, P.; Torregiani, E.; Gabetta, B.; Nizzola, R.; Bombardelli, E. *J.Chem.Soc.Perkin Trans*, **1993**, 1563
(36) Wahl, A. " Chimie des Taxoïdes. Hémisynthèse d'analogues structuraux du Taxotère®. Relations Structure-Activité", Ph.D. Dissertation, Université Paris XI Orsay, **1993**
(37) Georg, G.I.; Gollapudi, S.R.; Grunewald, G.L.; Gunn, C.W.; Himes, R.H.; Rao, B.K.; Liang, X-Z.; Mirhom, Y.W.; Mitscher, L.A.; Vander Velde, D.G.; Ye, Q-M. *Biorg.Med.Chem.Lett.*, **1993**,*3*, 1345
(38) Appendino, G.; Barbonbi, L.; Gariboldi, P.; Bombardelli, E.; Gabetta, B.; Viterbo, D. *J.Chem.Soc.Chem.Commun.*, **1993**, 1587
(39) Fuji, K.; Tanaka, K.; Li, B.; Shingu, T.; Sun, H.; Taga, T. *J.Nat.Prod.*, **1993**, *56*, 1520
(40) Chen, S-H.; Huang, S.; Wei, J.; Farina, V. *Tetrahedron*, **1993**, *49*, 2805
(41) Magri, N.F.; Kingston, D.G.I. *J.Org.Chem.*, **1986**, *51*, 797
(42) Samaranayake, G.; Neidigh, K.A.; Kingston, D.G.I. *J.Nat.Prod.*, **1993**, *56*, 884
(43) Chen, S-H.; Farina, V.; Wei, J-M.; Long, B.; Fairchild, C.; Mamber, S;W.; Kadow, J.F.; Vyas, D.; Doyle, T.W. *Biorg.Med.Chem.Lett.*, **1994**, *4*, 479
(44) Guéritte-Voegelein, F.; Sénilh, V.; David, B.; Guénard, D.; Potier, P. *Tetrahedron* **1986**, *42*, 4451
(45) Dubois, J.; Guénard, D.; Guéritte-Voegelein, F.; Guedira, N.; Potier, P.; Gillet, B.; Beloeil, J-C. *Tetrahedron*, **1993**, *49*, 6533
(46) Guéritte-Voegelein, F.; Mangatal, L.; Guénard, D.; Potier, P.; Guilhem, J.; Cesario, M.; Pascard, C. *Acta Cryst.***1990**,*C46*, 781
(47) Appendino, G.; Varese, M.; Gariboldi, P.; Bombardelli, E.; Gabetta, B.; *Tetrahedron Lett* , **1994**, 2217

RECEIVED September 13, 1994

Chapter 15

Recent Advances in the Chemistry and Structure–Activity Relationships of Paclitaxel

David G. I. Kingston

Department of Chemistry, Virginia Polytechnic Institute and State University, Blacksburg, VA 24061–0212

Recent studies on the chemistry of taxol are described. Side chain chemistry includes the conversion of cephalomannine to taxol, and the attachment of the taxol side chain to baccatin III. Chemistry of the northern hemisphere includes deoxygenation at C-10 and C-7, and chemistry of the southern hemisphere includes deacetylation at C-4 and debenzoylation and reacylation at C-2; this last transformation has yielded several derivatives with significantly improved tubulin-assembly activity as compared with taxol. A summary of structure-activity relationships of taxol is given.

The effectiveness of taxol (**1**) as a chemotherapeutic agent against ovarian and breast cancer has sparked a worldwide study of its synthesis, biosynthesis, mechanism of action, nad chemistry, and recent work in the chemical areas has been reviewed on several occasions, both by ourselves (*1-4*) and others (*5-9*). In our group we have elected to focus on the chemistry and structure-activity relationships of taxol, and this review will describe results since the publication of our recent reviews (*1-4*). The work will be presented in the form of a tour of the taxol structure, beginning with the side chain, and proceeding on to the northern hemisphere, and then on to the southern hemisphere. The tour will conclude with a summary of the known structure-activity relationships of taxol.

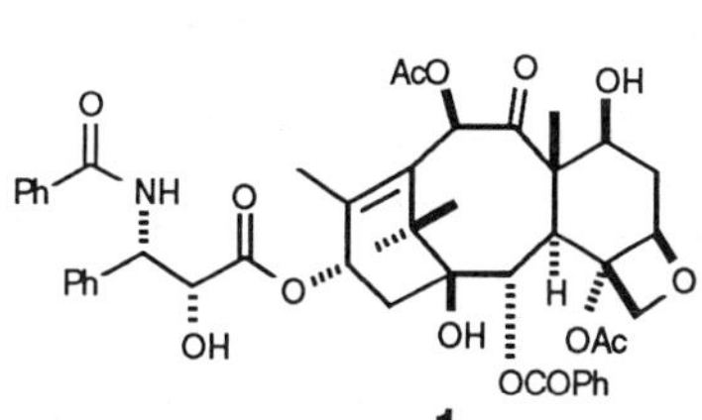

1

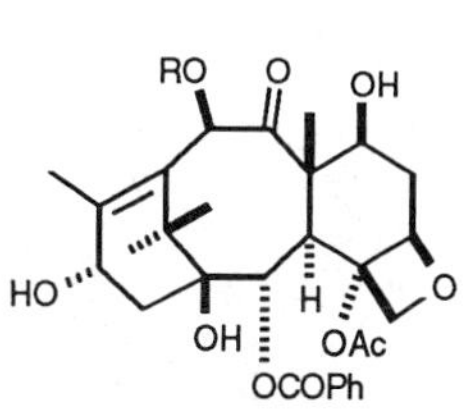

2 R = Ac
3 R = H

NOTE: Paclitaxel is the generic name for Taxol, which is now a registered trademark.

0097–6156/95/0583–0203$08.00/0

The Side Chain

The partial synthesis of taxol by attachment of the β-phenylisoserine side chain to a suitably protected baccatin III (**2**) has been the subject of intense study, since this conversion offers the capability of preparing taxol from the much more readily available 10-deacetylbaccatin III (**3**), and thus avoiding the supply problems inherent in the production of taxol from the bark of the relatively scarce and slow-growing western yew (*10*). Notable achievements in this area have been made by Greene (*11-13*), Holton (*14*), Ojima (*15*), Georg (*16*), and Commercon (*17*), and Holton's β-lactam process has been selected by Bristol-Myers Squibb for the synthesis of taxol from 10-deacetylbaccatin III. One aspect of the partial synthesis of taxol that has been overlooked, however, is that of its synthesis from cephalomannine (**4**). Cephalomannine occurs with taxol in *T. brevifolia* and other *Taxus* species, and it can be a major component of the mixture in some situations. A method to convert cephalomannine to taxol would thus be of interest as a means of extending the usefulness of the direct isolation process.

The conversion of cephalomannine to taxol could in principle be achieved by reductive removal of the side chain (*18*), followed by reacylation by one of the methods described above (*11-17*). The possibility of a direct conversion intrigued us, however, because it presented the challenge of hydrolysing an amide selectively in the presence of various ester linkages. In the event, the problem was solved by the reaction scheme shown below. Osmylation of cephalomannine as previously described (*19*), followed by periodate cleavage, yielded the ketoamide **5** in excellent yield. Benzoylation of **5** gave the 2'-benzoyl derivative **6** in quantitative yield, and treatment of **6** with *o*--phenylenediamine and acid selectively cleaved the ketoamide group and allowed intramolecular O→N acyl transfer to occur, yielding taxol (**1**) in good yield. This process thus provides an efficient and high-yield pathway for the conversion of cephalomannine to taxol.

The Oxazoline Route. Although the conversion described above is effective for the preparation of taxol from cephalomannine, the preparation of taxol from baccatin III remains the most important method for the semisynthesis of this compound. We thus elected to investigate this conversion. Our interest was sparked by the observation that treatment of taxol with triphenylphosphine and carbon tetrachloride converts it to a mixture of the *cis*- and *trans*-oxazolines **7** and **8**, with the isomer with inverted stereochemistry at C-2' (**7**), being the major product (Scheme 2). Hydrolysis of the minor product **8** with refluxing O.1N HCl yielded taxol in good yield. The minor product **8** could be prepared (as its 7-triethylsilyl derivative) in excellent yield by reacting the known oxazoline **10** (20) with 7-(triethylsilyl)baccatin III (**9**) in the presence of DCC and PP. The coupled product **11** was formed in 90% yield, and hydrolysis as before gave taxol in 75% yield (Scheme 2). This process thus provides an efficient method for the synthesis of taxol from 10-deacetylbaccatin III, and it will probably be comparable with other methods in terms of overall efficiency and yield once it has been optimized (*21*).

The Northern Hemisphere

The northern hemisphere of taxol consists of the region from C-12 to C-6, and comprises the C-11 (*12*) double bond, the C-10 acetoxyl group, the C-9 carbonyl group, and the C-7 hydroxyl group. We have earlier shown that the C-11 (*12*) double bond is resistnat both to hydrogenation (22) and to oxidation (*19*), and in our present studies we elected to investigate the significance of the oxygen functions at C-10 and C-7.

The C-10 Acetoxyl group. The C-10 acetoxyl group was removed in a six-step sequence starting from 10-deacetylbaccatin III (**3**) (Scheme 3). Treatment of **3** with triethylsilyl chloride yielded 10-deacetyl-7-(triethylsilyl) baccatin III (**12**), which was

Scheme 1

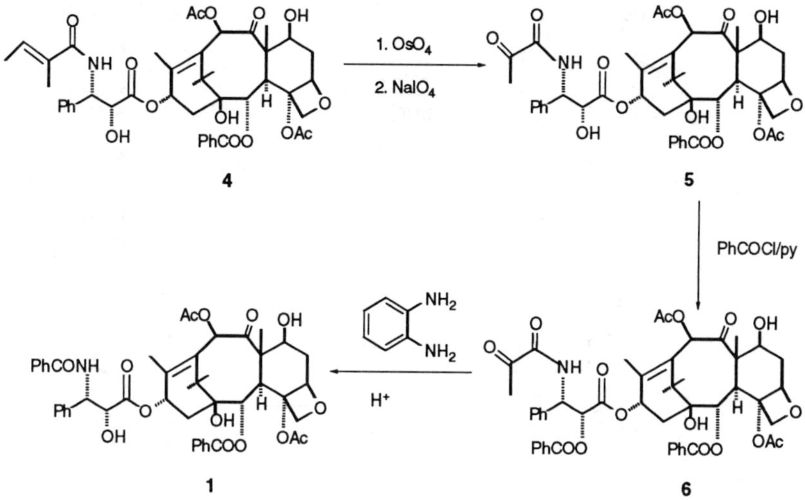

Scheme 2

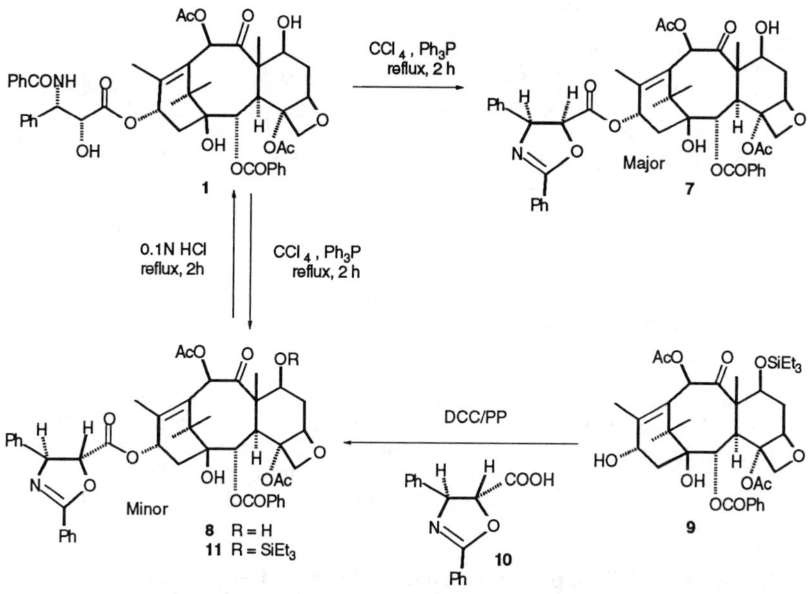

converted to its 10-(S-methylthiocarbonyl) derivative **13** with sodium hydride, carbon disulfide, and methyl iodide. Attachment of a protected C-13 side chain by Commercon's method (*17*) yielded the xanthate **14**, which was deoxygenated by Barton-McCombie reduction (*23*) to yield the 10-deacetoxy derivative **15**. Hydrolysis followed by benzoylation then gave 10-deacetoxytaxol (**16**) in 11% overall yield from **3** (*24*).

Scheme 3

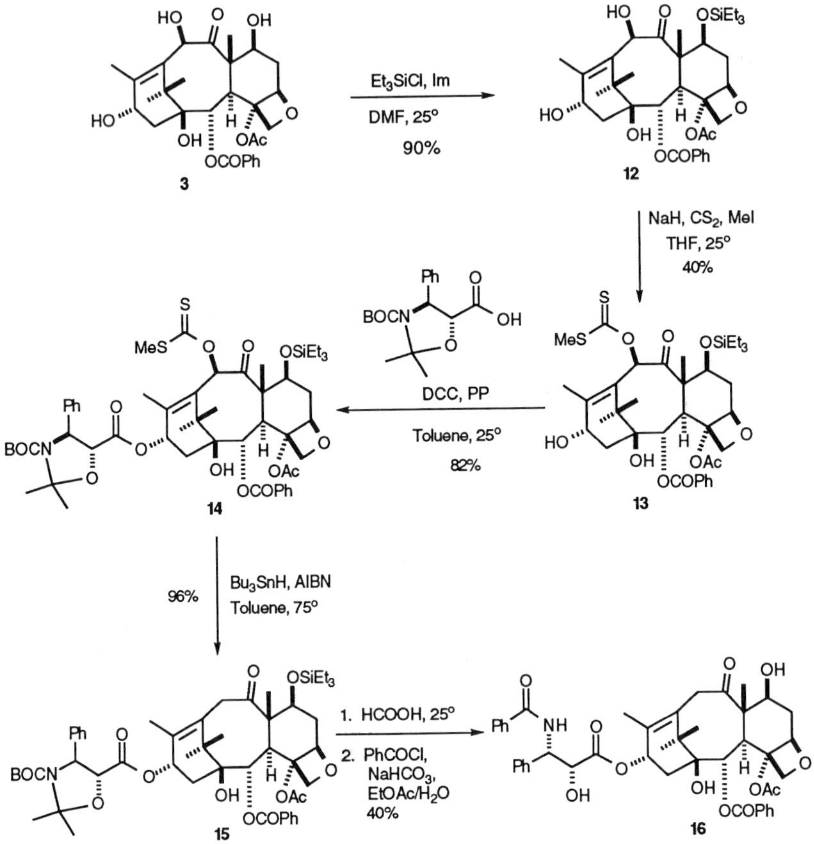

The cytotoxicity of **16** was determined in the P-388 lymphocytic leukemia cell line, and it was found to be identical to that of taxol. It is worth noting that other investigators have also prepared 10-deacetoxytaxol by different routes; thus Chen and his co-workers deoxygenated taxol through an unexpected elimination reaction (*25*), while recently Holton has shown that taxol can readily be deoxygenated at C-10 in one step with samarium diiodide (*26*).

The C-7 Hydroxyl Group. Deoxygenation studies at C-7 yielded some interesting chemistry, leading to a method for epimerization of the C-7 hydroxyl group in both directions (*27*). Thus treatment of taxol (**1**, Scheme 4) with sodium hydride in dry THF cleanly converted it to 7-epitaxol (**17**) in about 85% yield. The reverse process could be achieved by conversion of 7-epitaxol to its 2'-triethylsilyl derivative **18** and thence to its 7-(*S*-methylthiocarbonyl) derivative **19**. During this conversion the C-7 position is epimerized back to the normal configuration, since the 7-*epi*-hydroxyl group is in a very crowded environment and cannot form a xanthate derivative. Hydrolysis of the xanthate **19** by treatment with tributtin hydride and AIBN in wet tributyltin toluene gave 2'-(triethylsilyl)taxol **20**, which could readily be hydrolyzed back to taxol. The conversion of 7-*epi*taxol **17** to taxol **1** was achieved in 62% overall yield.

Scheme 4

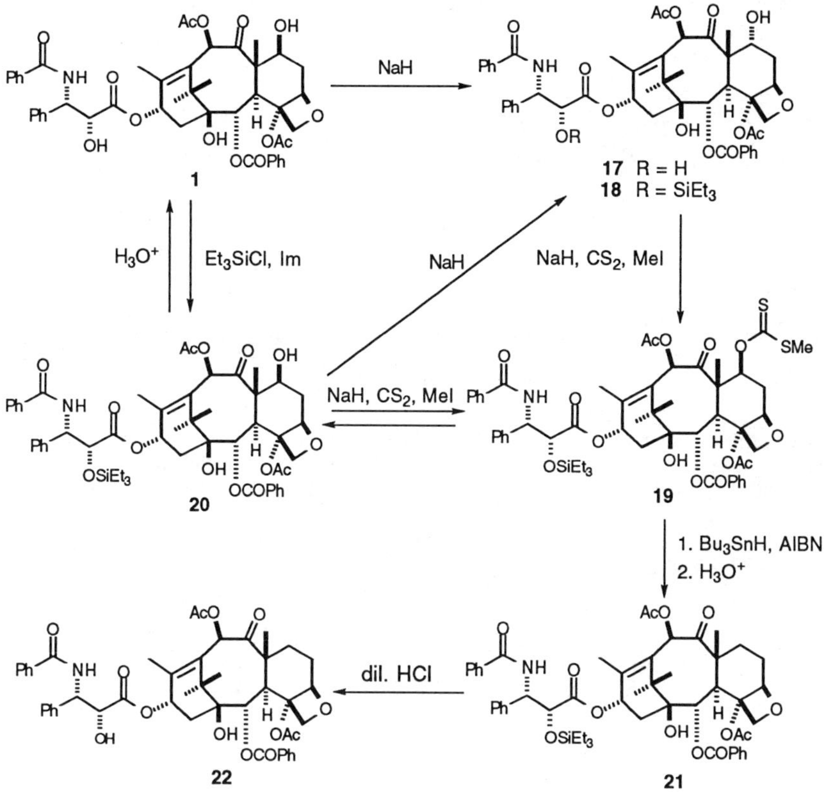

Reprinted with permission from ref. 27. Copyright 1993

7-Desoxytaxol was prepared in good yield by deoxygenation of the xanthate **19** under standard conditions, to give the 7-desoxyderivative **21**. Hydrolysis of **21** then gave 7-desoxytaxol, **22**. This compound was somewhat more active than taxol in the P-388 cytotoxicity assay (*27*), but it was only as active as taxol in the HCT116 cytotoxicity assay (*28*).

The Southern Hemisphere

Changes to the northern hemisphere appear to make little difference to the bioactivity of taxol, but the same is not true of the southern hemisphere. Here, relatively minor changes in the structure of the molecule can have profound consequences on its activity, as described below.

The Oxetane Ring. The first indication that the southern hemisphere was important to the activity of taxol came from studies that we carried out on the oxetane ring

(*29*). Taxol analogues in which the oxetane ring was opened, such as the D-*seco*taxol **23**, were uniformly inactive in both tubulin-assembly and cytotoxicity assays. As a follow-up to this work, we chose to examine the effect of the formation of a tetrahydrofuran ring in place of the oxetane ring. Both we (*21*) and others (*30,31*) had discovered that baccatin III rearranges to an isomeric tetrahydrofuran derivative such as **24** on treatment with certain acidic or basic reagents. The effect of such a modification on the activity of taxol was unknown, however.

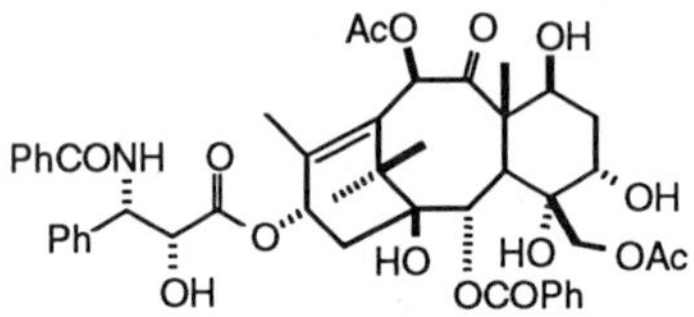

23

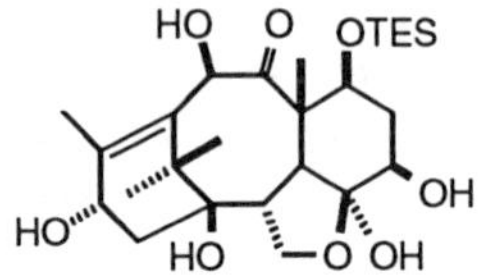

24

Treatment of taxol with di-*t*-butyl dicarbonate and DMAP under forcing conditions gave a mixture of products from which the tri-*t*-BOC derivative **25** could be isolated in reasonable yield. Treatment of this product with LiOH gave the rearranged product **26**, in which the 2-benzoate group had been hydrolyzed and the resulting alkoxide ion had attacked the oxetane ring. Benzoylation with benzoic acid and DCC, followed by removal of the BOC groups with formic acid, then yielded the isotaxol derivative **27**, in which the oxetane ring has been replaced by a tetrahydrofuran ring (Scheme 5). This compound was less active than taxol in the P-388 cytotoxicity assay. This result thus confirms and extends our earlier observations (*29*) on the necessity of the oxetane ring for the activity of taxol.

Scheme 5

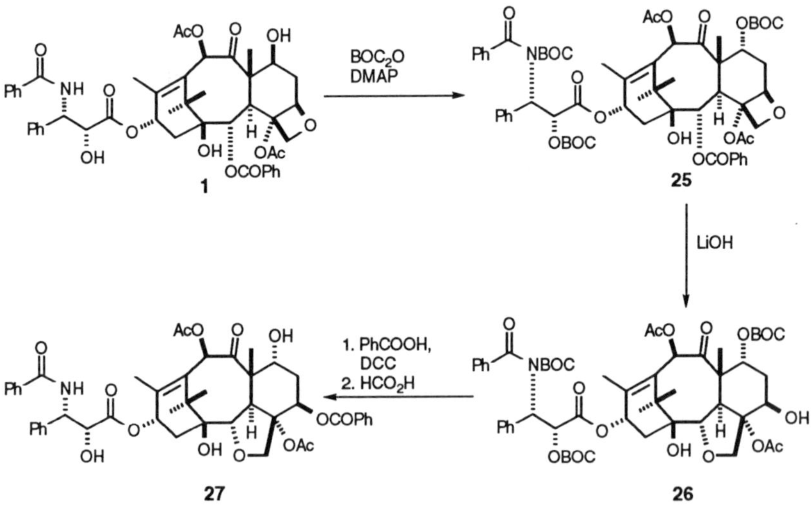

The C-4 Acetate. Given that the oxetane ring is required, the question then arose as to the importance of the C-4 acetate group. This question was resolved by the development of various selective methods for the hydrolysis of this group. In one method, 7-(triethylsilyl)baccatin III (**9**) was converted to its 4, 10-dideacetyl derivative **28** in 72% by treatment with potassium-*t*-butoxide. Coupling with Commercon's side chain (*17*) gave the product **29**, which was acetylated to give the corresponding 10-acetate **30**. Deprotection with formic acid and benzoylation then gave 4-deacetyl taxol (**31**) in 46% yield from **30** and 16% overall yield from **9** (Scheme 6) (Neidigh et al., *Tetrahedron Lett.*, in press).

A second process for the preparation of **31** proved to be even more efficient. Treatment of taxol with t-butyldimethylsilyl chloride followed by triethylsilyl chloride yielded the diprotected derivative **32**, which was converted to its 2-debenzoyl-4-deacetyl derivative **33** in 69% yield on treatment with benzyltrimethylammonium methoxide in methanol. Rebenzoylation ($PhCO_2H$, DCC, DMAP), followed by deprotection gave 4-deacetyltaxol **31** in 42% overall yield from taxol (Scheme 7) (Neidigh et al., *Tetrahedron Lett.*, in press).

The bioactivity of **31** proved to be slightly less than that of taxol in the P-388 cytotoxicity assay, suggesting that the C-4 acetate group only makes a small contribution to the activity of taxol.

Deoxygenation at C-2. Removal of the benzoyloxy group at the C-2 position has been accomplished in at least two ways. In one approach, 2-debenzoyltaxol was prepared in a nine-step sequence from baccatin III by Chen et al. (*33*). In the second approach, we prepared 1-benzoyl-2-debenzoyloxytaxol from taxol in three steps by the route shown in Scheme 8.

Scheme 6

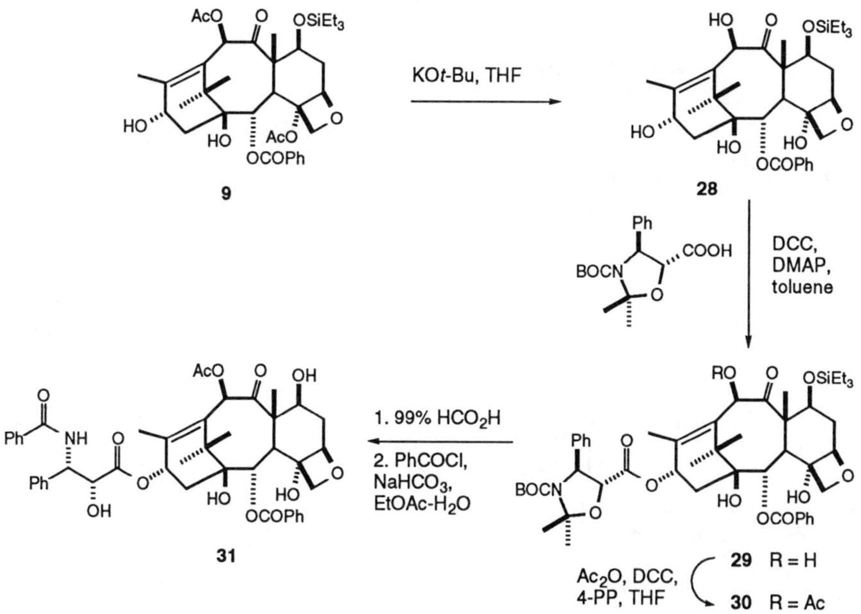

Scheme 7

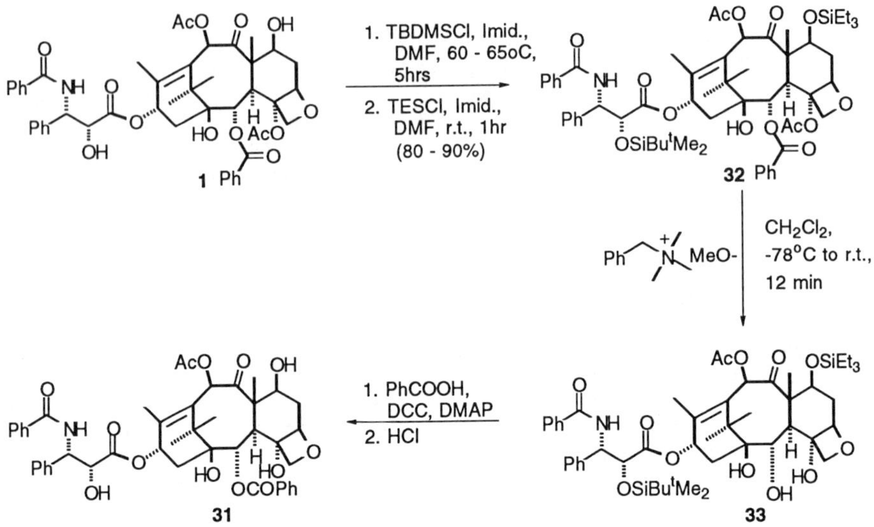

Scheme 8

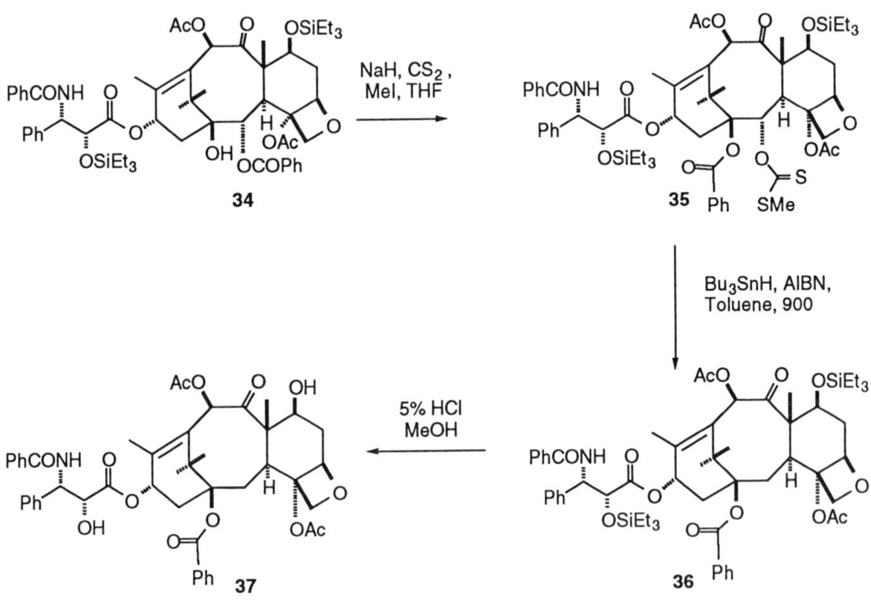

2',7-Di(triethylsilyl)taxol (**34**) was treated with carbon disulfide, sodium hydride, and methyl iodide in an attempt to prepare the 1-(S-methylthiocarboxyl) derivative. To our surprise, the product obtained was 1-benzoyl-2-debenzoyl-2-(S-methylthiocarboxyl)taxol **35**, in which the C-2 benzoate group had undergone an acyl migration to the C-1 position. The structure of **35** was confirmed by HMBC experiments, which showed three-bond correlations from the SMe protons to the thiocarbonyl carbon, and from the C-2 proton to the thiocarbonyl carbon. Deoxygenation of **35** in the usual way then gave the 2-desoxy derivative **36**, which was deprotected to yield 1-benzoyl-2-debenzoyloxytaxol (**37**). This product, like 2-debenzolyoxy taxol itself (*32*), was much less cytotoxic than taxol, suggesting the need for a benzoyl or other aroyl group at the 2-position of taxol for bioactivity. It should also be noted that the baccatin III analog of compound **37** has recently been prepared by Chen and his co-workers, although the mechanism proposed for the conversion differed slightly from that described above (*33*).

Preparation of 2-Aroyl-2-debenzoyltaxol Analogs. Given the significance of the 2-benzoate group of taxol described above, it became of interest to prepare various 2-aroyl-2-debenzoyl analogs. One method for accomplishing this was by a modification of the pathway of Scheme 5. If the hydrolysis of **25** with LiOH was carried out under carefully controlled conditions, oxetane ring-opening could be avoided and the resulting 2-debenzoyltaxol could be rebenzoylated and deprotected to yield 2-aroyl-2-debenzoyltaxol analogs.

A better approach, however, became available through studies on phase-transfer catalysis of taxol hydrolysis (*34*). Treatment of 2',7-di(triethylsilyl)taxol (**34**) with sodium hydroxide in benzene:dichloromethane with aqueous NaOH and a PTC catalyst yielded the 2-debenzoyl derivative **38** in 65-75% yield (Scheme 9). Reacylation of **38** under controlled conditions, followed by deprotection, then gave 2-aroyl-2-debenzoyltaxol analogs of general structure **40**, where ArCO is any desired aroyl group.

Scheme 9

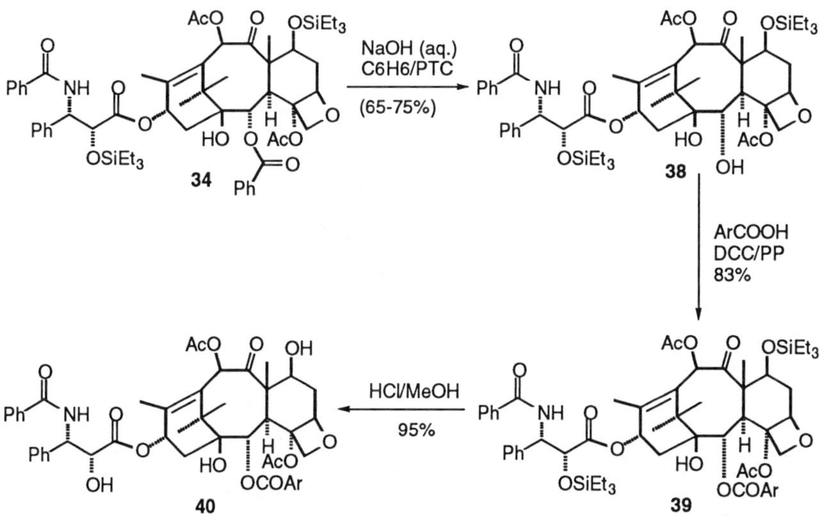

Interestingly, the bioactivities of analogs of general structure **40** are highly variable, depending on the nature of the aroyl group (Table I). Analogs with a *para*-substituted ring were uniformly much less active than taxol, except for the small *p*-fluoro substituent. Analogs with *meta*-substituted rings were, however, *often more cytotoxic than taxol.* This increased activity was particularly marked with the *m*-chloro, *m*-azido, and *m*-methoxy substituted analogs.

Table I. Cytotoxicity of 2-Aroyl-2-debenzoyltaxol Analogs **40**.

Substituent on Ar	*Cytotoxicity in the P-388 assay, ED_{50}/ED_{50} (taxol)* ortho	meta	para
Cl	0.01	0.0014	150
CN	N.D.	0.33	28
N_3	N.D.	0.002	175,000
OCH_3	N.D.	0.001	N.D.
NO_2	N.D.	0.3	8.3
NH_2	N.D.	1500	N.D.
CF_3	N.D.	15	28
F	0.07	0.35	0.5

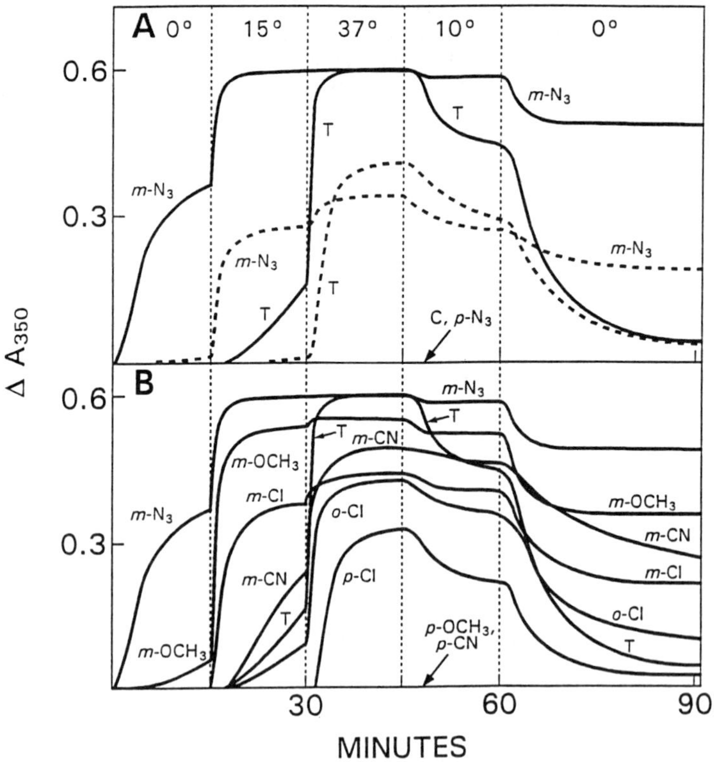

Figure 1. Effects of Taxol and 2-Aroyltaxols on Tubulin Polymerization
Reproduced with permission from ref. 34. Copyright 1994. American Chemical Society.

Additional studies of these exciting new analogs were carried out by our colleague Dr. E. Hamel at the National Cancer Institute. In these studies (35), the three anlaogs referred to above were all found to promote the assembly of tubulin to microtubules more efficiently than taxol (Table II). In addition, the *m*-azido and, to a small extent, the *m*-methoxy analogs assembled tubulin under conditions where taxol was unable to do so (Figure 1). These data thus confirm and extend the results of Table I, and offer an exciting glimpse of possible enhancements of taxol's activity by carefully chosen structural modifications.

Table II. Comparison of Taxol and C-2-Derivatives on Assembly Rates at 15°C and Extent of Assembly at 37°C

Compound	10 μM drug *Maximum Rate at 15°C* [1] *(Relative to Taxol)* *munits* ΔA_{350}*/ min*	10 μM drug *Extent* [2] ΔA_{350}	40 μM drug *Maximum Rate at 15°C* [1] *(Relative to Taxol)* *munits* ΔA_{350}*/ min*	40 μM drug *Extent* [2] ΔA_{350}
Taxol (1)	1.2 (1)	0.38 (1)	23 (1)	0.60 (1)
6a (*m*-N_3)	119 (99)	0.34 (0.9)	NM [3]	0.60 (1)
6e (*m*-OCH_3)	51 (43)	0.37 (1)	271 (12)	0.54 (0.9)
6g (*m*-Cl)	28 (23)	0.38 (1)	128 (6)	0.44 (0.7)
6c (*m*-CN)	1.3 (1.1)	0.40 (1.1)	33 (1.4)	0.51 (0.9)
6i (*o*-Cl)	0.7 (0.6)	0.25 (0.7)	9.3 (0.4)	0.43 (0.7)
6h (*p*-Cl)	0 (0)	0.15 (0.4)	0 (0)	0.32 (0.5)
6f (*p*-OCH_3)	0 (0)	0 (0)	0 (0)	0 (0)
6d (*p*-CN)	0 (0)	0 (0)	0 (0)	0 (0)
6b (*p*-N_3)	0 (0)	0 (0)	0 (0)	0 (0)

[1]The maximum assembly rate at 15°C was determined from the experiments shown in Fig. 1, together with additional experiments performed with analogs at 10 μM that are not presented here. Individual experiments were performed with 4 samples in the spectrophotometers, with a dwell time at each position of 5 sec. About 0.42 min. elapsed between successive readings at each position. The maximum interval increase in reading was used to calculate the maximum rate. [2]Maximum extent of assembly was determined from the same experiments, subtracting the initial turbidity reading at 0°C from the final reading at 37°C. [3]NM, not meaningful, since the reaction at 0°C was so extensive (see Fig. 1).

SOURCE: Reprinted with permission from ref. 34. Copyright 1994.

Summary and Conclusions

The work that we and others have done on structure-activity correlations for taxol can be summarized by the chart of Figure 2 (36). Although much has been accomplished in this area, much yet remains to be done, and it is expected that continuing studies will yield further insights into the chemistry and bioactivity of this fascinating molecule. Ultimately, it seems highly likely that taxol analogs with significantly improved clinical bioactivities will be developed through research such as we and others have carried out.

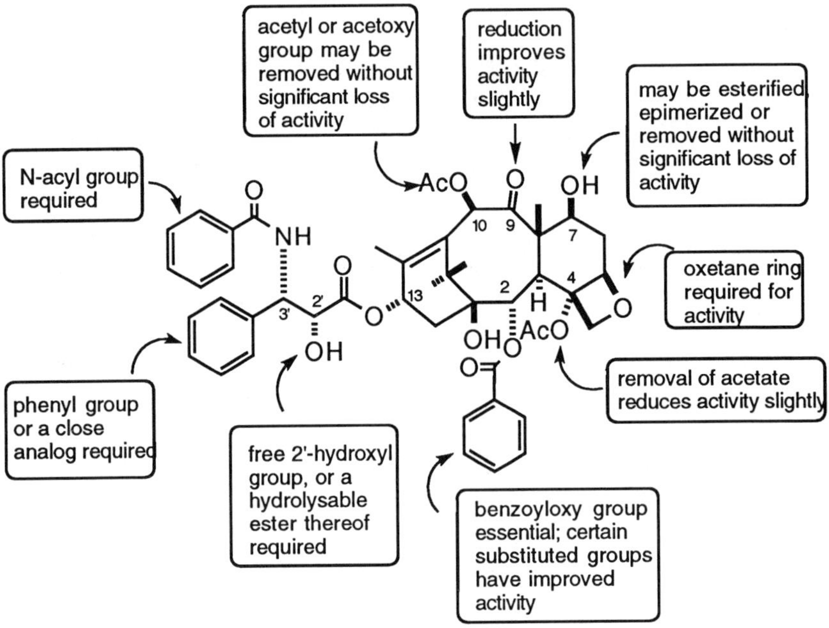

Figure 2. Structure-activity relationships of taxol

Reproduced with permission from ref. 35. Copyright 1994.

Acknowledgments

The work from my laboratory has been carried out with the help of an able and dedicated group of co-workers. The recent work described in this report was carried out by Dr. A. G. Chaudhary, Dr. M. D. Chordia, Dr. M. M. Gharpure, Dr. A. A. L. Gunatilaka, Mr. K. A. Neidigh, and Mr. (now Dr.) J. M. Rimoldi. Biological work was carried out by Dr. E. Hamel and his co-workers at the National Cancer Institute. To all of these associates I owe a great debt of gratitude, since the work described is really theirs and not mine. Financial support from the National Cancer Institute (Grants CA 55131 and CA 48974) and Bristol-Myers Squibb is gratefully acknowledged, and Dr. K. Snader (National Cancer Institute) is thanked for a gift of crude taxol-containing fractions from *T. brevifolia*.

Literature Cited

(1) Kingston, D. G. I.; Samaranayake, G. ; Ivey, C. A. *J. Nat. Prod.,* **1990** , *53:*, 1-12. .

(2) Kingston, D. G. I. *Pharmac. Ther.* **1991**, *52*, 1-34.

(3) Kingston, D. G. I.; Molinero, A. A.; Rimoldi, J. M. In *Progress in the Chemistry of Organic Natural Products*; W. Herz; G. W. Kirby; R. E. Moore; W. Steglich and C. Tamm, Ed.; Springer-Verlag: New York, 1993; Vol. 61; pp 1-206.

(4) Kingston, D. G. I. In *Human Medicinal Agents from Plants*; A. D. Kinghorn and M. F. Balandrin, Ed.; American Chemical Society: Washington, D. C., 1993; Vol. 534; pp 138-148.

(5) Suffness, M.; Cordell, G. A. In *The Alkaloids*; A. Brossi, Ed.; Academic Press: Orlando, 1985; Vol. 25; pp 1-369.

(6) Blechert, S.; Guenard, D. In *The Alkaloids, Chemistry and Pharmacology*; A. Brossi, Ed.; Academic Press, Inc.: San Diego, 1990; Vol. 39; pp 195-238.

(7) Guénard, D.; Guéritte-Voegelein, F.; Potier, P. *Acc. Chem. Res.* **1993**, *26*, 160-167.

(8) Nicolaou, K. C.; Dai, W.-M.; Guy, R. K. *Angew. Chem. Int. Ed. Engl.* **1994**, *33*, 15-44.

(9) Swindell, C. S. In *Studies in Natural Products Chemistry*; Atta-ur-Rahman, Ed.; Elsevier Science Publishing: New York, 1993; Vol. 12; pp 179-231.

(10) Cragg, G. M.; Schepartz, S. A.; Suffness, M.; Grever, M. R. *J. Nat. Prod.* **1993**, *56*, 1657-1668.

(11) Denis, J.-N.; Greene, A. E.; Guénard, D.; Guéritte-Voegelein, F.; Mangatal, L.; Potier, P. *J. Am. Chem. Soc.* **1988**, *110*, 5917-5919.

(12) Denis, J.-N.; Kanazawa, A. M.; Greene, A. E. *Tetrahedron Lett.* **1994**, *35*, 105-108.

(13) Kanazawa, A. M.; Denis, J.-N.; Greene, A. E. *J. Org. Chem.* **1994**, *59*, 1238-1240.

(14) Holton, R. A. Eur. Pat. Appl. EP 400,971 1990: 05 December (Chem. Abstr. 1991 164568q).

(15) Ojima, I.; Habus, I.; Zhao, M.; Zucco, M.; Park, Y. H.; Sun, C. M.; Brigaud, T. *Tetrahedron* **1992**, *48*, 6985-7012.

(16) Georg, G. I.; Cheruvallath, Z. S.; Harriman, G. C. B.; Hepperle, M.; Park, H. *Bioorg. Med. Chem. Lett.* **1993**, *3*, 2467-2470.

(17) Commerçon, A.; Bezard, D.; Bernard, F.; Bourzat, J. D. *Tetrahedron Lett.* **1992**, *33*, 5185-5188.

(18) Magri, N. F.; Kingston, D. G. I.; Jitrangsri, C.; Piccariello, T. *J. Org. Chem.,* **1986**, *51*, 3239-3242.

(19) Kingston, D. G. I.; Gunatilaka, A. A. L.; Ivey, C. A. *J. Nat. Prod.* **1992**, *55*, 259-261.

(20) Gou, D.; Liu, Y.; Chen, C. *J. Org. Chem.* **1993**, *58*, 1287-1289.

(21) Kingston, D. G. I.; Chaudhary, A. G.; Gunatilaka, A. A. L.; Middleton, M. L. *Tetrahedron Lett.* **1994**, *35*, 4483-4484.

(22) Samaranayake, G.; Neidigh, K. A.; Kingston, D. G. I. *J. Nat. Prod.* **1993**, *56*, 884-898.

(23) Barton, D. H. R.; McCombie, S. W. *J. Chem. Soc., Perkin Trans. I* **1975**, 1574-1585.

(24) Chaudhary, A. G.; Kingston, D. G. I. *Tetrahedron Lett.* **1993**, *34*, 4921-4924.

(25) Chen, S.-H.; Fairchild, C.; Mamber, S. W.; Farina, V. *J. Org. Chem.* **1993**, *58*, 2927-2928.

(26) Holton, R. A.; Somoza, C.; Chai, K.-B. *Tetrahedron Lett.* **1994**, *35*, 1665-1668.

(27) Chaudhary, A. G.; Rimoldi, J. M.; Kingston, D. G. I. *J. Org. Chem.* **1993**, *58*, 3798-3799.

(28) Chen, S.-H.; Huang, S.; Kant, J.; Fairchild, C.; Wei, J.; Farina, V. *J. Org. Chem.* **1993**, *58*, 5028-5029.
(29) Samaranayake, G.; Magri, N. F.; Jitrangsri, C.; Kingston, D. G. I. *J. Org. Chem.* **1991**, *56*, 5114-5119.
(30) Farina, V.; Huang, S. *Tetrahedron Letters* **1992**, *33*, 3979-3982.
(31) Wahl, A.; Guéritte-Voegelein, F.; Guénard, D.; Le Goff, M.; Potier, P. *Tetrahedron* **1992**, *48*, 6965-6974.
(32) Chen, S.-H.; Wei, J.-M.; Farina, V. *Tetrahedron Lett.* **1993**, *34*, 3205-3206.
(33) Chen, S.-H.; Huang, S.; Gao, Q.; Golik, J.; Farina, V. *J. Org. Chem.* **1994**, *59*, 1475-1484.
(34) Chaudhary, A. G.; Gharpure, M. M.; Rimoldi, J. M.; Chordia, M. D.; Gunatilaka, A. A. L.; Kingston, D. G. I.; Grover, S.; Lin, C. M.; Hamel, E. *J. Am. Chem. Soc.* **1994**, *116*, 4097-4098.
(35) Kingston, D. G. I. *Trends Biotechnol.* **1994**, *12*, 222-227.

RECEIVED August 8, 1994

Chapter 16

Medicinal Chemistry of Paclitaxel

Chemistry, Structure–Activity Relationships, and Conformational Analysis

Gunda I. Georg[1], Geraldine C. B. Harriman[1], David G. Vander Velde[2], Thomas C. Boge[1], Zacharia S. Cheruvallath[1], Apurba Datta[1], Michael Hepperle[1], Haeil Park[1], Richard H. Himes[3], and Lalith Jayasinghe[4]

[1]Department of Medicinal Chemistry, University of Kansas, Lawrence, KS 66045
[2]NMR Laboratory, University of Kansas, Lawrence, KS 66045
[3]Department of Biochemistry, University of Kansas, Lawrence, KS 66047
[4]Oread Laboratories, 1501 Wakarusa Drive, Lawrence, KS 66047

Chemical, conformational and structure-activity studies of taxol and related taxanes are detailed. Semisynthetic methodology for the preparation of taxol and related analogues with modified C-13 phenylisoserine side chains was developed and analogues, modified at the C-3' phenyl group and the *N*-benzoyl group, were prepared. 3'-Cyclohexyl and 2-cyclohexylcarbonyl taxol analogues and C-13 side chain homologated derivatives were synthesized. Methods for the selective hydrolysis of all ester groups in baccatin III and the conversion of 4-deacetylbaccatin III to 4-deacetyltaxol are reported. Reduction of taxanes with samarium diiodide provided 10-deacetyl derivatives as well as 9-dihydrotaxanes. Conformational analysis of taxol and other bioactive derivatives demonstrated the formation of hydrophobically clustered conformations in aqueous solvents.

The discovery by the Potier group that 10-deacetylbaccatin III (**4**) can be isolated in significant quantities from a regenerable source, the needles of the European yew tree *Taxus baccata* L., was the most significant finding in the attempt to secure the long term supply of the anticancer agent taxol (**1**) through semisynthesis (Fig. 1) (*1*). Extraction of the fresh needles yields **4** in amounts of up to 1g/kg, which is about ten times the amount of taxol isolated from the bark (0.1g/kg). It is of importance to note that the needles are a fully regenerable source and that their harvest does not threaten the survival of the yew species. The availability of **4** also facilitated semisynthetic studies directed at the elucidation of the taxol pharmacophore (*2,3*).

Synthesis and Biological Evaluation of C-13 Chain Modified Taxol Analogues

Since the C-13 *N*-benzoyl-3-phenylisoserine side chain of taxol is of crucial importance for taxol's cytotoxicity (*4*), efficient methodology for the asymmetric synthesis of the C-13 side chain **2** and its attachment to baccatin III required development (*5*).

NOTE: Paclitaxel is the generic name for Taxol, which is now a registered trademark.

0097–6156/95/0583–0217$08.00/0

The Semisynthesis of Taxol: Development of an Enantioselective Method for the Preparation of 3-Phenylisoserine. Our approach focused on the recognition that (3*R*,4*S*)-3-hydroxy-4-phenyl-2-azetidinone (**6**) could serve as a practical precursor for (2*R*,3*S*)-3-phenylisoserine (**5**) (Fig. 2). With this in mind, we developed an enantioselective approach to this β-amino acid via the ester enolate-imine cyclocondensation (Scheme 1). This chemistry, which has been under study in our

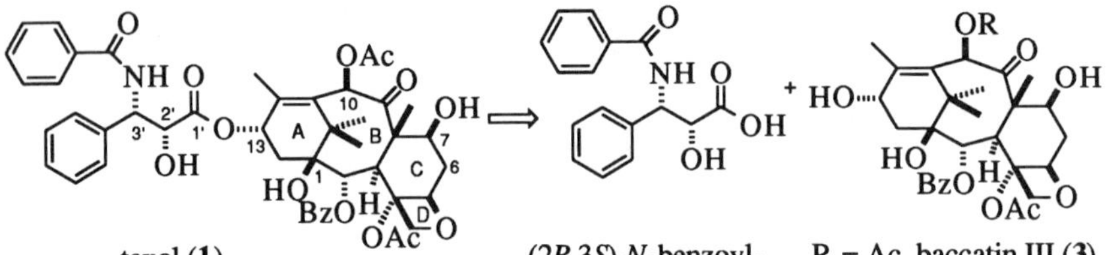

taxol (**1**) (2*R*,3*S*)-*N*-benzoyl-3-phenylisoserine (**2**) R = Ac, baccatin III (**3**) R = H, 10-deacetylbaccatin III (**4**)

Figure 1. Structures of taxol (**1**), *N*-benzoyl-3-phenylisoserine (**2**), baccatin III (**3**) and 10-deacetylbaccatin III (**4**).

laboratory since 1984 (*6,7*), provides β-lactams with high stereoselectivity (*8*). The advantage of this method lies in its flexibility with regard to the synthesis of 3'-phenyl analogues. Substitution of the *N*-trimethylsilyl benzaldimine in Scheme 1 with other aldehyde imines provides a facile method for the synthesis of 3' modified phenylisoserines.

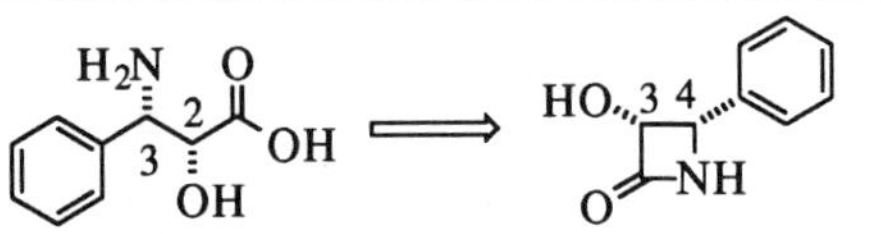

(2*R*,3*S*)-3-phenylisoserine (**5**) (3*R*,4*S*)-3-hydroxy-4-phenyl-2-azetidinone (**6**)

Figure 2. Structures of **5** and **6**.

The application of the ester enolate-imine cycloconden-sation to the synthesis of **5** was simultaneously explored by us (*9,10*) and the Ojima group (*9,11*).

An evaluation of several chiral auxiliaries in the ester enolate-imine cyclocondensation reaction revealed that Whitesell's and Oppolzer's (shown in Scheme 1) chiral auxiliaries provide β-lactam **8** and related analogues with excellent enantioselectivity (*9-12*).

Scheme 1

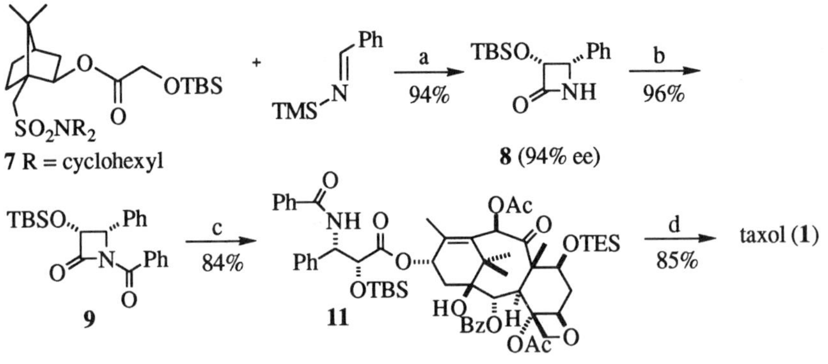

(a) LDA, THF, -78 °C, 4 h. (b) PhCOCl, Et_3N, DMAP, CH_2Cl_2, 0 °C, 1 h. (c) 7-Triethylsilylbaccatin III (**10**), NaH, THF, 0 to 35 °C, 2 h. (d) HF, pyridine, 25 °C, 5 h.

Lithiation of chiral glycolate **7** followed by addition of *N*-trimethylsilyl benzaldimine afforded the taxol side chain precursor **8** in 94% yield and 94% ee (*12*). For the attachment of **8** to baccatin III we employed the efficient acylation method developed by the Holton group (*13*). They recognized the potential of utilizing *N*-acyl-β-lactams as powerful acylating agents for the sterically hindered hydroxyl group at C-13 of baccatin III. Benzoylation of **8** utilizing a standard acylation protocol gave activated electrophilic imide **9**. Reaction of **9** with the sodium alkoxide of 7-triethylsilylbaccatin III (**10**) cleanly produced the coupling product **11** in 84% yield. Fluoride assisted removal of the silyl ethers resulted in the synthesis of taxol in 85% yield. The semisynthesis of taxol as outlined in Scheme 1 represents the shortest semisynthesis of taxol reported to date (*12*). Thus the semisynthesis utilizing the Holton/Georg/Ojima protocol is one of the most efficient methods for the semisynthesis of taxol and related analogues.

Synthesis and Biology of 3'-Aryl Taxol Analogues. With this chemistry in hand, the application to the syntheses of a variety of taxanes was undertaken. The methodology afforded us the versatility to utilize a variety of aldimines for β-lactam synthesis to ultimately functionalize the C-3' position of the phenylisoserine side chain. In a systematic approach to the development of these novel analogues, we employed the Topliss Operational Scheme (*14,15*) to explore the electronic, lipophilic, and steric parameters involved in the binding of these functionalities to their cellular targets.

Following the protocol described above for the synthesis of taxol (Scheme 1), the analogues shown in Fig. 3 and Fig. 4 were synthesized in good overall yields. The compounds were evaluated in a microtubule assembly assay and for their cytotoxicity against B16 melanoma cells (Fig. 3 and 4) (*10,16-20*).

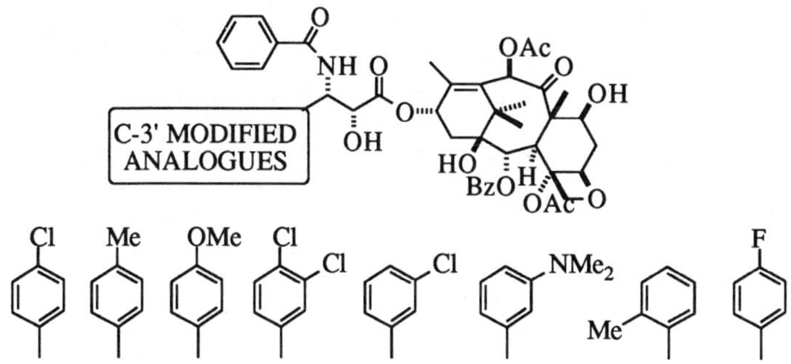

A =	1.9	2.4	0.51	7.1	4.4	4.6	5.1	1.1
B =	2.2	3.0	1.0	40	6.7	5.8	>33	1.2

A = $ED_{50}/ED_{50(taxol)}$, microtubule assembly
B = $ED_{50}/ED_{50(taxol)}$, cytotoxicity against B16 melanoma cells

Figure 3. Structures and biological activities of substituted 3'-phenyl taxol analogues.

Our studies demonstrated that no systematic change of bioactivity occurs as a result of aromatic substituent value alteration (Hansch π, δ, E_S). However, several of the derivatives demonstrated activity similar to taxol in both assays. The most potent analogues in this series the 4-ClPh, 4-MePh, 4-MeOPh and 4-FPh analogues carry a substituent at the para-position of the 3'-phenyl group, whereas derivatives with reduced bioactivity are substituted at the meta and ortho positions. In addition to substituted 3'-phenyl analogues we also prepared 3'-heteroaromatic taxol analogues (Fig. 4) (*21*). All derivatives displayed better activity than taxol in the microtubule assembly assay. Of interest is the observation that all of the pyridyl derivatives displayed excellent activity in the microtubule assembly assay but differed greatly in their cytotoxicity against B16 melanoma cells. The 2-pyridyl derivative displayed slightly better activity than taxol, but the 3-pyridyl derivative had greatly reduced cytotoxicity against B16 melanoma. Of the two furyl derivatives, the 2-furyl analogue showed excellent activity in both assays.

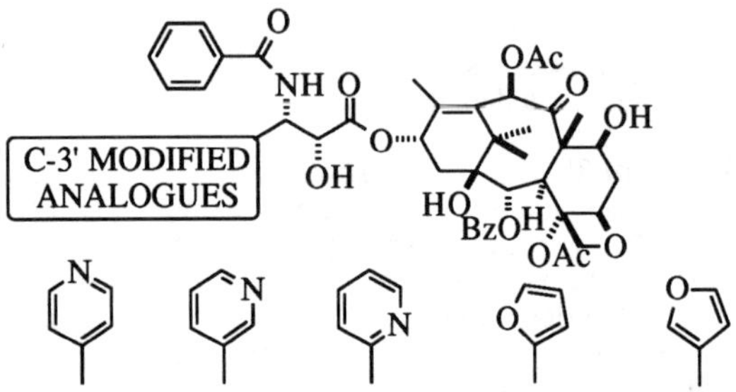

A =	0.42	0.5	0.69	0.85	0.9
B =	1.3	27	0.78	0.31	3.3

A = $ED_{50}/ED_{50(taxol)}$, microtubule assembly
B = $ED_{50}/ED_{50(taxol)}$, cytotoxicity against B16 melanoma cells

Figure 4. Structures and biological activities of 3'-heteroaromatic taxol analogues.

Development of a Convergent Synthetic Method for the Preparation of *N*-Acyl Taxol Analogues. For the synthesis of *N*-benzoyl modified taxanes via the β-lactam route, the acylation of β-lactam **8** with a variety of acid chlorides and subsequent reaction with 7-triethylsilylbaccatin III (**10**) as shown in Scheme 1 would provide the desired analogues. However, a more efficient and convergent route was desired. To accomplish this, a synthesis of *N*-debenzoyltaxol (**14**) was developed (Scheme 2) (*22*). Precursor **8** was acylated with di-*tert*-butyl dicarbonate to afford the activated imide **12**. The hydroxyl group at the C-13 position of 7-triethylsilylbaccatin III (**10**) was then acylated with **12** to afford the protected *N*-debenzoyltaxol precursor **13**. Removal of the protecting groups afforded *N*-debenzoyltaxol (**14**) in good yield. This versatile intermediate was then subjected to Schotten-Baumann acylation which produced a variety of *N*-substituted derivatives **15** in one step from common intermediate **14** (*22-24*). Utilizing this versatile intermediate, we synthesized and evaluated aromatic (Fig. 5) (*24*), heteroaromatic (Fig. 6) (*25*) and aliphatic *N*-acyl taxol analogues (Fig. 7) (*22*).

Aromatic and Heteroaromatic *N*-Acyl Analogues. The synthesis of the aromatic *N*-benzoyl taxol analogues (Fig. 5) was guided again by the Topliss Operational Scheme (*14,15*). Evaluation of the bioactivity of these analogues revealed a relatively flat response to changes of substituent constants (Hansch π, δ, E_S) in both tests. A relatively narrow range of activities ($ED_{50}/ED_{50(taxol)}$ = 0.55 to 6.0) was observed in the microtubule assembly assay. However, correlation of the aromatic substituent constants with the data from the microtubule assembly assay revealed that

Scheme 2

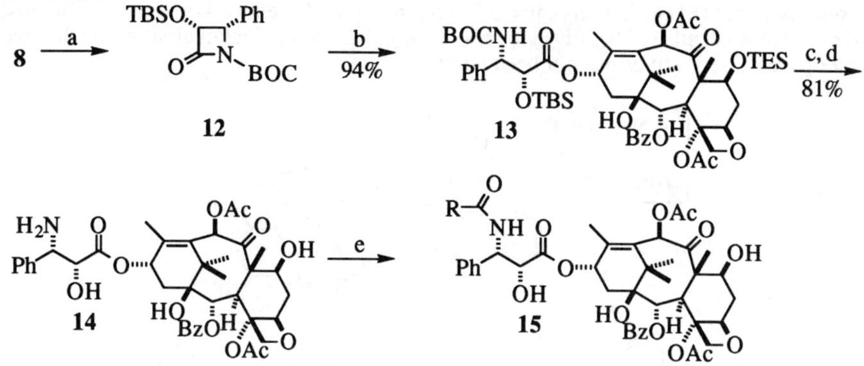

(a) $(BOC)_2O$, Et_3N, DMAP, THF, 25 °C, 1 h. (b) **10**, NaH, THF, 0 to 25 °C. (c) HF, pyridine, 25 °C, 5 h. (d) CF_3CO_2H, CH_2Cl_2, 25 °C, 30 min. (e) RCOCl, EtOAC, H_2O, $NaHCO_3$, 25 °C, 5 min. R = Aromatic, aliphatic or aliphatic RO.

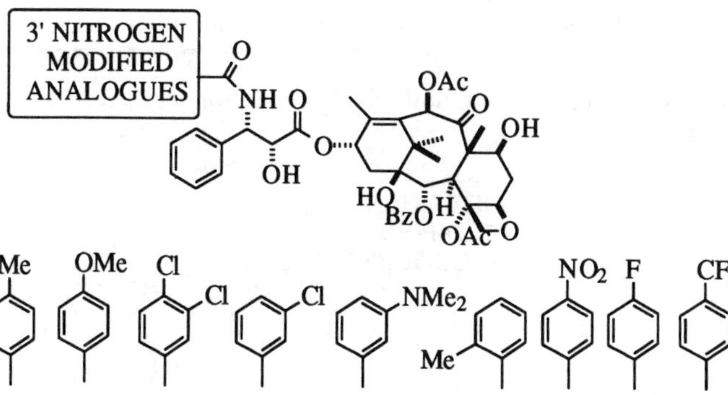

A =	2.4	1.6	0.55	2.1	2.0	1.4	1.9	2.0	1.2	6.0
B =	1.5	1.4	1.3	11	1.8	1.6	7.5	21	4.3	18

A = $ED_{50}/ED_{50(taxol)}$, microtubule assembly
B = $ED_{50}/ED_{50(taxol)}$, cytotoxicity against B16 melanoma cells

Figure 5. Structures and biological evaluation of aromatic *N*-acyl taxol analogues.

potency increased slightly with a decrease of π values (lipophilicity). This trend was unique to the microtubule assembly and was not observed in the cytotoxicity test against B16 melanoma cells, indicating that uptake and metabolism probably play a role in the cytotoxicity of these analogues. None of the derivatives had higher cytotoxicity than taxol against B16 melanoma cells, but several of them were similar in activity. The least active compounds were the 3,4-Cl_2Ph-, the 4-CF_3Ph- and the 4-NO_2Ph-derivatives (*18*).

Among the heteroaromatic analogues the 3-furyl derivative was the most active compound, possessing very good activity in both tests (Fig. 6) (*25*). Surprisingly it was found that the nicotinoyl analogue displayed very good activity in the microtubule assembly assay ($ED_{50}/ED_{50(taxol)} = 0.63$) but had greatly reduced cytotoxicity against B16 melanoma cells ($ED_{50}/ED_{50(taxol)}$ = >200). The biological evaluation of the two other pyridyl derivatives has not been completed.

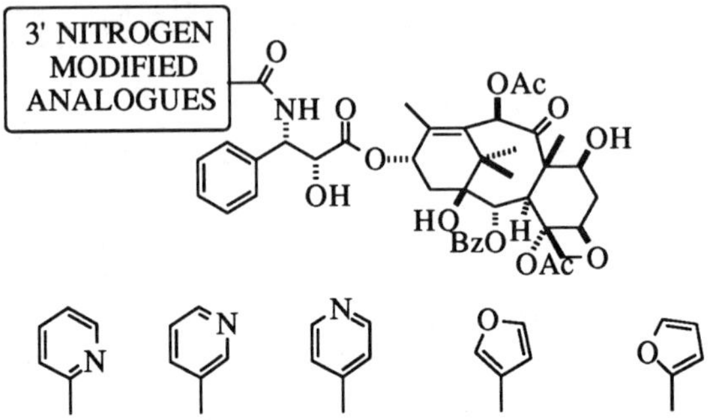

A =	3.5	0.63	2.0	0.36	0.78
B =	n. d.	> 200	n. d.	0.50	1.0

A = $ED_{50}/ED_{50(taxol)}$, microtubule assembly
B = $ED_{50}/ED_{50(taxol)}$, cytotoxicity against B16 melanoma cells

Figure 6. Structures and biological evaluation of heteroaromatic *N*-acyl taxol analogues.

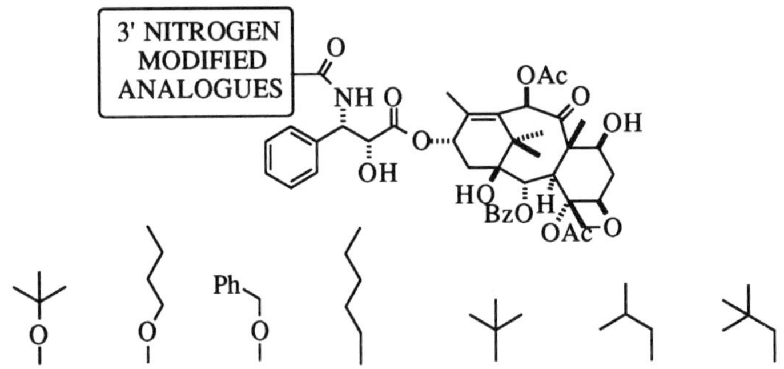

A =	0.55	0.80	3.3	0.32	2.6	1.2	0.74
B =	0.23	0.60	4.0	0.51	22	2.7	3.1

A = $ED_{50}/ED_{50(taxol)}$, microtubule assembly
B = $ED_{50}/ED_{50(taxol)}$, cytotoxicity against B16 melanoma cells

Figure 7. Structures and biological evaluation of aliphatic *N*-acyl and *N*-oxycarbonyl taxol analogues.

Aliphatic *N*-Acyl and *N*-Oxycarbonyl Analogues. We also investigated aliphatic *N*-acyl and oxycarbonyl taxol analogues (Fig. 7) (*22*). Surprisingly, the *N*-

hexanoyl derivative ($ED_{50}/ED_{50(taxol)}$ = 0.32) was slightly more active than the *N*-BOC (10-acetyltaxotere) analogue ($ED_{50}/ED_{50(taxol)}$ = 0.55) in the microtubule assembly. However, it showed slightly reduced activity in the B16 melanoma cytotoxicity test in comparison to 10-acetyltaxotere (*26*). Of interest is the high activity of the *N*-(*n*-butoxycarbonyl) analogue which was more active than taxol in both assays. The *N*-benzyloxycarbonyl derivative had reduced activity in both tests. In the aliphatic *N*-acyl series it appears that branching of the aliphatic chain is detrimental to cytotoxicity. The *N*-pivaloyl derivative displayed reduced ability to induce microtubule assembly and was the least cytotoxic agent in this series against B16 melanoma cells ($ED_{50}/ED_{50(taxol)}$ = 22).

Synthesis and Biology of 3'-Cyclohexyl and 2-Cyclohexyl Taxol Analogues

In light of our recent results that a number of bioactive taxol analogues adopt similar hydrophobically clustered conformations (see later discussion) (*27*), we synthesized representative bioisosteric taxol analogues to further probe the taxol pharmacophore. Since the 3'-phenyl group and the 2-benzoate are part of the proposed hydrophobic cluster, we were interested in exploring whether a replacement of the aromatic rings with cyclohexyl groups would influence bioactivity and conformational properties (*28,29*). 3'-Cyclohexyl-3'-dephenyltaxol (**19**) was synthesized from β-lactam **8** and 7-triethylsilylbaccatin III (**10**) as shown in Scheme 3. Hydrogenation of the phenyl group in β-lactam **8** was followed by benzoylation of the β-lactam nitrogen of **16** and subsequent coupling of imide **17** to **10**. Deprotection of **18** provided **19** in good overall yield.

Scheme 3

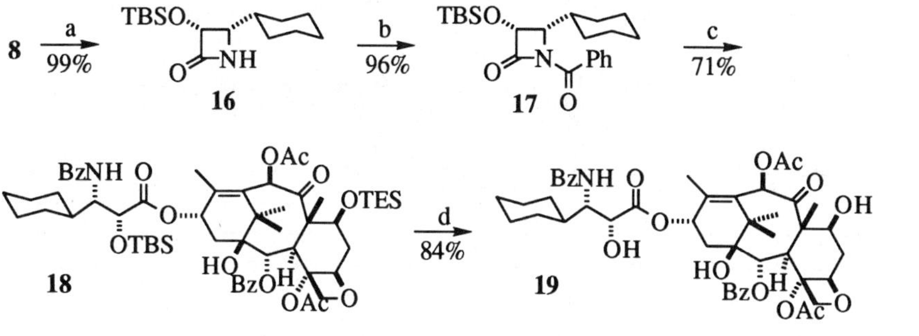

(a) 3% Pt/C, H_2 (55 psig), EtOAc, 12 h. (b) Benzoyl chloride, Et_3N, DMAP, CH_2Cl_2, 0 °C, 1 h. (c) **10**, NaH, THF, 0 to 35 °C, 2 h. (d) HF, pyridine, 25 °C, 2 h.

10-Acetyl-2-cyclohexylcarbonyl-2-debenzoyltaxotere **22** was prepared from 7-triethylsilylbaccatin III (**10**) by reduction of the 2-benzoyl group (Scheme 4). Acylation with β-lactam **12** and removal of the protecting groups from coupled product **21** provided the desired 2-cyclohexyl analogue **22**. A third derivative, in which all three phenyl groups of taxol are converted to cyclohexyl moieties was obtained in one step via hydrogenation of taxol (Scheme 5).

The three cyclohexyl analogues **19, 22** and **23** displayed better activity than taxol in the microtubule assembly and showed potent cytotoxicity against B16 melanoma cells (Table I). These results indicate that none of the aromatic groups in the taxol molecule is of crucial importance for microtubule assembly or cytotoxicity against B16 melanoma cells.

Scheme 4

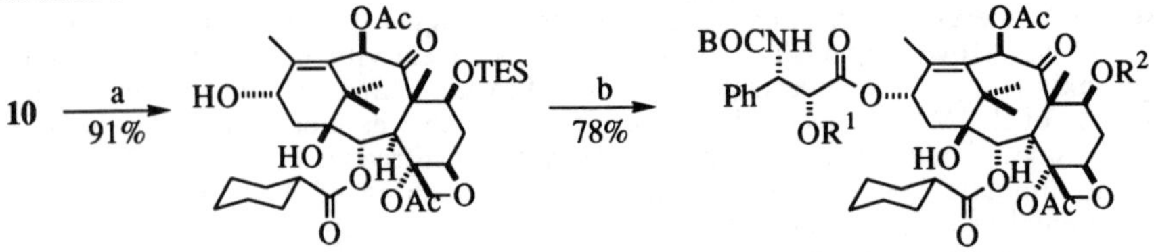

20

21 R^1 = TBS; R^2 = TES
22 R^1 = R^2 = H (84%) c

(a) 3% Pt/C, H_2 (30 psig), EtOAc, 12 h. (b) NaH, THF, **12**, 0 to 25 °C, 1.5 h. (c) HF, pyridine, 3 h.

Scheme 5

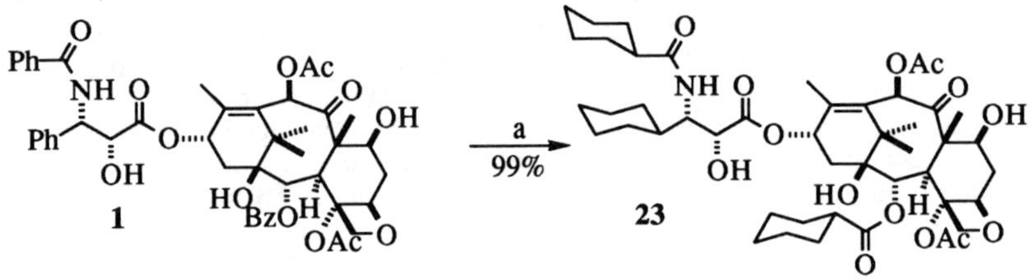

1

23

(a) 3% Pt/C, H_2 (55 psig), EtOAc, 24 h.

Table I. Biological Assays for Cyclohexyl Taxol Analogues 19, 22 and 23

compound	microtubule assembly ED_{50}/ED_{50}(taxol)	B16 melanoma ED_{50}/ED_{50}(taxol)
19	0.29	0.91
22	0.67	1.1
23	0.47	1.6

Synthesis and Biology of Homotaxotere and 10-Deacetylhomotaxol

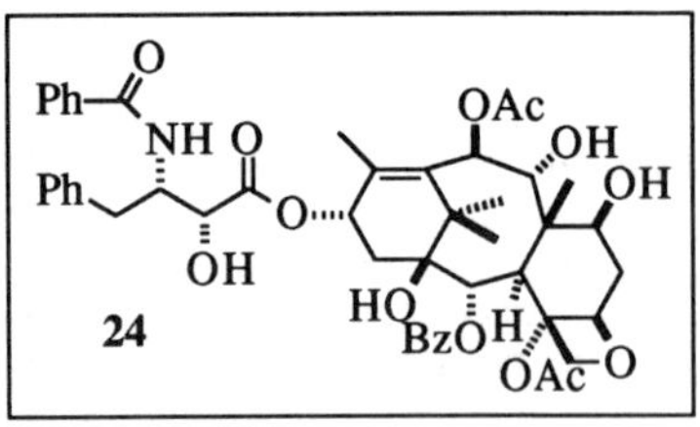

24

Additional side chain analogues of taxol and taxotere were prepared to further probe the taxol pharmacophore and to investigate the relationship between biological activity and C-13 side chain conformations. We were interested to learn what influence homologation of the C-13 taxol side chain might have on bioactivity and conformational equilibria (see later discussion) of these analogues (*30*).

The target compounds **25** and **26** (Scheme 6) were prepared via a modification of the Green/Commercon taxotere semisynthesis (*31*), utilizing phenylglycine as the starting material (*32*).

Biological evaluation demonstrated that analogues **25** and **26** had very poor ability to induce microtubule formation (ED50/ED50(taxol) = >27), thus indicating that homologation is detrimental to bioactivity. It is of interest that a related homologue **24**, prepared by the Abbott group, was also found to be an inactive compound (*33*). Neither **24** nor **25** assumes taxol-like conformations in aqueous solvents.

Scheme 6

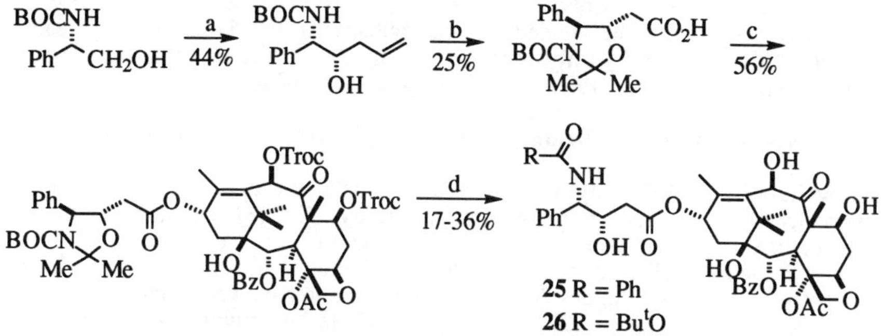

(a) i. $(COCl)_2$, DMSO, -78 to 0 °C, 2 h; ii. $H_2C{=}CH{-}CH_2MgBr$, Et_2O, CH_2Cl_2, 25 °C, 2 h. (b) i. $Me_2C(OMe)_2$, PPTS, toluene, 80 °C, 3 h, 57%; ii. $NaIO_4$, $RuCl_3$, $NaHCO_3$, H_2O, CH_3CN, CCl_4, 25 °C, 48 h, 43%. (c) 7,10-Ditroc-baccatin III, DCC, DMAP, toluene, 80 °C, 2 h. (d) i. HCO_2H, 25 °C, 4 h, 73%; ii PhCOCl, $NaHCO_3$, H_2O, EtOAc, 25 °C, 20 min, 79% or $(BOC)_2O$, $NaHCO_3$, H_2O, THF, 25 °C, 4 h, 43%; iii. Zn, AcOH, MeOH, 54 to 63%.

Selective Methods for the Hydrolysis of the Baccatin III Ester Functionalities

In order to investigate the contributions of the ester moieties of taxol to bioactivity and to obtain taxanes with additional sites for further structure activity studies, we decided to investigate the selective deesterification of baccatin III. Whereas most previous deacylation studies (*34-36*) resulted in mixtures of 2-, 4- and 10-deacylated products we were able to develop completely selective methods for the deacylation at all three positions (Scheme 7) (*37,38*).

Scheme 7

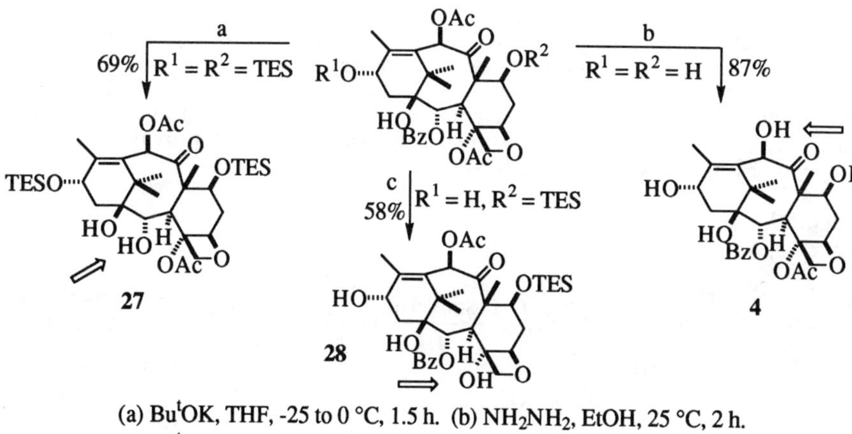

(a) Bu^tOK, THF, -25 to 0 °C, 1.5 h. (b) NH_2NH_2, EtOH, 25 °C, 2 h. (c) Bu^tOK, THF, -25 to 0 °C, 45 min.

Debenzoylation at C-2 was achieved by treating 7,13-bis(triethylsilyl)baccatin III with potassium *tert*-butoxide. The mechanism probably involves the formation of an oxyanion at the neighboring C-1 hydroxyl group, which assists in the hydrolysis of the 2-benzoate to form **27**. The same strategy, the *in situ* generation of a neighboring oxyanion, was successfully applied to the selective removal of the 4-acetyl group. Treatment of 7-triethylsilylbaccatin III with potassium *tert*-butoxide yielded the desired 4-deacetylated baccatin III derivative **28** as the sole hydrolysis product. Presumably, the oxyanion, generated at C-13 is in close proximity to the 4-acetyl group, assisting in the hydrolysis of this group. We also found that hydrolysis of baccatin III with hydrazine yielded 10-deacetylbaccatin III (**4**) as the only reaction product. This selective hydrolysis is presumably due to the fact that the 10-acetyl group is sterically less hindered than the acyl groups at positions 2 and 4.

Synthesis and Biology of 4-Deacetyltaxol and 4-Deacetyl-10-acetyltaxotere

The availability of 4-deacetyl derivative **28** provided the opportunity to probe the importance of the 4-acetyl group for taxol bioactivity. The key step in the synthesis of 4-deacetyltaxol (**30**) and 4-deacetyl-10-acetyltaxotere (**31**) is the coupling between 4-deacetyl-7-triethylsilylbaccatin III (**28**) and Commercon's oxazolidine carboxylic acid **29** (*31*) (Scheme 8). Analogues **30** and **31** were found to have very poor activity in the microtubule assembly assay ($ED_{50}/ED_{50(taxol)}$ = >27). This finding demonstrated that the 4-acetyl group is of critical importance for taxol bioactivity (*39,40*).

Scheme 8

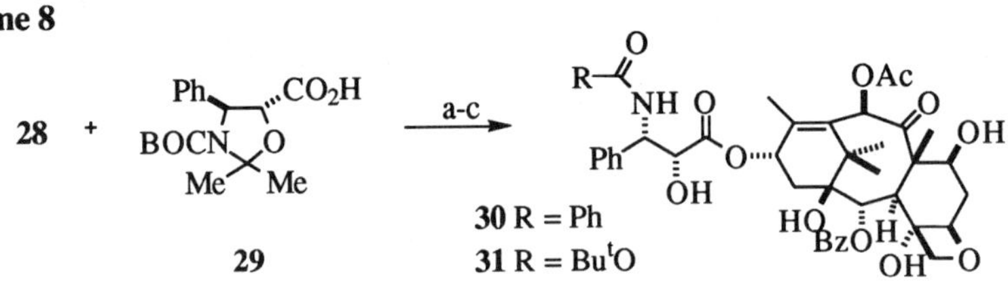

(a) DCC, DMAP, toluene, 70 °C, 45 min, 73%. (b) HCO_2H, 25 °C, 6 h, 70%. (c) PhCOCl, $NaHCO_3$, H_2O, EtOAc, 25 °C, 15 min, 68% or $(BOC)_2O$, $NaHCO_3$, THF, 25 °C, 4 h, 62%.

Preparation and Biology of 10-Deacetoxytaxol and 9-Dihydrotaxotere

To further probe the taxol pharmacophore we initiated deoxygenation studies and also explored the reduction of the C-9 carbonyl group. We have found that treatment of taxol or baccatin III with SmI_2 effects deoxygenation at C-10 (Scheme 9). This reaction is highly chemoselective and no protecting groups are needed for this transformation. The reaction is completed in five minutes and provides the target compound **32** in high yield (91%) (*41,42*).

Scheme 9

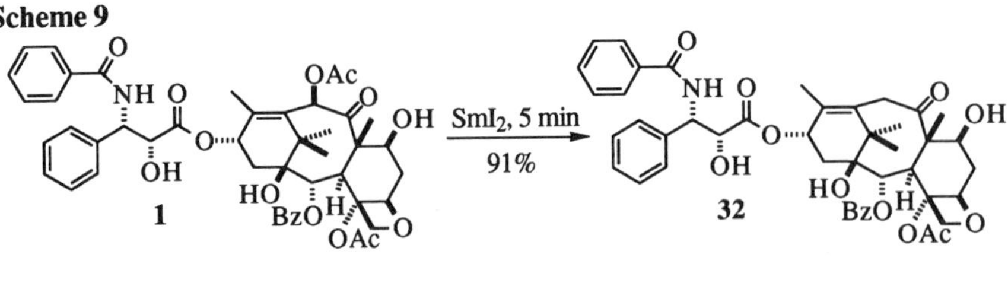

Because the mechanism of the SmI_2-mediated deoxygenation involves the C-9 carbonyl group we reasoned that it might be possible to use the same reagent for the reduction of the C-9 carbonyl group. Previous attempts to reduce the C-9 carbonyl group had demonstrated the remarkable stability of this group to a variety of reducing agents (*34,43*). When taxol was treated with SmI_2 for 12 h, reduction of the C-9 carbonyl occurred to yield 10-deacetoxy-9-dihydrotaxol **33** (Scheme 10) (*44*). Since the acetoxy functionality is a good leaving group, this moiety is lost before the C-9 carbonyl group is reduced.

Scheme 10

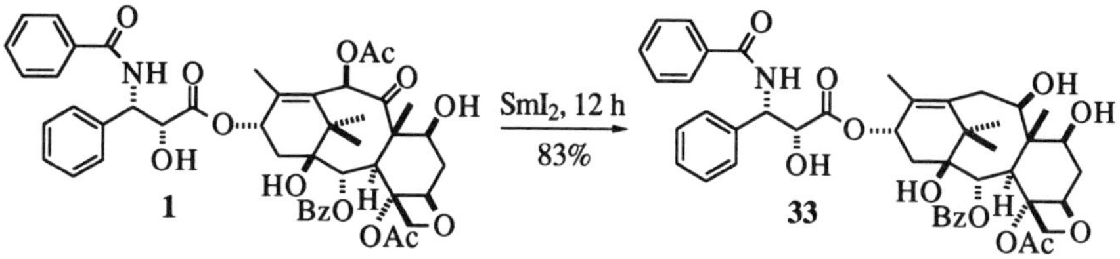

When the same reaction was carried out with taxotere (**34**) as the substrate, a mixture of 9-dihydrotaxotere (**35**) and 10-dehydroxy-9-dihydrotaxotere (**36**) was obtained (*44,45*). Since taxotere carries a C-10 hydroxyl group instead of a 10-acetyl group, elimination of the hydroxy group occurred at a rate slow enough to allow reduction at the carbonyl group (isolation of **35**) before hydroxide elimination.

Scheme 11

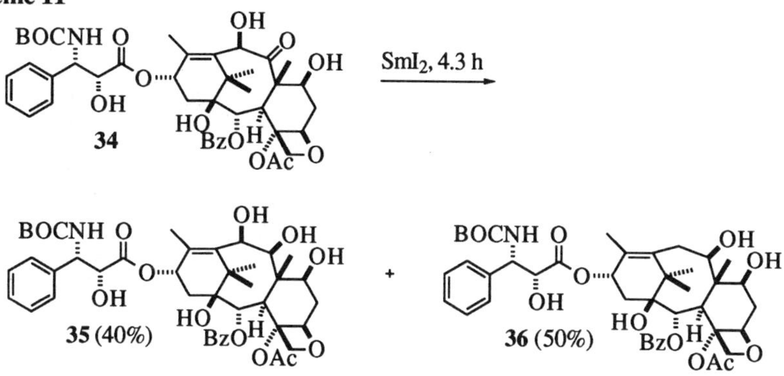

The four analogues were evaluated for their activity in the microtubule assembly assay and for their cytotoxicity against B16 melanoma cells (Table II).

Table II. Biological Evaluation of Taxanes 32-36

compound	microtubule assembly ED_{50}/ED_{50}(taxol)	B16 melanoma ED_{50}/ED_{50}(taxol)
taxol (**1**)	1	1
taxotere (**34**)	0.45	0.41
32	0.51	1.0
33	1.4	14
35	0.97	1.1
36	1.7	1.8

10-Deacetoxytaxol (**32**) was as active as taxol in both assays, indicating that this group is not important for biologial activity. 10-Deacetoxy-9-dihydrotaxol (**33**) had a only a slightly decreased ability to stimulate microtubule assembly compared to taxol but was significantly (14 times) less cytotoxic than taxol against B16 melanoma cells. The related 10-deacetoxy-9-dihydrotaxotere analogue **36** displayed similar data for microtubule assembly as **33** but was significantly more cytotoxic than **33**. Apparently, the BOC group in the side chain of **36** compensates for the decrease in activity caused by the changes in the diterpene moiety. 9-Dihydrotaxotere **35** displayed activity similar to taxol, but had reduced activity in both tests in comparison to taxotere (**34**). The synthesis of the 9α-hydroxy analogue of **35** was recently reported (*46*). This analogue displayed good microtubule assembly properties (ED_{50}/ED_{50}(taxol) = 1.3) and was more active than taxol in an *in vivo* test (mouse model of B16F10 ascites tumor). These and our studies suggest that reduction of the C-9 carbonyl to an α- or β-hydroxyl group does not significantly alter the bioactivity of taxotere analogues.

Conformational Studies of Taxol and Analogues

The taxane diterpene ring system predominantly occupies a single conformation which has been well characterized in the solid state (*47*) and in solution (*2*). The A ring is essentially locked in a boat conformation; the B ring is in a chair-boat conformation; the C ring assumes an envelope-like conformation distorted by the strained D ring fused to it. Alternative taxane conformations are only produced by substantial skeletal rearrangement, e.g. D ring opening, saturation of the 11-12 double bond, or A/B ring contraction. The C-13 side chain, on the other hand, is highly flexible, rapidly samples alternative conformations (as seen in molecular dynamics simulations and assumed to hold true in solution), and its preferred conformation(s) depend on the medium. The most important torsion appears to be around the C2'-C3' bond, whose three low-energy rotamers (with values of H2'-C2'-C3'-H3' near 60°, -60° and 180°) show very different populations which change in different solvents. In the taxotere crystal structure, and in nonpolar solutions, the dominant conformer **I** (Fig. 8) is the one with the 60° torsion (*48,49*). A VCD study (*48*) suggests a smaller population of the -60° conformer **II** is also present under these conditions. For both of these *gauche* conformers, the value of $^3J_{2',3'}$ is expected to be small; the averaged value in nonpolar solution is ca. 2.5 Hz. Conformer **III**, with a 180° torsion, becomes extensively populated in polar solutions (DMSO, water, or mixtures of the two; all giving very

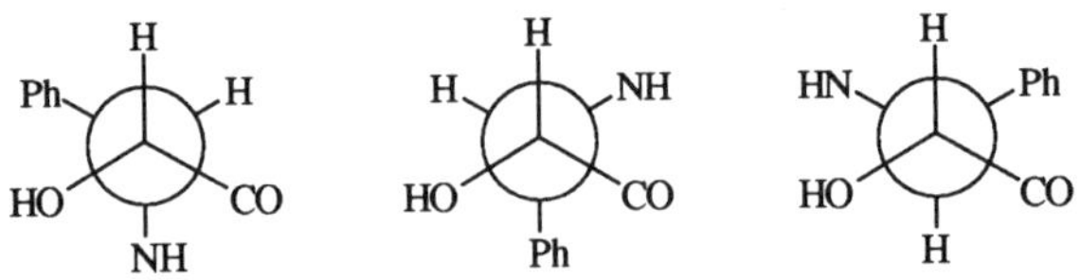

I II III

Figure 8. Newman projections (C2'-C3') of taxol C-13 side chain conformers.

similar spectra). This is evidenced by the increase of $^3J_{2',3'}$ to 7-8 Hz (*48*), and the appearance of NOE's between the 3'-phenyl and 2-benzoyl protons in taxol and taxotere (2-benzoyl *ortho* and *meta* to 3'-phenyl *meta* and *para*) (*27*). Coincident with the NOE's are changes in the chemical shifts, particularly for the 2-benzoyl *meta* and *para* protons and the 3'-phenyl *para* proton. All of these observations are consistent with an energetically favorable orientation of the aromatic rings, as shown in Fig. 9. There are also strong NOE's to the 4-acetyl methyl group from both aromatic rings

(*27*). Deletion of any of the three groups is known to decrease activity sharply (*39,50,51*). The clustering of these three groups has been described as a nonpeptidic example of "hydrophobic collapse" (*27*).

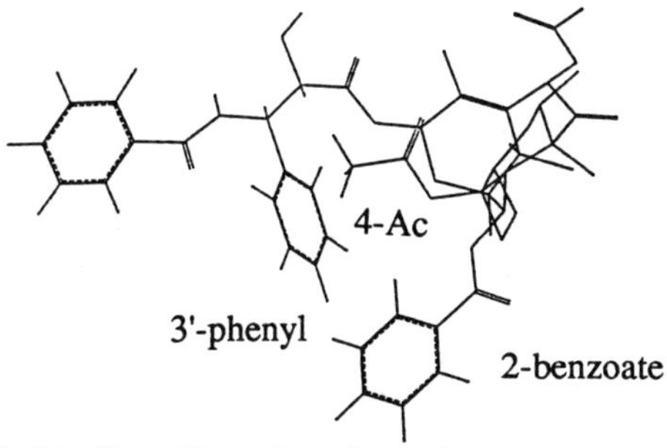

Figure 9. Hydrophobically collapsed conformation of taxol in aqueous solution.

The potential contributors to the solvent-dependent conformational preferences of the sidechain are: (1) intramolecular hydrogen bonds. Hydrogen bonds between the 1'-carbonyl, 2'-hydroxyl, and 3'-amide stabilize the 60° conformer in the solid state (*47*) and have also been detected in solution (*52*), but these likely become less important in hydrogen-bonding solvents; (2) hydrophobic interactions, which will increase in importance as intramolecular hydrogen bonding decreases; and (3) aromatic-aromatic interactions. These are worth distinguishing from purely hydrophobic interactions, since they are highly directional, involving both pi and sigma components (*53*) and could potentially stabilize a particular conformer in either polar or nonpolar solvents. By studying the conformational preferences of various taxol analogues in both polar and nonpolar solvents, the magnitude of these contributions can be appreciated. Since the proper conformation for binding is likely to be one that is low in energy and appreciably populated in aqueous solution, insight may also be gained into what the conformational requirements are for biological activity. (The highly flexible sidechain may, of course, change its conformation while bound because of specific interactions with groups at the active site.)

We have performed NMR conformational studies on numerous taxanes, both active and inactive, in polar and nonpolar solvents (*54*). The best results for polar solvents have been obtained in DMSO/water (3:1 v/v) mixtures at temperatures (typically -20° C in our experiments) well below the freezing point of water. Under these conditions, the combination of reduced molecular motion and high solvent viscosity make taxanes mimic the behavior of large molecules and allow the efficient use of the NOESY experiment. The dielectric constant of this solvent mixture also increases as the temperature is decreased, approaching the value of water at room temperature (*55*). This promotes hydrophobic interactions in the taxanes to a degree similar to pure water, while giving sufficient solubility for the NMR experiments without the necessity to attach ionizable functional groups to the taxanes.

The taxanes we have analyzed (*54*) have fallen into three classes: (1) active compounds with aromatic rings at the 2 and 3' positions, which all show conformational properties similar to taxol: conformer **III** becomes increasingly populated in polar solution, with increased values of $J_{2',3'}$ and NOE's between the aromatic rings. Besides taxol and taxotere, we have examined the 3'-(2-furyl) and 3'-(2-pyridyl) analogues (Fig. 4) which also belong in this class. In the heterocyclic analogues, it can be seen from the NOE intensities that the heteroatom faces the outside of the cluster formed with the 2-benzoyl and 4-acetyl groups. Reduction of the 2-benzoyl ring to a cyclohexyl ring gives a highly active compound (**22**, Scheme 4) with similar conformational properties.

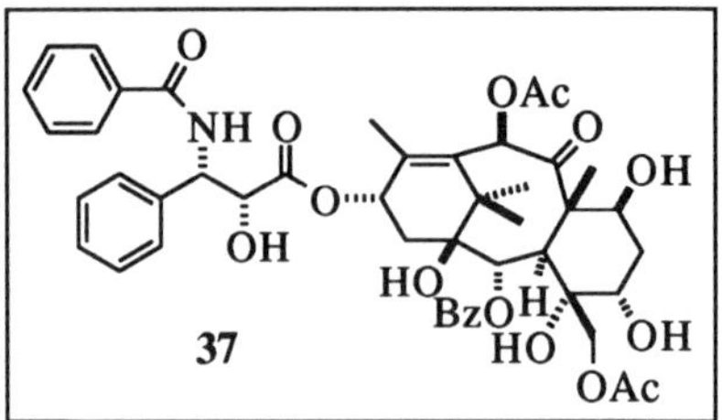

37

A second class of compounds, carrying aromatic rings at the 2 and 3' positions, are virtually inactive as a result of structural modifications and lack of one or more of the solvent-dependent conformational properties seen in the active compounds. These include two taxol homologues **24** and **25** with a methylene unit inserted in the side chain, which prevents the association of the two aromatic rings.

Also included in this group is the reaction product of taxol with Meerwein's reagent, compound **37,** in which the D ring has been opened and the 4-acetyl group has migrated to the less hindered 20-position (*56*).

Besides the displacement of the 4-acetyl, conformational changes in the A, B, and C ring resulting from the opening of the D ring effectively prevent the association of the aromatic rings. The centrality of the 4-acetyl group is also pointed out by the inactive 4-deacetyl analogue **30** of taxol, in which there is no impedance to the association of the aromatic rings, and NOE's between them are still observed; however, the characteristic aromatic chemical shift changes with solvent are not evident, which we interpret to mean that the rings are not constrained to their specific relative orientation as seen in the active compounds.

A third class contains 3'-alkyl compounds which do not extensively populate conformer **III** in polar solution and which show a different type of interaction between the 2- and 3'- substituents. 3'-Cyclohexyl and tricyclohexyl analogues **19** and **23** of taxol were found to belong to this group. Values of $J_{2',3'}$ remain small in polar solution, and no NOE's were observed between the 2 and 3' substituents in these compounds. There is a chemical shift change of one of the 2-methylenes in the 3'-cyclohexyl analogue in polar solution, suggesting there is still some tendency for the two rings to approach each other in conformer **I**, which is expected to be the dominant conformer, although they do not come within NOE distance of each other. A large (10 Hz) coupling constant between H-3' and H-1" on the cyclohexyl ring, and a relatively weak NOE is consistent with a dominantly *trans* conformation around this bond. In a model based on conformer **I** with this orientation, the cyclohexyl ring is observed to be about midway between the positions of the 3'-phenyl substituent in conformers **I** and **III**, which may help to account for the lack of solvent-dependent conformational change observed for these compounds. These results suggest that the enhanced stability of conformer **III** in taxanes with aromatic 2 and 3' substituents, and the specific orientation of those rings which is observed in polar solution, do result from the aromaticity of the 3'-substituent. Further conformational studies on other 3'-alkyl taxanes are in progress.

Abbreviations: Ac = acetyl; BOC = *tert*-butoxycarbonyl; Bz = benzoyl; DCC = 1,3-dicyclohexylcarbodiimide; DMAP = 4-dimethylaminopyridine; ED = effective dose; Et = ethyl; Me = methyl; n. d. = not determined; PPTS = pyridinium *p*-toluenesulfonate; TBS = *tert*-butyldimethylsilyl; TES = triethylsilyl; THF = tetrahydrofuran; TMS = trimethylsilyl; Troc = 2,2,2-trichloroethoxycarbonyl.

Acknowledgements: The authors wish to thank the following agencies for their financial support: The National Cancer Institute (CA 52790, CA 55141, CA 55160), Oread Laboratories, The Kansas Health Foundation for postdoctoral fellowships to G. C. B. Harriman and Z. S. Cheruvallath, The Scientific Education Partnership of the Marion Merrell Dow Foundation for postdoctoral fellowships to G. C. B. Harriman and Thomas C. Boge. We also would like to thank Larry L. Klein from Abbott Laboratories for providing us with a sample of compound **24** for NMR studies.

References

(1) Denis, J.-N.; Greene, A. E.; Guénard, D.; Guéritte-Voegelein, F.; Mangatal, L.; Potier, P. *J. Am. Chem. Soc.* **1988**, *110*, 5917-5919.

(2) For review: Georg, G. I.; Boge, T. C.; Cheruvallath, Z. S.; Clowers, J. S.; Harriman, G. C. B.; Hepperle, M.; Park, H. In *Taxol: Science and Applications*; Suffness, M., Ed.; CRC: Boca Raton, FL, 1994; (in press).

(3) For review: Suffness, M. *Annu. Rep. Med. Chem.* **1993**, *28*, 305-314.

(4) Wani, M. C.; Taylor, H. L.; Wall, M. E.; Coggon, P.; McPhail, A. T. *J. Am. Chem. Soc.* **1971**, *93*, 2325-2327.

(5) For review: Georg, G. I.; Ali, S. M.; Zygmunt, J.; Jayasinghe, L. R. *Exp. Opin. Ther. Pat.* **1994**, *4*, 109-120.

(6) Georg, G. I. *Tetrahedron Lett.* **1984**, *25*, 3779-3782.

(7) Georg, G. I.; Kant, J.; Gill, H. J. *J. Am. Chem. Soc.* **1987**, *109*, 1129-1135.

(8) Hart, D. J.; Ha, D.-C. *Chem. Rev.* **1989**, *89*, 1447-1465.

(9) Ojima, I.; Habus, I.; Zhao, M.; Georg, G. I.; Jayasinghe, L. R. *J. Org. Chem.* **1991**, *56*, 1681-1683.

(10) Georg, G. I.; Cheruvallath, Z. S.; Himes, R. H.; Mejillano, M. R.; Burke, C. T. *J. Med. Chem.* **1992**, *35*, 4230-4237.

(11) Ojima, I.; Habus, I.; Zhao, M.; Zucco, M.; Park, Y. H.; Sun, C. M.; Brigaud, T. *Tetrahedron* **1992**, *48*, 6985-7012.

(12) Georg, G. I.; Cheruvallath, Z. S.; Harriman, G. C. B.; Hepperle, M.; Park, H. *Bioorg. Med. Chem. Lett.* **1993**, *3*, 2467-2470.

(13) Holton, R. A.; Kim, H.-B.; Somoza, C.; Liang, F.; Biediger, R. J.; Boatman, P. D.; Shindo, M.; Smith, C. C.; Kim, S.; Nadizadeh, H.; Suzuki, Y.; Tao, C.; Vu, P.; Tang, S.; Zhang, P.; Murthi, K. K.; Gentile, L. N.; Liu, J. H. *J. Am. Chem. Soc.* **1994**, *116*, 1599-1600.

(14) Topliss, J. G. *J. Med. Chem.* **1972**, *15*, 1006-1011.

(15) Topliss, J. G. *J. Med. Chem.* **1977**, *20*, 463-469.

(16) Georg, G. I.; Cheruvallath, Z. S.; Himes, R. H.; Mejillano, M. R. *Bioorg. Med. Chem. Lett.* **1992**, *2*, 295-298.

(17) Georg, G. I.; Cheruvallath, Z. S.; Himes, R. H.; Mejillano, M. R. *Bioorg. Med. Chem. Lett.* **1992**, *2*, 1751-1754.

(18) Georg, G. I.; Cheruvallath, Z. S.; Harriman, G. C. B.; Hepperle, M.; Park, H.; Himes, R. H. *Bioorg. Med. Chem. Lett.* **1994**, (in press).

(19) Several 3'- and *N*-aryl taxol analogues have been claimed in a patent: Holton, R. A.; Hossain, N.; Beidiger, R. J.; Kim, S. European Patent Application **1992**, 0534709A1.

(20) 3'-Furyl and 3'-thienyl taxol analogues have been claimed in a patent: Holton, R. A.; Nadizadeh, H.; Beidiger, R. J. **1993**, European Patent Application 0534708A1.

(21) Georg, G. I.; Harriman, G. C. B.; Hepperle, M.; Himes, R. H. *Bioorg. Med. Chem. Lett.* **1994**, *4*, 1381-1384.

(22) Georg, G. I.; Boge, T. C.; Cheruvallath, Z. S.; Harriman, G. C. B.; Hepperle, M.; Park, H.; Himes, R. H. *Bioorg. Med. Chem. Lett.* **1994**, *4*, 335-338.

(23) Georg, G. I.; Harriman, G. C. B.; Park, H.; Himes, R. H. *Bioorg. Med. Chem. Lett.* **1994**, *4*, 487-490.

(24) Georg, G. I.; Boge, T. C.; Cheruvallath, Z. S.; Harriman, G. C. B.; Hepperle, M.; Park, H.; Himes, R. H. *Bioorg. Med. Chem. Lett.* **1994**, *4*, 1825-1830.

(25) Georg, G. I.; Clowers, J. S.; Harriman, B. G. C.; Hepperle, M.; Himes, R. H. **1994**, (unpublished results).

(26) *N*-Hexanoyl-*N*-debenzoyltaxol was recently isolated from cell cultures of *Taxus baccata*: Ma, W.; Park, G. L.; Gomez, G. A.; Nieder, M. H.; Adams, T. L.; Aynsley, J. S.; Sahai, O. P.; Smith, R. J.; Stahlhut, R. W.; Hylands, P. J.; Bitsch, F.; Shackleton, C. *J. Nat. Prod.* **1994**, *57*, 116-122.

(27) Vander Velde, D. G.; Georg, G. I.; Grunewald, G. L.; Gunn, K.; Mitscher, L. A. *J. Am. Chem. Soc.* **1993**, *115*, 11650-11651.

(28) Boge, T. C.; Himes, R. H.; Vander Velde, D. G.; Georg, G. I. *J. Med. Chem.* **1994**, (in press).

(29) Similar results were reported by the Ojima group: Duclos, O.; Zucco, M.; Ojima, I.; Bissery, M.-C.; Lavelle, F. *Abstracts of Papers*, 207th National Meeting of the American Chemical Society, San Diego, CA; American Chemical Society: Washington, DC, 1994; MEDI 86.

(30) Jayasinghe, L. R.; Datta, A.; Zygmunt, J.; Ali, S. M.; Vander Velde, D. G.; Georg, G. I. *J. Med. Chem.* **1994**, (in press).

(31) Commerçon, A.; Bézard, D.; Bernard, F.; Bourzat, J. D. *Tetrahedron Lett.* **1992**, *33*, 5185-5188.

(32) Denis, J.-N.; Correa, A.; Greene, A. E. *J. Org. Chem.* **1991**, *56*, 6939-6942.

(33) Klein, L. L.; Maring, C. J.; Li, L.; Yeung, C. M.; Thomas, S. A.; Grampovnik, D. J.; Plattner, J. J. *Abstracts of Papers*, 207th National Meeting of the American Chemical Society, San Diego, CA; American Chemical Society: Washington, DC, 1994; MEDI 143.

(34) For review: Kingston, D. G. I. *Pharmacol. Ther.* **1991**, *52*, 1-34.

(35) For selective removal of the C-2 benzoate of baccatin III see: Chen, S.-H.; Farina, V.; Wei, J.-M.; Long, B.; Fairchild, C.; Mamber, S. W.; Kadow, J. F.; Vyas, D.; Doyle, T. W. *Bioorg. Med. Chem. Lett.* **1994**, *4*, 479-482.

(36) For selective removal of the C-2 benzoate of taxol see: Chaudhary, A. G.; Gharpure, M. M.; Rimoldi, J. M.; Chordia, M. D.; Gunatilaka, A. A. L.; Kingston, D. G. I.; Grover, S.; Lin, C. M.; Hamel, E. *J. Am. Chem. Soc.* **1994**, *116*, 4097-4098.

(37) Datta, A.; Jayasinghe, L. R.; Georg, G. I. *J. Org. Chem.* **1994**, *59*, 4689-4690.

(38) Georg, G. I.; Datta, A.; Hepperle, M. **1994**, (unpublished results).

(39) Datta, A.; Jayasinghe, L.; Georg, G. I. *J. Med. Chem.* **1994**, (in press).

(40) The synthesis of compound **30** was also reported by Kingston: Kingston, D. G. *Abstracts of Papers*, 207th National Meeting of the American Chemical Society, San Diego, CA; American Chemical Society: Washington, DC, 1994; MEDI 145.

(41) Georg, G. I.; Cheruvallath, Z. S. *J. Org. Chem.* **1994**, *59*, 4015-4018.

(42) Holton, R. A.; Somoza, C.; Chai, K. B. *Tetrahedron Lett.* **1994**, *35*, 1665-1668.

(43) The synthesis of 9-dihydrotaxol possessing an α-hydroxy group at C-9 from 13-acetyl-9-dihydrobaccatin III was reported: Klein, L. L. *Tetrahedron Lett.* **1993**, *34*, 2047-2050.

(44) Georg, G. I.; Cheruvallath, Z. S.; Himes, R. H. **1994**, (unpublished results).

(45) The synthesis of derivative **35** was recently reported: Pulicani, J.-P.; Bourzat, J.-D.; Bouchard, H.; Commerçon, A. *Tetrahedron Lett.* **1994**, *35*, 4999-5002.

(46) Grampovnik, D. J.; Maring, C. J.; Klein, L. L.; Li, L.; Thomas, S. A.; Yeung, C. M.; Plattner, J. J. *Abstracts of Papers*, 207th National Meeting of the American Chemical Society, San Diego, CA; American Chemical Society: Washington, DC, 1994; MEDI 102.

(47) Guéritte-Voegelein, F.; Guénard, D.; Mangatal, L.; Potier, P.; Guilhem, J.; Cesario, M.; Pascard, C. *Acta Cryst.* **1990**, *C46*, 781-784.

(48) Williams, H. J.; Scott, A. I.; Dieden, R. A.; Swindell, C. S.; Chirlian, L. E.; Francl, M. M.; Heerding, J. M.; Krauss, N. E. *Tetrahedron* **1993**, *49*, 6545-6560.

(49) Cachau, R. E.; Gussio, R.; Beutler, J. A.; Chmurney, G. N.; Hilton, B. D.; Muschik, G. M.; Erickson, J. W. *Intl. J. Supercomput. Appl.* **1994**, *8*, 24-34.

(50) Swindell, C. S.; Krauss, N. E.; Horwitz, S. B.; Ringel, I. *J. Med. Chem.* **1991**, *34*, 1176-1184.

(51) Chen, S.-H.; Wei, J.-M.; Farina, V. *Tetrahedron Lett.* **1993**, *34*, 3205-3206.

(52) Baker, J. K. *Spectroscopy Letters* **1992**, *25*, 31-48.

(53) Hunter, C. A.; Sanders, J. M. *J. Am. Chem. Soc.* **1990**, *112*, 5525-5534.

(54) Vander Velde, D. G.; Boge, T. C.; Datta, A.; Georg, G. I.; Harriman, G. C. B.; Hepperle, M.; Mitscher, L. A.; Jayasinghe, L. R. **1994**, (unpublished results).

(55) Douzou, P.; Petsko, G. *Adv. Protein Chem.* **1984**, *36*, 245-361.

(56) Samaranayake, G.; Magri, N. F.; Jitrangsri, C.; Kingston, D. G. I. *J. Org. Chem.* **1991**, *56*, 5114-5119.

RECEIVED August 30, 1994

Chapter 17

Practical Semisynthesis and Antimitotic Activity of Docetaxel and Side-Chain Analogues

A. Commerçon, J. D. Bourzat, E. Didier, and François Lavelle

Rhône-Poulenc Rorer, Centre de Recherches de Vitry Alfortville, 13 Quai Jules Guesde, 94403 Vitry sur Seine, France

Docetaxel (Taxotere®) and a variety of semisynthetic side-chain analogs have been prepared and evaluated for their potency in inhibiting microtubules disassembly and for their antitumor activity in *in vitro* and *in vivo* experimental models. Their partial syntheses were achieved using stereoselective approaches. Different protection/deprotection strategies have been investigated. Structure-activity relationship studies demonstrate that biological activity is very dependent on the position and nature of substituents on the aromatic ring of 3'-modified-phenyl analogs. New carbamates have also been synthesized. Yet *tert*-butoxycarbonyl remains the substituent of choice for the 3'-nitrogen atom. Among all the new taxoids reported here, 3'-*para*-fluoro-docetaxel was identified as one of the most powerful analogs of docetaxel.

The structure of the natural antitumor agent paclitaxel (Taxol®, **1**) was first established by Wani, Wall *et al.* in 1971 (*1*). This diterpene, extracted from the bark of the Western Yew, *Taxus Brevifolia* Nutt (Taxaceae) (*2*), has proved highly cytotoxic against a wide number of cancer cell lines in *in vitro* and *in vivo* experimental models (*3*).

In the early 1980's, as the supply of paclitaxel for clinical evaluation was becoming scarce, Potier's group, i.e. Guéritte-Voegelein, Guénard *et al.*, at Gif-sur-Yvette started the partial synthesis of new taxane diterpenoids (taxoids) from 10-deacetyl-baccatin III **2**, an abundant constituent of the needles of the European yew species *Taxus baccata* L. (*4*). Using conveniently O-protected derivatives of baccatin III and 10-deacetyl-baccatin III (such as **3**) as key precursors (*5*), their semisynthetic work led to the discovery of the new and biologically potent taxoid docetaxel (Taxotere®, **4**) (*6*). The renewable source of 10-deacetyl-baccatin III **2** makes docetaxel easily available whereas the paclitaxel used in clinical trials has been harvested so far from the bark of yew trees, a process that unfortunately is fatal to the tree.

The unprecedented mechanism of action of docetaxel and paclitaxel has been extensively studied by Horwitz *et al.* (*7*) and Andreu *et al.* (*8*) who observed that both

0097–6156/95/0583–0233$08.00/0

compounds act by promoting tubulin assembly into stable microtubules. Clinical trials of both taxoids are currently underway utilizing Cremophor EL-ethanol for paclitaxel and polysorbate for docetaxel as formulation solvents. Paclitaxel and docetaxel clinical activities have been reported against many types of tumors such as advanced breast, ovarian and non-small cell lung cancers (*3,9*). Paclitaxel has received FDA registration approval for the treatment of metastatic ovarian cancer and breast cancer after failure of first line therapy.

A wide number of analogs of docetaxel and paclitaxel have already been prepared by different pharmaceutical and academic groups (*10*). Furthermore, two total syntheses of paclitaxel by Nicolaou *et al.* (*11*) and by Holton *et al.* (*12*) have been very recently reported. These major achievements from these two groups not only illustrate the present know-how of organic chemists in synthesizing complex structures but also may open the way to new and thus far inaccessible analogs.

Structure-activity relationships

The structural differences between paclitaxel and docetaxel are a *tert*-butoxycarbonyl (Boc) group instead of a benzoyl group on the nitrogen atom at C-3' on the side chain and an hydroxyl function instead of an acetate at the 10-position of the diterpene moiety (figure 1). These structural modifications lead to an increase of cytotoxicity in certain experimental models (*13*). These results suggested the possibility of further improvement by introducing, for instance, new side-chain modifications.

Structure-activity relationships of taxoids have already been reviewed (*10*) and our present knowledge in this area can be outlined as depicted in figure 2. The C-13 phenylisoserine side-chain and the diterpene moiety of paclitaxel are both crucial for biological activity. Thus baccatin III and its derivatives without the phenylisoserine side-chain at C-13 are neither active in the tubulin assay nor cytotoxic (*14*).

Studies on the diterpene moiety showed that the oxetane ring is essential for biological activity (*15*). Structural and molecular modelling studies show that this 4-membered ring is involved in a conformational lock of the diterpene skeleton and the C-13 side-chain through a pseudo chair conformation of ring C (*6, 16*). Paclitaxel and docetaxel rearrangement products possessing contracted (*17*) or cleaved (*18*) A-rings are significantly less bioactive products. B-ring contracted analogs have been reported as maintaining antitumor activity (*19*). A wide number of modifications can be introduced at the 7-position without significant loss of activity as noted with epimerization and acylation (acetyl, glutaryl, phenylalanyl, alanyl, N,N-dimethyl-glycyl, etc.) products

Figure 1. Structures of docetaxel, paclitaxel and baccatin III derivatives.

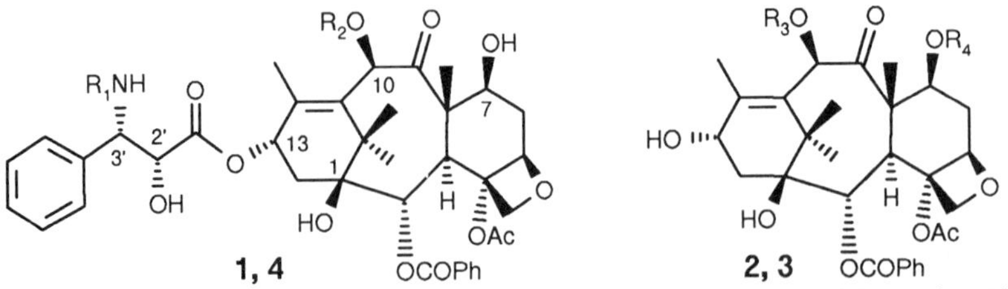

1, R_1=PhCO, R_2=Ac (paclitaxel) **2**, R_3=H, R_4=H (10-deacetyl-baccatin III)

4, R_1=tBuOCO, R_2=H (docetaxel) **3**, R_3=R_4=Troc (Cl_3CCH_2OCO)

(*20*). All of these C-7-modified derivatives showed activity comparable to paclitaxel in the microtubule disassembly assay but have generally slightly reduced cytotoxicity. Oxidation at C-7 is known to reduce bioactivity (*21*). Interestingly enough, 7-deoxy-paclitaxel exhibited *in vitro* activity similar to paclitaxel while 7,10-dideoxy-paclitaxel proved to be slightly less cytotoxic (*22, 23*).

Figure 2. Modifications influencing the cytotoxicity of paclitaxel.

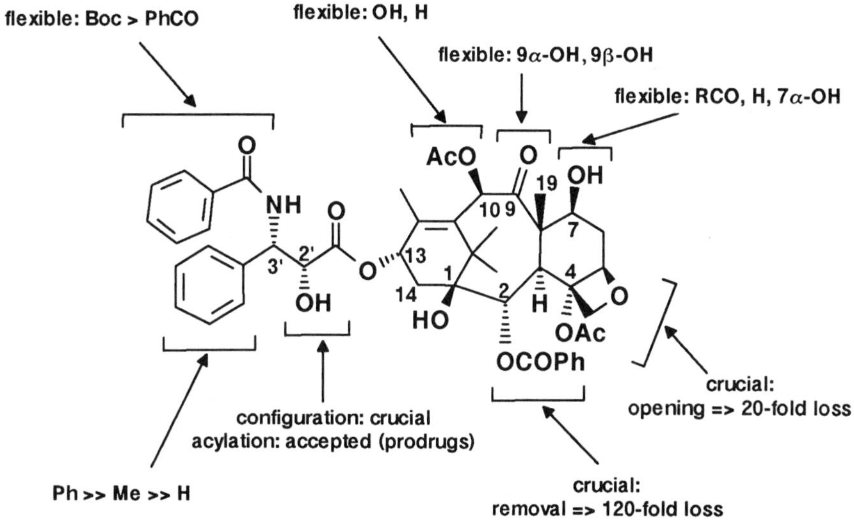

As reported by Klein with 9α-hydroxy-paclitaxel (*24*) and researchers at Rhône-Poulenc Rorer with 9α-and 9β-hydroxy-docetaxel (*25*), cytotoxicity of these reduced compounds is similar to paclitaxel and docetaxel respectively. Kingston *et al.* (*26*), Chen *et al.* (*27*), Holton *et al.* (*28*) as well as our own group (*25*) have all reported different approaches for preparing 10-deoxy-paclitaxel and 10-deoxy-docetaxel which were found to have comparable antitumor activity with respect to paclitaxel and docetaxel. Recently our group noted that 19-hydroxy-docetaxel, obtained from natural 10-deacetyl-19-hydroxy-baccatin III, has activity similar to docetaxel (*29*). Furthermore Chen *et al.* observed in the paclitaxel series an unprecedented rearrangement product possessing a cyclopropane moiety involving carbons C-7, C-8 and C-19 (*30*). This constrained analog of paclitaxel exhibits cytotoxicity equivalent to paclitaxel. Thus the top part of the diterpene moiety, that is positions 7, 9, 10 and 19, tolerate a wide variety of substituents. This allows us to assume that this region of taxoids may not play a crucial role in microtubule binding or, to some extent, to cytotoxicity. 2-Debenzoyloxy-paclitaxel as well as 2-debenzoyl-paclitaxel showed little *in vitro* cytotoxicity (*31*), while, very recently, different groups observed that modified benzoyl groups at C-2 have to be appropriately substituted to retain activity (*32*). These results indicate that the C-2 benzoate moiety plays an important role in the binding of paclitaxel to its receptor and, as a matter of fact, to cytotoxicity. Recently, it has been found that new taxoids derived from 14β-hydroxy-10-deacetylbaccatin III have activity close to paclitaxel and docetaxel (*33*). Optimum activity in the series was achieved with 14-β hydroxydocetaxel-1,14-carbonate.

Perhaps more than anyone else, Potier, Guéritte-Voegelin, Guénard *et al.* have focused on the study of structure-activity relationships of side-chain modified taxoids (*34*). They observed that the regio- and stereochemistry of each hetero atom at the 2'- and 3'-positions are crucial for retaining biological activity (*34*). The hydroxyl group at C-2' is critical since 2'-deoxy-paclitaxel is nearly 70-fold less cytotoxic than paclitaxel while isosteric 2'-deoxy-2'-fluoro-paclitaxel and 2'-O-methyl-paclitaxel proved to be nearly 100-fold and 200-fold less active respectively than the natural product (*35*). Introduction of an acyl group at 2' such as acetyl or hydrophilic acyl groups reduced activity in the tubulin assay but many of these derivatives retain activity *in vivo* or in cell-based assays since they are very likely to act as prodrugs (*36*). Replacement of the 3'-phenyl group of paclitaxel with methyl or hydrogen drastically lowers cytotoxicity (*34, 37*). The importance of an aryl group at C-3' was emphasized in Rhône-Poulenc Rorer (*38*) and Florida State University (*39*) patents as well as by the work of other groups (*40*) with the preparation of a wide number of very active analogs. Klein *et al.* (*19a*) observed that 3'-dephenyl-3'-isobutyl-9-dihydro-paclitaxel is highly cytotoxic, while Holton *et al.* patented 3'-dephenyl-3'-isobutenyl-paclitaxel (*41*) and 3'-dephenyl-3'-cyclohexyl-paclitaxel (*42*) as highly cytotoxic new taxoids. Our group (*43*) as well as Holton *et al.* (*39*) and more recently Georg *et al.* (*44*) reported that heteroaromatic groups at C-3' such as thienyl, furyl or pyridyl retain cytotoxicity.

Regarding modifications of the nitrogen atom at C-3', a free amino group as well as deletion of the amine function lead to less active analogs (*34*). Replacement of the 3'-benzoyl group of paclitaxel with other acyl groups such as tigloyl (cephalomanine), tosyl, butyryl and substituted benzoyl gives access to equally or less active compounds (*34, 45*) whereas introduction of a 3'-*tert*-butyloxycarbonyl substituent (docetaxel) leads to a more active compound in experimental models (*13*).

These preliminary results suggested that other modifications at C-3' might further improve the antitumor efficacy. We report herein some of our results regarding the stereoselective semisynthesis of docetaxel and new taxoids with either a 3'-modified-phenyl ring or a 3'-N-modified-carbamate moiety, along with their biological activity.

Chemistry.

Semisynthetic taxoid work has generated a demand for new stereoselective preparations of isoserine-type structures. This demand has been met during the last three years by a wide number of papers describing new approaches to phenylisoserinates or structurally related β-lactams (*46*). Our own efforts in this area have provided different stereoselective approaches.

Stereoselective approaches to 3'-modified-phenyl taxoids. Our first method used an aldol reaction as the key step (Scheme 1) (*47*). Asymmetric aldol reaction of the boron enolate of the chiral compound **5**, with differently substituted benzaldehydes led to the expected bromohydrins **6** as single isomers in moderate to good yields. Treatment of **6** with lithium ethoxide concomitantly cleaved the chiral auxiliary and formed epoxides **7**. Sodium azide ring opening gave, after hydrogenolysis and acylation, the corresponding ethyl phenylisoserinates **8**. At this point, an oxazolidine-type protection was introduced to increase the reactivity of the corresponding acid and minimize the risk of epimerization in the esterification process (*47*). Cyclic protection was achieved using 2-methoxy-propene. Subsequent alkaline hydrolysis afforded acids **9** in nearly quantitative overall yields. Esterification with O-diprotected baccatin III derivative **3** gave esters **10**. Deprotection under acidic conditions, N-acylation with $(Boc)_2O$ and reductive O-deprotection with zinc in acetic acid led to compounds **12** in satisfactory yields. This methodology was used to prepare a wide number of analogs **12** modified in the *ortho*, *meta* and *para* positions (R = Me, F, Cl, etc.).

Scheme 1. Approach based on a stereoselective aldol reaction

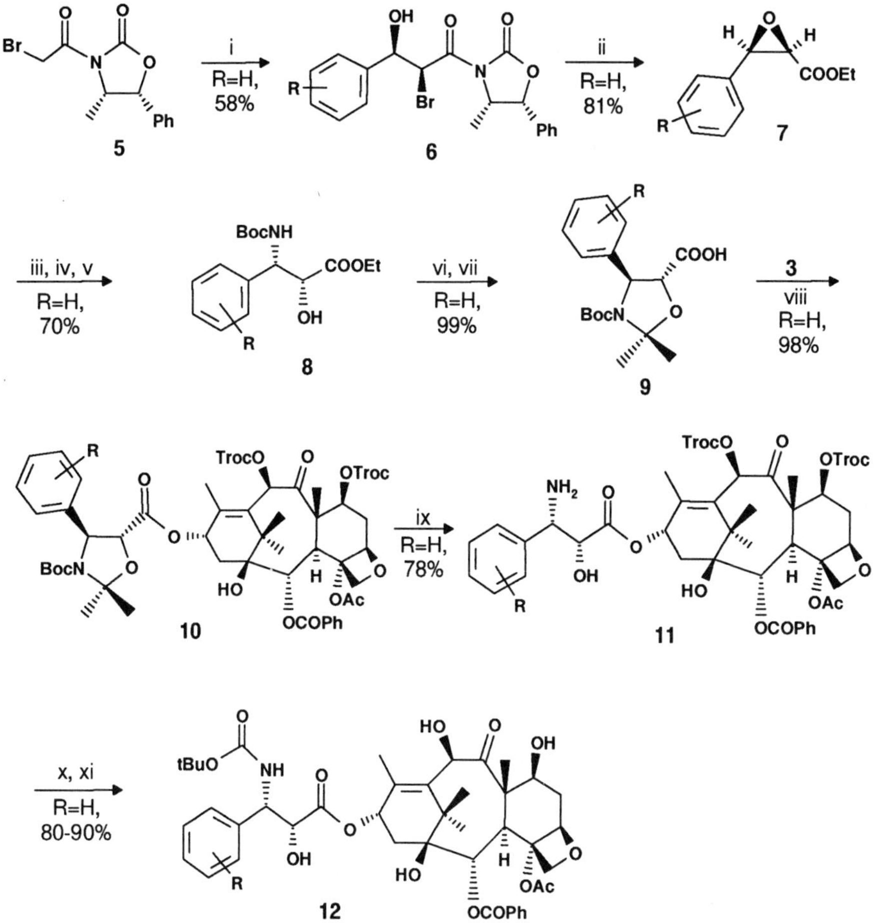

Reagents: i) **5** (1 equiv.), Et_3N (1.4 equiv.), Bu_2BOTf (1.2 equiv.), CH_2Cl_2, -70°C to 20°C, 2h then R-C_6H_4CHO (1 equiv.), -78°C to 0°C, 1h. ii) EtOLi, THF, -75°C to 15°C, 15min. iii) NaN_3, EtOH, NH_4Cl, 60°C, 8h. iv) H_2 (1 atm.), Pd/C (10%), AcOEt. v) $(Boc)_2O$, Na_2CO_3, CH_2Cl_2, 20°C. vi) $CH_2=C(CH_3)OCH_3$ (8 equiv.), PPTS, toluene, 80°C, (dist.). vii) LiOH, EtOH, H_2O, 20°C then H_3O^+. viii) **3** (1 equiv.), **9** (1.5 equiv.), DCC (1.6 equiv.), DMAP (0.5 equiv.), toluene, 80°C, 2h. ix) HCOOH, 20°C, 4h. x) $(Boc)_2O$, CH_2Cl_2, $NaHCO_3$, 20°C. xi) Zn, AcOH, MeOH, 60°C, 1h.

As the approach via the aldol reaction has several limitations (aldehydes bearing a basic nitrogen atom sometimes presented difficulties in the initial condensation reaction and, in the third step, regioselectivity of the oxirane opening is controlled by the substituents of the phenyl ring), we looked for other stereoselective approaches to phenylisoserines. Another way to prepare phenylisoserine derivatives is via opening of a corresponding β-lactam (*48*). Furthermore Holton (*49*) and Ojima (*50*) have shown that N-acyl β-lactams are very helpful in the esterification of baccatin derivatives at C-13. Extensive efforts at generating new stereoselective methods for the preparation of β-lactams quickly followed (*46*). Our own efforts at utilizing β-lactam chemistry as an expeditious way to generate isoserines after hydrolytic opening are outlined below (Scheme 2) (*51*). Staudinger [2+2] cycloaddition of readily available benzylidene-imines **13** with acetoxyketene generated *in situ* led to β-lactams **14a** and **14b** as a mixture of diastereomers, in approximately 50% d.e. After acetate hydrolysis the corresponding hydroxy-lactams were obtained. At this step (or in some cases at the previous one), we were generally able to separate the major diastereomer **15** by a single crystallization. After ring opening and hydrogenolysis, the desired phenylisoserine derivatives **16** were isolated in good overall yield. This latter compound was taken on to acid **9** as described in scheme 2.

Our approach has proven general enough to access a series of analogs modified on the phenyl ring at C-3' particularly at the *para*-position (R = F, $N(Me)_2$, etc.).

Scheme 2. Preparation of phenylisoserines using the Staudinger reaction.

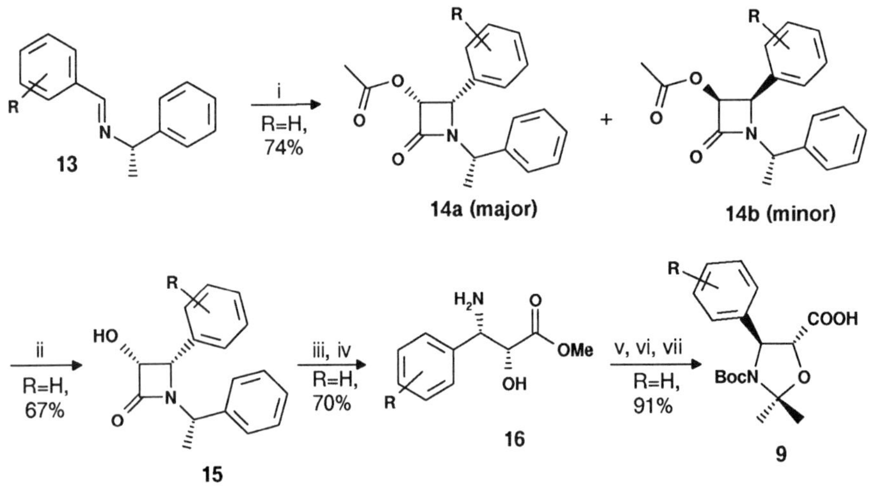

Reagents: i) $AcOCH_2COCl$, $CHCl_3$, Et_3N, 0°C, 1.5h then 20°C, 3h. ii) KOH/H_2O (1M), THF, 0°C, 1h or NH_3, 0°C, 1h then crystallisation from AcOEt. iii) HCl (6N)/MeOH (5/1), reflux 20h or HCl gas, MeOH, 40°C, 2.5h. iv) H_2 (345 psi), Pd/C (3%), MeOH/AcOH (3/1), 65°C, 4h or H_2 (15 psi), $Pd(OH)_2$ (20%), MeOH/AcOH (20/1), 20°C, 18h. v) $(Boc)_2O$, Na_2CO_3, CH_2Cl_2, 20°C. vi) $CH_2{=}C(CH_3)OCH_3$ (8 equiv.), PPTS, toluene, 80°C, (dist.). vii) LiOH, EtOH, H_2O, 20°C then H_3O^+.

In the early stage of our studies on structure-activity relationships, it was very important to know if the distance of the phenyl ring to the isoserine chain could be modified. Thus the corresponding compound, i.e. 3'-dephenyl-3'-benzyl-docetaxel **21**, was prepared by yet another route, as depicted in scheme 3, but still utilizing the same protection/deprotection sequence (Bourzat, J.D.; Commerçon, A., unpublished data). Benzylisoserinate **18** was obtained in several steps from N-Boc-phenylalaninol **17**. After isopropylidene protection and alkaline hydrolysis the derived acid **19** was esterified with the 7,10-diTroc baccatin derivative **3** to give the 13-O-esterified baccatin derivative **20**. Final deprotection, reacylation and cleavage of the Troc groups afforded the desired analog **21**. 3'-Dephenyl-3'-benzyl-docetaxel **21** proved to be 150-fold less active than docetaxel in the tubulin assay (*4*) and inactive in the different *in vitro* cytotoxicity assays (IC50 (P388): 8 μg/ml) showing that the distance of the phenyl group on the isoserine-chain was quite important.

Scheme 3. Preparation of 3'-dephenyl 3'-benzyldocetaxel

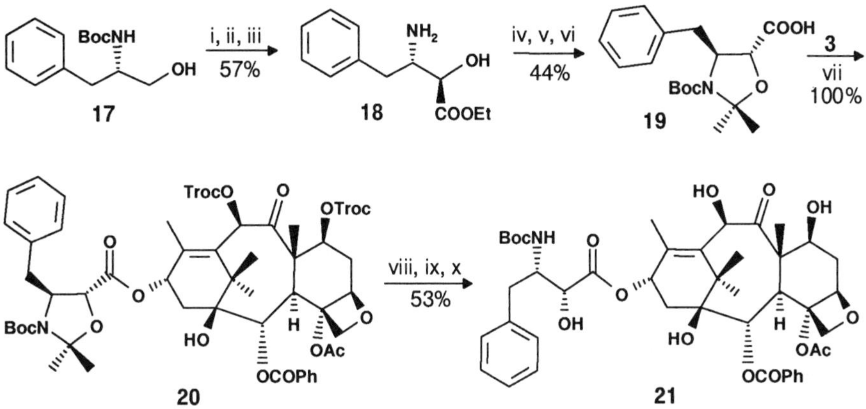

Reagents: i) $C_5H_5N{\cdot}SO_3$ (3 equiv.), Et_3N, DMSO, CH_2Cl_2, -70°C. ii) Me_3SiCN, $MgBr_2$, CH_2Cl_2, 26h, -20°C, then chromatogr. iii) EtOH, HCl, H_2O, 20°C. iv) $(Boc)_2O$, CH_2Cl_2, $NaHCO_3$, 20°C, 16h. v) $CH_2{=}C(CH_3)OCH_3$ (8 equiv.), PPTS, toluene, 80°C (dist.). vi) LiOH, EtOH, H_2O, 20°C, 3h then H_3O^+. vii) **3** (1 equiv.), **19** (1.5 equiv.), DCC (1.6 equiv;), DMAP (0.5 equiv.), toluene, 80°C, 2h. viii) HCOOH, 20°C, 4h. ix) $(Boc)_2O$, CH_2Cl_2, $NaHCO_3$, 20°C, 20h. x) Zn, AcOH, MeOH, 60°C, 35 min.

Protective groups improvements. One of the problems that we constantly faced in all of the above synthetic schemes involved the side-chain protection: cleavage of the isopropylidene protection of compounds **10** occurs under acidic conditions but with removal of the Boc moiety. Such a cleavage is desired when Boc replacement analogs are to be prepared (*vide infra*) but is of course less desirable when the Boc group is to remain on the molecule. In order to retain the Boc moiety, we investigated a large number of other acidic conditions but none afforded a chemoselective opening of the oxazolidine ring. These results led us to examine other 2-modified oxazolidines as protecting groups.

Our idea was to modify the 2-position with a group such as trichloromethyl, which might give, under reducing conditions, both cleavage of the cycle and removal of the 7,10-diTroc groups on the baccatin moiety.

Although trichloroacetaldehyde (chloral) is known to react with β-aminoalcohols to prepare 2-trichloromethyl-oxazolidines (*52*), to our knowledge, the use of such derivatives as protective groups of β-aminoalcohols has never been reported.

In the event (scheme 4)(*53*), cyclization of N-Boc phenylisoserine **22** with chloral in the presence of pyridinium *para*-toluenesulfonate (PPTS) in refluxing toluene yielded the expected oxazolidine derivatives as a mixture of diastereomers (d.e. 30%) but unfortunately with removal of the Boc group. We subsequently found that we could directly obtain these trichloromethylene-protected isoserinates **24** by treating phenylisoserinate **23** with chloral under the same conditions. The corresponding acids **25** were then prepared by alkaline hydrolysis. The nitrogen atom of these oxazolidines proved to be weakly basic. Thus we were able to directly esterify **25** with the O-diprotected baccatin derivative **3** affording the corresponding esters in good yield. Simultaneous cleavage of the trichloroethylidene protection and the Troc protective groups was performed with zinc in acetic acid at room temperature to give amino-taxoid **27** which was acylated to give docetaxel **4**.

Although in this case the initial goal of finding a selective protection with respect to the Boc was not fulfilled, this new approach represents a very expeditious semisynthetic access to docetaxel and side-chain modified analogs.

Scheme 4. 2-Trichloromethyl-oxazolidine-type protection

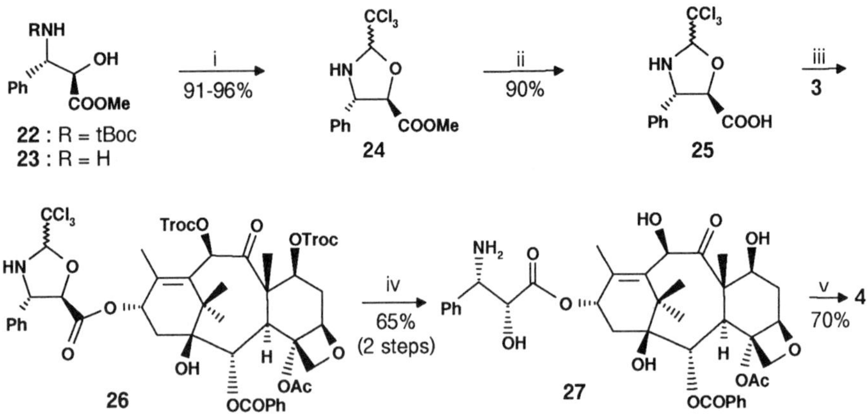

Reagents: i) from **22**: CCl_3CHO (20 equiv.), PPTS, toluene (dist.), flash chromatography in AcOEt/cyclohexane 1/3, 91% (d.e.=30% by NMR); from **23**: CCl_3CHO (4 equiv.), PPTS, toluene (dist.), 96% (d.e.=35% by HPLC); ii) KOH or $LiOH{\cdot}H_2O$ (1.1 equiv.), MeOH, H_2O, HCl 1N, >90%; iii) **3** (1 equiv.), **25** (1.8 equiv.), DCC (1.03 equiv.), DMAP (0.2 equiv.), toluene, 25°C (3h); iv) **26**, Zn (10 equiv.), AcOH (40 equiv.), AcOEt, 25°C (16h), 65% by HPLC from **3**; v) **27**, Di-*tert*-butyldicarbonate (1.2 equiv.), MeOH, 25°C (15h), 70% by HPLC.

Another idea was to introduce at the 2-position of the oxazolidine a substituent, such as an alkoxy, a phenyl or a mono- or dimethoxy-phenyl group, which might favor the hydrolytic cleavage of the ring without removal of the Boc group. We tested different aryl and alkoxy groups at the 2-position of the oxazolidine-protection (*54*). In considering the resulting yields, products stability and ability to give crystallized compounds, we selected the mono- or dimethoxy-phenyl groups. As an example, we report in scheme 5 the preparation of docetaxel with *para*-methoxy-phenyl as the activating group.

N-Boc phenylisoserine methyl ester reacts under kinetic control with *para*-methoxy-benzaldehyde dimethylacetal to give the corresponding 2-monosubstituted-1,3-oxazolidine-carboxylates **28a** and **28b** in 70% d.e. allowing the isolation of the major diastereomer **28b** after a single crystallization. Alkaline hydrolysis of **28b** followed by coupling reaction with the 7,10-O-diprotected baccatin derivative **3** afforded the corresponding esterification product **30** in nearly quantitative overall yield and without detectable epimerization at the 5-position of the oxazolidine ring. Acid-mediated

oxazolidine cleavage of the baccatin ester **30** was conducted with *para*-toluenesulfonic acid (PTSA) in methanol without removal of the Boc to give compound **31**.
Thus, this methodology fulfilled our original goal of maintaining the Boc group during the oxazolidine cleavage. It has also proved general and easily applicable to the preparation of docetaxel and side-chain modified analogs.
These oxazolidine-type protections have different advantages: compared to their open forms, the cyclic acids are very stable and reactive in the esterification reaction. Furthermore, in the cyclic-protected phenylisoserine, phenyl and carboxylic functions enjoy a trans disposition, which limits the risk of isomerization at C-2' during the esterification reaction. This 2'-epimerization possibility has also recently been reported by Greene *et al.* using trichloromethoxymethyl as protecting group at C-2' under the same esterification conditions (*55*).

Scheme 5. 4-Methoxyphenyl-oxazolidine-type protection

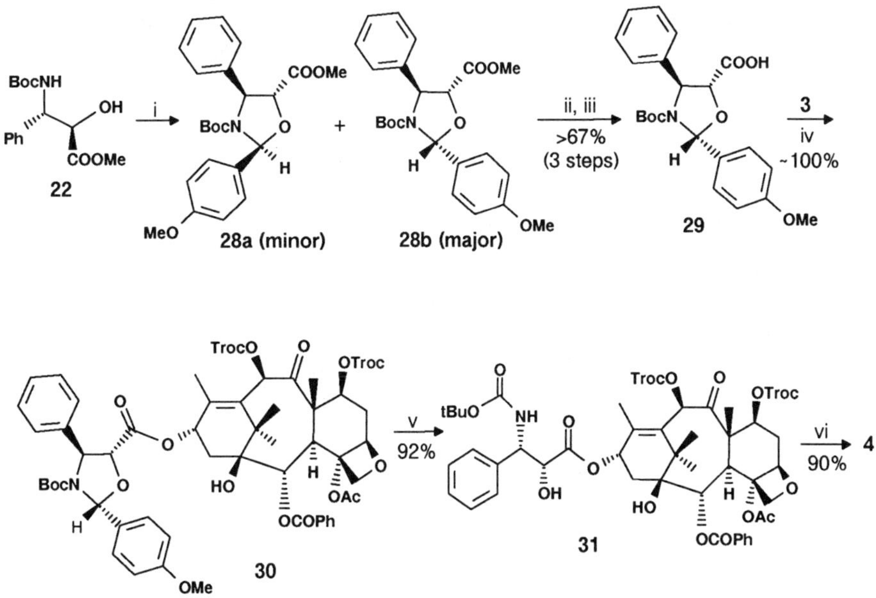

Reagents: i) **22**, 4-MeO-PhCH(OCH$_3$)$_2$ (1.1 equiv.), PPTS, toluene (dist.), 0.5-2h. ii) cryst. toluene/cyclohexane. iii) KOH or LiOH·H$_2$O (1.1 equiv.), MeOH, H$_2$O then HCl 1N. iv) **3** (1 equiv.), **29** (1.7equiv.), DCC (1.06 equiv.), DMAP (0.2 equiv.), toluene, 25°C, 2h. v) PTSA (1 eq.), MeOH, 25°C, (2h). vi) Zn (11 equiv.), AcOH (40 equiv.), AcOEt, 30°C, 3h.

Surprisingly enough, we observed limitations of such cyclic protections in the coupling of the corresponding enantiomeric or diastereomeric forms of acids **9** or **29** with the baccatin derivative **3**: partial to complete epimerization at 2' occurred in the esterification process (Bourzat, J.D.; Commerçon, A., unpublished data). This was also independently observed by Greene *et al.* (*55*).

Preparation of 3'-N-modified taxoids. Partial synthesis of new carbamate and amide derivatives of the amino group at C-3' have been easily realized using the free amino intermediate **32** (Scheme 6). Acylation was achieved with a wide range of acylating agents. For instance new carbamates were prepared from the free amino derivative with halocarbonates (when available or stable), or *para*-nitrophenyl carbonates. We also applied Senet's method which makes use of very reactive 1,2,2,2-tetrachloroethyl-carbonates (*56*). Application to our case proved to be general since we rarely faced difficulties in introducing bulky substituents at the N-3' position.

Scheme 6. General access to new carbamate analogs of docetaxel.

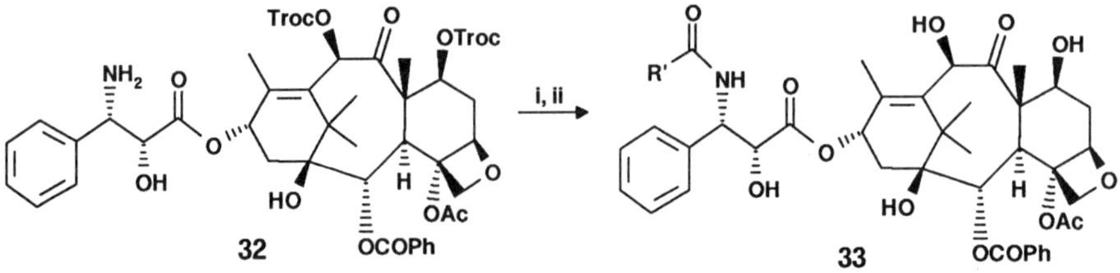

32 33

Reagents: i) method A: ClCOR', $NaHCO_3$, AcOEt, H_2O, 20°C, 15min; method B: $(R'CO)_2O$, $NaHCO_3$, CH_2Cl_2, 20°C, 24h; method C: $R'COOCHClCCl_3$, pyridine, CH_2Cl_2, 20°C, 16h; method D: $R'COOC_6H_4$-(*p*-NO_2), pyridine, CH_2Cl_2, 20°C. ii) Zn (11 equiv.), AcOH (40 equiv.), MeOH, 60°C, 1h..

Biological results and discussion.

Results of inhibition of the disassembly process of microtubules at 4°C (*4*) for docetaxel and its side-chain analogs as well as their *in vitro* cytotoxicity (P388 leukemia cell line) (*13*) and *in vivo* antitumor activity (B16 melanoma grafted in mice) (*57*) are reported in Tables I and II.

Table I : Biological activities of 3'-modified-phenyl analogs of docetaxel

Entry	Substituent R	Tubulin assay IC50/IC50 (**1**)	P388 IC50 μg/ml	B16 Melanoma mg/Kg/inj	B16 Melanoma %T/C	B16 Melanoma NCI Score
1	H (**4**)	0.64	0.04	13.4	0	++[d]
2	*o*-CH_3[a]	2.2	0.8	30	35	+
3	*m*-CH_3[a]	2.8	0.7	30	42	-
4	*p*-CH_3[a]	1.3	0.04	15.5	8	++
5	*o*-F[a]	1.1	0.04	10.8	0	++
6	*m*-F[a]	1	0.1	30	42	-
7	*p*-F[b]	0.9	0.03	20	6	++[e]
8	*m*-OMe[a]	3	0.7	18	88	-
9	*p*-OMe[a]	0.7	0.05	23	0	++
10	*p*-Cl[a]	1	0.05	18	0	++
11	*p*-I[c]	2.2	0.3	35	88	-
12	*p*-NMe_2[b]	3.5	0.65	20	39	+

[a]Prepared using scheme 1. [b]Prepared using scheme 2. [c]Obtained by another method (Bourzat, J-D. *et al.*, in preparation). [d]Log cell kill: 3.3. [e]Log cell kill: 3.2

Regarding the 3'-modified-phenyl analogs (Table I), one observes a rather good correlation between the microtubule disassembly inhibitions and both the cellular and the *in vivo* antitumor activities. The results obtained in this study demonstrate the influence of the substituent position on biological activity. We detected that introduction of a substituent at the *meta*-position (entries 3, 6, 8,) significantly decreases the level of cytotoxicity in experimental models. This lowering of activity is not apparently dependent on the electronic properties of the substituent (electron-withdrawing groups such as fluorine, or electron-donating groups such as methyl and methoxy). The presence of a substituent in the *ortho*-position looks acceptable as long as it is quite small (F versus Me, entries 2, 5). Modifications at the *para*-position are better tolerated although with modulated activity. Thus electronic effects are not detectable at *para* (entries 4, 9, 10) while bulky and basic substituents cause a loss in activity (entries 11, 12). Based on the *in vivo* results (log cell kill: 3.2), 3'-*para*-fluoro-docetaxel (entry 7) is a very powerful analog of docetaxel.

Concerning 3'-N-modified analogs, biological results (Table II) clearly demonstrate the superiority of the Boc group (entry 1) over other primary (entries 3, 4), secondary (entries 5, 6) or other tertiary (entries 7, 8, 9) alkyl carbamates. Isosteric forms of *tert*-butyl carbamate were prepared such as the C-isostere obtained from *tert*-butylacetyl chloride (entry 2). None of these derivatives exhibits higher cytotoxicity than docetaxel.

Table II : Biological activities of 3'-modified-carbamate analogs of docetaxel

Entry	Substituent R'	Tubulin assay IC50/IC50(1)	P388 IC50 µg/ml	B16 Melanoma mg/Kg/inj	%T/C	NCI Score
1	O-*tert*-Bu (**4**)	0.64	0.04	13.4	0	++[e]
2	CH_2-*tert*-Bu[a]	1.3	0.5	38	107	-
3	OMe[a]	0.7	6	50	68	-
4	O-*n*-Bu[a]	1.1	0.3	50	64	-
5	O-*iso*-Pr[a]	1.2	0.06	22	4	++[f]
6	$OCH(Et)_2$[c]	0.5	0.08	20	49	-
7	$OC(Me)_2Et$[b]	0.6	0.04	38	18	+
8	$OC(Me)_2CH_2Cl$[c]	0.6	0.07	20	30	+
9	$OC(Me)_2CH_2CN$[d]	0.65	0.45	20	16	+

[a]Prepared using method A. [b]Prepared using method B. [c]Prepared using method C. [d]Prepared using method D. [e]Log cell kill: 3.3; [f]Log cell kill: 1.4.

Concluding remarks

In conclusion, our work has contributed to the ongoing search for facile and practical preparations of phenylisoserinates and phenyl-modified phenylisoserinates. These methodologies led to the synthesis of docetaxel and docetaxel analogs modified at the 3'-position (phenyl ring or substituent of the nitrogen atom). Preliminary results clearly demonstrate that the *para*-position of the phenyl ring can be modified with retention of activity while *tert*-butylcarbamate remains thus far one of the best substituents for antitumor efficacy. Among all the taxoids disclosed, 3'-*para*-fluoro-docetaxel proved to be the most powerful analog of docetaxel.

Acknowledgements

We would like to thank Drs. H. Bouchard, E. Fouque, T. Hart, P. Ni and I. Taillepied as well as M. Alasia, D. Bézard, F. Bernard, E. Bouley, M-F. Marzin, C. Massey, P. Nemecek and C. Souder for their chemical contributions. We are also grateful to Dr. M. Vuilhorgne and his collaborators for structural analyses, Drs. M-C. Bissery, C. Combeau, J-F. Riou, P. Vrignaud and their collaborators for biological evaluation and Dr. C.J. Burns for critical reading of this manuscript. We also gratefully acknowledge collaborations with Drs. F. Guéritte-Voegelin, D. Guénard and Profs. P. Potier, A. Greene and I. Ojima. Finally we are also indebted to Drs. J-M. Mas, D. Bernard and P. Léon for fruitful discussions.

Literature cited

1. Wani, M.C.; Taylor, H.L.; Wall, M.E.; Coggon, P.; McPhail, A.T. *J. Am. Chem. Soc.* **1971**, *93*, 2325.
2. For a recent review on historical and ecological aspects of paclitaxel, see: Appendino, G. *Fitoterapia* **1993**, *54*, suppl. al N.1, 5.
3. a) Rowinsky, E.K.; Cazenave, L.A.; Donehower, R.C. *J. Natl. Cancer Inst.* **1990**, *82*, 1247. b) Rowinsky, E.K.; Donehower, R.C. *Pharm. Ther.* **1991**, *52*, 35.
4. Chauvière, G.; Guénard, D.; Picot, F.; Sénilh, V.; Potier, P. *C. R. Séances Acad. Sci., (série 2),* **1981**, *293*, 501.
5. Guéritte-Voegelein, F.; Sénilh, V.; David, B.; Guénard, D.; Potier, P. *Tetrahedron* **1986**, *42*, 4451.
6. Guénard, D.; Guéritte-Voegelein, F.; Potier, P. *Acc. Chem. Res.* **1993**, *26*, 160.
7. a) Schiff, P.B.; Fant, J.; Horwitz, S.B. *Nature* **1979**, *277*, 665. b) Ringel, I.; Horwitz, S.B. *Journal of the Natl. Cancer Inst.* **1991**, *83*, 288.
8. a) Diaz, J.F.; Andreu, J.M. *Biochemistry* **1993**, *32*, 2747. b) Diaz, J.F.; Menendez, M.; Andreu, J.M. *Biochemistry* **1993**, *32*, 10067.
9. a) Lavelle, F. *Curr. Opin. Invest. Drugs* **1993**, *2 (6)*, 627. b) Rothenberg, M.L. *Curr. Opin. Invest. Drugs* **1993**, *2 (12)*, 1269. c) Pazdur, R.; Kudelka, P.A.; Kavanagh, J.J.; Cohen, P.R.; Raber, M.N. *Cancer Treatment Reviews* **1993**, *19*, 351.
10. For recent reviews on the topic, see: a) Kingston, D.G.I. *Progress in the Chemistry of Organic Natural Products* **1993**, 1. b) Nicolaou, K.C.; Dai, W-M.; Guy, R.K. *Angew. Chem. Int. Ed. Engl.* **1994**, *33*, 15. c) Georg, G.I.; Ali, S.M.; Zygmunt, J.; Jayasinghe, L.R. *Exp. Opin. Ther. Patents* **1994**, *4*, 109.
11. Nicolaou K.C.; Yang, Z.; Liu, J.J.; Ueno, H.; Nantermet, P.G.; Guy, R.K.; Claiborne, C.F.; Renaud, J.; Couladouros, E.A.; Paulvannan, K.; Sorensen, E.J. *Nature* **1994**, *367*, 630.
12. a) Holton, R.A.; Somoza, C.; Kim, H-B.; Liang, F.; Biediger, R.J.; Boatman, P.D.; Shindo, M.; Smith, C.C.; Kim, S.; Nadizadeh, H.; Suzuki, Y.; Tao, C.; Vu, P.; Tang, S.; Zhang, P.; Murthi, K.K.; Gentile, L.N.; Liu, J.H. *J. Am. Chem. Soc.* **1994**, *116*, 1597. b) Holton, R.A.; Kim, H-B.; Somoza, C.; Liang, F.; Biediger, R.J.; Boatman, P.D.; Shindo, M.; Smith, C.C.; Kim, S.; Nadizadeh, H.; Suzuki, Y.; Tao, C.; Vu, P.; Tang, S.; Zhang, P.; Murthi, K.K.; Gentile, L.N.; Liu, J.H. *J. Am. Chem. Soc.* **1994**, *116*, 1599.
13. Riou, J-F.; Naudin, A.; Lavelle, F. *Biochem. Biophys. Res. Commun.* **1992**, *187*, 164.
14. Lataste, H.; Sénilh, V.; Wright, M.; Guénard, D.; Potier, P. *Proc. Natl. Acad. Sci. U.S.A.* **1984**, *81*, 4090.
15. Samaranayake, G.; Magri, N.F.; Jitragsri, C.; Kingston, D.G.I. *J. Org. Chem.*, **1991**, *56*, 5114.
16. Guéritte-Voegelein, F.; Guénard, D.; Mangatal, L.; Potier, P.; Guilhem, J.; Cesario, M.; Pascard, C.; *Acta Cryst.* **1990**, *C46*, 781.

17. Wahl, A.; Guéritte-Voegelein, F.; Guénard, D.; Le Goff, M-T.; Potier, P. *Tetrahedron*, **1992**, *48*, 6965.
18. Ojima, I.; Fenoglio, I.; Park, Y.H.; Sun, C-M.; Appendino, G.; Pera, P.; Bernacki, R.J. *J. Org. Chem.* **1994**, *59*, 515.
19. a) Klein, L.L. *207th Amer. Chem. Soc. Natl. Meeting* San Diego (USA) **1994**. b) Klein, L.L.; Maring, C.J.; Li, L.; Yeung, C.M.; Thomas, S.A.; Grampovnik, D.J.; Plattner, J.J.; Henry, R.F. *J. Org. Chem.* **1994**, *59*, 2370.
20. a) Ringel, I.; Horwitz, S.B. *J. Pharmacol. Exp. Ther.* **1987**, *242*, 692. b) Huang, C.H.O.; Kingston, D.G.I.; Magri, N.F.; Samaranayake, G. *J. Nat. Prod.*, **1986**, *49*, 665. c) Mellado, W.; Magri, N.F.; Kingston, D.G.I.; Garcia-Arenas, R.; Orr, G.A.; Horwitz, S.B.; *Biochem. Biophys. Res. Commun.* **1984**, *124*, 329.
21. Kingston, D.G.I. *Pharmacol. Ther.* **1991**, *52*, 1.
22. Chaudhary, A.G.; Rimoldi, J.M.; Kingston, D.G.I. *J. Org. Chem.* **1993**, *58*, 3798.
23. a) Chen, S-H.; Huang, S.; Kant, J.; Fairchild, C.; Wei, J.; Farina, V. *J. Org. Chem.* **1993**, *58*, 5028.b) Chen, S-H.; Wei, J-M.; Vyas, D.M.; Doyle, T.W.; Farina, V. *Tetrahedron Lett.* **1993**, *34*, 6845.
24. Klein, L. *Tetrahedron Lett.* **1993**, *34*, 2047.
25. a) Commerçon, A. *207th Amer. Chem. Soc. Natl. Meeting* San Diego (USA), March 1994. b) Pulicani, J-P.; Bourzat, J-D.; Bouchard, H.; Commerçon, A. *Tetrahedron Lett.* **1994**, *in press*.
26. Chaudhary, A.G.; Kingston, D.G.I. *Tetrahedron Lett.* **1993**, *34*, 4921.
27. Chen, S.H.; Fairchild, C.; Mamber, S.W.; Farina, V. *J. Org. Chem.* **1993**, *58*, 2927.
28. Holton, R.A.; Somoza, C.; Chai, K-B. *Tetrahedron Lett.* **1994**, *35*, 1665.
29. Margraff, R.; Bézard, D.; Bourzat, J-D.; Commerçon, A. *Bioorg. Med. Chem. Lett.* **1994**, *4*, 233.
30. Chen, S-H.; Huang, S.; Wei, J.; Farina, V. *J. Org. Chem.* **1993**, *58*, 4520.
31. Chen, S.H.; Wei, J.M.; Farina, V. *Tetrahedron Lett.* **1993**, *34*, 3205.
32. a) Chen, S-H.; Farina, V.; Wei, J-M.; Long, B.; Fairchild, C.; Mamber, S.W.; Kadow, J.F.; Vyas, D.; Doyle, T.W. *Bioorg. Med. Chem. Lett.* **1994**, *4*, 479. b) Chaudhary, A.G.; Gharpure, M.M.; Rimoldi, J.M.; Chordia, M.D.; Gunatilaka, A.A.L.; Kingston, D.G.I. *J. Am. Chem. Soc.* **1994**, *116*, 4097. c) Ojima, I.; Duclos, O.; Zucco, M.; Bissery, M-C.; Combeau, C.; Vrignaud, P.; Lavelle, F. *J. Med. Chem.* **1994**, *in press*.
33. Ojima, I.; Park, Y.H.; Sun, C-M.; Fenoglio, I.; Appendino, G.; Pera, P. *J. Med. Chem.* **1994**, *37*, 1408.
34. Guéritte-Voegelein, F.; Guénard, D.; Lavelle, F.; Le Goff, M-T.; Mangatal, L.; Potier, P. *J. Med. Chem.*, **1991**, *34*, 992.
35. Kant, J.; Huang, S.; Wong, H.; Fairchild, C.; Vyas, D.; Farina, V. *Bioorg. Med. Chem. Lett.* **1993**, *3*, 2471.
36. Prodrugs of paclitaxel have been reviewed by Nicolaou (*10b*) and Georg (*10c*).
37. Swindell, C.S.; Krauss, N.E.; Horwitz S.B.; Ringel, I. *J. Med. Chem.* **1991**, *34*, 1176.
38. Bourzat, J-D.; Commerçon, A.; Paris, J-M. *PCT Int. Appl.* WO 92/09 589, **1992**.
39. Holton, R.A. *PCT Int. Appl.* WO 93/06 079, **1993**.
40. a) Monsarrat, B.; Mariel, E.; Cros, S.; Garès, M.; Guénard, D.; Guéritte-Voegelein, F.; Wright, M. *Drug Metab. Dispos.* **1990**, *18*, 895. b) Georg, G.I.; Cheruvallath, Z.S.; Himes, R.H.; Mejillano, M.R. *Bioorg. Med. Chem. Lett.* **1992**,

2, 1751. c) Georg, G.I.; Cheruvallath, Z.S.; Himes, R.H.; Mejillano, M.R. *Bioorg. Med. Chem. Lett.* **1992**, *2*, 295. d) Georg, G.I.; Cheruvallath, Z.S.; Himes, R.H.; Mejillano, M.R.; Burke, C.T. *J. Med. Chem.* **1992**, *35*, 4230.

41. Holton, R.A.; Hossain, N. *US Pat. Appl.* 5 284 864, **1994**.
42. Holton, R.A.; Hossain, N. *US Pat.* Appl. 5 284 865, **1994**.
43. Bouchard, H.; Bourzat, J-D.; Commerçon, A. *in press*.
44. Georg, G.I.; Harriman, G.C.B.; Hepperle, M.; Himes, R.H. *Bioorg. Med. Chem. Lett.* **1994**, *4*, 1381.
45. a) Miller, R.W.; Powel, R.G.; Smith, C.R. Jr.; Arnold, E.; Clardy, J. *J. Org. Chem.* **1981**, *46*, 1469. b) McLaughlin, J.L.; Miller, R.W.; Powel, R.G.; Smith, C.R. Jr. *J. Nat. Prod.* **1981**, *44*, 312. c) Georg. G.I.; Cheruvallath, Z.S.; Himes, R.H.; Mejillano, M.R. *Bioorg. Med. Chem. Lett.* **1992**, *2*, 1751. d) Georg. G.I.; Cheruvallath, Z.S.; Himes, R.H.; Mejillano, M.R. *Bioorg. Med. Chem. Lett.* **1992**, *2*, 295. e) Georg. G.I.; Cheruvallath, Z.S.; Himes, R.H.; Mejillano, M.R. *J. Med. Chem.* **1992**, *35*, 4230.
46. Approaches to the preparation of paclitaxel and docetaxel side-chains have already been reviewed by several groups (*10*). For a new recent survey, see: Boa, A.N.; Jenkins, P.R.; Lawrence, N.J. *Contemporary Organic Synthesis*, **1994**, *1*, 47.
47. Commerçon, A.; Bernard, F.; Bézard, D.; Bourzat, J-D. *Tetrahedron Lett.* **1992**, *33*, 5185.
48. Ojima, I.; Park, Y.H.; Sun, C.M.; Brigaud, T.; Zhao, M. *Tetrahedron Lett.* **1992**, *33*, 5737.
49. Holton, R.A. *Eur. Pat. Appl. EP 400,971*, **1990**.
50. a) Ojima, I.; Habus, I.; Zhao, M.; Georg, G.I.; Jayashinge, L.R. *J. Org. Chem.* **1991**, *56*, 1681. b) Ojima, I., Sun, C.M.; Zucco, M.; Park, Y.H.; Duclos, O.; Kuduk, S. *Tetrahedron Lett.* **1993**, *34*, 4149. c) Ojima, I.; Habus, I.; Zhao, M.; Georg, G.I.; Zucco, M.; Park, Y.H.; Sun, C.M.; Brigaud, T. *Tetrahedron* **1992**, *48*, 6985.
51. Bourzat, J.D.; Commerçon, A. *Tetrahedron Lett.* **1993**, *34*, 6049.
52. a) Steiner, B.; Sasinkova, V.; Koos, M.; Repas, M.; Koenigstein, J. *Chem. Pap.* **1988**, *42*, 817. b) Ruske, W.; Hartmann, I. *J. Prakt. Chem.* **1962**, *18*, 146.
53. Didier, E.; Fouque, E.; Commerçon, A. *Tetrahedron Lett.* **1994**, *35*, 3063.
54. Didier, E.; Fouque, E.; Taillepied, I.; Commerçon, A. *Tetrahedron Lett.* **1994**, *35*, 2349.
55. Denis, J.N.; Kanazawa, A.M.; Greene, A.E. *Tetrahedron Lett.* **1994**, *35*, 105.
56. Barcelo, G.; Senet, J-P.; Sennyey, G. *J. Org. Chem.* **1985**, *50*, 3951.
57. Bissery, M-C.; Guénard, D.; Guéritte-Voegelein, F.; Lavelle, F. *Cancer Res.* **1991**, *51*, 4845.

RECEIVED August 19, 1994

Chapter 18

Paclitaxel Structure–Activity Relationships and Core Skeletal Rearrangements

Shu-Hui Chen and Vittorio Farina

Central Chemistry, Bristol-Myers Squibb, 5 Research Parkway, Wallingford, CT 06492–7660

Abstract: A number of novel paclitaxel (Taxol®) analogs that contain modifications at the C_2, C_7 and C_{10} position have been prepared during the course of a taxol SAR study. These analogs include deoxygenated, fluorinated and C_7-C_{19} cyclopropane paclitaxel derivatives. Several very interesting skeletal rearrangements within the taxol and baccatin core were also discovered while developing analog syntheses. The details of the chemistry and the biological evaluation of these analogs will be discussed.

Paclitaxel (Taxol®) **1**, a highly oxygenated diterpenoid isolated from the bark of *Taxus brevifolia* (*1*), has recently been approved for the treatment of advanced ovarian cancer (*2*). In the ongoing clinical trials involving breast and lung cancer, taxol has also demonstrated clinical potential (*3*). Paclitaxel functions by arresting cell replication in the mitotic phase of the cell cycle. These effects are attributed to paclitaxel's ability to polymerize tubulin into highly stable microtubules (*4*). The discovery of paclitaxel's unique antimitotic mechanism and its clinical impact have stimulated intensive research efforts that are aimed at designing novel analogs with improved biological profiles (*5*).

The structure of paclitaxel **1** and its semisynthetic side chain analog, docetaxel (Taxotere®) **2**, are shown in Figure 1. Figure 2 contains the structures of baccatin III **3** and 10-desacetylbaccatin III (10-DAB), **4**, important precursors of the clinically active taxanes. 10-DAB, a natural product from *Taxus baccata* (*6*), is being used by Bristol-Myers Squibb as a starting material for the semisynthesis of paclitaxel **1**. Likewise, Rhône-Poulenc Rorer utilizes the same compound as their starting material in the semisynthesis of docetaxel **2**. It should be pointed out that Taxotere® **2** is more potent than Taxol® **1** in certain *in vitro* assays and *in vivo* models, and is currently in clinical trials in Europe and Japan (*7*).

Our involvement in the study of paclitaxel's SAR began four years ago when relatively little was known about the subject (*8*). The initial goal of our research was to systematically modify each of the functional groups on the core, such as, the C_2 benzoate (*9*), C_{10} acetate (*10*), C_7 hydroxyl group (*11*), and oxetane ring (*12*), and then to subsequently evaluate the *in vitro* activity of these modified analogs. The initial results from our core SAR study thus form the topics of this article and they are: (i) deoxygenation at C_2, C_7 and C_{10}; (ii) fluorination studies at C_7; (iii) regioselective oxetane ring opening chemistry.

0097–6156/95/0583–0247$08.00/0

Figure 1

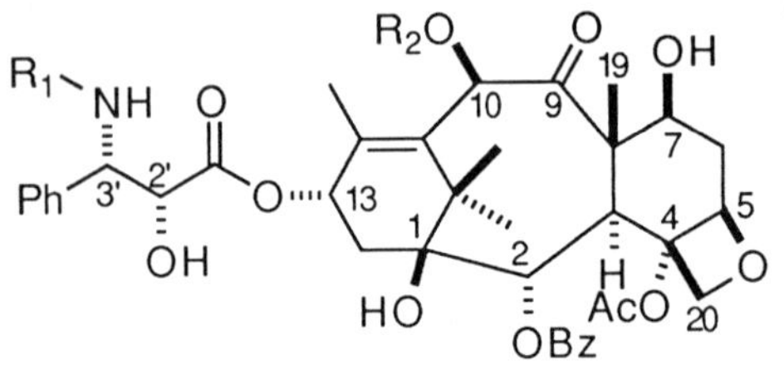

Paclitaxel 1 R_1=Bz, R_2=Ac

Docetaxel 2 R_1=Boc-t, R_2=H

Figure 2

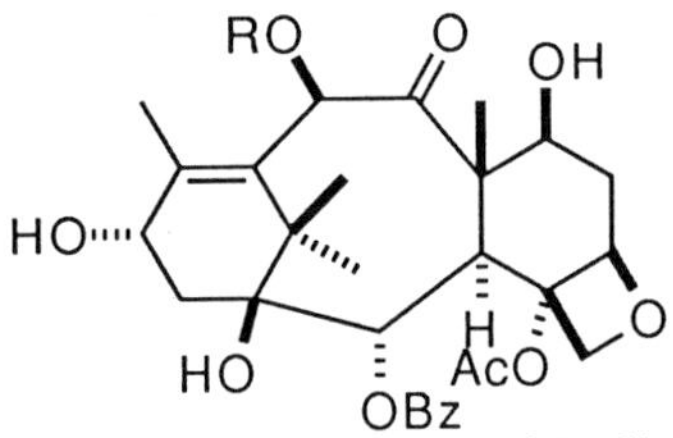

Baccatin III 3 R=Ac

10-desacetylbaccatin 4 R=H

Figure 1 & 2: Structures of Paclitaxel, Docetaxel, Baccatin III and 10-DAB.

Scheme 1

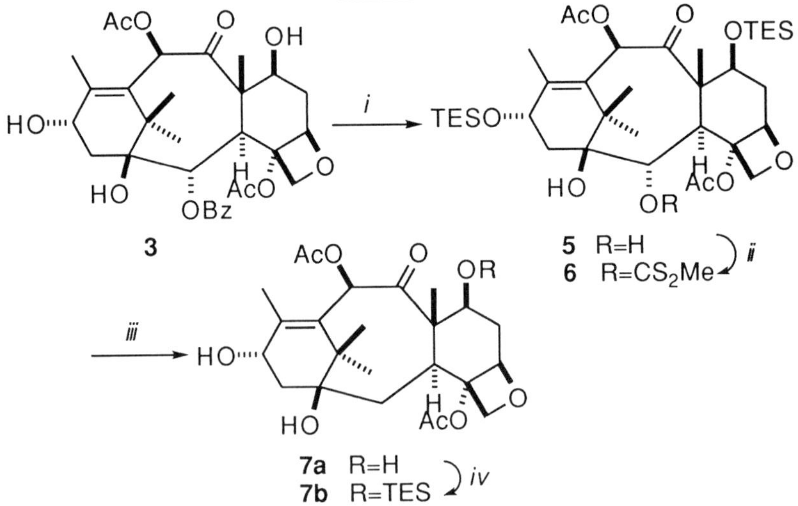

Conditions: (i) TESCl/imidazole/DMF/r.t., (90%); then Red-Al/THF/0°C, (84%); (ii) NaH/THF+CS_2/MeI, (61%); (iii) Bu_3SnH/AIBN/Toluene/90°C, (89%); then TBAF/THF/0°C, (85%); (iv) TESCl/imidazole/DMF/0°C, (83%).

Results and Discussions

(i) Deoxygenation chemistry:

Our systematic deoxygenation studies began with the C_2 debenzoylation of baccatin III **3** as shown in Scheme 1 (*9*). We envisioned that the xanthate **6** would be the ideal precursor for C_2 deoxy baccatin **7a**, when utilized in the Barton deoxygenation reaction. C_2 xanthate **6** can be prepared from baccatin **3** *via* the corresponding C_2 alcohol **5**. Regioselective C_2 debenzoylation was initially attempted by hydrolysis with sodium methoxide in methanol but was unsuccessful. However, after considerable experimentation, we were pleased to discover that treatment of 7,13-bisTES baccatin with Red-Al in THF provided a clean C_2 debenzoylation reaction (*9b*). The notable selectivity observed in this reaction is presumably due to the precoordination of reducing agent with the C_1 hydroxyl group which results in the observed directed reduction. The $C_{1,2}$ diol **5** was then treated with sodium hydride, carbon disulfide and methyl iodide, to afford the C_2 xanthate **6**. Upon Barton deoxygenation, xanthate **6** was smoothly converted to the desired C_2 deoxybaccatin **7a**.

As can be seen in Scheme 2, the C_7 xanthate **8** was prepared from baccatin III **3** for use as an intermediate in the preparation of both C_7 deoxybaccatin and $C_{7,10}$ dideoxy baccatin (*11*). It was necessary to introduce the C_{13} triethylsilyl group during the course

Scheme 2

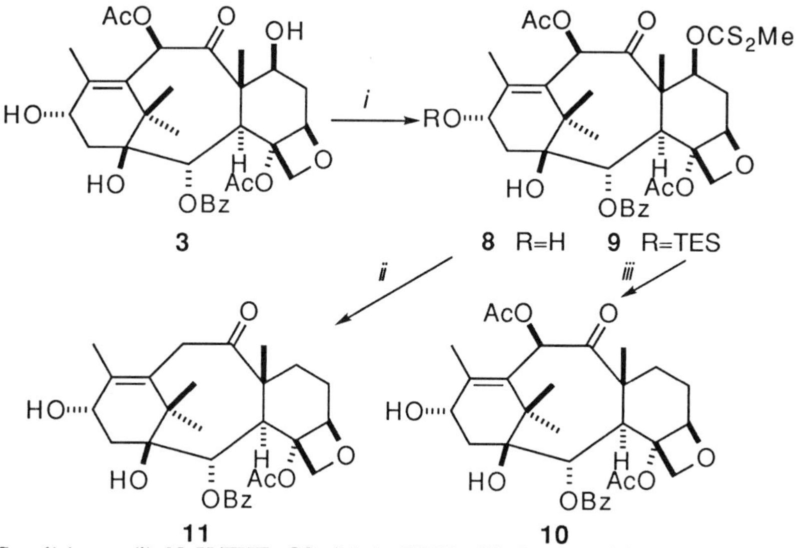

Conditions: (i) NaH/THF+CS_2/MeI, (57%); (ii) for **8** to **11**: 6 eq. Bu_3SnH/AIBN /Toluene/100°C, (85%); (iii) for **8** to **9**: TESCl/imidazole/DMF/r.t., (86%); for **9** to **10**: Bu_3SnH/AIBN/Toluene/85°C, (83%); then TBAF/THF/0°C, (73%).

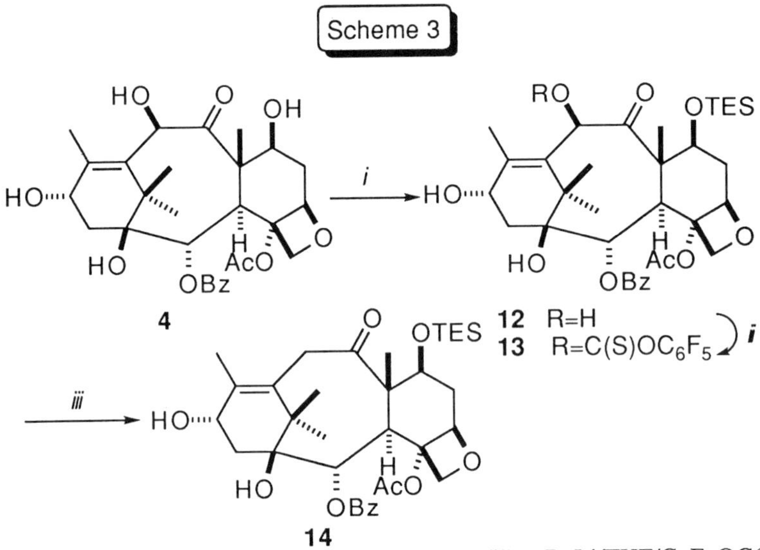

Conditions: (i) TESCl/imidazole/DMF/0°C, (89%); (ii) n-BuLi/THF/$C_6F_5OC(S)Cl$/-40°C, (74%); (iii) Bu_3SnH/AIBN/Toluene/90°C, (99%).

of the C_7 deoxygenation study in order to provide adequate solubility for the deoxygenation reaction. Reduction of xanthate **9** with Bu_3SnH/AIBN in toluene at 85°C gave the desired C_7 deoxy baccatin **10**. Interestingly, reduction of xanthate **8** with a large excess of Bu_3SnH/AIBN (6-8 equiv.) in toluene at elevated temperature (110°C) afforded 7,10-dideoxy baccatin **11** (*10b*). Not surprisingly, using identical conditions, C_7 deoxybaccatin **10** could be converted to **11** in high yield. Precedent for this type of reductive removal of the C_{10} acetate is contained in reports describing the reductive cleavage of certain activated (e.g. allylic, α-keto) benzoates (*10b*).

Scheme 4

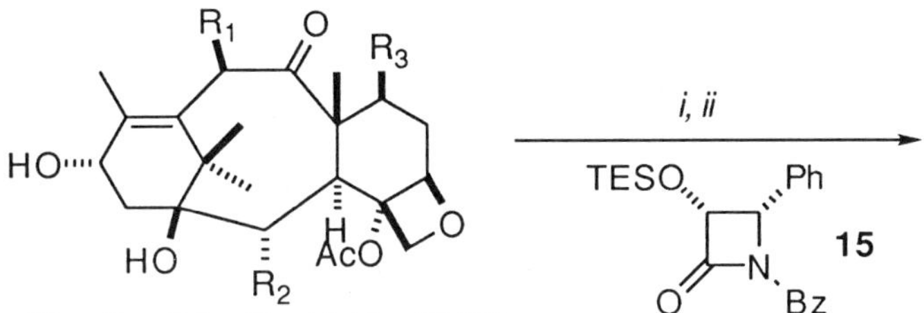

7b R_1=OAc, R_2=H, R_3=OTES
10 R_1=OAc, R_2=OBz, R_3=H
11 R_1=R_3=H, R_2=OBz
14 R_1=H, R_2=OBz, R_3=OTES

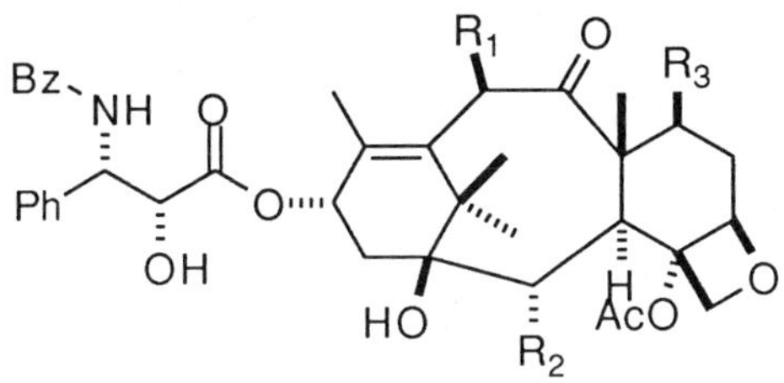

2-deoxytaxol **16** R_1=OAc, R_2=H, R_3=OH
7-deoxytaxol **17** R_1=OAc, R_2=OBz, R_3=H
7,10-dideoxytaxol **18** R_1=R_3=H, R_2=OBz
10-deoxytaxol **19** R_1=H, R_2=OBz, R_3=OH

Conditions: (i) LHMDS/THF/-40°C, then **15**; (ii) Pyridine/48% HF/CH_3CN/5°C. For **7b** to **16**, (63%); for **10** to **17**, (53%); for **11** to **18**, (30%); for **14** to **19**, (78%).

We recently reported the first synthesis of 10-deoxy taxol *via* the use of a novel $C_{10\text{-}11}$/ $C_{12\text{-}18}$ dienone intermediate (*10a*). Alternatively, 10-deoxy taxol can be prepared from 10-deoxy baccatin **14**, whose synthesis is shown in Scheme 3. Selective C_7 silylation (TESCl/imidazole/DMF) and subsequent C_{10} acylation (n-BuLi/ $C_6F_5OC(S)Cl$/THF) of 10-DAB **4** afforded the C_{10} thiocarbonate **13**. This compound was subsequently converted to 10-deoxy baccatin **14** in quantitative yield using Barton's deoxygenation procedure (*11a*).

The paclitaxel side chain was then readily attached to the four deoxy baccatin derivatives which were in hand using Holton's protocol (*13*). As shown in Scheme 4, treatment of these deoxy baccatins (**7b**, **10**, **11**, **14**) with LHMDS followed by β-lactam

15 (*14*) provided the corresponding four deoxy paclitaxel derivatives (**16, 17, 18, 19**) in fair to excellent yield, after standard desilylation.

The deoxy paclitaxel analogs were evaluated first in a tubulin polymerization assay (*15*) followed by a cytotoxicity assay (*16*) against a human colon cancer cell line. As shown in Table 1, 7-deoxypaclitaxel **17** and 10-deoxypaclitaxel **19** were as potent as the parent paclitaxel in both *in vitro* assays. 7,10-dideoxypaclitaxel **18** is less potent in the cytotoxicity assay, while 2-deoxytaxol **16** is essentially inactive in both assays. In the case of C7-deoxytaxol **17** and C7,10-dideoxytaxol **18**, the results from the tubulin assay do not correlate well with the results from the cytotoxicity assay. The above biological data seems to indicate that the C_7 and C_{10} functional groups are not absolutely essential for *in vitro* cytotoxicity, while the C_2 benzoate is indeed involved in the intimate receptor binding.

Table 1: Tubulin and cytotoxicity data of four deoxygenated taxol analogues:

Compound#	*Compound name*	*Tubulin Polymer. init. rate ratio*	*IC_{50} (nM) HCT 116*
1	**Taxol**	100%	4.0
16	2-deoxytaxol	0%	483
17	7-deoxytaxol	23%	4.0
18	7,10-dideoxytaxol	33%	29.0
19	10-deoxytaxol	143%	6.0

While investigating an alternative synthesis of 7,10-dideoxy baccatin **11**, we came across a series of very interesting free radical rearrangements (*11*). As shown in Scheme 5, reduction of 7-xanthate **20** with Bu_3SnH gave the desired 7,10-dideoxy baccatin **11** and its tetracyclic isomer **21**. The ratio of **11/21** was *ca.* 2:1. With the intention of preventing the rearrangement and of preparing **11** in higher yield, a more reactive hydride donor, Ph_3SnH was chosen as the reducing agent (*17*). Consistent with this hypothesis, we felt that the use of relatively less reactive reducing agent, tris(trimethylsilyl)silane (*18*), should lead to more extensive rearrangement.

Scheme 5

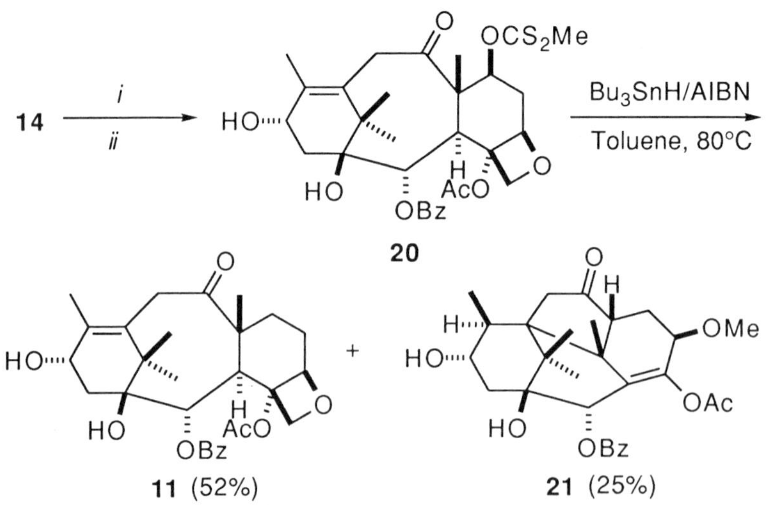

Conditions: (i) TBAF/THF/0°C, (77%); (ii) NaH/CS_2/THF, then MeI, (77%).

As shown in Scheme 6, the use of Ph_3SnH led not only to a slightly higher ratio of products **11/21**, but also to the formation of the C_7 methyl ether **22**. The mechanism for the formation of **22** was in part reported by Barton (*19*) and more recently by Pradham (*20*) and Bowman (*21*). Reduction of **20** with $(TMS)_3SiH$, as predicted, resulted in the formation of products such as **23** and **24** which arose from more extensive rearrangement reactions. The structures of compounds **21**, **23** and **24** were established on the basis of extensive NMR analysis (*11b*).

Scheme 6

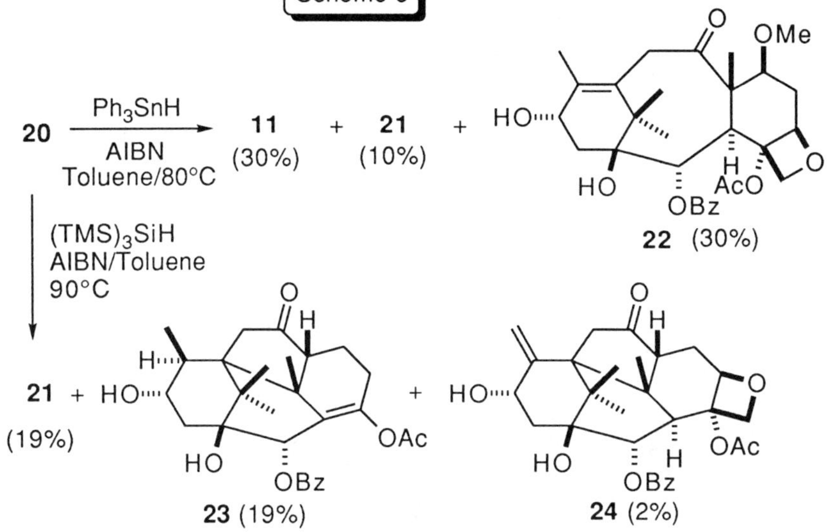

The formation of these rearranged products from xanthate **20** can be readily rationalize by invoking a cascade of radical rearrangements. (Scheme 7). The initially formed β-keto radical **25** (*22*), can isomerize *via* alkoxy radical **26** to the isomeric radical **27**. This places the C_8 carbon radical in close proximity to the C11-12 double bond, which results in the formation of **28** via a 5-exo cyclization (*23*). This radical is apparently too hindered to be quenched by hydride and thus rearranges further. The major pathway for radical **28** is a remote intramolecular hydrogen abstraction from C_3 to afford **29**, as described previously in our recently published paper on photochemical rearrangement of paclitaxel (*24*). Radical **28** also disproportionated to give the minor product **24** when $(TMS)_3SiH$ was employed as the hydride donor.

Radical **29** is evidently also sterically hindered, and undergoes an unusual oxetane fragmentation reaction, analogous to the one recently "clocked" by Nonhebel and Walton (*25*). These authors measured a rate of opening of 8.9×10^2 sec^{-1}, a rate much slower than that of the oxiranylmethyl radical, and even slower than that of the cyclobutylmethyl radical (*26*). This suggests that it is the hindered nature of the radical-bearing carbon in **29** that prevents trapping by hydride and accounts for the formation of **21**. When the reducing agent employed was tributyltin deuteride, the product was specifically labled only at the C_5 methoxy group. However, in the case of $(TMS)_3SiH$ reduction, in addition to **21**, alkoxymethyl radical fragmentation (*27*), with loss of formaldehyde, led to allylic radical **31**, presumably due to the slow trapping of **30** by the rather unreactive silane reagent. After this cascade of six intramolecular rearrangements, radical **31** was finally quenched by silane to give compound **23**.

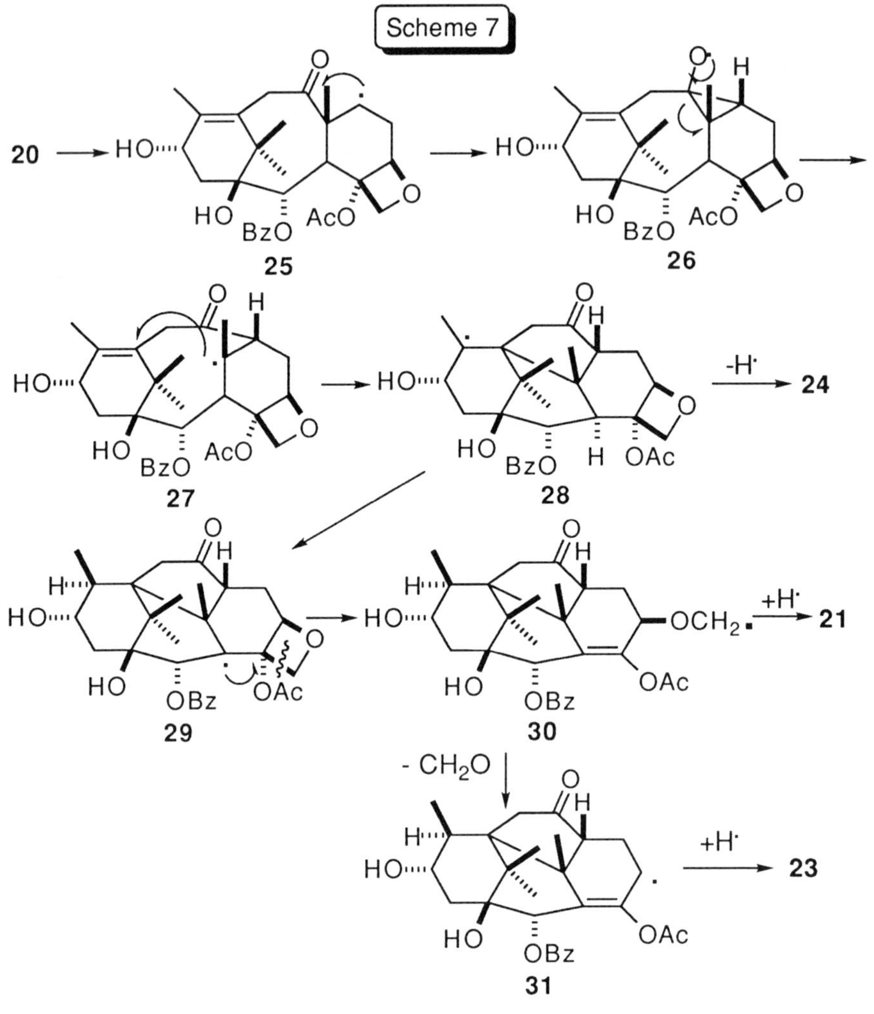

(ii) Fluorination chemistry:
Incorporation of fluorine atoms into bioactive molecules has become an important area of medicinal research. With this in mind, we decided to prepare 7-fluorotaxol for biological evaluation (*28*). Treatment of $C_{2'}$-Cbz taxol **32a** with two eqivalents of diethylaminosulfur trifluoride (DAST) in dichloromethane provided a 55% yield of 7-epi-fluorotaxol **33a**, as well as 32% of the 7,8-cyclopropane analog **34a**. These two products were isolated *via* semipreparative HPLC. Reacting **32a** with four equivalents of DAST afforded the A-ring contracted fluorinated or cyclopropane-bearing taxol analogues, **35a** and **36a**. Standard hydrogenation of these products with palladium on charcoal yielded compounds **33b**, **34b**, **35b** and **36b**. (Scheme 8)

With 7-α-fluorotaxol **33b** in hand, we then turned our attention to the preparation of its C_7 epimer, 7-β-fluorotaxol. Towards this end, 2'-Cbz-7-epi taxol **32b** was chosen as the starting material. To our great surprise, treatment of **32b** with two to four equivalents of DAST in dichloromethane led only to a mixture of cyclopropane-bearing

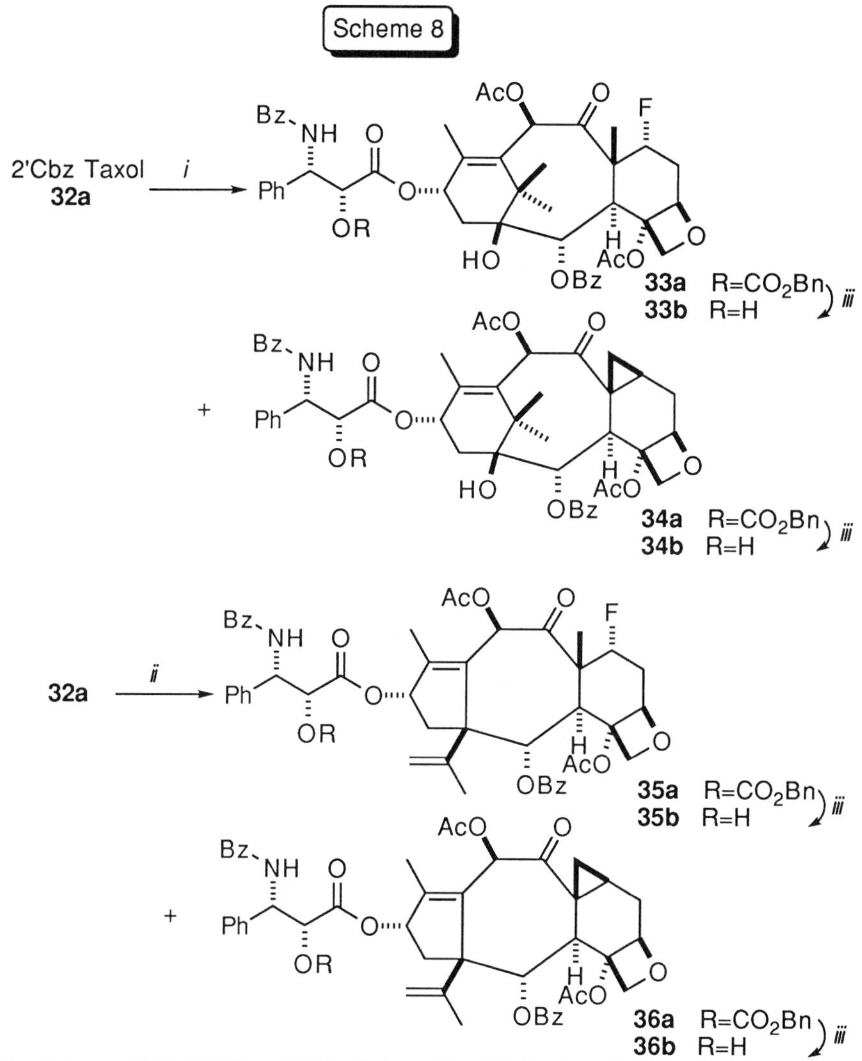

Conditions: (i) DAST(2 eq.)/CH_2Cl_2/r.t., **33a** (55%), **34a** (32%); (ii) DAST(4 eq.)/ CH_2Cl_2/r.t., **35a** (57%), **36a** (38%); (iii) H_2/Pd/C/EtOAc, **33b** (88%), **34b** (90%), **35b** (82%), **36b** (88%).(Reproduced with permission from reference 28)

derivatives, **34a** and **36a**. Further treatment of this mixture with excess DAST resulted in a single product, **36a**, in 80-90% yield. The structure assignment of **36a** was based on the mass spectrum (loss of two molecules of water), and ^{1}H NMR spectrum. In particular, the ^{1}H NMR spectrum of **36a** failed to show the signal arising from the C_{19} methyl group. This had apparently been replaced by the signal of an aliphatic methylene group. Standard COSY and HETCOR experiments indicated that the structure contained a cyclopropane ring. This 7,8-cyclopropane ring bearing baccatin core in **34a** was later confirmed by X-ray structure analysis (*30*).

As shown in Scheme 9, the absence of 7-β-fluorotaxol from the fluorination of 7-epimer **32b** is quite surprising, however, it can be understood by involking corner-

Scheme 9

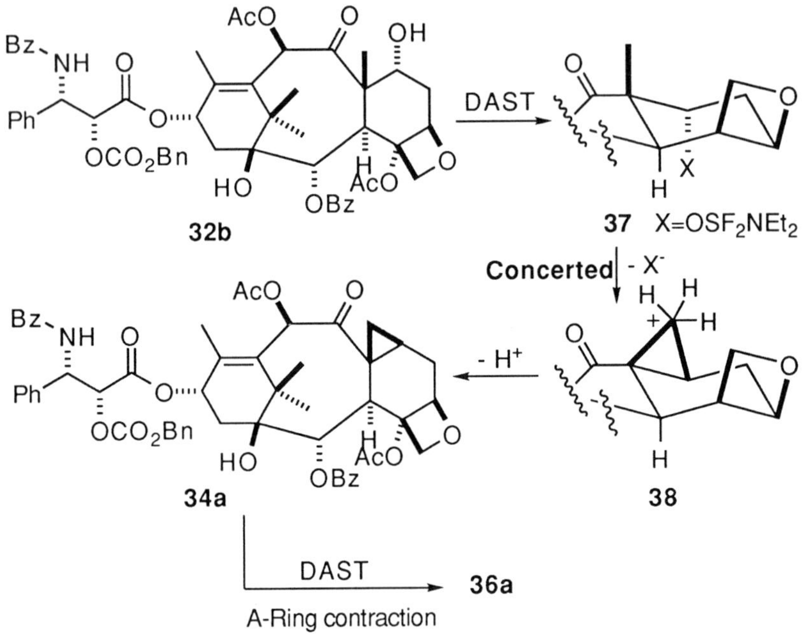

Scheme 10

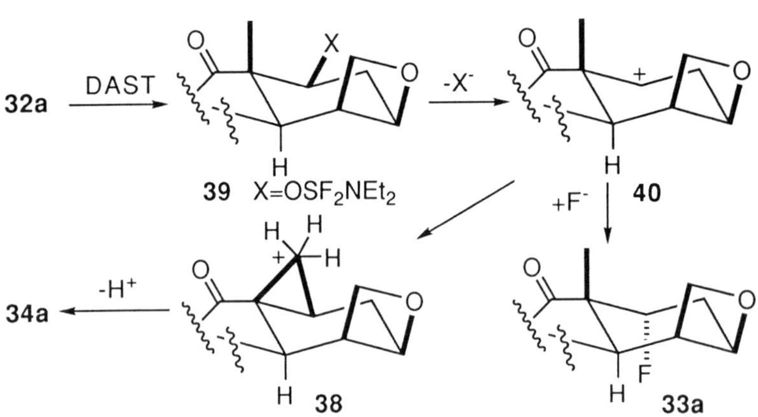

protonated cyclopropane intermediate, **38**, which then suffers the loss of a proton to yield **34a**, and thereafter, the A-ring contracted congener **36a** by loss of another molecule of water. Our proposed intermediate **38** is precedented by well-studied *enzymatic* counterparts identified in the production of cycloartenol from squalene (*31*), and in the biosynthesis of cyclopropanated sesquiterpenes (*32*). We believe several factors may contribute to this unprecedented type of cyclopropanation reaction. (A) Fluorine displacement *via* S_N2 mechanism in **37** is severely hindered by the C_{19} methyl and oxetane ring; (B) *Trans* di-axial geometry between the C_{19} methyl group and C_7 leaving group is ideal for concerted C_{19} methyl participation; (C) The usual 1,2 Wagner-Meerwein type methyl migration is totally supressed due to the presence of the C_9 keto group which diafavors the formation of a positive charge at its α position (*33*).

In sharp contrast to the C_7 epimer series, as shown in Scheme 10, the fluorination of **32a** is believed to proceed through carbonium ion **40**, which may suffer fluoride attack from the less hindered α face, to yield product **33a**. C_{19} methyl participation apparently also takes place, leading to a second intermediate **38**, which provides 7,8-cyclopropane taxol analog **34a** after the loss of one proton (*28*). The switch of reaction mechanism (from concerted process to carbocation intermediate) clearly indicates that the*trans* di-axial relationship between C_{19} methyl and C_7 leaving group is crucial for the concerted process observed in Scheme 9.

All of these products were submitted for biological evaluation. The detailed results are summarized in the table below. As can be seen in Table 2, C_7 fluoropaclitaxel **33b** and 7,8-cyclopropane of paclitaxel **34b** are only 2-3 fold less potent than the parent in the cytotoxicity assay. In contrast, two A-ring contracted analogues, **35b** and **36b**, suffer significant loss of potency in both tubulin and the *in vitro* cytotoxicity assay. It is quite encouraging to see that the 7-epi-fluorotaxol **33b** performs better than taxol in a standard tubulin polymerization assay. The significance of this observation is under further study.

Table 2: Biological data for 7,8-cyclopropane and 7-fluorotaxol derivatives:

Compound#	*Compound name*	*Tubulin Polymer. init. rate ratio*	*IC_{50} (nM) HCT 116*
33b	7-epi-fluorotaxol	140%	11
35b	7-epi-fluoro-nortaxol	28%	78
34b	7,8-cyclopropanetaxol	56%	8
36b	7,8-cyclopropane-nortaxol	5%	820

(iii) Oxetane ring opening chemistry:

Our initial SAR effort was dedicated to the understanding the role that oxetane ring plays in the bioactivity of taxol (*12*). Although oxetane ring opening chemistry was studied previously by Kingston and Potier, a method for the regioselective oxetane ring opening was still elusive. Kingston reported that treatment of paclitaxel with acetyl chloride led to a product similar to **41a** in which the oxetane ring is opened and the A ring is contracted (*12b*).

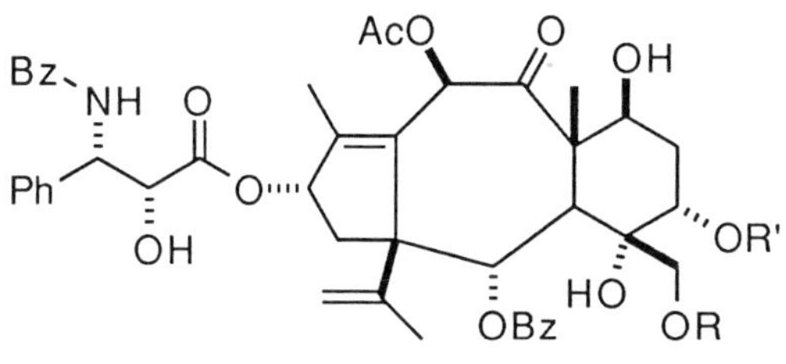

41a R=Ac, R'=H
41b R=H, R'=Ac

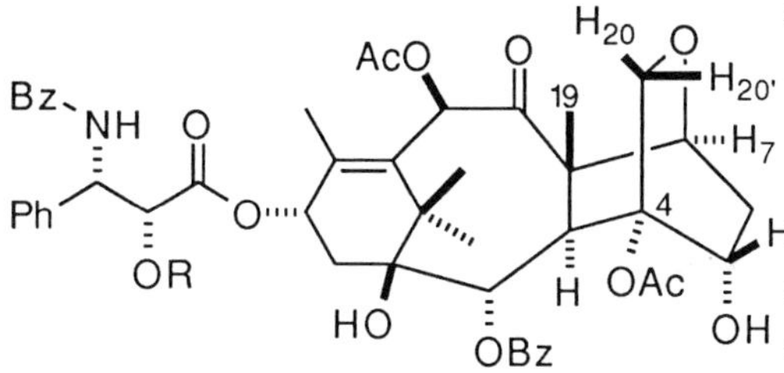

42a R=CO_2Bn
42b R=H

Similar unavoidable A ring contraction was also observed by Potier in their oxetane ring chemistry upon the use of hydrochloric acid and trifluoroacetic acid (*12c*). With the aim of avoiding this undesired A ring contraction in the oxetane ring opening reaction, we undertook a comprehensive study dealing with the reactivity of paclitaxel with Lewis acids in aprotic media.

Treatment of paclitaxel with trimethylsilyl bromide or boron tribromide in dichloromethane led only to a mixture of the two A ring contracted products, **41a** and **42b**, which are similar to those reported by Kingston (*12b*). Despite this initial setback, we next turned to a milder Lewis acid with the hope of preventing A ring contraction. It was then found that reacting 2'Cbz-taxol **32a** with tin tetrachloride gave high yields of a 1:4 ratio of the two desired products **43a** and **43b**, without concomitant A ring contraction. (Scheme 11) Titanium tetrachloride and boron trifluoride can also effect the same regiospecific oxetane ring opening reaction; however, a reverse ratio of **43a/43b** was obtained with $TiCl_4$, whereas, a 1:1 mixture of **43a** and **43b** was obtained when BF_3 was employed as the Lewis acid. Interestingly, prolonged treatment of **32a** with tin tetrachloride in dichloromethane yielded 15-25% of a novel bicyclo C ring derivative, **42a**, in addition to the major product **43b**. The structure of **42a** was established after extensive NMR characterization. In particular, proton-proton and carbon-carbon connectivities obtained by 2-D NMR were in agreement with the structure proposed.[12a] In addition, a three bond coupling was observed between H-7 and C-20, thus establishing the presence of the ether linkage. Strong NOEs were obtained between the C-19 methyl hydrogens and H-20, and H-5 and H-20', thus confirming the stereochemistry at C-5. Compound **42a** was further converted to **42b** *via* standard hydrogenation in high yield (*12a*).

Scheme 11

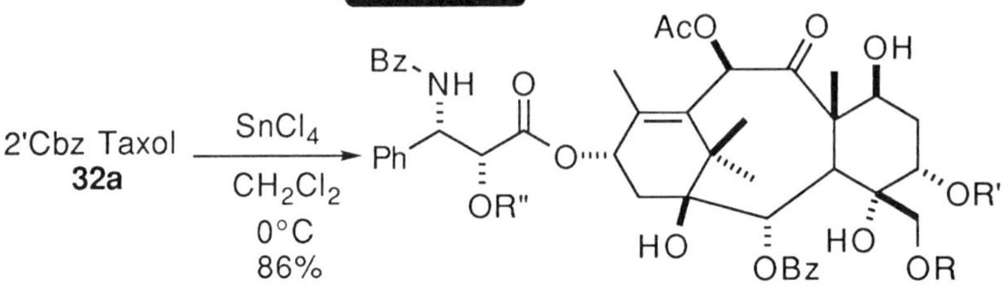

43a R=Ac, R'=H, R"=CO_2Bn
43b R=H, R'=Ac, R"=CO_2Bn
44a R=Ac, R'=H, R'=H
44b R=H, R'=Ac, R'=H

H_2/Pd/C
EtOAc
91%

Scheme 12

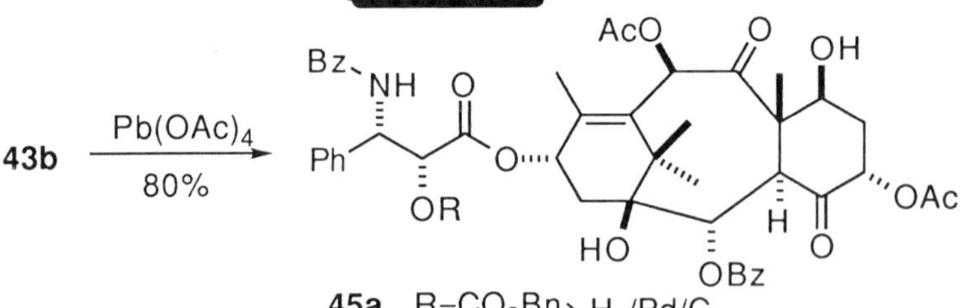

45a R=CO_2Bn
45b R=H

H_2/Pd/C
89%

In order to further confirm the vicinal diol functionality in **43b**, the major product **43b** was treated with lead tetraacetate in acetonitrile. As we expected, the C_4 keto derivative **45a** was obtained in 80% yield. Upon further hydrogenation, **45b** was prepared in 89% yield for the SAR study.(see Scheme 12)

The mechanism for the oxetane ring opening can be rationalized by invoking Winstein's classical study on neighbouring acetyl group participation in solvolysis reactions (*34*). As shown in Scheme 13, the most likely first step is the complexation of the Lewis acid with the oxetane oxygen which is likely the most basic atom in the molecule. The acetate group at C_4 is positioned for backside attack onto C_5 leading to acetoxonium ion **46**, which can then rearranged, reversibly, to a second acetoxonium ion **50**, *via* orthoester **49**. Upon "silica gel" quench (direct chromatography), oxonium ion **47** was converted to hemi-orthoester **48**, which was further unraveled to give **43b**, the C_5 acetoxy group bearing product. Likewise, oxonium ion **50** gave the corresponding C_{20} acetoxy bearing product **43a**, after silica gel quench. The formation of **42a** can be explained by assuming that the C ring conformation in **50** can flip to a boat, a conformation in which the C_7 hydroxy group can attack C_{20}, to permanently lock the C ring in a boat conformation (*12a*).

Scheme 13

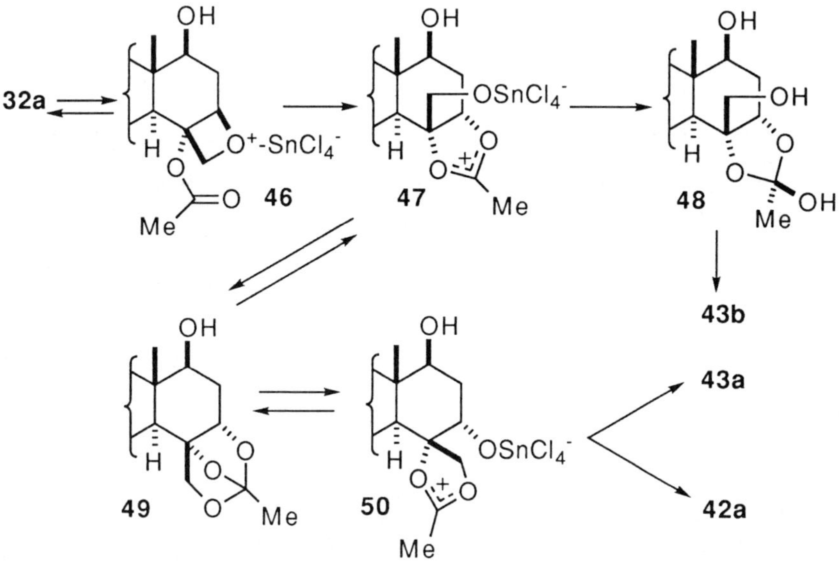

Table 3 shows the results of biological evaluation for **42b**, **44a** and **45b**. All of three analogs lacking the oxetane ring failed to polymerize tubulin under standard conditions. Compared with taxol, these analogues were also not cytotoxic. These data seem to suggest that oxetane ring is indeed needed for the biological activity.

Table 3: Biological evaluation for compounds ***42b***, ***44a***, *and* ***45b***.

Compound#	*Compound name*	*Tubulin Polymer. initi. rate ratio*	*IC_{50} (nM) HCT 116*
42b	bicyclo-taxol analog	0%	>78
44a	non-oxetane taxol	0%	220
45b	4-keto taxol analog	0%	3100

Conclusion

In summary, our SAR results clearly suggest that removal of C_{10} acetate and deoxygenation or fluorination at C_7 position do not affect the biological activity significantly; whereas, deletion of C_2 benzoate and opening of oxetane ring completely destroy the antitumor activity. In addition to the SAR work, we have also discovered several novel reactions and mechanisms, such as concerted cyclopropanation, oxetane ring fragmentation.

Acknowledgments

I would like to thank Dr. S. Huang for many complicated structure assignments, Drs. D.M. Vyas, T.W. Doyle for many valuable insights and suggestions. I am grateful to Ms. J.M. Wei for her preliminary work. I would also like to thank Drs. S.W. Mamber, B. Long, C. Fairchild for performing the tubulin assay and the cytotoxicity assay. I am grateful to Dr. J.F. Kadow and Dr. P. Scola for proof reading.

References and Notes

* To whom correspondence should be addressed.

† Current address: Department of Process Research, Boehringer Ingelheim Pharmaceuticals. 900 Ridgebury Road, Ridgefield, CT 06877.

¶ Taxol® is a registered trademark of Bristol-Myers Squibb company.

(1). Wani, M.C.; Taylor, H.L.; Wall, M.E.; Coggon, P.; Mcphail, A. *J. Am. Chem. Soc.* **1971**, *93*, 2325.

(2). Rowinsky, E.K.; Cazenave, L.A.; Donehower, R.C.; *J. Nat. Cancer Inst.* **1990**, *82*, 1247.

(3). McGuire, W.P.; Rowinsky, E.K.; Rosenshein, N.B.; Grumbine, F.C.; Ettinger, D.S.; Armstrong, D.K.; Donehower, R.C. *Ann. Int. Med.* **1989**, *111*, 273.

(4). Schiff, P.B.; Horwitz, S.B. *Proc. Natl. Acad. Sci. U.S.A.* **1980**, *77*, 1561.

(5). For total synthesis of taxol, see: Holton, R.A.; Somoza, C.; Kim, H.B.; Liang, F.; Biediger, R.J.; Boatman, P.D.; Shido, M.; Smith, C.C.; Kim, S.; Nadizadeh, H.; Suzuki, Y.; Tao, C.; Vu, P.; Tang, S.; Zhang, P.; Murthi, K.K.; Gentile, L.N.; Liu, J.H.*J. Am. Chem. Soc.* **1994**, *116*, 1597. Also, Nicoloau, K.C.; Yang, Z.; Liu, J.J.; Ueno, H.; Natermet, P.G.; Guy, R.K.; Claiborne, C.F.; Renaud, J.; Couladouros, E.A.; Paulvannan, K.; Sorensen, E.J. *Nature* **1994**, 367.

(6). Senilh, V.; Blechert, S.; Colin, M.; Guénard, D.; Picot, F.; Potier, P.; Varenne, P. *J. Nat. Prod.* **1984**, *47*, 131.

(7). Ravdin, P.M. In Stony Brook Symposium on Taxol and Taxotere; Stony Brook, NY, May 14-15, 1993; Abstracts p. 18.

(8). Review: Kingston, D.G.I. *Pharm. Ther.* **1991**, *52*, 1.

(9). (a) Chen, S.H.; Wei, J.M.; Farina, V. *Tetrahedron Lett.* **1993**, *34*, 3205. (b) Chen, S.H.; Farina, V.; Wei, J.M.; Long, B.; Fairchild, C.; Mamber, S.W.; Kadow, J.F.; Vyas, D.M.; Doyle, T.W. *Bioorg. & Med. Chem. Lett.* **1994**, *4*, 479.

(10). (a) Chen, S.H.; Fairchild, C.; Mamber, S.W.; Farina, V. *J. Org. Chem.* **1993**, *58*, 2927. (b) Chen, S.H.; Wei, J.M.; Vyas, D.M.; Doyle, T.W.; Farina, V. *Tetrahedron Lett.* **1993**, *34*, 6845.

(11). (a) Chen, S.H.; Huang, S.; Kant, J.; Fairchild, C.; Wei, J.M.; Farina, V. *J. Org. Chem.* **1993**, *58*, 5028. (b) Chen, S.H.; Huang, S.; Gao, Q.; Golik, J.; Farina, V. *J. Org. Chem.* **1994**, *59*, 1475.

(12). (a) Chen, S.H.; Huang, S.; Wei, J.M.; Farina, V. *Tetrahedron* **1993**, *49*, 2805. (b) Samaranayake, G.; Magri, N.F.; Jitrangsri, C.; Kingston, D.G.I. *J. Org. Chem.* **1991**, *56*, 5114. (c) Wahl, A.; Guéritte-Voegelein, F.; Guénard, D.; Le Golf, M.; Potier, P. *Tetrahedron* **1992**, *48*, 6965.

(13). Holton, R.A. Presented at the 203rd Meeting of the American Chemical Society, San Francisco, 1991, Abstract ORGN 0355.

(14). Ojima, I.; Habus, I.; Zhao, M.; Zucco, M.; Park, Y.H.; Sun, C.M.; Brigaud, T. *Tetrahedron* **1992**, *48*, 6985.
(15). Tubulin polymerization rate of analogue relative to that of taxol. For details, see: Swimdell, C.S.; Krauss, N.E.; Horwitz, S.B.; Ringel, I.*J. Med. Chem.* **1991**, *34*, 1176.
(16). IC_{50} measures drug concentration required for the inhibition of 50% cell proliferation. For details, see: Scudiero, D.A.; Shoemaker, R.H.; Paull, K.D.; Monks, A.; Tiernet, S.; Nofziger, T.H.; Currens, M.J.; Seniff, D.; Boyd, M.R. *Cancer Res.* **1988**, *48*, 4827.
(17). Friedrich, E.C.; Holmstead, R.L. *J. Org. Chem.* **1971**, *36*, 971.
(18). Chatgilialoglu, C. *Acc. Chem. Res.* **1992**, *25*, 188.
(19). Barton, D.H.R.; Crich, D.; Lobberding, A.; Zard, S.Z. *Tetrahedron* **1986**, *42*, 2329.
(20). Pradham, S.K.; Sitabkham, S. *Tetrahedron Lett.* **1993**, *34*, 5615.
(21). Bowman, W.R.; Brown, D.S.; Burns, C.A.; Marples, B.A.; Zaidi, N.A. *Tetrahedron* **1992**, *48*, 6883.
(22). Beckwith, A.L.J.; Duggan, P.J. *J. Chem. Soc. Perkin Trans II*, **1992**, 1777 and reference therein.
(23). Griller, D.; Ingold, K.U. *Acc. Chem. Res.* **1980**, *13*, 317.
(24). (a) Chen, S.H.; Combs, C.M.; Hill, S.E.; Farina, V.; Doyle, T.W. *Tetrahedron Lett.* **1992**, *33*, 7679. (b) Chen, S.H.; Farina, V.; Huang, S.; Gao, Q.; Golik, J.; Doyle, T.W. submitted to *Tetrahedron.*
(25). Laurie, D.; Nonhebel, D.C.; Suckling, C.J.; Walton, J.C. *Tetrahedron* **1993**, *49*, 5869.
(26). Beckwith, A.L.J.; Westwood, S.W. *J. Am. Chem. Soc.* **1988**, *110*, 2565.
(27). Steenken, S.; Schuchmann, H.P.; von Sonntag, C. *J. Phys. Chem.* **1975**, *79*, 763.
(28). Chen, S.H.; Huang, S.; Farina, V. On The Reaction of Taxol With DAST. *Tetrahedron Lett.* **1994**, *35*, 41.
(29). Chen, S.H.; Huang, S., Wei, J.M.; Farina, V. *J. Org. Chem.* **1993**, *58*, 4520.
(30). Gao, Q. unpublished result.
(31). Clayton, R.B. In Aspects of Terpenoid Chemistry and Biochemistry; Goodwin, T.W. Ed.; Academic Press: New York, 1971; P 1.
(32). Cane, D.E.; Sohng, J.K.; Willard, P.G. *J. Org. Chem.* **1992**, *57*, 844.
(33). Creary, X. *J. Am. Chem. Soc.* **1981**, *103*, 2463. Also, Creary, X. *J. Org. Chem.* **1979**, *44*, 3938.
(34). Roberts, R.M.; Corse, J.; Boschan, R.; Seymour, D.; Winstein, S. *J. Am. Chem. Soc.* **1958**, *80*, 1247 and references therein.

RECEIVED October 7, 1994

Chapter 19

Syntheses and Structure–Activity Relationships of New Taxoids

Iwao Ojima[1], Young Hoon Park[1], Ivana Fenoglio[1], Olivier Duclos[1], Chung-Ming Sun[1], Scott D. Kuduk[1], Martine Zucco[1], Giovanni Appendino[2], Paula Pera[3], Jean M. Veith[3], Ralph J. Bernacki[1], Marie-Christine Bissery[4], C. Combeau[4], P. Vrignaud[4], J. F. Riou[4], and François Lavelle[4]

[1]Department of Chemistry, State University of New York at Stony Brook, Stony Brook, NY 11794–3400
[2]Dipartimento di Scienza e Tecnologia del Farmaco, Università di Torino, 10125 Torino, Italy
[3]Department of Experimental Therapeutics, Grace Cancer Drug Center, Roswell Park Cancer Institute, Elm and Carlton Streets, Buffalo, NY 14263
[4]Rhône-Poulenc Rorer, Centre de Recherches de Vitry Alfortville, 13 Quai Jules Guesde, 94403 Vitry sur Seine, France

A series of new taxoids are synthesized from 14β-hydroxy-10-deacetylbaccatin III (14-OH-DAB). These new taxanes possess strong cytotoxicities against human cancer cell lines, and at least one of them possesses excellent antitumor activity *in vivo*. *Pseudo*-taxoids bearing *N*-acylphenylisoserine side chain at C-14 are synthesized, which are less active, but retain a certain level of cytotoxicity. Novel nor-seco-paclitaxel and docetaxel analogs are synthesized, which retain a certain level of activity despite the destruction of the A ring. New analogs bearing cyclohexyl groups at the C-3' and/or C-2 positions are synthesized and their cytotoxicity examined. The results clearly indicate that phenyl group at C-3' or C-2 is not a requisite for biological activity. 3'-Isobutenyl and 3'-isobutyl analogs of docetaxel show excellent activity against a drug-resistant cancer cell lines.

Taxol (paclitaxel), a complex diterpene isolated from the bark of the western yew (*Taxus brevifolia*) is currently considered the most exciting lead in cancer chemotherapy (*1-3*). Taxotere (docetaxel), a semisynthetic analog, also has shown great promise (*4*). Paclitaxel and docetaxel possess high cytotoxicity and strong antitumor activity against different cancers which have not been effectively treated by existing antitumor drugs (*5,6*). Paclitaxel has been approved by the FDA for the treatment of advanced ovarian cancer (December, 1992) and breast cancer (April, 1994), and is currently in phase II and III clinical trials for lung and other cancers (*2*). Docetaxel is currently in phase II and III clinical trials in United States, Europe, and Japan (*6*).

A recent report on clinical trials of paclitaxel and docetaxel, however, has disclosed that these highly effective drugs have a number of undesired side effects as well as multi-drug resistance (MDR) (*1,5,7*). Therefore, it is very important to develop new anticancer drugs which have less undesirable side effects, better pharmacological

0097–6156/95/0583–0262$08.00/0

properties, and/or activity spectra against various tumor types different from those of these two drugs.

Recently, a novel taxane diterpenoid, 14β-hydroxy-10-deacetylbaccatin III (14β-OH-DAB), was isolated from the needles of *Taxus wallichiana Zucc.* and other plant parts (*8*). Because of an extra hydroxyl group at the C-14 position, 14β-OH-DAB has proven to possess much higher water solubility than the usual 10-deacetylbaccatin III (DAB) which is currently used for the practical production of paclitaxel and docetaxel as mentioned above. Therefore, the new antitumor taxanes derived from 14β-OH-DAB can be expected to have substantially improved water solubility, bioavailability, and hydrophobicity-related drug resistance (*7*). These improved pharmacological properties may well be related to the modification of undesirable toxicity and activity spectra against different cancer types. In fact, recent reports on the related "hydroxy-taxoids", 9-dihydrodocetaxel (*9*) and 19-hydroxydocetaxel (*10*), show quite promising results.

Syntheses of New Taxoids (*11*). A series of new taxoids were synthesized from 14β-OH-DAB using a highly efficient and practical coupling protocol based on the *β-Lactam Synthon Method* developed in our laboratory (*13,14*). Thus, the C-13 side chain precursors, (3*R*,4*S*)-1-acyl-3-(EEO)-4-phenylazetidin-2-ones (**4-EE**) (EE = ethoxyethyl) with extremely high enantiomeric purity, were obtained through our efficient chiral ester enolate – imine cyclocondensation method (*12-14*) in four steps in 78-80% overall yields. The 7,10-ditroc-14β-OH-DAB (**3**, 75%) (troc = 2,2,2-trichloroethoxycarbonyl) and 7,10-ditroc-14β-OH-DAB-1,14-carbonate (**2**, 55%) were prepared by reacting 14β-OH-DAB with troc-Cl in pyridine (Scheme 1).

Scheme 1

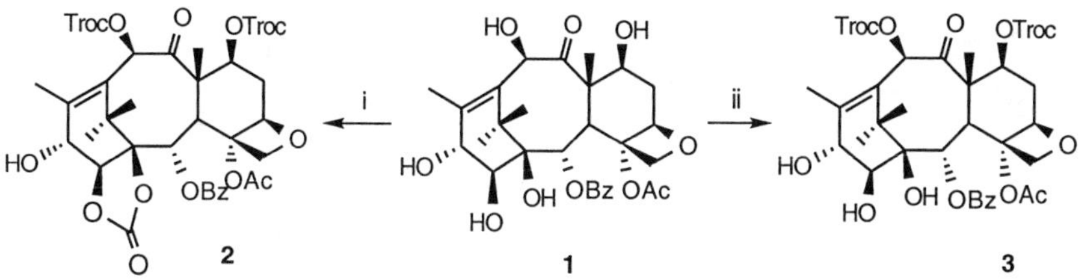

(i) Cl_3CCH_2OCOCl (6 eq.), py, 80 °C, 5 min.; (ii) Cl_3CCH_2OCOCl (4 eq.), py, 80 °C, 5 min.

The reaction of **3** with β-lactam **4a-EE** (1.2 equiv.) was carried out in the presence of 1.2 equiv. of NaHMDS at -40 °C for 30 min to give 7,10-ditroc-2'-EE-14-hydroxy-docetaxel-1,14-carbonate (**5a**) in 86% yield (Scheme 2). The coupling product **5a** was deprotected under modified Commerçon conditions (*15*) by treating with activated Zn in acetic acid and methanol at 40 °C for 9 h to give 14-hydroxy-docetaxel-1,14-carbonate (**SB-T-1011**) in 70% yield (Scheme 2). 14β-Hydroxy-taxol-1,14-carbonate (**SB-T-1012**) was obtained in the same manner through the coupling of **2** with β-lactam **4b-EE** (1.2 equiv.) (89% yield) followed by deprotection (70% yield) (Scheme 2).

The coupling of **3** with β-lactam **4a-EE** and subsequent deprotection were performed under similar conditions to give *pseudo*-docetaxel (**SB-T-1021**) in 50% overall yield (Scheme 3) (*11*). In the same manner, 10-deacetyl-*pseudo*-paclitaxel (**SB-T-1022**) was synthesized through the coupling of β-lactam **4b-EE** with **3**, followed by deprotection in 52% overall yield (Scheme 3) (*13*).

We have found that the reactivities of the four secondary hydroxyl groups of 14β-OH-DAB toward acylation decreases in the order C-7 > C-10 > C-14 > C-13, which appears to be in good agreement with the relative steric congestion at these positions. Accordingly, the coupling of the baccatin in the 13-position with an isoserine side chain precursor requires an appropriate protection of hydroxyl groups at the 7, 10 and 14 positions. As mentioned above, the selective protection of the 7- and 10-hydroxyl groups was easily carried out using troc-Cl to give **3**. However, the protection of the 14-

Scheme 2

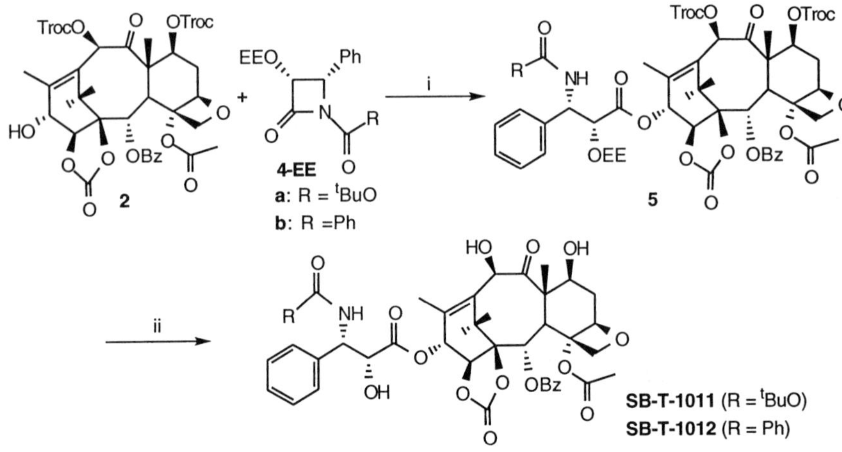

(i). NaHMDS, THF, -40 °C, 30 min.; (ii) Zn (activated), AcOH, MeOH, reflux, 2 h.

Scheme 3

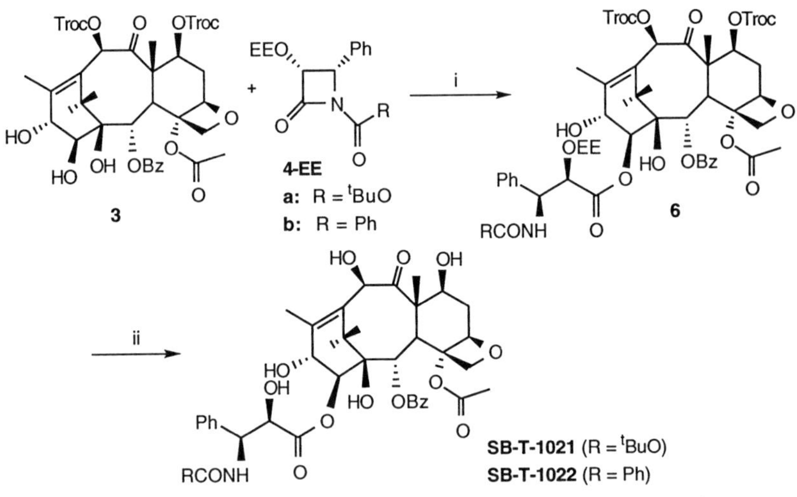

(i). NaHMDS, THF, -40 °C, 30 min.; (ii) Zn (activated), AcOH, MeOH, 40 °C, 9 h.

hydroxyl group was not straight forward, e.g., (i) an attempted protection with troc-Cl/pyridine resulted in the formation of 1,14-carbonate **2** (*vide supra*). (ii) the selective protection with triethylsilyl (TES) was successful, but the attempted coupling with a C-13 side chain precursor, a β-lactam **4a-EE** (*vide infra*), did not work because of steric hindrance caused by the introduction of TESO group adjacent to the 13-hydroxyl group. We successfully introduced an acetonide as a potential protecting group for the 1- and 14-hydroxyl groups which constitutes a cis-diol, giving 7,10-ditroc-14-OH-DAB-1,14-acetonide (**7**) in 89% yield (Scheme 4), but the attempted deprotection of this acetonide under acidic conditions led to the opening of the D ring. Thus, the acetonide turned out to be inappropriate as the protecting group for our purpose. Finally, we introduced an orthoformate to the 1- and 14-hydroxyl groups using triethyl orthoformate in the presence of *p*-toluenesulfonic anhydride, giving 7,10-ditroc-14-OH-1,14-ethoxymethylene-DAB

(**8**) in 93% yield (Scheme 4) (*16*). This protecting group was found to be quite appropriate for the coupling with the C-13 side chain precursor (*vide supra*) and subsequent deprotections to give the desired 14β-hydroxydocetaxel (**SB-T-1001**).

Scheme 4

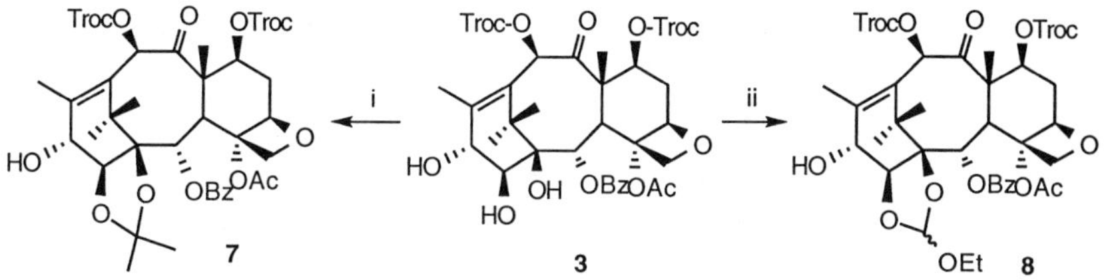

(i) $Me_2C(OMe)_2$, TsOH, RT, 30 h; (ii) $HC(OEt)_3$, Ts_2O, CH_2Cl_2, RT, 30 min.

The protected 14-OH-DAB **7** was coupled with enantiopure (3*R*,4*S*)-1-t-BOC-3-TESO-4-phenylazetidin-2-one (**4a-TES**) using our protocol (*14*), i.e., NaHMDS in THF at -40 °C for 45 min, to give the corresponding coupling product in 82% yield. Since the 1,14-acetonide moiety of this compound cannot be removed without affecting the D-ring, the TES group at the 2'-position and the troc groups at the 7 and 10 positions were removed sequentially by treating with 0.5*N* HCl – THF (1:4) and then with zinc in 0.5*N* HCl – THF (1:4), respectively to give 14β-hydroxydocetaxel-1,14-acetonide (**SB-T-1071**) in 72% overall yield (Scheme 5) (*16*).

Scheme 5

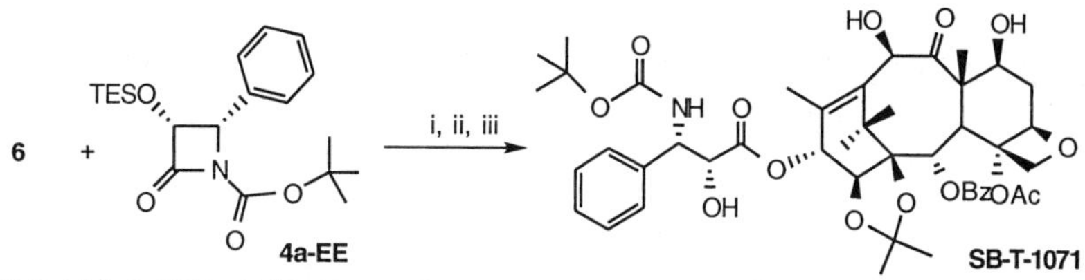

(i) NaHMDS, THF, -40 °C, 30 min; (ii) 0.5*N* HCl in EtOH, RT, 2 h; (iii) Zn, 0.5*N* HCl/THF, 0 °C, 20 min.

In the same manner, the protected 14-OH-DAB **8** was coupled with **4a-TES**, giving **9** in 78% yield. Subsequently, **9** was first treated with 0.5*N* HCl in EtOH to remove the TES group at the 2'-position (96%), followed by reacting with formic acid in dioxane at room temperature for 24 h, yielding 14-formyl-7,10-ditroc-14-hydroxydocetaxel, which was treated *in situ* with 1% aqueous $NaHCO_3$ in THF–MeOH (1:7:3) at room temperature for 5 h to give 7,10-ditroc-14-hydroxydocetaxel (73% for two steps). The troc groups of this compound at the 7 and 10 positions were deprotected by using zinc in 0.5*N* HCl – THF (1:2) at 0 °C for 30 min to give 14β-hydroxydocetaxel (**SB-T-1001**) in 73% yield (Scheme 6) (*16,17*).

Cytotoxicity and Microtubules Disassembly Inhibitory Activity of New Taxoids (*16*). Cytotoxicities of these new taxanes were evaluated *in vitro* against human cancer cell lines. As Table I shows, these new taxanes possess strong cytotoxicities against human breast, non-small cell lung, ovarian, and colon cancer cells. 14-OH-Docetaxel-1,14-carbonate (**SB-T-1011**) exhibits activity better than that of paclitaxel for non-small cell lung cancer (A549) and colon cancer (HT-29) cell lines. 14-OH-Docetaxel (**SB-T-1001**) possesses strong cytotoxicity in between docetaxel and

Scheme 6

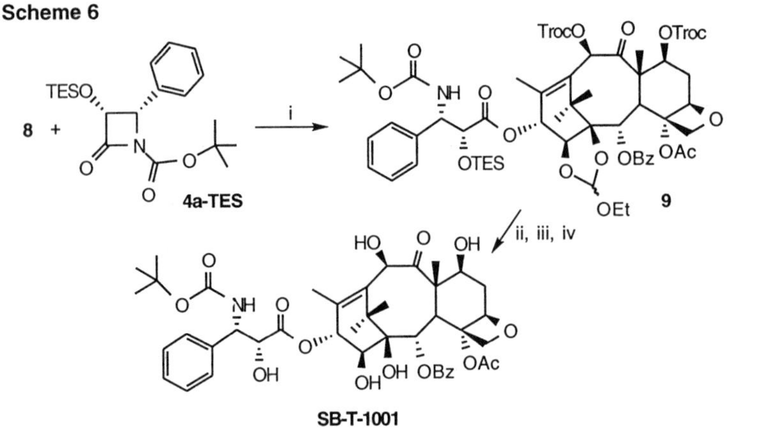

(i) 0.5*N* HCl, EtOH, RT, 2h; (ii)HCOOH/dioxane (1:1), RT, 24 h;
(iii) 1% $NaHCO_3$/ THF/MeOH (1:7:3), RT, 5 h; (iv) Zn, 0.5*N* HCl/THF (2:1), 0 °C, 30 min.

paclitaxel except for the activity against A549 human non-small cell lung cancer cell line (IC_{50} = 0.8 nM) which is slightly better than that of docetaxel. Although it is preliminary, this result is very encouraging from a viewpoint that structural modifications of docetaxel and paclitaxel can alter the activity spectrum against different tumor types, which may complement the existing anticancer drugs.

Table I. Cytotoxicities of New Taxoids from 14-OH-DAB (1) (IC_{50} nM)[a]

Taxoid	A121[a] (ovarian)	A549[a] (NSCLC)	HT-29[a] (colon)	MCF7[a] (breast)	Microtubules disassembly inhibition[b]
Paclitaxel	6.1	3.6	3.2	1.7	1.0T
Docetaxel	1.2	1.0	1.2	1.0	0.7T
SB-T-1001	3.3	0.8	2.1	1.9	0.8T
SB-T-1011	6.0	2.1	1.8	1.8	0.9T
SB-T-1012	105	33	24	11	3T
SB-T-1021	80	31	50	26	>10T
SB-T-1022	1044	512	248	269	>100T
SB-T-1071	46	18	21	34	3T

[a] The concentration of compound which inhibit 50% (IC_{50}) of the growth of human tumor cell line, A121 (ovarian carcinoma), A549 (non-small cell lung carcinoma), HT-29 (colon carcinoma), MCF7 (mammary carcinoma) after 72 h drug exposure according to the method developed by Skehan *et al.* (*18*). The data represent the mean values of at least three separate experiments. [b] IC_{50}/IC_{50}(paclitaxel).

14-OH-Docetaxel-1,14-acetonide (**SB-T-1071**), however, showed one order of magnitude weaker activity in both cytotoxicity and microtubules disassembly inhibitory activity. This can be ascribed to conformational change caused by the bulky acetonide moiety. In fact, molecular modeling study on the conformations of docetaxel, **SB-T-1001**, and **SB-T-1071** based on the SYBYL 6.0 program and NOESY analyses ($DMSO-D_2O$) clearly indicates the distortion in the "hydrophobic cluster" (*19*) of **SB-T-1071** (*16*). It should be noted that the introduction of 14β-hydroxyl group to docetaxel has only a little effect on the overall conformation of the molecule although there was a possibility that this hydroxyl group might have significantly changed the side chain conformation through hydrogen bonding. It is also worth mentioning that the modeled conformations of **SB-T-1001** and **SB-T-1011** show excellent overlap with that of paclitaxel rather than docetaxel (*16*).

Attachment of the *N*-acylphenylisoserine side chain at C-14 instead of the C-13 positions results in ca. 10-fold decrease in cytotoxicity (*11*). However, **SB-T-1021** still retains 10 nM level (IC_{50}) cytotoxicities against human cancer cells. This observation might provide us with a very important clue for an understanding of the active conformation of the paclitaxel series anticancer taxanes. For instance, the molecular modeling of **SB-T-1021**, starting from the X-ray structure of docetaxel (*20*), using the SYBYL 6.0 program reveals an unexpected excellent overlap of the energy minimized structure of **SB-T-1021** with that of docetaxel in which the two hydrophobic substituents, i.e., phenyl at C-3' and the *tert*-butoxy group, have exchanged their positions almost perfectly (*11*). This implies that the *pseudo*-taxoid **SB-T-1021** can mimic, to some extent, the binding conformation of docetaxel to the tubulin receptor although the identification of the "active conformation" of paclitaxel and docetaxel should await further investigation.

The *in vivo* antitumor activity of **SB-T-1011** has been evaluated against A151 human ovarian carcinoma xenograft in nude athymic mice, which shows an equivalent or slightly better activity than paclitaxel, causing total regression of the tumor (*11*).

Syntheses of Novel Nor-seco-taxoids (*21*). Studies on the structure-activity relationships (SAR) of paclitaxel and docetaxel have disclosed the essential role of the *N*-acylphenylisoserine moiety at C-13 as well as the oxetane ring (D-ring) for their strong anticancer activity (*3,4,22*). However, little is known of the SAR of the A and B rings of the baccatin framework. It is very important to clarify the minimum structural requirements for taxanes to exhibit anticancer activity by looking at simplified structure analogs of paclitaxel. Along this line, we have investigated the role of the A ring by synthesizing novel nor-seco analogs of docetaxel and paclitaxel in which the A ring is cleaved, but the B, C, and D rings are retained.

The novel nor-seco baccatin **10** was synthesized in 92% yield through oxidative cleavage of the A ring of 14-OH-DAB (**1**) with periodic acid (eq. 1). It should be noted that this oxidative cleavage is realized because of the presence of the hydroxyl group at C-14, i.e., this type cleavage is impossible for the usual 10-deacetylbaccatin III (DAB). The oxidative cleavage of **1** introduces a carbonyl to C-1 first, giving the corresponding 1-ketobaccatin, which immediately reacts with the hydroxyl group at C-10 *in situ* to form the hemiketal **10**.

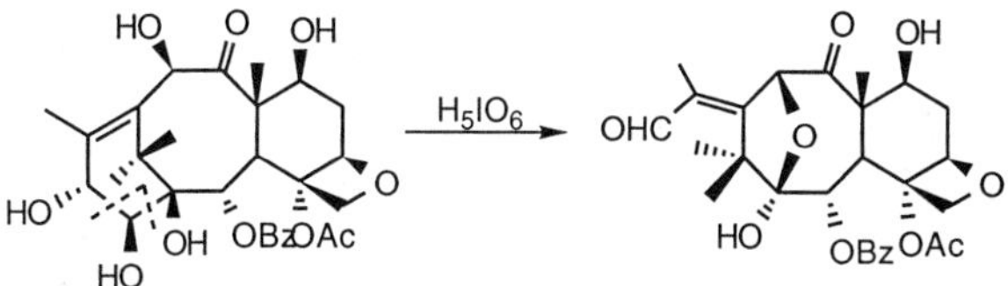

1 **10** (1)

The hydroxyl group of **10** at C-7 was selectively protected as triethylsilyl (TES) ether to give 7-TES-nor-seco-baccatin **11** in 76% yield (eq. 2). The aldehyde moiety of **11** was then reduced with sodium cyanoborohydride at pH 6 (*19*) to afford 7-TES-nor-seco-baccatin alcohol **12** in 80% yield (eq. 2).

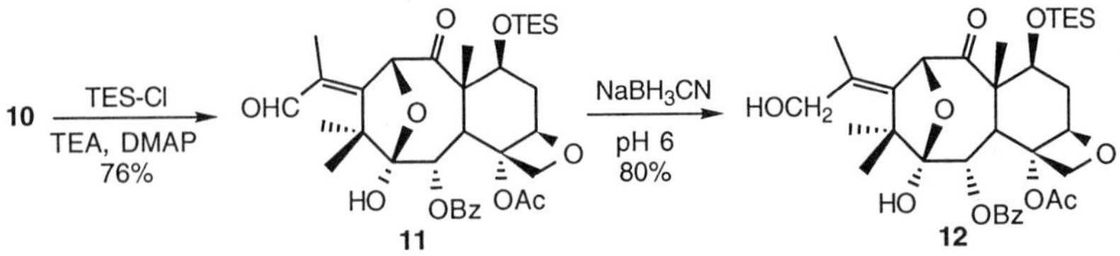

(2)

Finally, 7-TES-nor-seco baccatin alcohol **12** was coupled with β-lactams **4a**,β using our protocol (*vide supra*) to give novel nor-seco-docetaxel (**SB-T-2001**) (58%) and nor-seco-paclitaxel (**SB-T-2002**) (82%) in fairly good to excellent yields (Scheme 7). It is worthy of note that the removal of 7-TES group was not possible under the standard acidic conditions (0.5% HCl in MeOH at room temperature for 36 h) (*23*) and was only accomplished with the use of tetra-*n*-butylammonium fluoride (TBAF) in THF at -10 °C for 5 min.

Scheme 7

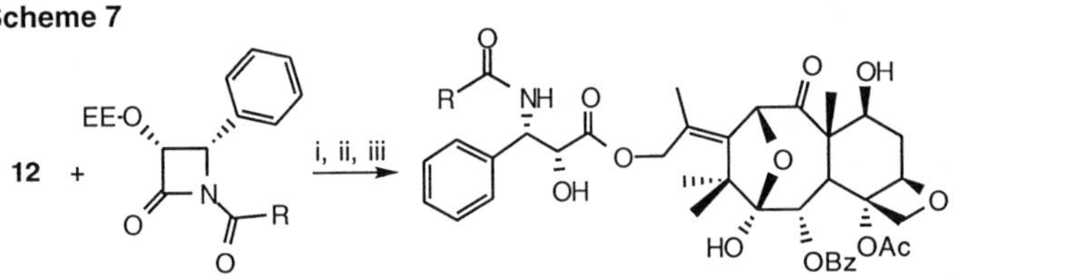

4a-EE: R = Ph
4b-EE: R = *t*-BuO

SB-T-2001 (R = Ph)
SB-T-2002 (R = *t*-BuO)

i) NaHMDS, THF, -40 °C, 30 min; ii) 0.5% HCl, RT, 36 h; iii) TBAF, THF. -10 °C, 5 min

Cytotoxicity of Nor-seco-taxoids (*21*). The cytotoxicity of these novel nor-seco taxoids were evaluated against human ovarian (A121), human non-small cell lung (A549), human colon (HT-29), and human breast (MCF7) cancer cell lines (*vide supra*). Both nor-seco analogs **SB-T-2001** and **SB-T-2002** showed 0.17–0.10 μM and 0.13–0.08 μM level IC_{50} values, respectively, which were 20-40 times weaker activities than paclitaxel in the same cell line assay (see Table I). It is worthy of note that **SB-T-2001** and **SB-T-2002** exhibit IC_{50} values of 0.36 and 0.47 μM, respectively, against an adriamycin-resistant human breast cancer (MCF7-R) cell line, i.e., these nor-seco analogs are only slightly weaker than the activities of paclitaxel (0.30 μM) against this particular drug resistant cancer cell.

The results obtained in this SAR study clearly indicate the importance of the A-ring for the strong cytotoxicity of paclitaxel and docetaxel. However, it is also very important to mention that these two reduced-structure analogs retain a certain level of cytotoxicity. The conformational analysis of **SB-T-2001** based on molecular modeling (SYBYL 6.0) indicates that this nor-seco-taxoid can take a folded structure similar to that of docetaxel or paclitaxel. However, (i) this analog lacks hydroxyl groups at C-1 and C-10, (ii) the C-16 methyl is at the concave face (α-face) of the molecule, and (iii) the C-13 side chain conformation is rather flexible. These changes may account for the reduced cytotoxicity of these novel nor-seco-taxoids.

Syntheses of Cyclohexyl Analogs of Docetaxel and Paclitaxel (*24,25*). In spite of extensive SAR studies on paclitaxel and docetaxel, little has been reported on the modification at the C-3' phenyl and the C-2 benzoate (*26*), which would clarify if phenyl groups at these positions are essential for antitumor activity. In order to introduce a cyclohexyl in place of the phenyl group at the C-3' of docetaxel, enantiomerically pure ß-lactam **13** obtained through our protocol (*12-14*) (*vide supra*) was hydrogenated on 5% Rh on carbon at 80 °C and 800 psi of hydrogen to give enantiomerically pure (3*R*,4*S*)-3-hydroxy-4-cyclohexyl-β-lactam (**14**) in 85% yield (Scheme 8). Protection of the 3-hydroxyl of **14** as ethoxyethyl (EE) ether (100 % yield) followed by *N*-acylation with di-*tert*-butyl dicarbonate in dichloromethane in the presence of triethylamine and a catalytic amount of 4-(dimethylamino)pyridine (DMAP) gave enantiomerically pure *N*-*tert*-butoxycarbonyl-β-lactam (**15**) in 91% yield (Scheme 8) (*24*).

Scheme 8

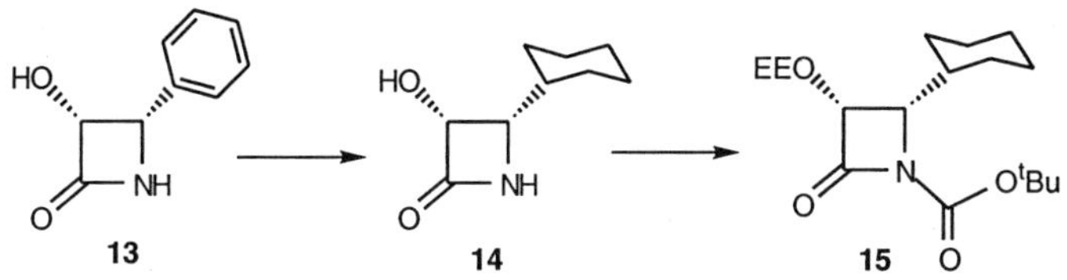

(i) H_2 (800 psi), 5% Rh-C, MeOH, 90 °C. (ii) EVE, *p*-TSA, 0 °C, CH_2Cl_2.
(iii) $^tBoc_2O$, Et_3N, DMAP, CH_2Cl_2

For the substitution of the 2-benzoate of docetaxel with 2-cyclohexane-carboxylate, we prepared the 2-cyclohexyl derivative (**16**) of 10-deacetylbaccatin III (DAB) through hydrogenation of DAB over 5% Rh on carbon at 35 °C and 500 psi of hydrogen, which gave **16** in quantitative yield (*27*). It is noteworthy that the Δ^{11} double bond remains intact under these conditions. Selective protection of the hydroxyl groups of **16** at C-7 and C-10 as 2,2,2-trichloroethoxycarbonate (troc) afforded 7,10-ditroc-2-hexahydro-DAB (**17**) in 74% yield (Scheme 9).

3'-Dephenyl-3'-(cyclohexyl)docetaxel (**SB-T-1031**) was synthesized through an efficient coupling of **15** with 7,10-ditroc-DAB (**18**) based on our coupling protocol (*13,14*) (*vide supra*), giving the product in 86% yield, followed by deprotection (*15*) (65% yield) (Scheme 10). In a similar manner, 2-(hexahydro)docetaxel (**SB-T-1041**) and 3'-dephenyl-3'-cyclohexyl-2-(hexahydro)-docetaxel (**SB-T-1051**) were synthesized through couplings of **4a-EE** with **17** and **15** with **17**, respectively, in moderate to fairly good yields (Scheme 10).

Scheme 9

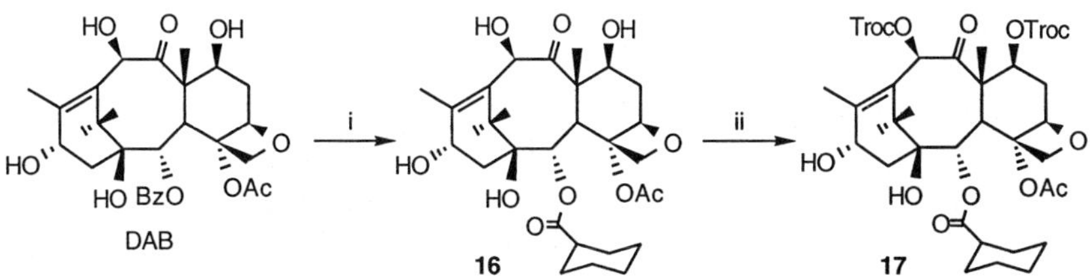

(i) H_2 (500 psi), 5%Rh-C, MeOH, 35 °C; (ii) TrocCl , pyridine, 80 °C.

Scheme 10

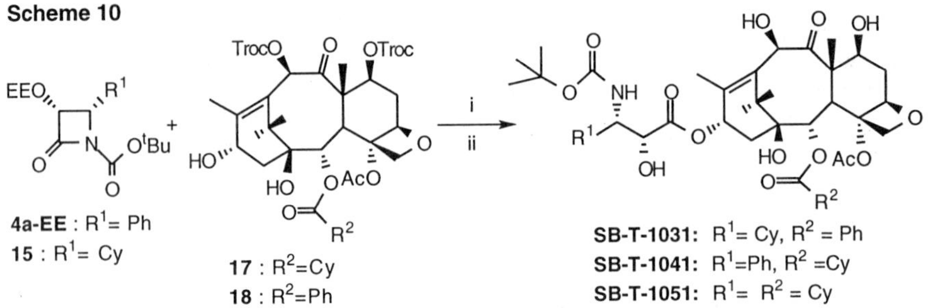

(i) NaHMDS, THF, -30 °C, 30 min.; (ii) Zn, AcOH, MeOH, 60 °C.

2-(Hexahydro)paclitaxel (**20**) was also synthesized through coupling of **4b-EE** with 7-TES-2-(hexahydro)baccatin III (**19**) which was obtained by hydrogenation of 7-TES-baccatin III over 5% Rh on carbon at 45 °C and 600 psi of hydrogen in methanol (85 %), followed by deprotection under the modified Grenoble–Gif conditions (*23*), i.e, 0.5% HCl in ethanol at 4 °C for 48 h (Scheme 11).

Scheme 11

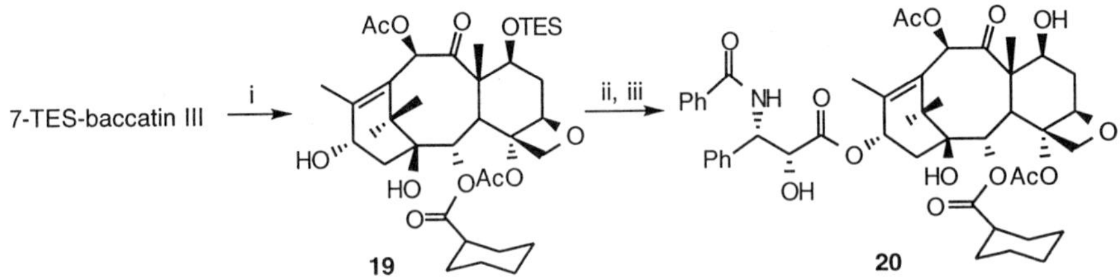

(i) H_2 (800 psi), 5% Rh-C, MeOH, 80 °C, 85%.; (ii) (a) NaHMDS, THF, 0 °C, (b) **4b-EE**, 0 °C, 80%. (iii) 0.5% HCl, 4 °C, 88%.

Microtubule Disassembly Inhibitory Activity and Cytotoxicity of the Cyclohexyl Analogs (*24*). Biological activities of the cyclohexyl analogs of docetaxel and 2-(hexahydro)paclitaxel (**20**) were evaluated in three assay systems, *i.e.*, inhibition of microtubule disassembly (*28*) and cytotoxicity against murine P388 leukemia cell lines as well as doxorubicin resistant P388 leukemia (P388/Dox) cell line (*29*). Results are summarized in Table II.

As Table II shows, the introduction of one cyclohexyl moiety to docetaxel virtually does not affect the tubulin binding ability, i.e., **SB-T-1031** (1.0Tt) and **SB-T-1041** (1.17Tt) (Tt = IC_{50} of docetaxel). Cytotoxicities of these analogs, however, appear to be more sensitive to the modification, i.e., **SB-T-1031** and **SB-T-1041** show ca. 8-fold and 12-fold weaker activities than docetaxel against P388 cell line, respectively. It is worthy of note that **SB-T-1031** is only 2 times less active than docetaxel against P388/Dox cell line. On the other hand, the substitution of two phenyl groups at the C-3' and the C-2 benzoate positions with two cyclohexyl groups (**SB-T-1051**) results in a substantial loss of activity in the microtubule disassembly inhibition (2.86Tt), but cytotoxicity (9.5Tt, P388; 2.4Tt, P388/Dox) is at the same level as those of **SB-T-1031** and **SB-T-1041**. 2-(Hexahydro)paclitaxel (**20**) shows much weaker activity than 2-(hexahydro)docetaxel (**SB-T-1041**) in the same assays.

Table II. Microtubule Disassembly Inhibitory Activity and *in vitro* Cytotoxicity of Cyclohexyl Analogs

Taxoid	Microtubule disassembly inhibitory activity[a] IC_{50}/IC_{50}(paclitaxel)	P388 cell line[b] IC_{50} (mg/mL)	P388/Dox[b] IC_{50} (mg/mL)
Docetaxel	0.70T	0.008	1.5
SB-T-1031	0.72T	0.063	3.1
SB-T-1041	0.85T	0.090	6.3
SB-T-1051	2.0T	0.076	3.6
20	1.7T	0.45	7.5

[a] IC_{50} represents the concentration of an agent leading to 50% inhibition of the rate of microtubule disassembly. IC_{50}(paclitaxel) is the IC_{50} value of paclitaxel in the same assay. In the same assay, the IC_{50} of paclitaxel is 0.015 mM. [b] IC_{50} represents the concentration that inhibits 50% of cell proliferation.

Although extensive conformational analyses of paclitaxel and docetaxel have been performed by NMR spectroscopy and molecular modeling studies (*19,30*), the binding conformations of these drugs are still to be determined. Nevertheless, it has been believed that phenyl groups at the C-3' and C-2 positions play an important role in the microtubule-binding process through their hydrophobic interactions with microtubules (*19,30*). In this context, it is worthy of note that **SB-T-1031**, **SB-T-1041**, **SB-T-1051**, and **20** are active without such a phenyl moiety at C-3' and/or C-2 (*25*). *These results clearly indicate that phenyl or an aromatic group at C-3' or C-2 is not a requisite for strong binding to the tubulin receptor* (*24*).

As mentioned above, the bis-cyclohexyl analog **SB-T-1051** exhibits unexpectedly weaker inhibitory activity for tubulin disassembly than mono-cyclohexyl analogs **SB-T-1031** and **SB-T-1041** in that the observed activity is much lower than that anticipated based on additivity of substituent effect. This fact may imply that the two cyclohexyl moieties of **SB-T-1051** are spatially in close proximity to render undesired change in the binding conformation although *in vitro* cytotoxicity of **SB-T-1051** does not appear to reflect this change. Our molecular modeling study on these cyclohexyl analogs using the SYBYL 6.0 program on the basis of the "hydrophobic collapse" conformation in aqueous media proposed by Vander Velde *et al.* (*19*) strongly supports this implication.

Antitumor Activity of the Cyclohexyl Analogs *in vivo* (*24*). The *in vivo* antitumor activity of **SB-T-1031**, **SB-T-1041**, and **SB-T-1051** were evaluated against B16 melanoma in $B6D2F_1$ mice. The antitumor activity is expressed by the T/C (%) value in which T is the median tumor weight of tolerated mice and C is the median tumor weight of control mice (*31*). Measurements were made when C is approximately equal to 1 g. Taxoids (0.4 mL/mouse) were administered intravenously (i.v.) on days of 5, 7, and 9. Results are as follows: **SB-T-1031**, T/C = 38% (20 mg/Kg/day); **SB-T-1041**, T/C = 76% (32.2 mg/Kg/day); **SB-T-1051**, T/C = 79% (12.4 mg/Kg/day). The results clearly indicate that **SB-T-1031** is only marginally active at the maximal *tested* dose, but **SB-T-1041** (at the maximal *tested* dose) and **SB-T-1051** (at the maximal *tolerated* dose) are inactive *in vivo*. Under the same conditions, docetaxel, showed excellent activity, i.e., T/C = 0% at 20 mg/Kg/day. The observed significant loss of activity *in vivo* could be ascribed to faster metabolism, e.g., oxidations by P450, faster excretion, or other bioavailability problems, but it is apparent that further investigation is necessary to deduce any conclusion.

Syntheses of 3'-Alkyl and 3'-Alkenyl Analogs of Docetaxel (*32*). The discovery of the fact that 3'-phenyl is not essential for the biological activity of paclitaxel and docetaxel at least *in vitro* prompted us to synthesize new 3'-alkyl and 3'-alkenyl analogs of docetaxel. Since we had developed efficient routes to (3*R*,4*S*)-4-alkyl-3-hydroxyazetidin-2-ones as well as (3*R*,4*S*)-4-alkenyl-3-hydroxyazetidin-2-ones with high enantiomeric purity on the basis of the chiral ester enolate – imine cyclocondensation (*33*), the corresponding 1-[t]BOC-β-lactams **21a,b** and **22a,b**, respectively, were readily synthesized in good yields in the same manner as **4** and **15** (*vide supra*).

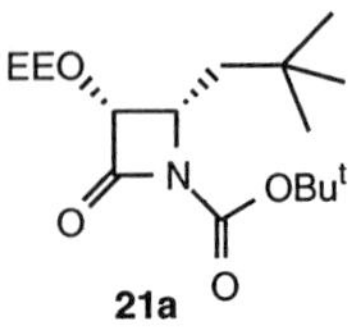
21a

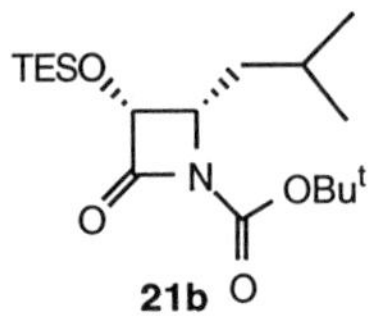
21b

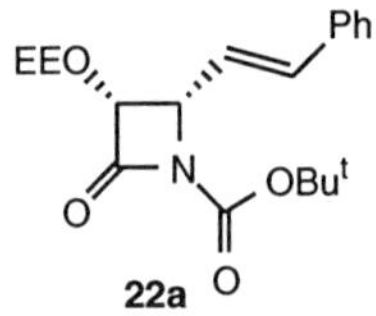
22a

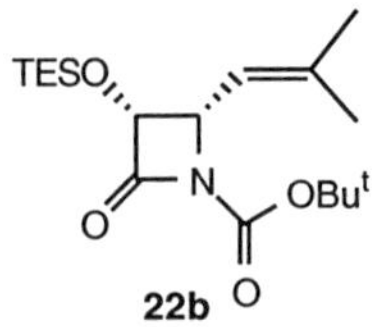
22b

The couplings of these β-lactams with 7,10-ditroc-DAB (**18**) and 7-TES-baccatin III were carried out using our protocol (*14*) using NaHMDS as the base followed by deprotection (*vide supra*) to give the corresponding 3'-dephenyl-3'-alkyl- and 3'-alkenyldocetaxels in good overall yields (Chart 1) (*32*).

Chart 1

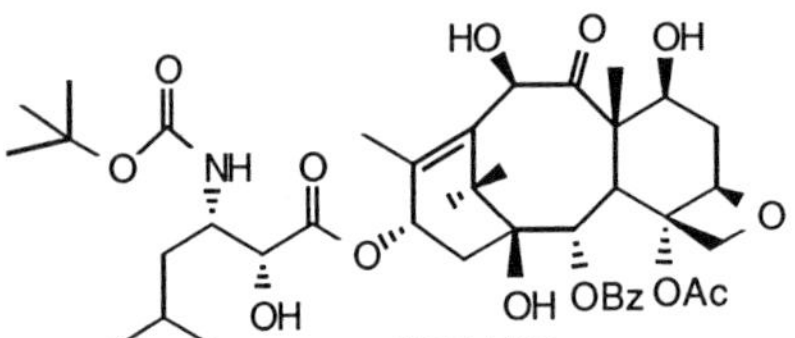
SB-T-1101

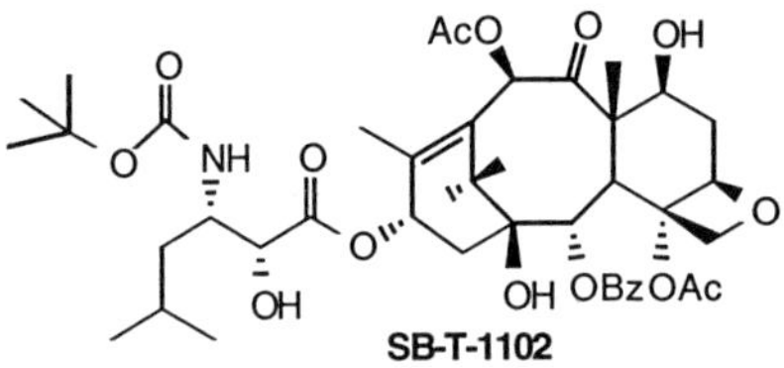
SB-T-1102

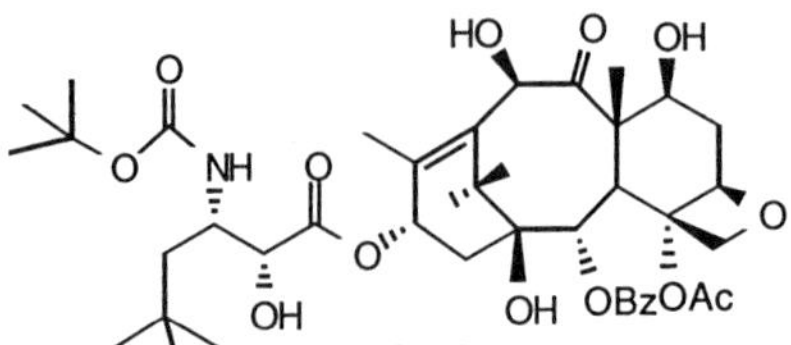
SB-T-1111

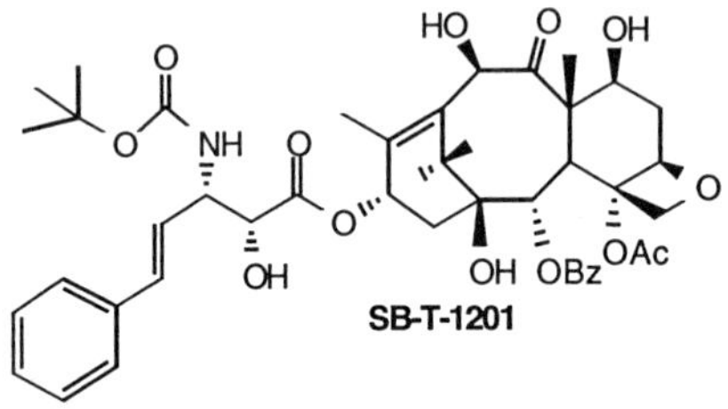
SB-T-1201

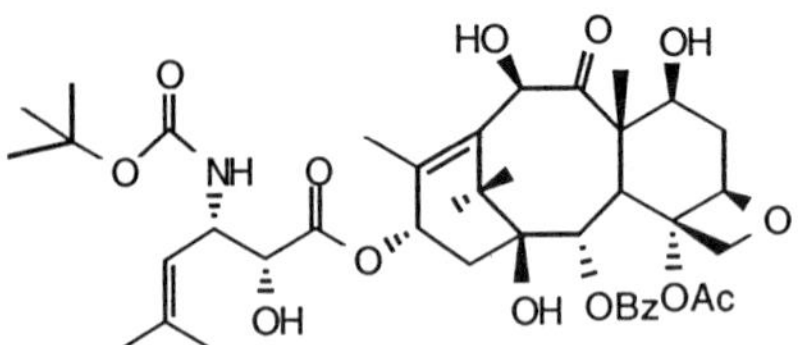
SB-T-1211

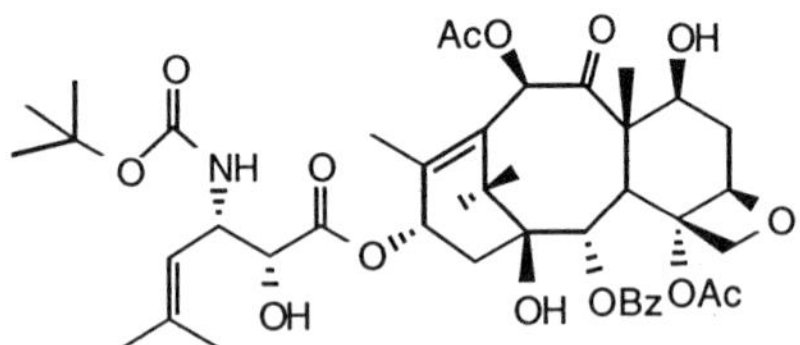
SB-T-1212

Microtubule Disassembly Inhibitory Activity and Cytotoxicity of the 3'-Alkyl and 3'-Alkenyl Analogs (*32*). Biological activities of four 3'-alkyl and 3'-alkenyl analogs of docetaxel were evaluated in three assay systems, *i.e.*, inhibition of microtubule disassembly and cytotoxicity against murine P388 leukemia cell lines as well as doxorubicin resistant P388 leukemia (P388/Dox) cell line (*vide supra*). As Table III shows, the 3'-isobutyl analog **SB-T-1101** and the 3'-isobutenyl analog **SB-T-1211** possess excellent activities comparable to docetaxel (*34*). It is also obvious that the activity is very sensitive to the bulkiness of the 3'-substituents.

Table III. Microtubule Disassembly Inhibitory Activity and *in vitro* Cytotoxicity of 3'-Alkyl and 3'-Alkenyl Analogs

Taxoid	Microtubule disassembly inhibitory activity[a] IC_{50}/IC_{50}(paclitaxel)	P388 cell line[b] IC_{50} (mg/mL)	P388/Dox[b] IC_{50} (mg/mL)
Docetaxel	0.70T	0.008	1.5
SB-T-1101	0.78T	0.0096	1.25
SB-T-1201	1.45T	0.22	6.25
SB-T-1111	1.45T	0.1	5.5
SB-T-1211	0.64T	0.01	1.85

[a] IC_{50} represents the concentration of an agent leading to 50% inhibition of the rate of microtubule disassembly. IC_{50}(paclitaxel) is the IC_{50} value of paclitaxel in the same assay. In the same assay, the IC_{50} of paclitaxel is 0.015 mM. [b] IC_{50} represents the concentration that inhibits 50% of cell proliferation.

The cytotoxicity of 3'-alkyl and 3'-alkenyl analogs derived from **18** and 7-TES-baccatin III were evaluated against human ovarian (A121), human non-small cell lung (A549), human colon (HT-29), and human breast (MCF7) cancer cell lines (*vide supra*). In addition to these cell lines, the activity against an adriamycin-resistant human breast cancer cell line (MCF7-R) was also examined. As Table IV shows, 3'-isobutenyl and 3'-isobutyl analogs, especially 3'-isobutenyl analog **SB-T-1212**, exhibits excellent cytotoxicity. *It is noteworthy that **SB-T-1102** and **SB-T-1212** show one order of magnitude better activity than paclitaxel and docetaxel against MCF7-R. This finding may provide significant information for the development of newer antitumor agents effective against drug-resistant tumors.*

Table IV. Cytotoxicities (IC_{50} nM)[a] of 3'-Alkyl and 3'-Alkenyl Analogs from 18 and 7-TES-baccatin III

Taxoid	A121[a] (ovarian)	A549[a] (NSCLC)	HT-29[a] (colon)	MCF7[a] (breast)	MCF7-R[a,b]
Paclitaxel	6.1	3.6	3.2	1.7	300
Docetaxel	1.2	1.0	1.2	1.0	235
SB-T-1101	1.9	0.7	0.5	0.8	107
SB-T-1102	3.8	0.98	3.2	4.0	36
SB-T-1212	0.46	0.27	0.63	0.55	12

[a] See the footnote of Table I. [b] MCF7-R = mammary carcinoma cells 180 fold resistant to adriamycin.

Antitumor Activity of the 3'-Alkyl and 3'-Alkenyl Analogs *in vivo* (32). The *in vivo* antitumor activity of **SB-T-1101** and **SB-T-1211** was evaluated against B16 melanoma in B6D2F1 mice. Taxoids (0.4 mL/mouse) were administered intravenously (i.v.) on days of 5, 7, and 9 (*vide supra*). Results are as follows: SB-T-**1101**, T/C = 5% (20 mg/Kg/day), time for median tumor to reach 1,000 mg (days) = 26.03, $\log_{10}$ cell kill = 1.97; **SB-T-1211**, T/C = 8% (12.4 mg/Kg/day), time for median tumor to reach 1,000 mg (days) = 27.54, $\log_{10}$ cell kill = 2.25. The results clearly indicate that both analogs are very active *in vivo*, and their activities are equivalent to that of docetaxel in the same assay.

Consequently, it is apparent that 3'-alkyl and 3'-alkenyl analogs of docetaxel and paclitaxel have opened a new avenue for the development of newer and potent taxoid antitumor agents, especially against drug-resistant tumors.

Acknowledgments. This research was supported by grants from the National Institutes of Health (GM417980), the Center for Biotechnology at Stony Brook which is sponsored by the New York State Science & Technology Foundation, Rhône-Poulenc Rorer, and the National Cancer Institute (CA13038). A generous support from Indena, SpA is also gratefully acknowledged. The authors would like to thank Dr. Ezio Bombardelli, Indena, SpA for providing them with 14β-OH-DAB. They extend sincere thanks to Dr. Daniel Guénard, Institut de Chimie des Substances Naturelles, CNRS, Gif-sur-Yevette, for the microtubule disassembly assay. They also thank Dr. Alain Commerçon of Rhône-Poulenc Rorer for his helpful discussions. One of the authors (I. F.) is grateful to Universitá degli Studi di Torino for a postdoctoral fellowship. They also thank Dr. David G. Vander Velde, University of Kansas, for the 1D and 2D NMR analyses of **SB-T-1001** and **SB-T-1071**, and for his helpful discussions regarding the "hydrophobic cluster" conformations.

Literature Cited

1. Rowinsky, E. K.; Onetto, N.; Canetta, R. M.; Arbuck, S. G. *Seminars in Oncology* **1992**, *19*, 646-662.
2. Suffness, M. In *Annual Reports in Medicinal Chemistry*; J. A. Bristol, Ed.; Academic Press: San Diego, 1993; Vol. 28; pp Chapter 32, 305-314.
3. Kingston, D. G. I. *Pharmac. Ther.* **1991**, *52*, 1-34 and references cited therein.
4. Guénard, D.; Guéritte-Vogelein, F.; Potier, P. *Acc. Chem. Res.* **1993**, *26*, 160-167.
5. Seidman, A. D. *Stony Brook Symposium on Taxol and Taxotère, May 14-15, 1993, Stony Brook , NY, Abstracts* pp 14-16.
6. Ravdin, P. M. *Stony Brook Symposium on Taxol and Taxotère, May 14-15, 1993, Stony Brook , NY, Abstracts* p 18.
7. Horwitz, S. B. *Stony Brook Symposium on Taxol and Taxotère, May 14-15, 1993, Stony Brook , NY, Abstracts* pp 23-24.
8. Appendino, G.; Gariboldi, P.; Gabetta, B.; Pace, R.; Bombardelli, E.; Viterbo, D. *J. Chem. Soc., Perkins Trans 1* **1992**, 2925-2929.
9. Grampovnik, D. J.; Maring, C. J.; Klein, L. L.; Li, L.; Thomas, S. A. *The 207th American Chemical Society National Meeting, March 13-17, 1994, San Diego, Abstracts* MEDI 102.
10. Margraff, R.; Bézard, D.; Bourzat, J. D.; Commerçon, A. *Bioorg. Med. Chem. Lett.* **1994**, *4*, 233-236.
11. Ojima, I.; Park, Y. H.; Sun, C. M.; Fenoglio, I.; Appendino, G.; Pera, P.; Bernacki, R. J. *J. Med. Chem.* **1994**, *37*, 1408-1410.
12. (a) Ojima, I.; Habus, I.; Zhao, M.; Georg, G. I.; Jayasinghe, R. *J. Org. Chem.* **1991**, *56*, 1681-1684. (b) Ojima, I. *U. S. Patent* **1994**, 5,294,737.
13. (a) Ojima, I.; Habus, I.; Zhao, M.; Zucco, M.; Park, Y. H.; Sun, C. M.; Brigaud, T. *Tetrahedron* **1992**, *48*, 6985. See also (b) Holton, R. A. *Eur. Pat. Appl.* EP 400,971, 1990; (c) Georg, G. I.; Cheruvallath, Z. S. *J. Med. Chem.* **1992**, *35*, 4230-4237.
14. (a) Ojima, I.; Sun, C. M.; Zucco, M.; Park, Y. H.; Duclos, O.; Kuduk, S. D. *Tetrahedron Lett.* **1993**, *34*, 4149-4152. (b) Ojima, I.; Zucco, M.; Duclos, O.; Kuduk, S. D.; Sun, C. M.; Park, Y. H. *Bioorg. Med. Chem. Lett.* **1993**, *3*, 2479-2482.
15. Commerçon, A.; Bezard, D.; Bernard, F.; Bourzat, J. D. *Tetrahedron Lett.* **1992**, *33*, 5185-5188.
16. Ojima, I.; Fenoglio, I.; Park, Y. H.; Pera, P.; Bernacki, R. J. *Bioorg. Med. Chem. Lett.*, **1994**, *4*, 1571-1574.
17. Kant, J.; Farina, V.; Fairchild, C.; Kadow, J. F.; Langley, D. R.; Long, B. H.; Rose, W. C.; Vyas, D. M. *Bioorg. Med. Chem. Lett.*, **1994**, *4*, 1565-1570.

18. Skehan, P.; Streng, R.; Scudierok D.; Monks, A.; McMahon, J.; Vistica, D.; Warren, J. T.; Bokesch, H.; Kenney, S.; Boyd, M. R. *J. Nat. Cancer Inst.* **1990**, *82*, 1107-1112.
19. Vander Velde, D. G.; Georg, G. I.; Grunewald, G. L.; Gunn, C. W.; Mitscher, L. A. *J. Am. Chem. Soc.* **1993**, *115*, 11650-11651.
20. Guéritte-Vogelein, F.; Guénard, D.; Mangatal, L.; Potier, P.; Guilhem, J.; Cesario, M.; Pascard, C. *Acta Cryst.* **1990**, *C46*, 781.
21. Ojima, I.; Fenoglio, I.; Park, Y. H.; Sun, C. M.; Appendino, G.; Pera, P.; Bernacki, R. J. *J. Org. Chem.* **1994**, *59* , 515-517.
22. Swindell, C. S.; Krauss, N. E. *J. Med. Chem.* **1991**, *34*, 1176-1184.
23. Denis, J.-N.; Greene, A. E.; Guénard, D.; Guéritte-Vogelein, F.; Mangatal, L.; Potier, P., *J. Am. Chem. Soc.* **1988**, *110*, 5917-5919.
24. Ojima, I.; Duclos, O.; Zucco, M.; Bissery, M.-C.; Combeau, C.; Vrignaud, P.; Riou, J. F.; Lavelle, F. *J. Med. Chem.* **1994**, *37*, 2602-2608.
25. The results were presented at *the 207th American Chemical Society National Meeting, San Diego, March 13-17, 1994*; Duclos, O.; Zucco, M.; Ojima, I.; Bissery, M.-C.; Lavelle, F. *Abstracts* MEDI 86. At the same meeting, Professor Gunda I. Georg presented similar results on the cyclohexyl analogs of paclitaxel and docetaxel; Georg, G. I. *Abstracts* MEDI 146.
26. For 2-(hexahydro)paclitaxel (**20**) see also Chen, S.-H.; Farina, V.; Wei, J.-M.; Long, B.; Fairchild, C.; Mamber, S. W.; Kadow, J. F.; Vyas, D.; Doyle, T. W. *Bioorg. Med. Chem. Lett.* **1994**, *4*, 479-482.
27. Kingston *et al.* reported that baccatin III could be hydrogenated over platinum: Samaranayake, G.; Neidigh, K. A.; Kingston, D. G. I. *J. Nat. Prod.* **1993**, *56*, 884-898.
28. Lataste, H.; Sénilh, V.; Wright, M.; Guénard, D.; Potier, P. *Proc. Natl. Acad. Sci. U.S.A.* **1984,** *81*, 4090-4094.
29. Riou, J. F.; Naudin, A.; Lavelle, F. *Biochem. Biophys. Res. Commun.* **1992**, *187*, 164-170.
30. (a) Dubois, J.; Guénard, D.; Guéritte-Voegelein, F.; Guedira, N.; Poyier, P.; Gillet, B.; Beloeil, J. C. *Tetrahedron* **1993**, *49*, 30, 6533-6544. (b) Williams, H. J.; Scott, A. I.; Dieden, R. A.; Swindell, C. S.; Chirlian, L. E., Francl, M. M.; Krauss, N. E. *Tetrahedron* **1993**, *49*, 30, 6545-6560.
31. Geran, R. I.; Greenberg, N. H.; MacDonald, M. M.; Schumacher, A. M.; Abbott, B. J. *Cancer Chemother. Rep.* **1972**, *Part 3*, 1-103.
32. Ojima. I.; Duclos, O.; Kuduk, S. D.; Slater, J. C.; Simonot, B.; Bissery, M.-C.; Lavelle, F.; Veith, J.; Bernacki, R. To be published.
33. Ojima, I.; Park, Y. H.; Sun, C. M.; Zhao, M.; Brigaud, T. *Tetrahedron Lett.* **33**, 5739-5742 (1992).
34. (a) A recent patent claims that 3'-isobutenyl analog of paclitaxel shows excellent *in vitro* cytotoxicity: Holton, R. A.; Nadizadeh, H. *U. S. Patent* **1994**, 5,284,864. (b) Klein, L. L. *The 207th American Chemical Society National Meeting, March 13-17, 1994, San Diego, Abstracts* MEDI 103.

RECEIVED September 7, 1994

Chapter 20

Chemistry and Antitumor Activity of 9(*R*)-Dihydrotaxanes

L. L. Klein, L. Li, C. M. Yeung, C. J. Maring, S. A. Thomas, D. J. Grampovnik and J. J. Plattner

Anti-Infective Division, Abbott Laboratories, One Abbott Park Road, Abbott Park, IL 60064–3500

The 9(R)-dihydrotaxanes are a new family of semisynthetic antitumor agents which show great promise as a second generation class of antimicrotubule agents. These compounds have increased water solubility and stability as compared to taxol and also exhibit excellent activity in tumor models. Other advantages of the 9(R)-hydroxyl group can be found in its use as an additional site for chemical modification towards the preparation of new derivatives. Furthermore, its effect on the surrounding functionalities allows for access to novel chemistry and ring systems from this taxane template.

The discovery of new analogs of the antitumor agent taxol which exhibit broader spectrum, enhanced in vivo activity, or improved water solubility will be important in the continued evolution of this class of agents. The 9(R)-dihydrotaxanes show great promise as a second generation class of agents in that they have greater solubility and stability and also show excellent in vivo activity in several solid tumor models.

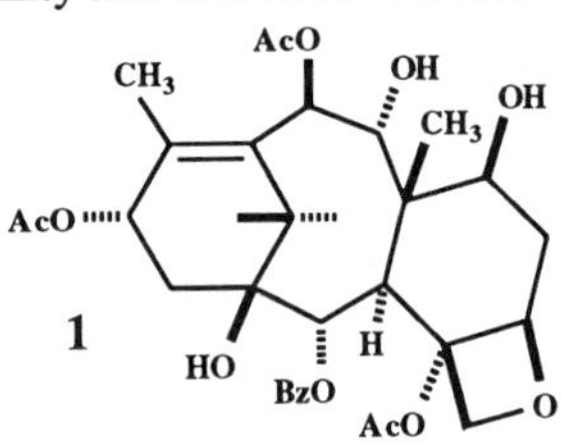

1

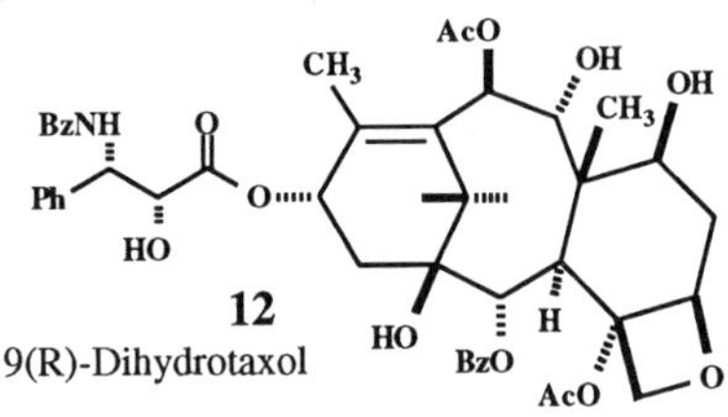

12
9(R)-Dihydrotaxol

The isolation of the novel component, 13-acetyl-9(R)-dihydrobaccatin III (**1**) was reported by several laboratories in 1992 (*1-3*). This component differs in structure from taxol in two respects: 1) the C-13 phenyl isoserinate sidechain which is required for activity is replaced by an acetate; 2) the C-9 carbonyl is replaced by an α-hydroxyl group. It is found as a constituent in *Taxus canadensis*, a bush common to the northeastern United States and southern Quebec (*4*), but has also been found in other species such as *Taxus chinensis* in China. Since **1** exists primarily in the needles, access to this component is quite similar to that of baccatin III from *Taxus baccata* in that the source is a cultivatable and renewable resource. The quantity of **1** available from this plant varies from 0.08-0.3% (dried needles); however, a simple extractive workup and recrystallization can give access to this component without the need for any chromatography (Gunawardana, G., Abbott Labs., unpublished data).

NOTE: Paclitaxel is the generic name for Taxol, which is now a registered trademark.

0097–6156/95/0583–0276$08.00/0

The x-ray structure of **1** shows exceptional overlap with that of the corresponding baccatin III derivatives due to the fact that the C-9 carbonyl in the latter series bisects the hydrogen/hydroxyl sp3 center in **1** (*1*). Several advantages and opportunities afforded by this novel structure are apparent: 1) the presence of the C-9 hydroxyl serves as an additional site for modifications; 2) the presence of this hydroxyl serves to increase the water solubility of these analogs; 3) the lack of a C-9 carbonyl serves to stabilize the system relative to the base-catalyzed C-7 epimerization (*5*).

Chemistry of 13-Acetyl-9(R)-dihydrobaccatin III (1)

In order to ascertain the effect of the C-9 hydroxyl on the antitumor activity of this system, we needed to remove the C-13 acetyl group present in **1** and reacylate this center with an appropriate sidechain. While many ways of conducting the latter reaction have been reported, no method for selective removal of the C-13 acetate was known prior to our work.

Selective Deacetylation at C-13. Initial deacetylation attempts involving hydrolytic deacylation succeeded in rapid and selective removal of C-10 acetate in good yield; however, further hydrolysis afforded only complicated mixtures involving non-selective deacylation of C-2, C-4, and C-13 acyl groups. Deacetylation of the 7,9-acetonide **2** obtained under standard conditions (Scheme 1) only slowed the C-10 deacylation but failed to enhance selectivity at the C-13 center. Under more vigorous conditions the C-2 benzoate

Scheme 1

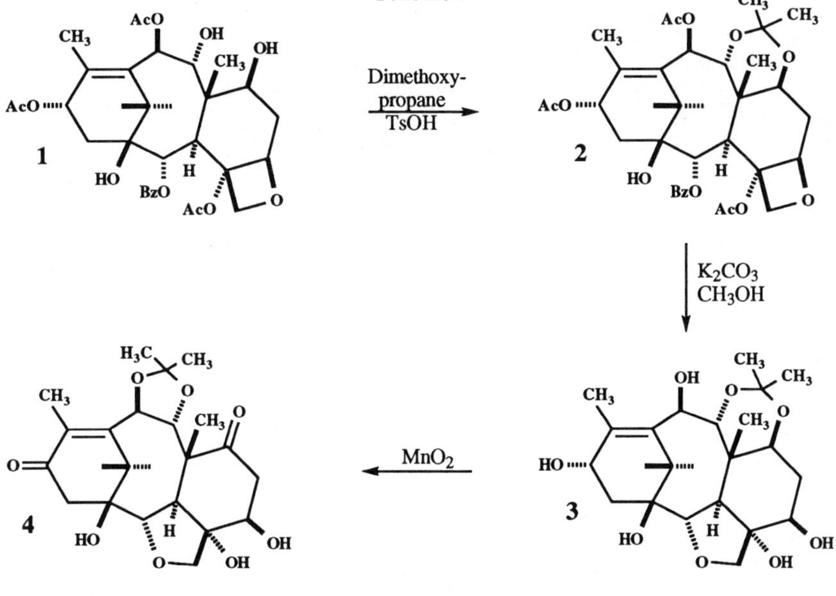

also undergoes hydrolysis followed by facile ring opening of the oxetane ring with concomitant formation of the tetrahydrofuran sideproduct **3**. This product was reported in the C-9 carbonyl series and has been shown to lead to inactive compounds (*6,7*). In this case, structure **3** was actually identified via characterization of its oxidation product **4** obtained by treatment of **3** with manganese dioxide (acetone, reflux). Under these conditions, this oxidation of the C-13 hydroxyl group takes place in conjunction with transfer of the 7,9-acetonide to the 9,10-position occurs thus allowing further oxidation at C-7.

In contrast to hydrolytic conditions, the use of carbanion reagents such as alkyl lithiums, Grignard reagents, and hydride reagents such as lithium triethylborohydride produced a different reaction profile such that the desired C-13 deacetylation product **5** predominated (*8*). Methyl lithium was found to be the reagent of choice giving >80% yield of the desired product from **1** without the need for protecting groups (Scheme 2) (Klein, L.L., Abbott Labs, unpublished data). This selective deacylation was found to be of general use, also applicable to the C-9 carbon Perhaps one reason for this selectivity can be related to another experiment which involved treatment of the acetonide **2** with lithium hexamethyldisilazide and allyl bromide (Scheme 3). The major product was the 13-(4-pentenoyl) analog **6** which arose from the enolization of the C-

Scheme 2

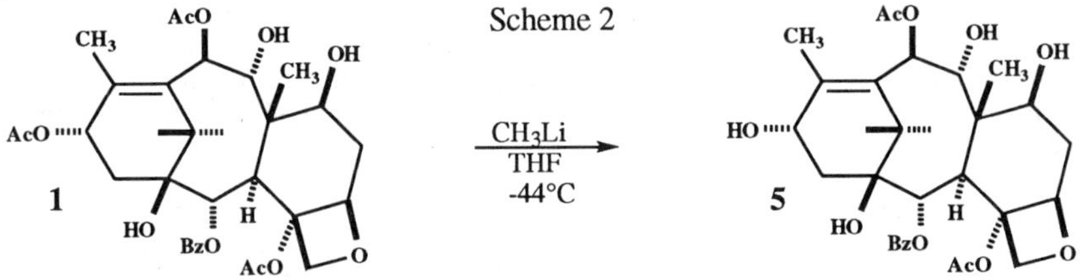

13 acetyl group. This result established that: 1) condensation of the C-13 enolate with electrophiles was viable. A similar reaction of this enolate with benzaldehyde led, after elimination, to the styryl derivative **7** containing the carbon skeleton of the taxol sidechain without requiring removal of this acetate; 2) decomposition of this enolate via a ketene elimination provides a mechanistic explanation for the deacetylation; however, it is still not clear why this presumably less accessible ester is the site for deprotonation.

One minor product isolated from the deacetylation of the acetonide **2** was the 4,13-dideacetyl analog **8**. The sidechain was attached to **8** under standard conditions which involved low temperature deprotonation of the C-13 hydroxyl group with lithium hexamethyldisilazide followed by treatment with the appropriate 4-substituted-3-(2-ethoxy)ethoxy-N-acyl-2-azetidinone **9** (*9*) giving good yields of the adduct (*10*).

Scheme 3

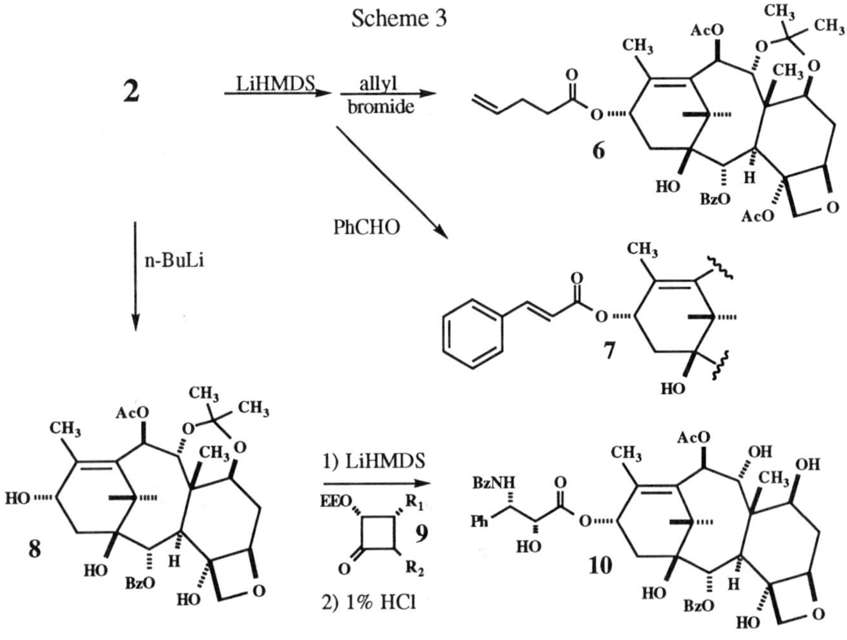

Deprotection of the ethoxyethyl protecting group and the acetonide afforded 4-deacetyl-9(R)-dihydrotaxol (**10**). This compound proved to be practically devoid of activity (Table I) in the tubulin assembly assay (*11*) and against the four tumor cell lines (*12*) which made up our routine testing protocol; thus, the C-4 position appeared to be very sensitive to change.

Synthesis of 9(R)-Dihydrotaxol (9-DHT). For the preparation of 9(R)-dihydrotaxol, the C-13 deacetylation was performed on isolate **1** and was followed by standard protection (triethylsilyl ether) of the C-7 hydroxyl group to give intermediate **11** (Scheme 4). Either sodium hydride at 25°C or lithium hexamethyldisilazide (LiHMDS) at -44°C was used to form the oxyanion of **11** which was reacted with lactam **9** (R_1 = Ph; R_2 = Bz). After deprotection under acidic conditions, 9(R)-dihydrotaxol (**12**) was produced. This synthetic sequence also allows for the preparation of the more soluble 10-deacetyl analogs in this series. Toward that end reaction of **11** with modified lactam **9** (R_1 = Ph; R_2 = Boc) in a similar manner produces an adduct which first undergoes quantitative basic hydrolysis followed by acidic deprotection. In this way 9(R)-dihydrotaxotere (**13**) was easily produced. The facile deacetylation of the C-10 acetate is not trivial in the C-9 carbonyl series and reflects the greater stability of the 9(R)-dihydro series. These products were shown to have excellent tubulin assembly activity and similar *in vitro* activity as compared to taxol and taxotere; therefore, these preliminary results establish that the C-9 carbonyl is not required for activity.

Scheme 4

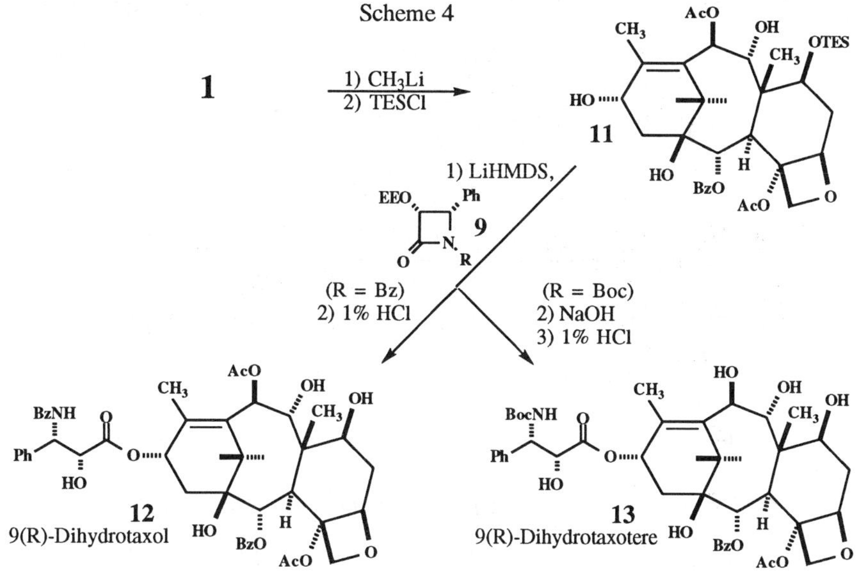

9(R)-Dihydrotaxol exhibited good efficacy versus the murine M109 solid tumor model (IP), delaying tumor growth to a greater extent than taxol. Compound **13** also showed good efficacy *in vivo* and was found to have 100-fold greater water solubility than taxol (**13**, 226 µg/mL; taxol 1-3 µg/mL). This allowed for a decrease in the amount of Cremophor EL vehicle required in the administrative solutions.

Table I. Tumor Cell Cytotoxicity of 9(R)-Dihydrotaxol Analogs

Compound	Tumor cell lines, IC_{50} (ng/mL)				Tubulin (μM)
	A549	*HT-29*	*B16F10*	*P388*	ED_{50} / ED_{50} taxol
Taxol	2.5-4.3	1.8-3.5	3.4-6.3	8.8-11	1.00
Taxotere	0.18	0.21	0.6	1.5	0.7
4-Deacetyl-9-dihydrotaxol (10)	>1000	8800	7900	>1000	>10
9-Dihydrotaxol (12)	16-22	6.4-9.6	25	49-57	0.86
9-Dihydrotaxotere (13)	0.26	0.65	0.4	2.8	1.31
10-Deacetyl-9-dihydrotaxol	11	1.9	39	140	0.83

Synthesis of 9DHT Analogs

Encouraged by the discovery that the 9(R)-dihydrotaxanes define an active template, the preparation of derivatives having increased solubility and/or potency was pursued. Modifications to both the sidechain and to the ring system were studied.

C-3' Side Chain Analogs. Analog work involved variation of the sidechain nitrogen substituent at C-3' via formation of amides, ureas, and carbamates; however,

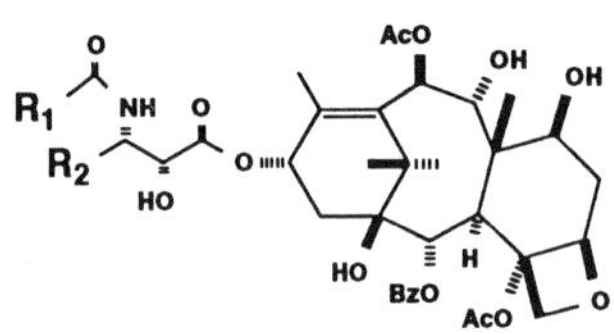

Table II. Tumor Cell Cytotoxicity of C-3' Alkyl Analogs

Entry	Compound R_1	R_2	Tumor cell lines, IC_{50} (ng/mL) *A549*	*HT-29*	*B16F10*	*P388*
1.	Ph **9(R)-DHT (12)**	Ph	16-22	6.4-9.6	25	49-57
2.	Ph	CH_3OCH_2	>100	79	>100	>100
3.	t-BuO		>100	>100	86	>100
4.	t-BuO	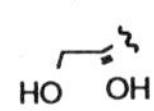	>100	>100	>100	>100
5.	t-BuO	CH_3	15	8.3	19	43
6.	t-BuO	$PhCH_2$	>100	>100	>100	>100
7.	t-BuO	Et	0.83	1.4	3.2	11
8.	t-BuO	$CH_2=CH$	1.1	1.8	4.4	16
9.	t-BuO	c-C_6H_{12}	1.2	5	5.7	14
10.	t-BuO	i-Bu	0.12	0.38	0.9	3.6

just as in the case for the C-9 carbonyl series, the activity exhibited by the t-butoxycarbonyl (Boc) group is optimum relative to those which have been tested (*13*). Work by others regarding the C-3' carbon substituent (phenyl) has involved subtle variation of substituents or heteroatoms on the aryl ring (*14*). Recent results have shown that this phenyl group is not required for the retention of activity (*15*), and in fact, groups such as C-3'-isobutyl (entry 10) have shown excellent *in vitro* activity (Table II). Inclusion of polar groups at this position such as the diol or the acetonide generally decreases the activity. Comparison of the activities of the methyl ether (entry 2) and simple carbon chain analogs such as ethyl (entry 7) show peculiar functional sensitivity. In contrast to the C-3' nitrogen substituent, the lack of activity of the C-3' benzyl homolog (entry 6) is most interesting reflecting high structural sensitivity at this position It was found that the C-3' benzyl analog in the C-9 carbonyl series is similarly inactive.

C-7/C-9 Analogs. Modifications at the C-7 and C-9 positions on the ring have shown high compatibility with retention of activity. For example 7,9-acetals, carbonates, and orthoesters were readily formed under standard conditions and reflect the pseudoequatorial positioning of these two groups on the ring backbone. In contrast to the sensitivity of the C-9 carbonyl series under basic conditions, the 9(R)-dihydro system can be treated directly with strong base in order to alkylate the C-7 and/or the C-9 hydroxyl groups. Alkylation of the C-9 (entry 7) and/or of the C-7 hydroxyl group

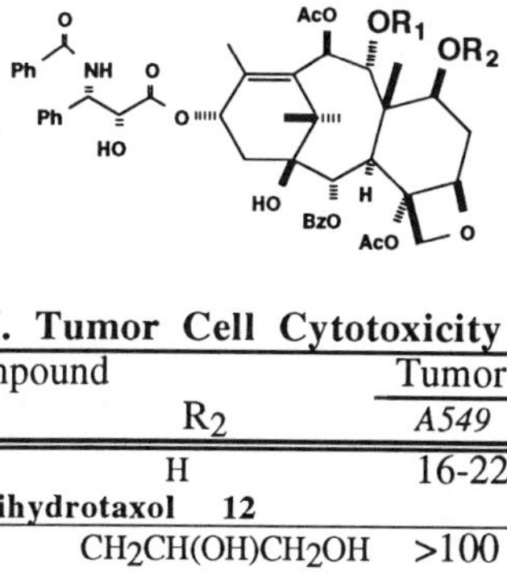

Table III. Tumor Cell Cytotoxicity of C-7,9 Analogs

Entry	Compound R1	R2	Tumor cell lines, IC_{50} (ng/mL) A549	HT-29	B16F10	P388
1.	H	H	16-22	6.4-9.6	25	49-57
	9-Dihydrotaxol 12					
2.	H	$CH_2CH(OH)CH_2OH$	>100	>100	>100	>100
3.	H	$CH_2CH_2NEt_2$	>100	>100	>100	>100
4.	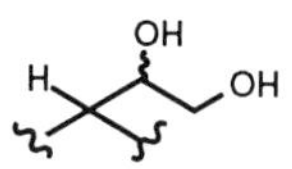		>100	79	90	>100
5.	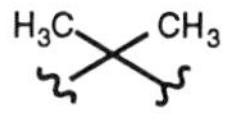		25	26	34	42
6.	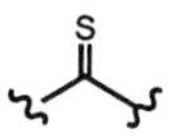		19	11	20	35
7.	CH_3	H	4.7	3.1	4.8	7.8
8.	H	CH_3	1.2	1.4	1.5	3.9
9.	H	$CH_2CH{=}CH_2$	1	1.2	2.7	5.3
10.	H	CH_3 (3'-NBoc)	0.27	0.15	0.2	0.6

(entry 8) is shown to increase potency (Table III). Further modification of the C-7 allyl group by adding polar substituents (entries 2,3) result in a loss of activity. The 3'-N-Boc analog in the 7-O-methyl series (entry 10) exhibits potent tubulin assembly activity ($ED_{50}/ED_{50taxol}$ = 0.35 mM) and extremely potent cytotoxicity. In general, masking of the C-7/9 hydroxyls leads to analogs of greater potency whereas addition of hydroxyl or amino groups in this quadrant produced compounds of lesser activity.

Ring B Rearrangement

Besides imparting greater stability and water solubility, the presence of the C-9 hydroxyl also allows for different chemistry to take place relative to the C-9 carbonyl parent structure. In the process of studying nucleophilic substitutions at the C-7 position, an interesting rearrangement took place upon activation of this center. Initial treatment of **1** with mesyl chloride formed the corresponding C-7 mesylate but this group was stable to subsequent attempts at displacement and under more vigorous conditions underwent decomposition; however, use of triflic anhydride for the same purpose led to formation of two new products. Through x-ray crystallographic analysis, these two structures were found to be novel rearrangement systems, **14** and **15** (*16*).

Scheme 5

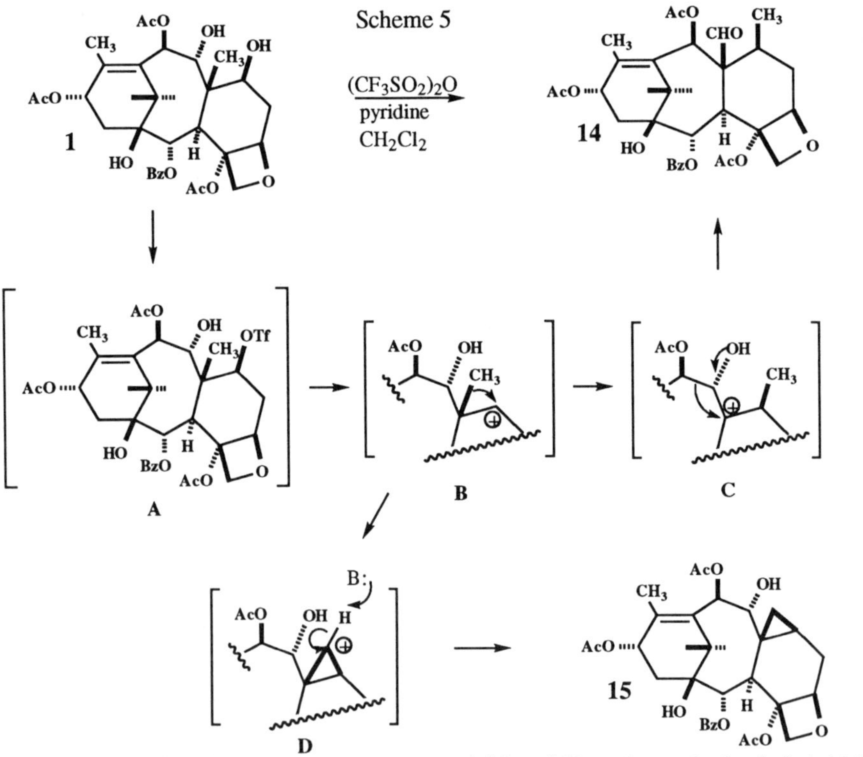

The novel ring-B rearrangement compound **14** could have been obtained via initial ionization of the C-7 triflate followed by C-19 methyl migration to give the more stable tertiary carbocation **C** (Scheme 5). This ion is structurally favored to undergo a pinacol rearrangement leading to product **14**. The cyclopropane product **15** may be derived by loss of a proton from the methonium ion **D** which in turn was formed from trapping of the C-7 carbocation by the C-19 methyl group. The presumed C-8 carbocation **C** would not be expected to form in the presence of a C-9 carbonyl, and in fact, this rearrangement has not been observed in the C-9 carbonyl system, whereas the cyclopropane product **15** which does not require this ion to form, has been reported in the carbonyl system (*17*). This novel ring-B rearrangement product is another example of the differences and opportunities afforded by the 9-dihydro system.

In a similar way, 9(R)-dihydrotaxol, appropriately protected at C-2', can also be directly rearranged to give the sidechain bearing compounds **16-18** (Scheme 6). It was

Scheme 6

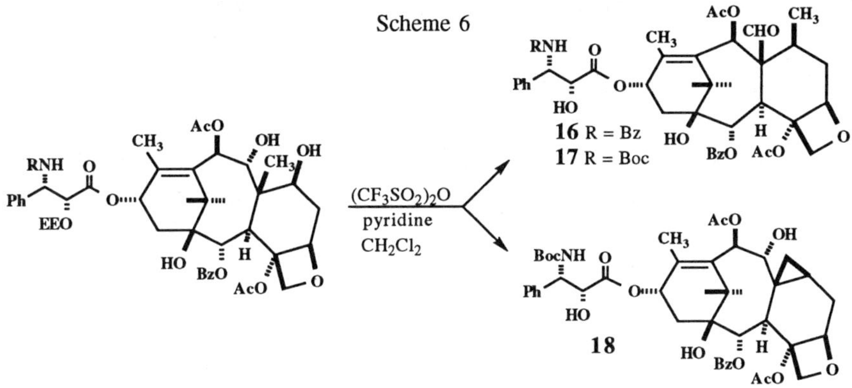

found that the corresponding N-benzoyl derivative **16** of this rearrangement product exhibited 10-30 fold less activity as compared to taxol; however, the corresponding N-Boc analog **17** showed in vitro activity more similar to taxol (Table IV). The structural overlap

Table IV. Tumor Cell Cytotoxicity of Rearrangement Analogs

Compound	Tumor cell lines, IC_{50} (ng/mL)			
	A549	*HT-29*	*B16F10*	*P388*
Taxol	2.5-4.3	1.8-3.5	3.4-6.3	8.8-11
9-Dihydrotaxol (12)	16-22	6.4-9.6	25	49-57
16	17	76	45	32
17	9.8	5.4	4.6	8.3
18	0.82	1.2	0.5	1.1

of the x-ray data from this novel system and that of the taxol ring system showed poor-to-fair overlap with the C-7 - C-10 "top" area whereas the C-1 - C-5 "bottom" of the molecule showed better alignment (*16*). In contrast, the cyclopropane structure **18** bearing close resemblance to the starting material shows excellent *in vitro* potency. Further chemical modification of the new aldehyde group in **14** was carried out via reduction to the C-19 alcohol with concomitant removal of the C-10 acetate. These and other modifications are currently under study.

Synthesis of 9-Deoxotaxol Analogs

Treatment of **1** with strong base followed by carbon disulfide and methyl iodide led to the 9-xanthate **19**, rather than the expected 7-xanthate (Scheme 7). If amine bases were used or reagents such as thiophosgene, thiocarbonyl diimidazole, or phenyl chlorothionoformate applied only the 7,9-thionocarbonate of **1** was obtained. Treatment of **19** with tributyltin hydride under standard conditions led to a good yield of 13-acetyl-9-deoxobaccatin III (**20**) (*18*). Following deacetylation at C-13 with methyl lithium and protection of the C-7 hydroxyl with triethylsilyl chloride, the sidechain was added to **21** as before and after deprotection, 9-deoxotaxol (**22**) was obtained.

Scheme 7

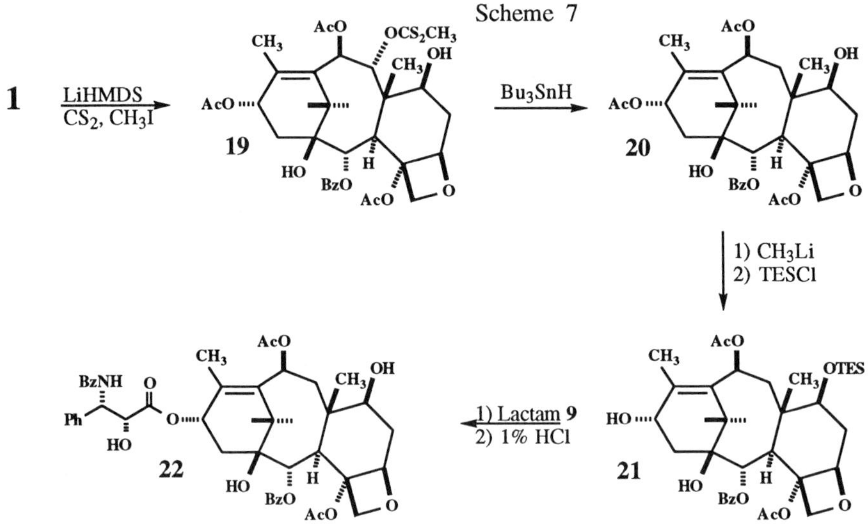

In order to methodically survey the effect of each of the oxygens, 13-acetyl-9-deoxobaccatin III (**20**) was again thioacylated and retreated with tributyltin hydride to give the 7,9-deoxygenated product, **23** (Scheme 8). In this case, treatment of **23** with methyllithium did not produce the typical selective C-13 deacylation result but instead led to a relatively nonselective deacylation of both C-13 and C-10 acetates to give the desired 13-deacetyl- (**24**) and the 10,13-dideacetyl-7,9-dideoxy (**25**) compounds in equal amounts. The dideacetyl product **25** could be selectively thioacylated to afford the C-10 thiocarbonylimidazolide which when treated with tributyltin hydride led to the 7,9,10-trideoxygenated derivative, **26**. Unfortunately, when the deoxygenated systems **24** and **26** were reacted with base and lactam as previously described for sidechain addition, little or no adduct was produced.

The fact that the deacetylation was nonselective, the standard sidechain acylation did not succeed, and the compounds showed poorer stability than their precursors to acid and base, suggested that the loss of the C-7/9 functionalities may also have affected the conformations of these molecules. An alternative method of acylation was applied via chemistry involving the acetonide sidechain precursor **27**. The corresponding Boc derivative of this sidechain precursor had been previously utilized for C-13 sidechain addition to the baccatin system (*19*). Instead of the Boc protecting group, the Cbz analog was prepared which allowed for final sidechain manipulation to occur under practically neutral conditions. Treatment of **24** or **26** with the Cbz acid **27** and DCC afforded the desired adducts under mild conditions. Hydrogenolysis of the Cbz group in the presence of aqueous methanol then affected hydrolysis of the incipient aminal. Quenching of the product amine *in situ* with benzoic anhydride produced the benzoyl sidechain compounds **28** and **29** in 61% and 55% yields, respectively (*18*).

Scheme 8

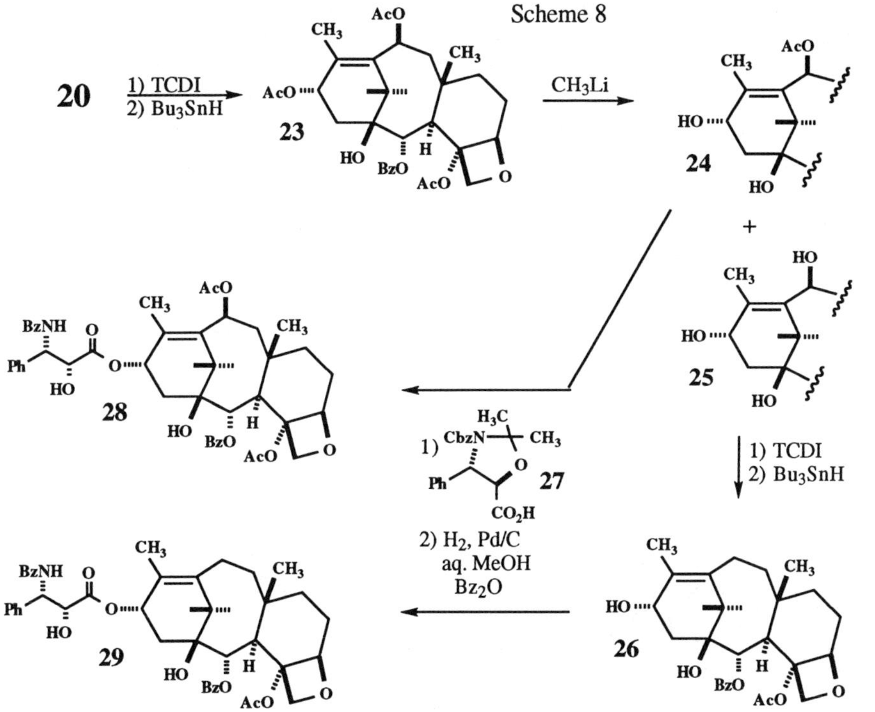

The *in vitro* results for 9-deoxo- **22**, 7-deoxy-9-deoxo- **28**, and 10-desacetoxy-7-deoxy-9-deoxotaxol **29** (Table V) clearly show that removal of the functionality at the C-9 or both the C-7 and C-9 positions has no affect on activity. Not until all three oxygen functionalities are excised is a diminishment of *in vitro* activity observed, and in this case a ten-fold loss of activity was measured for **29**. Nevertheless, even this trioxygenated compound retains significant cytotoxicity thereby delineating the relative unimportance of these functionalities for the antitumor activity of taxol.

Table V. Tumor Cell Cytotoxicity of 9-Deoxo Analogs

	Tumor cell lines, IC_{50} (ng/mL)			
Compound	*A549*	*HT-29*	*B16F10*	*P388*
Taxol	2.5-4.3	1.8-3.5	3.4-6.3	8.8-11
9(R)-Dihydrotaxol (12)	16-22	6.4-9.6	25	49-57
9-Deoxotaxol (22)	3.0	2.2	1.8	5.5
28	4.5	4.5	4.7	9.0
29	33	30	31	57

Attempted deoxygenation of the only remaining free hydroxyl group at C-1 led to a another novel rearrangement product, **30** (Scheme 9). Treatment of the acetonide derivative **2** with lithium hexamethyldisilazide followed by carbon disulfide and methyl iodide afforded a new xanthate, which NMR heteronuclear coupling studies showed to be **31**, the product resulting from xanthylation at C-2; therefore, the xanthylation had been preceded by a migration of the C-2 benzoate to the C-1 hydroxyl. Reduction of xanthate **31** led to a rearrangement of the skeleton to give novel compound **30**, for which a mechanistic explanation has been proposed. Formation of the C-2 radical **A** may lead to transannular attack of the C-11,12 double bond giving cyclobutane ring **B** which can then open to form a tertiary radical. Subsequent loss of the C-1 benzoyloxy radical leads to this novel ring system.

Scheme 9

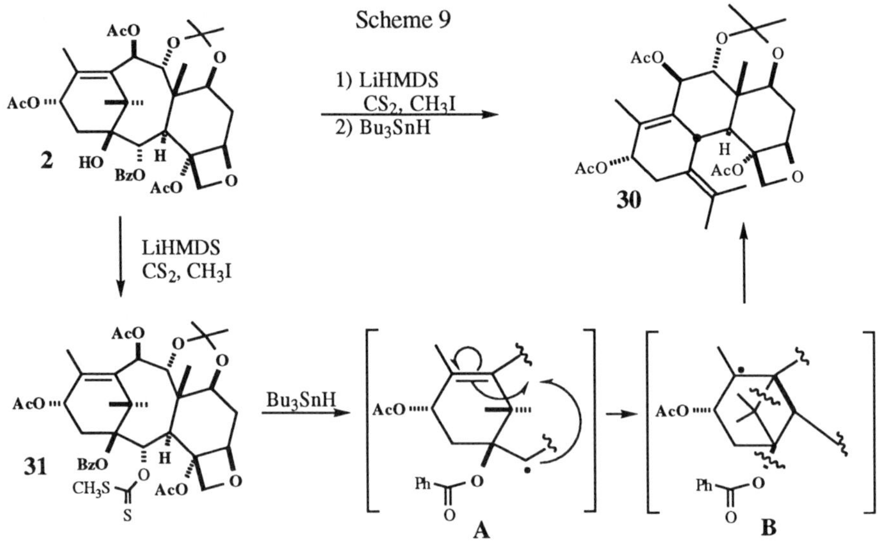

Acknowledgments

We are grateful to the isolation work by Dr. G. Gunawardana and M. O'Beirne. The NMR studies were performed by Dr.D. Whittern and X-ray analysis by R. Henry. Solubility measurements were performed by Dr. S. Krill, L. Cammack, and Dr. K. Garren. Activity studies were performed by D. Balli (in vitro), Dr. J. Alder(in vivo), Dr. R. Himes (U. of Kansas, microtubule assembly). Pharmacokinetic studies were performed by Dr. K. Marsh and E. McDonald. We also appreciate the helpful discussions with Dr. P. Lartey.

Literature Cited

1. Gunawardana, G.P.; Premachandran, U.; Burres, N.S.; Whittern, D.N.; Henry, R.; Spanton, S.; McAlpine, J.B. *J. Nat. Products*, **1992**, *55*, 1686-1689.
2. Zamir, L.O.; Nedea, M.E.; Belair, S.; Sauriol, F.; Mamer, O.; Jacqmain, E.; Jean, F.I.; Garneau, F.X. *Tetrahedron Lett.* **1992**, *33*, 5173-5176.
3. Zhang, S.; Chen, W.M.; Chen, Y.H. *Acta Pharm. Sinica*, **1992**, *27*, 268-272.
4. Martell, A.M. In *Shrubs and Vines for Northeastern Wildlife;* Gill, J.D. and Healy, W.M., Ed.; USDA Forest Service General Technical Report NE-9, Northeastern Forest Experiment Station, Upper Darby, PA, 1974, pp. 158-160.

5. Miller, R.W.; Powell, R.GZ.; Smith, C.R.; Jr. Arnold, R.; Clardy, J. *J. Org. Chem.* **1981**, *43*, 1469-1474.
6. Farina, V.; Huang, S. *Tetrahedron. Lett.* **1992**, *34*, 3979-3982.
7. Guenard, D.; Gueritte-Vogelein, F.; Wahl, A.; Potier, P.; Chemistry on 10-deacetylbaccatin III: Synthesis of new structural analogues of taxol and taxotere, 203rd ACS Meeting, San Francisco, CA, 1992; MEDI 033; Kingston, D.G.I. Studies on the Chemistry of Taxol, 203rd ACS Meeting, San Francisco, CA, 1992; MEDI 193.
8. Klein, L.L. *Tetrahedron Lett.*, **1993**, *34*, 2047-2050.
9. Ojima, I.; Habus, I.; Zhao, M.; Georg, G.I.; Jayasinghe, L.R. *J. Org. Chem.* **1991**, *56*, 1681-1683.
10. Ojima, I.; Zucco, M.; Duclos, O.; Kuduk, S.D.; Sun, C.M.; Park, Y.H. *Biorg. Med. Chem. Lett.* **1993**, *3*, 2479-2482.
11. Algaier, J.; Himes, R.H. *Biochim. Biophys. Acta*, **1988**, *954*, 235-243.
12. Mosmann, T.; *J. Immunol. Meth.*, **1983**, *65*, 55.
13. Maring, C.J.; Grampovnik, D.J.; Yeung, C.M.; Klein, L.L.; Li, L.; Thomas, S.A.; Plattner, J.J. *Bioorg. Med. Chem. Lett.* **1994**, *X*, X-X.
14. Georg, G.; Cheruvallath, Z.S.; Himes, R.H.; Mejillano, M.R. *Bioorg. Med. Chem. Lett.* **1992**, *2*, 1751-1754.
15. Li, L.; Thomas, S.A.; Klein, L.L.; Yeung, C.M.; Maring, C.J.; Grampovnik, D.J.; Plattner, J.J. *J. Med. Chem.* **1994**, *X*, X-X.
16. Klein, L.L.; Maring, C.J.; Li, L.; Yeung, C.M.; Thomas, S.A.; Grampovnik, D.J.; Plattner, J.J. *J. Org. Chem.* **1994**, *59*, 2370-2373.
17. Chen, S-H; Huang, S.; Wei, J.-M.; Farina, V. *J. Org. Chem.*, **1993**, *58*, 4520-4521.
18. Klein, L.L.; Yeung, C.M.; Li, L.; Plattner, J.J.*Tetrahedron Lett.*. **1994**, *X*, X-X.
19. Commercon, A.; Bourzat, J.D. *Tetrahedron Lett.* **1993**, *34*, 6049-6052.

RECEIVED August 8, 1994

Chapter 21

The Total Synthesis of Paclitaxel Starting with Camphor

Robert A. Holton, Carmen Somoza, Hyeong-Baik Kim, Feng Liang, Ronald J. Biediger, P. Douglas Boatman, Mitsuru Shindo, Chase C. Smith, Soekchan Kim, Hossain Nadizadeh, Yukio Suzuki, Chunlin Tao, Phong Vu, Suhan Tang, Pingsheng Zhang, Krishna K. Murthi, Lisa N. Gentile, and Jyanwei H. Liu

Dittmer Laboratory of Chemistry, Florida State University, Tallahassee, FL 32306

Abstract: The first total synthesis of the potent anti-tumor agent taxol is described. The synthesis proceeds from the commodity chemical camphor, which is available in either enantiomeric form, and builds the taxol molecule in a linear fashion. An important feature of the synthetic route is the utilization of conformational control to enable the functionalization of the eight membered B ring of taxol.

The total synthesis of the potent anti-tumor[1] agent taxol (**1**), isolated by Wall and Wani in 1971,[2] has stood for over twenty years as a major challenge for organic chemists. Taxol has been the subject of extensive chemical and biological studies, which have recently been summarized in several reviews,[3] and many synthetic approaches have been described.[3e,4]

Until now, our taxane research program has produced a synthesis of the taxane ring system,[5] a total synthesis of taxusin,[6a] and a (now commercialized) semisynthesis of taxol.[7] Here we describe the first total synthesis of taxol.[8a,b]

The facile epimerization of taxol at C-7[9] is well documented,[3e] and has been postulated to occur via a retroaldol-aldol process. We chose to pursue a synthetic strategy in which this stereocenter would be introduced at an early stage and carried throughout most of the synthesis in the absence of a C-9 carbonyl group, thereby avoiding the C-7 epimerization complication. Thus, our route to taxol proceeds retrosynthetically through C-7 protected baccatin III (**2**) to the tricyclic ketone **3**, which arises from C ring closure of a precursor **4**, properly functionalized at C-1, C-2, C-3, C-7, and C-8. Synthesis of this precursor would be made possible by conformational control of the eight membered B ring, via ketone **5**. Ketone **5** was projected to arise from an aldol condensation of the enolate of ketone **6**, the formation of which would be made possible by the conformational control exerted by the C-10α substituent, with an aldehyde carrying the carbons necessary for the eventual construction of the taxane C ring.

The synthesis begins with camphor, which is readily available in either enantiomeric form. Following the route provided by Buchi,[10] camphor can be converted to β-patchouline and its epoxide, commercially available[11] under the trade name "Patchino." Our synthetic studies had heretofore utilized the commercially available enantiomer of Patchino which leads ultimately to the unnatural enantiomers of

NOTE: Paclitaxel is the generic name for Taxol, which is now a registered trademark.

0097–6156/95/0583–0288$08.00/0

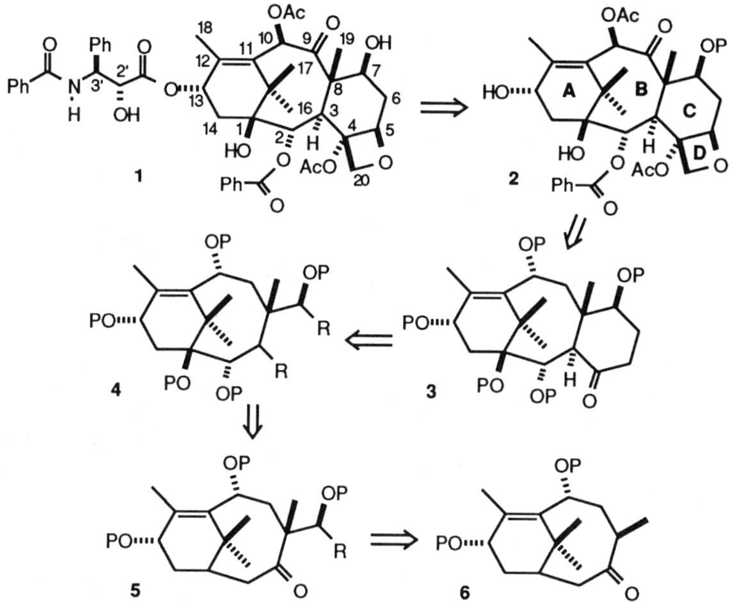

the taxane natural products. To provide access to material in the enantiomeric series leading to natural taxanes, we have modified the synthesis reported by Buchi so that Patchino can be prepared in multigram quantities from camphor without the need for chromatographic separation at any stage.

Patchino reacted, as reported by Buchi,[10] in the presence of boron trifluoride etherate to give the rearranged homoallylic tertiary alcohol **7**. This set the stage for discovery of the fragmentation of bicyclic epoxy alcohols pioneered in our laboratory nine years ago,[5,12] known as the "epoxy alcohol fragmentation," and the cornerstone of our syntheses of the taxane skeleton, taxusin, and now taxol. The epoxy alcohol fragmentation has enabled the synthesis of a variety of molecules having different substitution patterns on the bicyclo [5.3.1] skeleton, e.g., ketone **8**, generated in quantitative yield upon Sharpless epoxidation of alcohol **7**.

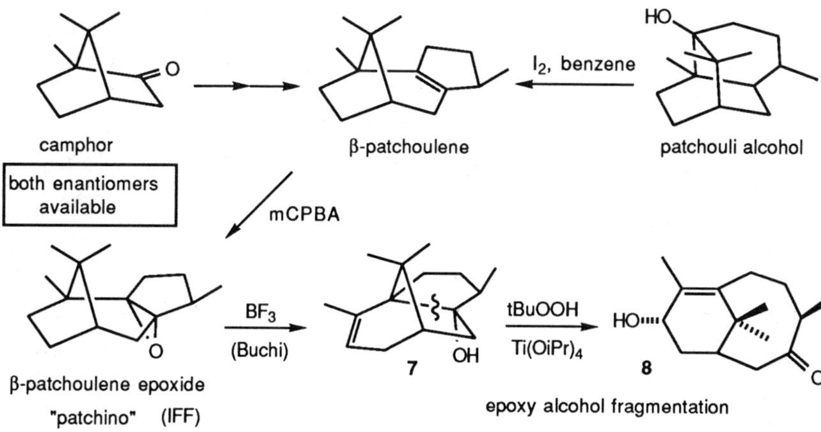

Scheme I

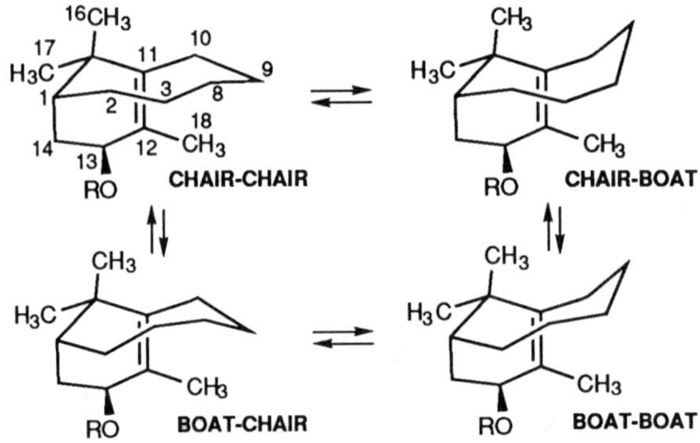

Spectroscopic studies[13] of compounds having the the bicyclo [5.3.1] skeleton have demonstrated that there are *four distinct conformations of this eight membered ring*, as shown in Scheme 1. For a given compound the equilibrium will shift to favor conformations which orient substituents toward the periphery of the B ring to minimize non-bonded interactions.

The combination of a C-10β alkoxyl group, a C-8β methyl group, and a C-3 ketone in this ring system shifts the equilibrium to strongly favor the chair-boat conformation. For example, ketone **8a**[13] was found to exist exclusively in the chair-boat conformation, and calculations[14] indicate that, in the case of **8a**, this conformation is favored by ca. 7 kcal/mol. This ketone was projected to be an intermediate in the synthesis of taxusin and other taxanes until we discovered that it does not undergo deprotonation at C-8α, presumably because of the approximately eclipsing alignment of the C-8α hydrogen and the C-3 carbonyl group. Therefore, to enable C-8α deprotonation (and subsequent aldol condensation), we chose to utilize a C-10α silyloxy substituent as a conformational control element.

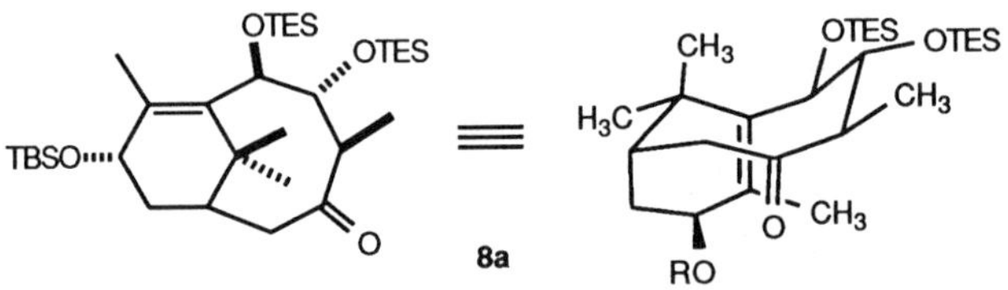

8a

As in the synthesis of taxusin,[6] patchino was converted to epoxide **9** in 98% yield, and **9** then rearranged under very specific conditions to provide diol **10** in 93% yield at 75% conversion. Silylation (TESCl, pyridine) of **10** was followed by epoxy alcohol fragmentation and protection at C-13 to give **11** in 93% overall yield. Although **11** was found to be in the chair-boat conformation, calculations indicate that, while the chair-boat conformer is lowest in energy, the chair-chair and boat-chair conformers are only ca. 2.5 kcal/mol less stable.[14] Presumably deprotonation of one of these other conformers at C-8 is facile.

The magnesium enolate[15] of ketone **11** ($HN(iPr)_2$, THF, MeMgBr, 25 °C, 3 h, then **11**, 1.5 h) underwent aldol condensation with 4-pentenal (THF, -23 °C, 1.5 h), and the crude product was directly protected (Cl_2CO, pyridine, CH_2Cl_2, -10 °C, 0.5 h, then ethanol, 0.5 h) to give ethyl carbonate **12**, a ca. 6:1 mixture of chair-chair and boat-chair conformers ($CDCl_3$),[16] slowly interconverting on the NMR time scale, in 75% yield. Hydroxylation at C-2 (**12**, LDA, THF, -35 °C, 0.5 h, then -78 °C, 1.0 mol equiv of (+)-camphorsulfonyl oxaziridine (for the enantiomer leading to taxol; (-)-camphorsulfonyl oxaziridine for the enantiomer leading to *ent*-taxol), 0.5 h)[17] gave

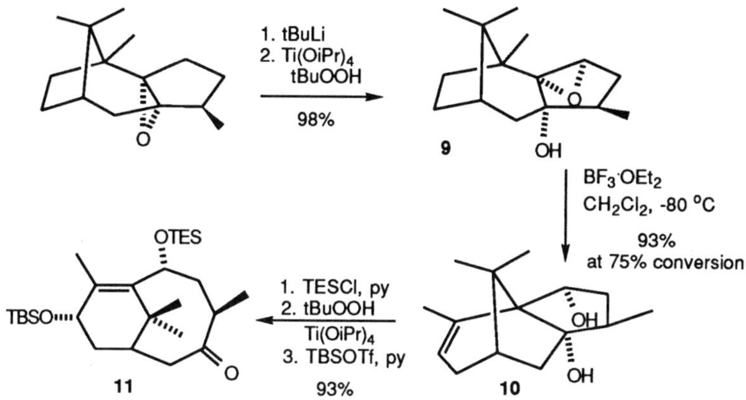

hydroxy carbonate **13** (chair-chair conformation) in 85% yield. Reduction of **13** from the periphery of the molecule[18] (20 mol equiv of RedAl, toluene, -78 °C, 6 h, then warm to 25 °C over 6 h) gave a triol which, without isolation, was converted to carbonate **14** (Cl_2CO, pyridine, CH_2Cl_2, -78 to 25 °C, 1 h, 97%). Carbonate **14** could be obtained directly from RedAl reduction of **13**, but complete reduction followed by regeneration of the cyclic carbonate was operationally easier and more efficient.

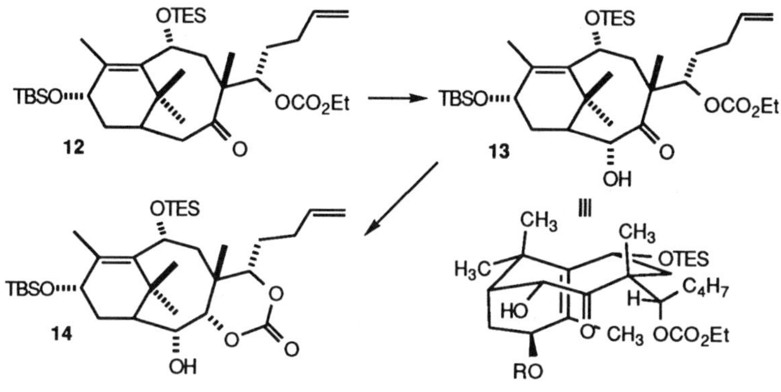

Our strategy for Synthesis of the C-1 through C-3 portion of taxol is illustrated in Scheme 2. Introduction of a second conformational control element, a sufficiently large epimerizable substituent at C-3α, would shift the equilibrium in favor of the boat-chair conformation. This conformation was expected to permit generation of the C-1, C-2 enolate of a C-2 ketone, which would undergo hydroxylation at C-1 followed by hydride reduction of the C-2 carbonyl group from the periphery to generate the C-2α alcohol. Finally, epimerization at C-3, facilitated by the hindered environment of the C-2α hydroxyl group, would return the B ring to the chair-chair conformation.

Thus, **14** underwent Swern oxidation to give C-2 ketone **15** in 95% yield. That **15** was still in the chair-chair conformation (apparently the C-3α oxygen substituent is not bulky enough to shift the equilibrium to favor the boat-chair conformation) was a matter of some concern. Treatment of **15** with 1.05 mol equiv of LTMP from -25 to -10 °C gave hydroxy lactone **16** in 90% yield. This remarkable result is analogous to the Chan rearrangement,[19] which, to our knowledge, has been used but once in

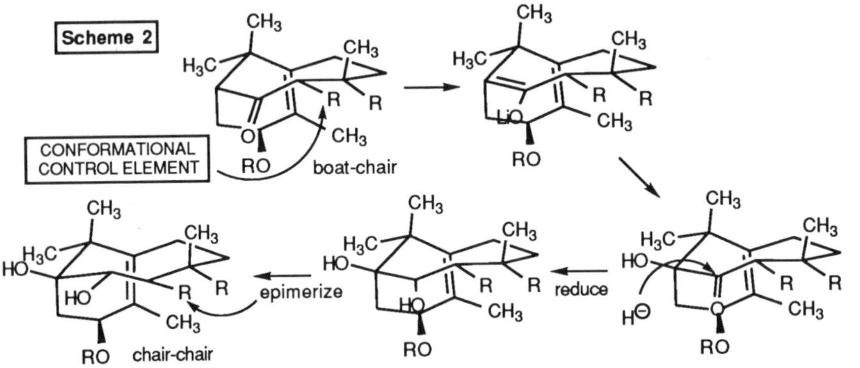

synthesis.[20] The formation of **16** is the first example of this reaction in a cyclic system, and this is also the first indication that the Chan rearrangement can, under some circumstances, be very stereoselective. We are further investigating this rearrangement.

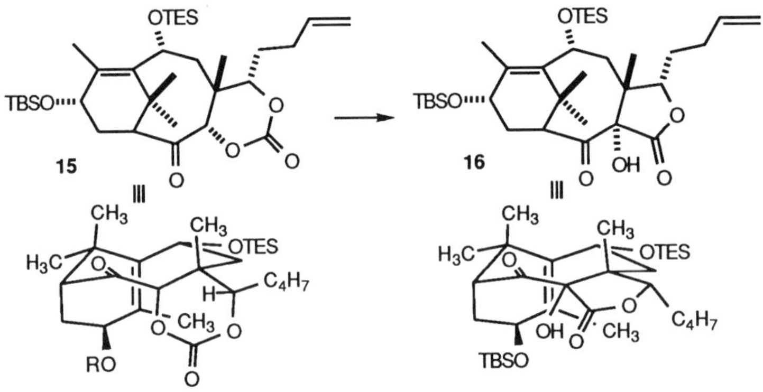

The chair-chair conformation of **16** aligns the C-3α hydroxyl group for facile reductive removal, and its samarium diiodide reduction led to the stable enol **17**, which, upon treatment with silica gel, was converted to a 6:1 mixture of cis- and trans-fused lactones **18**, from which the cis-fused lactone **18c** (boat-chair conformation) could be obtained by crystallization. Treatment of the trans-fused lactone **18t** with KOtBu in THF followed by quenching with acetic acid gave back **17**, and through this recycling **18c** was obtained in 91% yield from **16**. Attempts to generate and hydroxylate a dienolate from **17** were unsuccessful. Lactone **18t** was not deprotonated by LTMP at temperatures up to -10 °C, and was recovered unchanged. However, treatment of **18c** with 4 mol equiv of LTMP at -10 °C followed by addition of (±)-camphorsulfonyl oxaziridine (5 mol equiv) to the enolate at -40 °C gave 88% of **19c** along with 8% of its trans-fused isomer **19t**, which was formed upon chromatographic separation of the small amount (3%) of unreacted **18c**. It is remarkable that *deprotonation of* ***18c*** *with LTMP apparently occurs first, and perhaps only, at C-1*, even though the C-3 proton would normally be expected to be more acidic. Reduction of **19c** with RedAl[18] (THF, -78 °C, 1.5 h) followed by a basic work-up gave C-2α-hydroxy trans-fused lactone (88%) and 4% of **19t**, which could be converted to the C-2α-hydroxy trans-fused lactone almost quantitatively by samarium diiodide reduction.[21] The C-2α-hydroxy trans-fused lactone was quantitatively converted to carbonate **20**[22] by treatment with

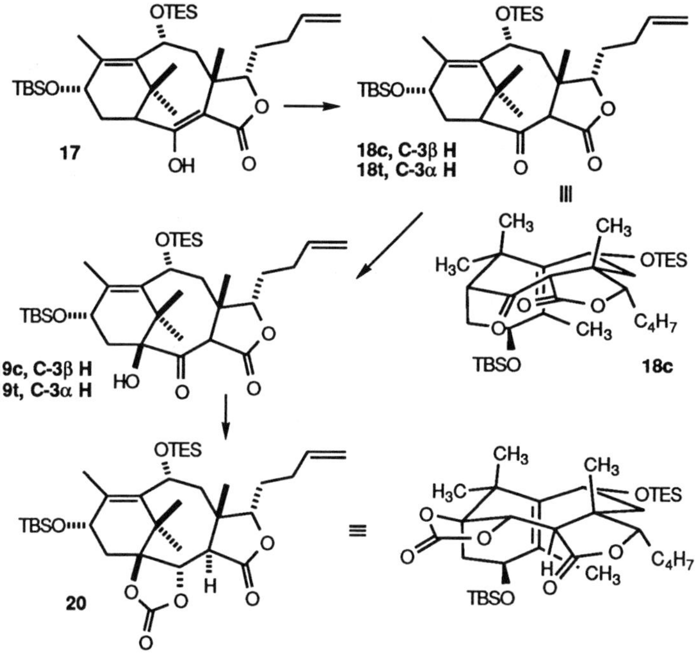

phosgene (10 mol equiv, pyridine, CH_2Cl_2, -23 °C, 0.5 h).

Therefore, as outlined here, patchino can be transformed to lactone carbonate **20** in fifteen steps and 36% overall yield. This synthetic sequence provides functionality at C-1, C-2, C-3, C-7, and C-8 as needed for a synthesis of taxol through careful conformational control of the bicyclo [5.3.1] eight membered ring. Conversion of **20** to taxol requires cyclization of the C ring, introduction of the oxetane D ring, and, finally, oxidation at C-9 followed by adjustment of the regio- and stereochemistry at C-9 and C-10.

Oxidative cleavage of the terminal olefin of **20** was carried out by ozonolysis in

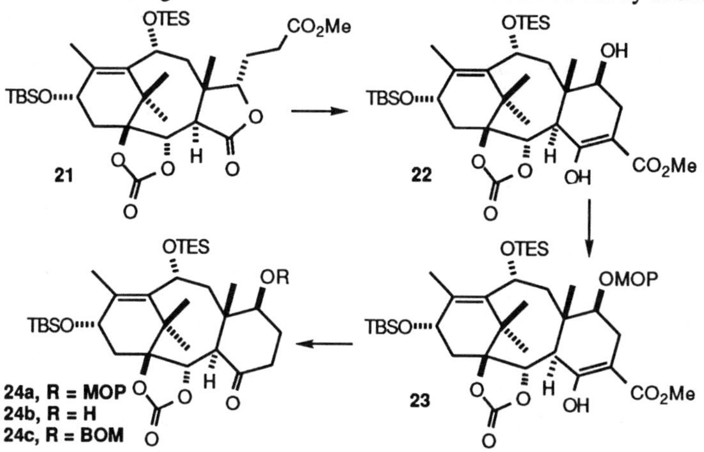

the presence of methanol and sodium hydroxide,[23] or, in somewhat higher yield (93%),by first ozonolysis to the aldehyde followed by oxidation to the acid ($KMnO_4$, KH_2PO_4)[24] and esterification with diazomethane to produce methyl ester **21**. Dieckman cyclization of **21**, following the protocol recently developed in our laboratory (LDA, THF, -78 °C, 0.5 h, then HOAc, THF),[25] gave the enol ester **22** in 93% yield at 90% conversion. Attempted decarbomethoxylation of **22** resulted in at least partial reversion to **21**, and **22** was therefore temporarily protected (pTsOH, 2-methoxypropene, 100%) to give **23**, which smoothly underwent decarbomethoxylation (PhSK, DMF, 86 °C, 3 h) to provide **24a**, or, if an acidic work up was employed, hydroxy ketone **24b**, in 92% yield.

To secure completely unambiguous structure confirmation, **24a** was converted to diol carbonate **27**, which was independently synthesized from baccatin III (**2**, P=H). Selective deprotection of **24a** with TBAF (1 mol equiv, THF, -10 °C, 6 h) gave alcohol **25**, which was then oxidized with tetrapropylammonium perruthenate[26] (TPAP(cat), NMO, molecular sieves, CH_2Cl_2, 25 °C, 1.5 h) to give ketone **26** in 86% overall yield from **24a**. Deprotection (HF, pyridine, CH_3CN) of **26** provided diol carbonate **27** in 96% yield.

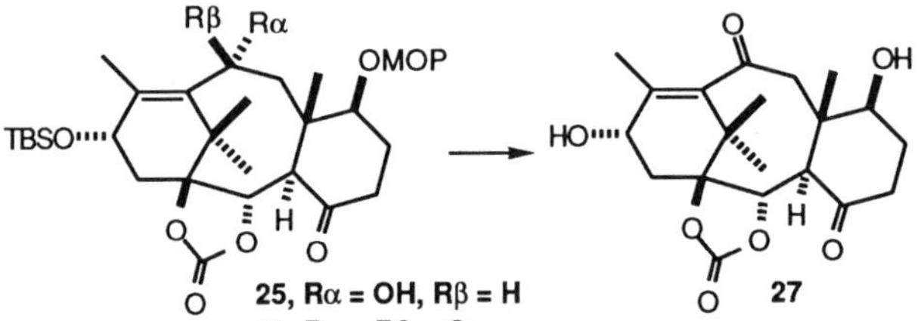

25, Rα = OH, Rβ = H
26, Rα = Rβ = O
27

The synthesis of **27** from baccatin III proceeded as follows.[27] Reduction of baccatin III with tetrabutylammonium borohydride (CH_2Cl_2, 25 °C, 60 h) provided the C-9β hydroxy derivative in 70% yield, and protection (TESCl, pyridine, 25 °C, 20 h) then gave the 7,13-(bis)-TES derivative **28** in 94% yield. Treatment of **28** with KH (THF, 25 °C, 1 h) followed by an acetic acid quench yielded a substance in which the acetyl group had migrated from C-10 to C-9, and subsequent oxidation (TPAP, NMO, CH_2Cl_2, 25 °C, 1 h) led to the C-10 ketone **29** in 66% yield from **28**. Reductive removal of the C-9 acetate (SmI_2, THF, 0 °C, 1 h) proceeded in 83% yield, and the C-13 TES group was then selectively removed (TBAF, THF, -10 °C, 10 h) to provide **30** in 85% yield. Hydroxy ketone **30** quantitatively underwent selective cleavage of the C-4 acetate with super-hydride (THF, -78 °C, 1 h) to give **31**, which underwent solvolysis of the C-2 benzoate (NaOMe), with formation of varying amounts of the C-2, C-20 tetrahydrofuran.[28] The resulting tetraol was converted to the C-1, C-2 cyclic carbonate[22] (Cl_2CO, pyridine, CH_2Cl_2, -78 to -10 °C, 1 h) **32** in 70% yield from **31**. Oxetane cleavage with TMSCl[29] (CH_2Cl_2, 25 °C, 1 h) provided C-5α-chloro triol **33**, which was converted to keto carbonate **34** with lead tetraacetate[30] (benzene, 25 °C, 10 min) in 90% yield from **32**. Finally, **34** underwent samarium diiodide reduction (THF, -78 °C, 5 min) followed by removal of the C-7 TES protecting group (HF, pyridine, acetonitrile, 25 °C, 3 h) to give **27**, identical with the material prepared from **24a**, in 70% yield from **34**. With the structure of **24** securely established, the remaining issues of the conversion of **24** to taxol were addressed.

We expected synthesis of the acetoxy oxetane to be difficult, and our expectations were fulfilled. We had attempted several non-traditional approaches to this functional array, and these had not been completely successful. In order to bring the total synthesis to fruition, we decided to pursue synthesis of the oxetane through

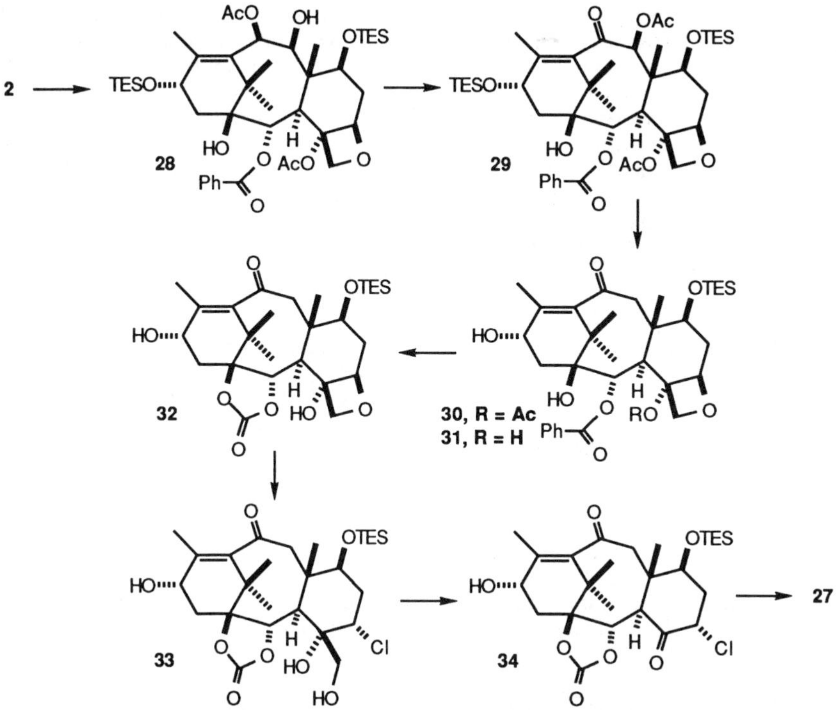

utilization of a nucleophilic displacement. Although several recent reports described successful oxetane syntheses in model systems, we undertook a model study to determine whether it would be more efficient to cyclize the oxetane by formation of the C-5-oxygen bond or by formation of the C-20-oxygen bond. To answer this question the simple model compounds **35** and **37** were prepared by a straightforward route that will be described elsewhere. The cyclization of **35** was attempted under a wide variety of conditions, and, after numerous tries, the maximum yield of oxetane **36** obtained was about 30%, and it was accompanied by a number of byproducts. On the other

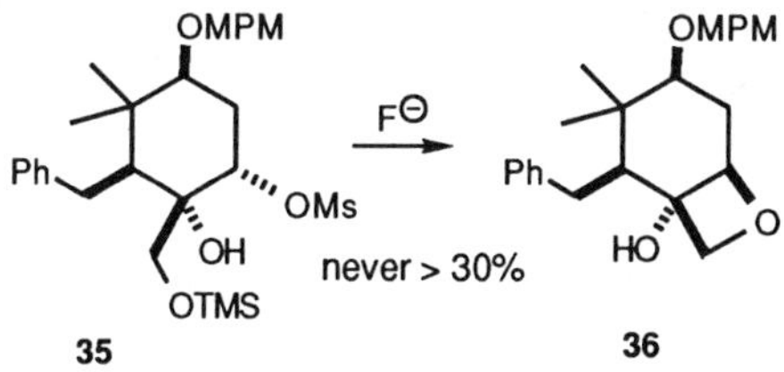

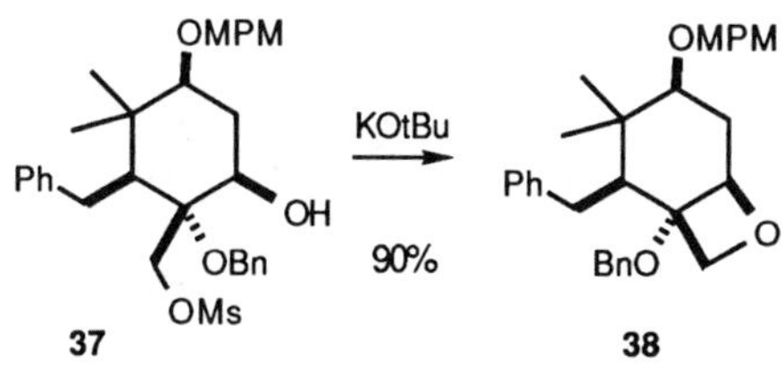

hand, **37**, in which the tertiary hydroxyl group was protected as its benzyl ether to prevent epoxide formation, cyclized readily to provide oxetane **38** in high yield.

Therefore, we set out to complete the synthesis of taxol by conversion of **24a** to the corresponding C-5β alcohol having a C-20 mesylate, and cyclization of the oxetane as in the conversion of **37** to **38**. However, the C-4 carbonyl group of **24** was found to be very hindered, and addition of C-20 to it in the presence of the carbonate failed with most nucleophilic reagents. For example, Wittig reagents completely failed to react with **24a**. Although the enol triflate could be prepared from **24a** in high yield, it failed to undergo oxidative addition to a transition metal (e.g., Pd (0)). Furthermore, all reagents that successfully underwent addition to C-4 did so from the α face, thus requiring introduction of C-20 in the β configuration by indirect means.

The C-4, C-5 enolate derived from ketone **24a** reacted with camphorsulfonyl oxaziridine to give predominantly the C-5β alcohol, which was temporarily protected as the TMS ether. Addition of methylmagnesium bromide occurred from the α face, and the TMS group was removed in an aqueous workup to give diol **39** in approximately 80% overall yield from **24a**. At this point, the C-7 MOP protecting group became extremely labile, and great care was required to prevent its partial loss in subsequent operations. The C-5 hydroxyl group was acetylated, and elimination of the tertiary hydroxyl group could be brought about through the action of thionyl chloride in pyridine to give allylic acetate **40** in about 55% yield from **39**, along with significant quanties of the isomeric C-4, C-5 olefin. Osmylation of **40** occurred from the α face to provide diol **41a** in about 90% yield. Several different routes were pursued with the intent of protecting the C-4α hydroxyl group of **41a**, and it soon became evident that, while this could be accomplished, it would require a lengthy and laborious effort. The C-20 mesylate **41b** could readily be prepared from **41a**, and as expected, it cyclized to

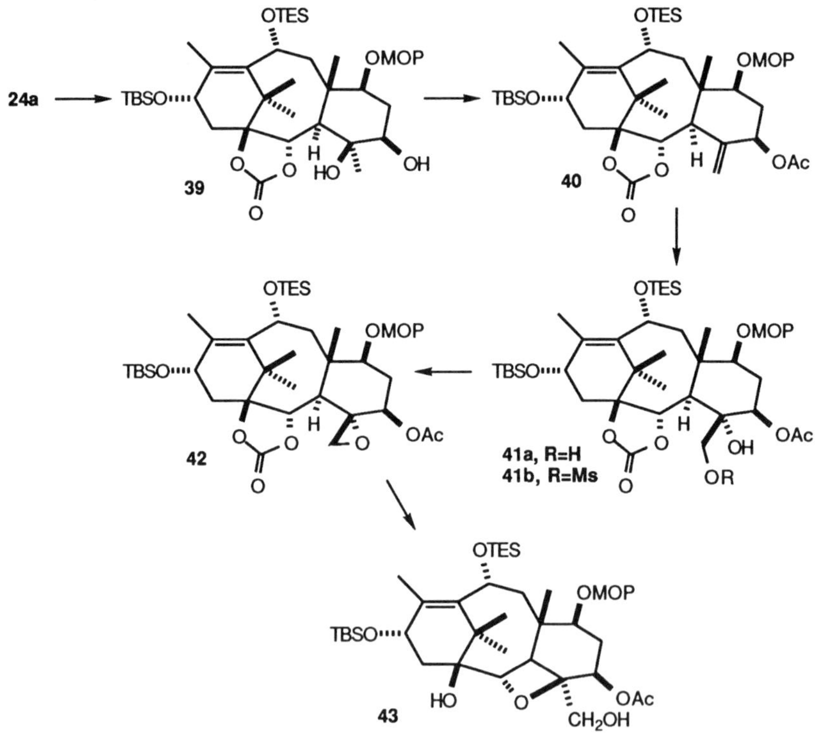

the epoxide **42** easily, even under relatively vigorous acetylation conditions. Epoxide **42** was refractory to intermolecular opening, and in one such attempt, treatment of it with lithium methyl mercaptide in DMF produced the C-2, C-4 oxetane **43** in high yield. The formation of **43**, presumably by carbonate opening followed by nucleophilic opening of the epoxide by the C-2 alkoxide, is one of the many examples of unwanted rearrangement encountered during the course of this work.

While this effort was progressing, it was discovered that, similar to the conversion of **32** to **33**, carbonate **45** was converted by trimethylsilyl bromide quantitatively to diol bromide **44**. Treatment of **44** with DBU in toluene at reflux or with diisopropyl ethyl amine at reflux produced oxetanol **45** in over 90% yield.[29a] Thus it appeared that we had been mislead by the simple model study described above, and that the alternate mode (i.e., C-20 alkoxide displacement of a C-5α leaving group) of oxetane cyclization would be effective.

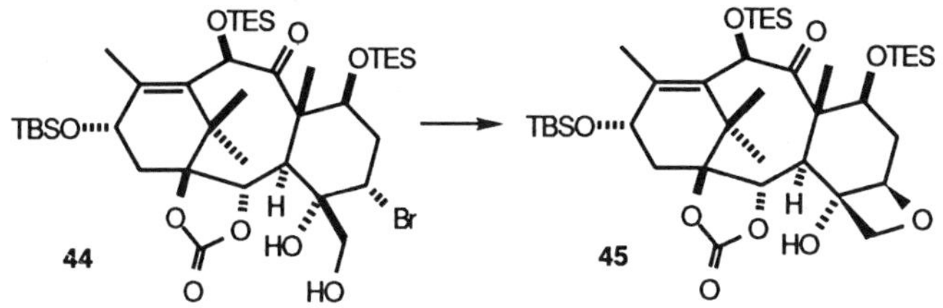

44 45

Since we had encountered difficulty in retaining the MOP group, a robust protecting group at C-7 which would survive the remainder of the synthesis was selected, and **24b** reacted with BOM chloride ($EtN(iPr)_2$, CH_2Cl_2, $(Bu)_4NI$, reflux, 32 h) to give C-7 BOM derivative **24c** in 92% yield. Remarkably, the TMS enol ether of **24c** (LDA, THF, TMSCl, -78 °C) underwent oxidation with mCPBA in hexane (25 °C, 5 h) to stereoselectively provide C-5α trimethylsilyloxy ketone **46** in 86% yield at 86% conversion. When the mCPBA oxidation was conducted in methylene chloride solution little stereoselectivity was observed. Addition of methylmagnesium bromide to **46**, a low yield (30-60%) reaction in ethereal or hydrocarbon solvents, proceeded well in methylene chloride solution (-45 °C, 15 h, 10 mol equiv of MeMgBr) to give tertiary alcohol **47** in 95% yield. Elimination of **47** was carried out with Burgess' reagent,[31] and acidic work up then provided allylic alcohol **48a** in 63% yield.

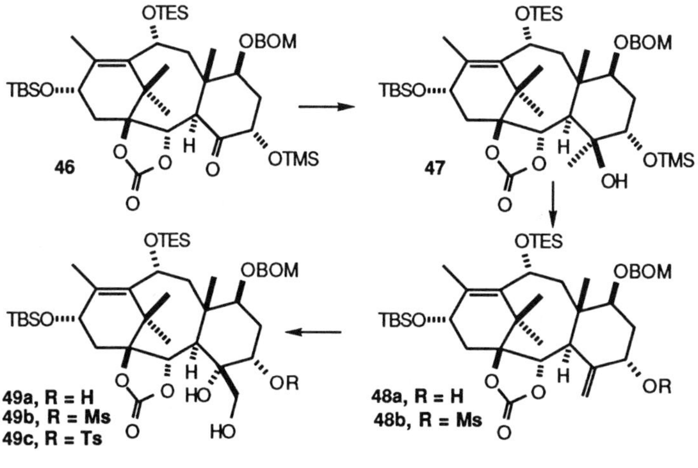

46 47

49a, R = H
49b, R = Ms
49c, R = Ts

48a, R = H
48b, R = Ms

Alcohol **48a** could be converted to either C-5α mesylate **49b** or C-5α tosylate **49c**. Mesylate **48b** was generated quantitatively (MsCl, pyridine) from alcohol **48a**, and osmylation of **48b** gave **49b** in 60-65% yield. Alternatively, osmylation (ether/pyridine, 0 °C, 12 h) of **48a** gave triol **49a** in 80% yield at 91% conversion. Triol **49a** was converted to tosylate **49c** through temporary protection of the C-20 hydroxyl group as the TMS ether (TMSCl, Et_3N, -78 °C), formation of C-5α tosylate and cleavage of the temporary TMS protecting group (LDA, TsCl, -35 °C, 3 h, then HOAc, 0 °C, 14 h) in 85% overall yield at 94% conversion without isolation of intermediates.

Either **49b** or **49c** underwent cyclization[29a,32] to oxetanol **50** (DBU, toluene, 105 °C, 2 h) in 80-85% yield. Acetylation of **50** (Ac_2O, pyridine, DMAP, 24 h, 25 °C) was difficult and proceeded in only 70-75% yield, but was followed by quantitative removal of the C-10 TES group (HF pyridine complex, CH_3CN, 0 °C, 11 h) to give **51**. Addition of phenyllithium[33] (2.1 mol equiv, THF, -78 °C, 10 min) to **51** to provide the C-2 benzoate was followed by TPAP oxidation (NMO, molecular sieves, CH_2Cl_2, 25 °C, 15 min), giving ketone **52** in 85% yield.

Oxidation at C-9 and rearrangement to the taxol C-9, C-10 regio- and stereochemistry, following the protocol established in our previous studies,[27] proceeded nicely. A THF solution of the enolate of **52** (4 mol equiv of KOtBu, THF, -78 to 0 °C, 0.5 h) was added to a suspension of benzeneseleninic anhydride (8 mol equiv, THF, 0 °C, 40 min) and the product was directly treated further with KOtBu (4 mol equiv, THF, -78 °C, 10 min). Direct acetylation of the product (Ac_2O, pyridine, DMAP, 20 h, 25 °C) provided 7-BOM-13-TBS baccatin III (**53a**) quantitatively. As the difficult acetylation of the C-4 hydroxyl group had already shown, the underside of the molecule was quite crowded, and removal of the C-13 TBS group was almost a problem. Treatment of **53a** with TBAF led to cleavage of the C-2 benzoate prior to removal of the TBS group. The TBS group could be removed with HF (pyridine, 90 °C, 12 h) to give **53b** in 65% yield along with several rearrangement products. Finally it was found that with TASF[34] (THF, 25 °C, 1 h) 7-BOM baccatin III (**53b**) was produced in 94% yield. Attachment of the C-13 side chain[7] proceeded uneventfully: the lithium alkoxide[35] of **53b** (LHMDS, THF) was treated with β-lactam **54** (THF, 0 °C, 1 h), the product was

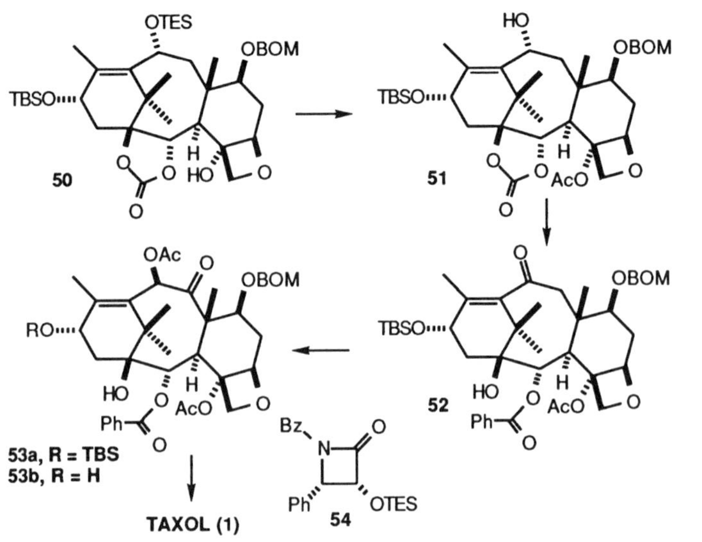

desilylated (HF, pyridine, CH_3CN, 0 °C, 1 h) and, finally, the C-7 BOM group was removed by hydrogenolysis (H_2, Pd/C, EtOH, reflux, 1 h) to give taxol in 93% yield from **53b**.

Herein we have described the first total synthesis of the anti-tumor agent taxol. It illustrates the use of conformational control in synthesis, and has provided a variety of uniquely challenging and interesting situations, many of which are the subject of ongoing investigations in our laboratory. The synthesis produces (-)-taxol[16a] from (-)-borneol, and *ent*-(+)-taxol[16b] from (-)-Patchino.[11] The overall yield of taxol from diol **10** is ca. 4-5%. We continue to search for modifications which will improve the efficiency of this synthesis and also provide a route to new, more active, taxol-type anti-tumor agents.

Acknowledgment: We thank the National Cancer Institute (CA 42031, CA 55131) and private donors to the Taxol Research Fund for financial support of this work.

REFERENCES AND NOTES

1. Reviews: (a) Rowinsky, E. K.; Onetto, N.; Canetta, R. M.; Arbuck, S. G. *Sem. Oncol.*, **1992**, *19*, 646. (b) Holmes, F. A.; Walters, R. S.; Theriault, R. L.; Forman, A. D.; Newton, L. K.; Raber, M. N.; Buzdar, A. U.; Frye, D. K.; Hortobagyi, G. N. *J. Natl. Cancer Inst..*, **1991**, *83*, 1797. (c) Slichenmyer, W. J.; Von Hoff, D. D. *Anti-Cancer Drugs*, **1991**, *2*, 519.
2. Wani, M. C.; Taylor, H. L.; Wall, M. E.; Coggon, P.; McPhail, A. T. *J. Am. Chem. Soc.*, **1971**, *93*, 2325.
3. (a) Suffness, M.; Cordell, G. A. in: *The Alkaloids. Chemistry and Pharmacology*; Brossi, A., Ed.; Academic Press: New York, 1985; vol. 25, p. 3. (b) Kingston, D. G. I.; Samaranayake, G.; Ivey, C. A. *J. Nat. Prod.*, **1990**, *53*, 1. (c) Kingston, D. G. I. *Pharm. Ther.*, **1991**, *52*, 1. (d) Guenard, D.; Gueritte-Voegelein, F.; Potier, P. *Acc. Chem. Res.*, **1993**, *26*, 160. (e) Kingston, D. G. I.; Molinero, A. A.; Rimoldi, J. M. *Prog. Chem. Org. Nat. Prod.*, **1993**, *61*, 1.
4. (a) Swindell, C. S. *Org. Prep. Proced. Int.* **1991**, *23*, 465. (b) For more recent references see: Swindell, C. S.; Chander, M. C.; Heerding, J. M.; Klimko, P. G.; Rahman, L. T.; Raman, J. V.; Venkataraman, H. *Tetrahedron Lett.*, **1993**, *34*, 4921.
5. Holton, R. A. *J. Am. Chem. Soc.*, **1984**, *106*, 5731.
6. (a) Holton, R. A.; Juo, R. R.; Kim, H. B.; Williams, A. D.; Harusawa, S.; Lowenthal, R. E.; Yogai, S. *J. Am. Chem. Soc.*, **1988**, *110*, 6558. (b) Holton, R. A.; Juo, R.-R.; Lowenthal, R. E. U. S. Patent 4,876,399 (**1989**).
7. (a) Holton, R. A. "Advances in Taxane Synthesis" Midwestern Regional ACS Meeting; June, 1989, Cleveland, Ohio. (b) Holton, R. A. "Progress in Taxol Synthesis" National ACS Meeting; April, 1992, San Francisco, Ca. (c) Holton, R. A., U. S. Patent 5,015,744 (**1991**); European Patent 0 400 971 (**1990**); (d) Holton, R. A., U. S. Patent 5,136,060 (**1992**); (e) Holton, R. A., U. S. Patent 5,175,315 (**1992**); (f) Holton, R. A., PCT Patent Application International Publication Number WO 93/06079 (**1993**); (g) Holton, R. A., U. S. Patent 5,229,526 (**1993**); (h) other patents pending. (i) Subsequent to these disclosures similar findings have been reported; see Ojima, I.; Sun, C. M.; Zucco,M.; Park, Y. H.; Duclos, O.; Kuduk, S. *Tetrahedron Lett.*, **1993**, *34*, 4149, and references cited therein. (j) For an alternate approach, see Commercon, A.; Bezard, D.; Bernard, F.; Bourzat, J. D. *Tetrahedron Lett.*, **1992**, *33*, 5185, and references contained therein.
8. a) Holton, R.A.; Somoza, C.; Kim, H.B.; Liang, F.; Biediger, R.J.; Boatman, P.D.; Shindo, M.; Smith, C.C.; Kim, S.; Nadizadeh, H.; Suzuki, Y.; Tao, C.; Vu,

P.; Tang, S.; Zhang, P.; Murthi, K.K.; Gentile, L.N.; Liu, J.H. *J. Am. Chem. Soc.*, **1994**, *116*, 1597. b) Holton, R.A.; Kim, H.B.; Somoza, C.; Liang, F.; Biediger, R.J.; Boatman, P.D.; Shindo, M.; Smith, C.C.; Kim, S.; Nadizadeh, H.; Suzuki, Y.; Tao, C.; Vu, P.; Tang, S.; Zhang, P.; Murthi, K.K.; Gentile, L.N.; Liu, J.H. *J. Am. Chem. Soc.*, **1994**, *116*, 1599. c) After our total synthesis was completed, a second total synthesis was reported; see: Nicolaou, K.C.; Yang, Z.; Liu, J.J.; Ueno, H.; Nantermet, P.G.; Guy, R.K.; Claiborne, C.F.; Renaud, J.; Couladouros, E.A.; Paulvannan, K.; Sorensen, E.J. *Nature*, **1994**, *367*, 630.

9. The taxane numbering system, as illustrated in (**1**), is used throughout.
10. Buchi, G.; MacLeod, Jr., W. D.; Padilla, J. *J. Am. Chem. Soc.* **1964**, *86*, 4438.
11. Patchino is a product of International Flavors and Fragrances, Inc. We are grateful to Dr. William Schrieber for bringing this fact to our attention and for providing us with a generous supply of Patchino.
12. (a) For application in the bicyclo [3.2.1] system, see reference 5. (b) For application in the bicyclo [2.2.1] system, see Holton, R. A.; Kennedy, R. M. *Tetrahedron Lett.*, **1984**, *25*, 4455. (c) For application in the bicyclo [3.1.1] system, see Wender, P. A.; Mucciaro, T. P. *J. Am. Chem. Soc.*, **1992**, *114*, 5878.
13. Solution conformations have been determined by ^{1}H nOe difference spectroscopy at 500 MHz: Holton, R. A.; Somoza, C.; Kim, H. B.; Juo, R. R.; Williams, A. D.; Harusawa, S.; Takemoto, Y.; Smith, C. C.; Gentile, L. N.; Liang, F. manuscript in preparation.
14. Calculations were performed using Sybyl molecular modeling software and the Tripos force field obtained from Tripos Associates, St. Louis, Missouri; see Clark, M.; Cramer III, R. D.; Opdenbosch, N. V. *J. Comp. Chem.* **1989**, *10*, 982.
15. Krafft, M. E.; Holton, R. A. *Tetrahedron Lett.*, **1983**, *24*, 1345.
16. The ratio of conformers depended on the nature of both the C-7 protecting group and the alkyl side chain.
17. (a) Davis, F. A.; Vishwakarma, L. C.; Billmers, J. M.; Finn, J. *J. Org. Chem.*, **1984**, *49*, 3241. (b) Davis, F. A.; Haque, M. S.; Ulatowski, T. G.; Towson, J. C. *J. Org. Chem.*, **1986**, *51*, 2402.
18. For a related example of carbonyl reduction in a taxane ring system, see Shea, K. J.; Higby, R. G.; Gilman, J. W. *Tetrahedron Lett.*, **1990**, *31*, 1221.
19. Lee, S. D.; Chan, T. H.; Kwon, K. S. *Tetrahedron Lett.*, **1984**, *25*, 3399.
20. White, J. D.; Vedananda, T. R.; Kang, M.; Choudhry, S. C. *J. Am. Chem. Soc.*, **1986**, *108*, 8105.
21. For an example of reduction of a remotely related C-2 ketone with sodium, see reference 12c.
22. The C-1, C-2 cyclic carbonate has been found to be an excellent protecting group; it offers some resistance to acid-promoted rearrangement and can be readily converted to the C-2 benzoate. Holton, R. A.; Kim, S.; Tao, C. manuscript in preparation. Patents pending.
23. (a) Sundaraman, P.; Walker, E. C.; Djerassi, C. *Tetrahedron Lett.*, **1978**, *19*, 1627. (b) Marshall, J. A.; Garofalo, A. W.; Sedrani, R. C. *Synlett*, **1992**, *8*, 643.
24. Abiko, A.; Roberts, J. C.; Takemasa, T.; Masamune, S. *Tetrahedron Lett.*, **1986**, *27*, 4537.
25. (a) For a similar cyclization reported to proceed in 13% yield, see Gardner, P. D.; Haynes, G. R.; Brandon, R. L. *J. Org. Chem.*, **1957**, *22*, 1206. (b) Holton, R. A.; Boatman, P. D.; Kim, H. B.; Murthi, K. K. manuscript in preparation.
26. Griffith, W. P.; Ley, S. V.; Whitcombe, G. P.; White, A. D. *J. Chem. Soc., Chem. Commun.*, **1987**, 1625.
27. Holton, R. A.; Somoza, C.; Smith, C. C.; Suzuki, Y. manuscripts in preparation. Patents pending.
28. (a) Farina, V.; Huang, S. *Tetrahedron Lett.*, **1992**, *33*, 3979. (b) Wahl, A.; Gueritte-Voegelein, F.; Guenard, D.; Le Goff, M. T.; Potier, P. *Tetrahedron*, **1992**, *48*, 6965.

29. (a) Holton, R. A.; Tao, C.; Shindo, M. manuscript in preparation. (b) See reference 30 for skeletal rearrangement of baccatin derivatives under similar conditions.
30. Chen, S. H.; Huang, S.; Wei, J.; Farina, V. *Tetrahedron*, **1993**, *49*, 2805.
31. Burgess, E. M.; Penton, Jr., H. R.; Taylor, E. A. *J. Org. Chem.*, **1973**, *38*, 26.
32. Ettouati, L., Ahond, A. Poupat, C.; Potier, P. *Tetrahedron*, **1991**, *47*, 9823.
33. Holton, R. A.; Kim, S.; Tao, C. manuscript in preparation. Patents pending.
34. (a) Noyori, R.; Nishida, I.; Sakata, J.; Nishizawa, M. *J. Am. Chem. Soc.*, **1980**, *102*, 1223. (b) Noyori, R.; Nishida, I.; Sakata, J. *J. Am. Chem. Soc.*, **1981**, *103*, 2106.
35. Holton, R. A. U. S. Patent 5,229,526 (**1993**); Holton, R. A. U. S. Patent 5,274,124 (**1993**).
36. (a) Identical in all respects with an authentic sample of natural taxol provided by Bristol-Myers Squibb Company. (b) $[\alpha]^{23}{}_D = +47^\circ$ (c 0.19, MeOH); identical in all other respects with an authentic sample of natural taxol.

RECEIVED August 19, 1994

Chapter 22

The Total Synthesis of Paclitaxel by Assembly of the Ring System

K. C. Nicolaou[1,2] and R. K. Guy[1]

[1]Department of Chemistry, The Scripps Research Institute, La Jolla, CA 92037
[2]Department of Chemistry, University of California, San Diego, CA 92037

Taxol is an important anticancer agent that is active in a number of cancers including breast and ovarian. The total synthesis of Taxol has been a goal for numerous synthetic groups over the past two decades. Presented herein is the total synthesis of the natural enantiomer of Taxol.

Taxol (Figure 1, **1**) is the lead compound in a new class of antitumor antibiotics called taxoids (*1-3*). Wani and Wall, at the Research Triangle Institute, isolated Taxol from the cytotoxic fraction of an alcoholic extract of the bark of the Pacific Yew, *Taxus brevifolia*. They reported its biological activity, isolation procedure, and structure in 1971 (*4*). Due primarily to a lack of understanding of Taxol's unique mode of action, the drug remained undeveloped until 1979 when, in a seminal series of papers (*5*), Horwitz's group revealed Taxol's molecular target: microtubules. They discovered that, unlike other microtubule drugs that depolymerize microtubules, Taxol enhances the polymerization of tubulin and stabilizes the resulting microtubules (*6*). They also found that cells treated with Taxol exhibit abnormal stable bundles of microtubules and are incapable of forming a normal mitotic spindle (*7*). Over the next decade, Taxol slowly became a molecule of interest to biologists as its unique properties found application in a wide range of fields (*1,3*).

Clinical development of Taxol began with Phase I trials in 1983. Acute hypersensitivity reactions delayed and almost completely derailed the initial trials (*8*). The adoption of long (24 h) infusion times and pretreatment of patients with antihistamines and steroids successfully controlled these problematic responses (*9*) Excitement about Taxol began its exponential rise in both the scientific and public communities in 1989 when Rowinsky and coworkers reported that Taxol induced a high rate of response in patients with drug resistant ovarian cancer (*10*). Since then, other studies have confirmed Taxol's usefulness with ovarian (*11-13*), breast (*14*), and lung cancers (*15-17*). Taxol also shows effect against skin cancers (*18,19*) and head and neck carcinoma (*20*). The principle clinical toxicity for Taxol is granulocytopenia (*21*) and the co-administration of granulocyte colony stimulating factor has shown promise as a method for mediation of this problem (*22*). Clinical evaluation of Taxol, including co-administration with Cisplatin (*23*), is ongoing. Early in 1993, the FDA approved Taxol for treatment of ovarian cancer (*24*). Another, similar, agent, Taxotere™, is under development by Rhone-Poulenc Rorer (*25*).

NOTE: Paclitaxel is the generic name for Taxol, which is now a registered trademark.

0097–6156/95/0583–0302$08.00/0

At the time (Jan. 1992) that this group undertook synthetic studies towards Taxol, there was serious question as to the ability of Nature to supply enough Taxol to meet clinical demands. *T. brevifolia* is a slowly maturing tree that grows, for the most part, enmeshed in old growth forests. Bristol-Meyers Squibb, who had acquired exclusive license for production of the drug from NIH, estimated that natural supplies would last for less than five years (*26*). Concomitantly, the public began to raise concerns about the ecological harm of its harvest (*27*). Fortunately, a semi-synthetic process developed by Holton (*28-30*) and Ojima (*31,32*) largely rendered this issue moot since this process began with the natural product 10-deacetylbaccatin III, isolated from the leaves of the European yew, *T. baccata*, by Hauser's method (*33*). Another semi-synthesis procedure, developed earlier by Potier and Greene (*34*), led to the discovery of Taxotere™.

Thus, our motivations included a need to provide a synthesis of Taxol that might lead to a practical route for its production. The imposing structure of Taxol and the opportunities for pushing the envelope of synthetic chemistry that it provided also challenged us greatly. Consequently, we undertook this synthesis with a greater than normal sense of urgency.

Choosing the Route

Topographic Analysis. Even a cursory examination of the structure of Taxol (Figure 1, **1**) reveals several intimidating features that could prove cruxes for any route chosen for the assault. First, the carbon skeleton contains four and eight membered rings, both notoriously difficult to form. To further increase the difficulty, the eight membered B ring includes two pseudo-axial methyl groups as well as a bridging ring juncture with a six membered ring that includes an alkene, formally forbidden by Bredt's rule. Densely packed with functionality, the molecule contains four tertiary centers, eleven stereocenters, four esters, a ketone, an amide, three free alcohols, and an oxetane ring all within a molecule with a molecular weight of only 850 daltons. Additionally, previous studies (*35*) by other groups had shown that the C7 alcohol was prone to epimerization through a retro-Aldol/Aldol mechanism. The sum of these elements was sufficiently challenging to have engaged many synthetic groups by the time we undertook the challenge in 1992.

Retrosynthesis. While establishing our strategy, we had a central goal secondary to the total synthesis: the route must be sufficiently flexible and practical to allow the preparation of many analogs unavailable from Nature or man's manipulation of the natural product.

Relying on the previous work of other groups as well as our own studies, we envisaged the late formation of the oxetane (D) ring (*36-38*), attachment of the side chain (*28-32*), and oxygenation/reduction of the C13 position (*39*). Thus, we perceived the problem as limited to the assembly of Taxol's ABC ring system in either its fully functionalized form or a form that would serve as its progenitor. Figure 2 displays our strategy. The presence of two pairs of vicinal oxygenation in the central B ring seemed to be a good starting point. To take full advantage of this arrangement, we retrosynthetically applied the McMurry pinacol coupling (*40,41*) transform to the C9-C10 bond. After some preliminary studies, we realized that a directed epoxidation/reductive opening retrosynthesis maneuver at C1-C14 gave a substrate that Shapiro coupling (*42*) of suitable synthons for rings A (**3**) and C (**4**) could readily assemble. This analysis left, as the remaining significant problem, the production of the six membered A and C rings. Faced with the need to control five stereocenters in the C-ring, we rapidly turned to the Diels-Alder transform as a possible means of setting four in one step. Indeed, application of this transform to both rings led to starting materials that were either commercially available (**5**, *43*) or known in the literature: **6** (*44*), **7** (*38*), and **8** (*45*).

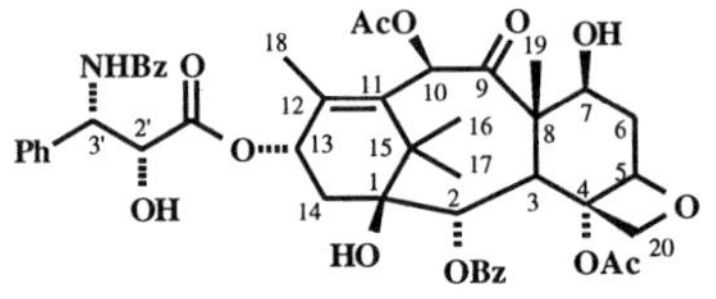

Figure 1. The Structure and Numbering Scheme of Taxol.

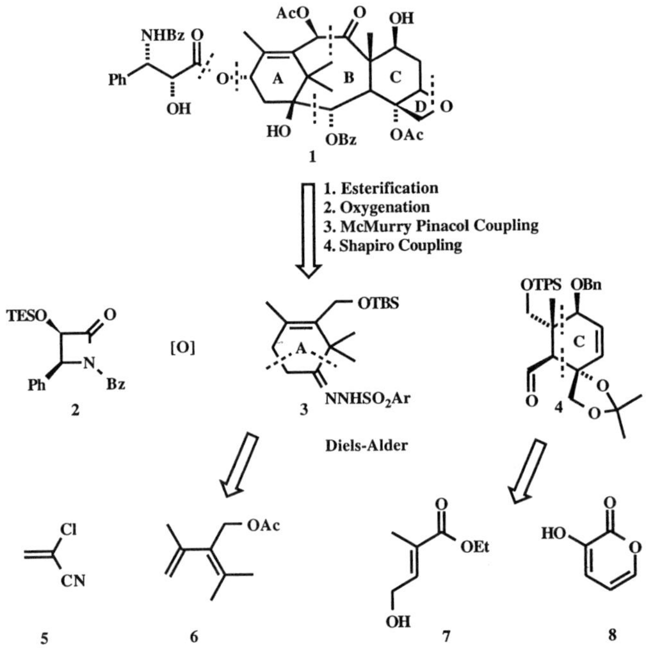

Figure 2. Retrosynthetic Analysis of Taxol.

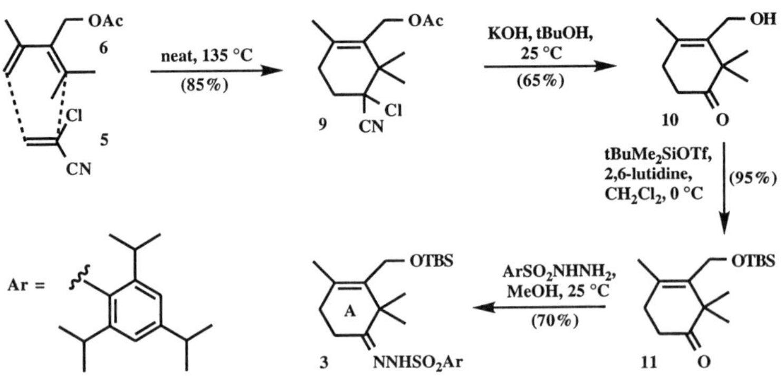

Figure 3. Synthesis of Taxol's A-Ring.

Potential Problems. This plan left three potential problems that we viewed as unanswerable by precedent. First, we worried that differentiation between the sterically less reactive C9 hydroxyl group and that of the electronically less reactive C10 would prove difficult. Expecting to be able to interchange the isomers of the keto-alcohol through enolization/protic trapping, we trusted Nature to have chosen the thermodynamically more stable isomer. Second, the crucial pinacol coupling seemed fraught with the seeds for disaster: two other groups had applied the methodology to similar substrates (*46,47*) and found only modest success. Relying upon improvements in the technology (*48-50*) and our own model studies (*51*) we hoped to build successfully upon these foundations. Lastly, the stereoselectivity of the Shapiro coupling seemed problematic. Since such reactions are often extremely diastereoselective, we decided to table our worries until that pitch became rough.

Assembling the Party

The A-Ring. With a plan in hand, it remained to reduce the expectations to practice. As shown in Figure 3, Diels-Alder reaction of diene **6** (*44*), neat in chloroacrylonitrile **5** (*43*) gave, as a single diastereomer, the cycloadduct **9** (*52*). Hydrolysis of the geminal cyanochloride under basic conditions (*53*) produced the corresponding ketone, **10**, with concomitant removal of the acetyl group. Treatment with tert-butyldimethylsilyl triflate reprotected the alcohol as the silyl ether. Finally, conversion of the ketone into its arylsulfonylhydrazone **3** (*54*) produced a substrate suitable for Shapiro reaction.

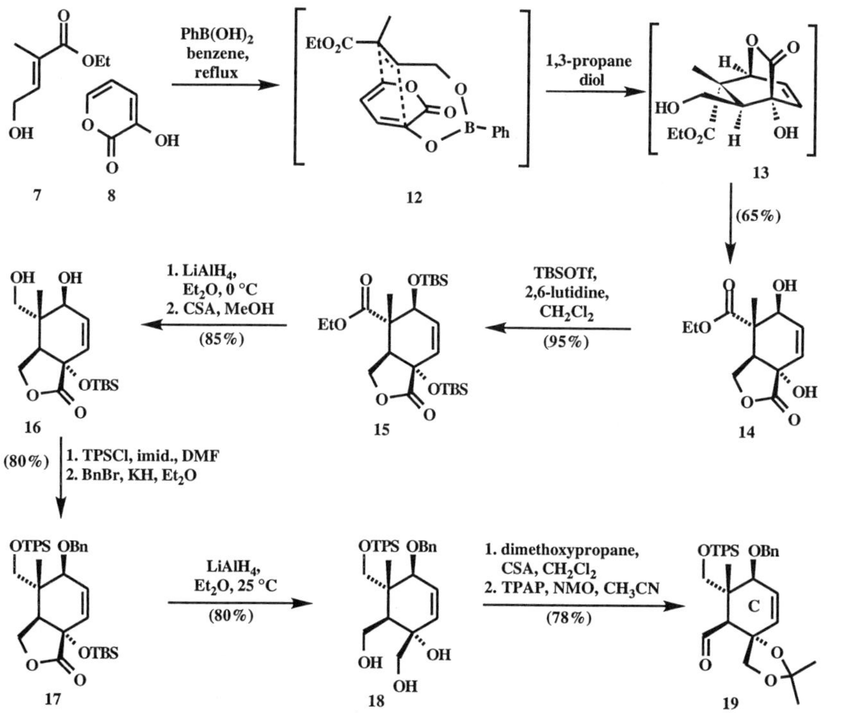

Figure 4. Synthesis of Taxol's C-Ring.

The C-Ring. Construction of the stereochemically demanding C-ring proceeded somewhat less straightforwardly, as shown in Figure 4. Initial attempts at cycloadditions gave exclusively the regioisomer antipodal to that desired. Propitiously, a report by *Narasaka, et. al* (*55*) that met our needs appeared in the literature during our hour of despair. Thus, formation of a temporary tether (**12**) between dienophile **7** (*38*) and diene **8** (*45*) by reaction with phenylboric acid under dehydrating conditions forced the reactants
into affording the desired adduct. Decomplexation with propane diol gave, presumably, the [2.2.2] cycloaddition product **13** that promptly rearranged, by transesterification under the acidic conditions, to the less strained fused [3.4.0] system **14**.

Reaction between tert-butyldimethylsilyl triflate and the two alcohols of **14** simultaneously protected them as their silyl ethers to give **15**. Treatment of this ester-lactone with lithium aluminum hydride then selectively reduced the *ester*. Obstruction of the lactone's Dünitz angles (*56*) by the substituents at C8 probably caused this inversion of chemoselectivity. Acid catalyzed removal of the less hindered silyl ether provided diol **16**. Selective silylation at the primary alcohol using tert-butyldiphenylsilyl chloride followed by benzylation at the secondary alcohol produced **17** in good yield. Exhaustive reduction of the lactone gave, with simultaneous removal of the tertiary silyl ether, triol **18**. Transketalization of the triol with dimethoxypropane yielded, after equilibration of isomers, almost exclusively the favored five-membered cyclic acetonide. Oxidation of the resulting primary alcohol using Ley's ruthenium system (*57*) cleanly gave the desired aldehyde **4**.

Climbing to Base Camp

At this point we reached the first of our major strategic goals for, as displayed in Figure 5 the addition of the vinyl anion of **3**, generated by the method of Martin (*58*), to the aldehyde **4** proceeded with the anticipated discrimination to give one diastereomer of the allylic alcohol **19**. Such exquisite stereocontrol can be attributed to control of the facial selectivity of the incoming nucleophile by a chelate formed between the oxygens of the acetonide and aldehyde and the lithium cation. Chelation controlled epoxidation following the work of Sharpless (*59*) afforded, as a single diastereomer, epoxide **20**. Reductive opening of this epoxide with lithium aluminum hydride proceeded with excellent regioselection to produce the diol **21**. Transformation of this diol to its cyclic carbonate **22**, through the action of phosgene and potassium hydride in ether, enforced our desire for the propinquity of the latent aldehydes. Synchronous removal of the silyl ethers using nucleophilic fluoride, followed by oxidation of the resulting alcohols, gave the dialdehyde **23** -- in our view suitably predisposed for successful McMurry reaction (*40,41,48-50*). Indeed, exposure of this substrate to the titanium reagent generated by the reduction of titanium (III) with zinc-copper couple gave the desired diol **24**, albeit in modest yield. Unfortunately, the desired reaction pathway faced stiff competition from other ketyl couplings, most notably the product of 1,4 addition to the allylic aldehyde. Nevertheless, the methodology provided a flexible and reasonably efficacious route to the ABC ring system, a base camp from which we could explore the lines that lay ahead.

Setting the High Camp

As shown in Figure 6, resolution of this diol proceeded by formation of the diastereomeric esters **25** from 1-(S)-(-)-camphanic chloride followed by chromatographic separation. X-ray crystallographic analysis of one isomer allowed the picking of the desired stereochemical orientation at C9. Hydrolysis of this ester gave **24** as a single enantiomer. We now faced a pitch that we perceived as potentially quite problematic. Allaying all of our fears, selective acetylation with acetic anhydride and DMAP provided,

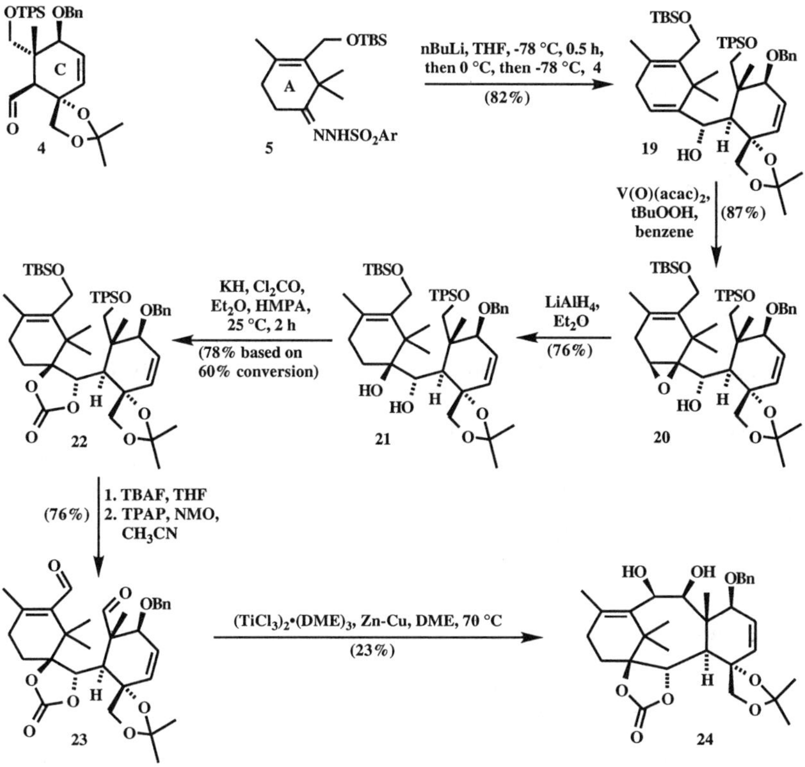

Figure 5. Synthesis of Taxol's ABC Ring System.

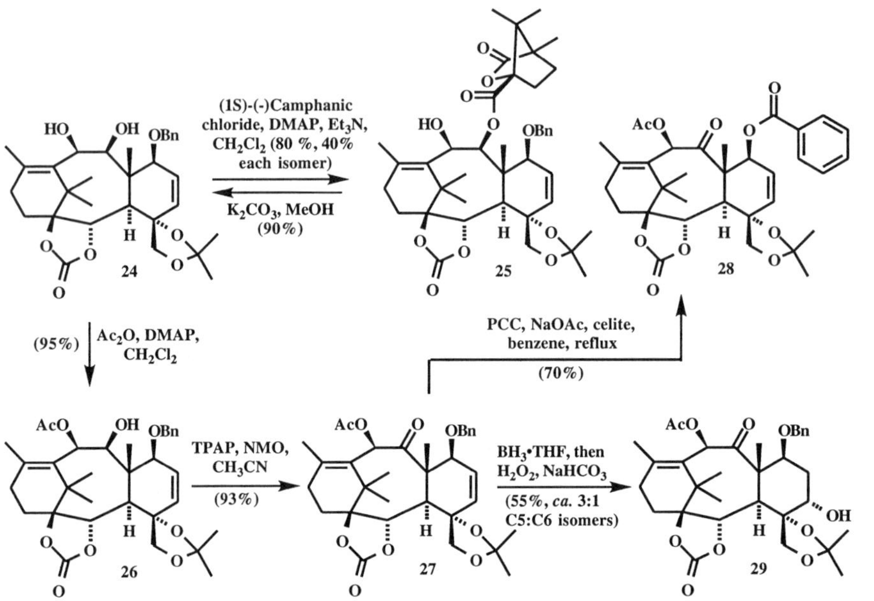

Figure 6. Functionalization of Taxol's ABC Ring System.

very cleanly, material assigned, based upon ^{1}H NMR, the structure **26**. Oxidation of the C10 alcohol gave what we believed to be the desired ketoacetate **27**. To our great joy, crystallographic analysis of the crystalline C7 benzoate **28**, available from **27** by benzylic oxidation with pyridinium chlorochromate, confirmed categorically our inference of this structure. With this crux behind us, we were at the high camp, ready for our assault on the summit.

The Summit Push

The next goal was installment of the oxetane ring. Hydroboration of the C-ring alkene, followed by oxidative workup, gave, with reasonable regioselectivity and excellent stereoselection, the alcohol **29**. Due to the proximity of other functionality, the conversion of this alcohol to the desired oxetane required traversing a somewhat circuitous route. As shown in Figure 7, exposure of **29** to acidic conditions removed the acetonide blocking group. Then, acetylation temporarily and regioselectively protected the resulting primary alcohol as its acetate to give **30**. In order to provide a system faithful to the chemistry developed in degradative studies (*60*) further manipulation exchanged the benzyl protecting group at C7 for a triethylsilyl ether to give **31**. Next, to prepare for the envisioned annulation, the acetyl group was now replaced with an acid labile trimethylsilyl ether and the secondary alcohol was selectively converted to the triflate, producing **32**. Gratifyingly, this material produced, in reasonable yield, the desired oxetane upon exposure to acidic, dehydrating conditions, giving **33**. Finally, acetylation of the tertiary alcohol gave **34**.

Conversion of this material to Taxol followed, as shown in Figure 8, smoothly. Opening of the cyclic carbonate with phenyllithium, a reaction that knew only rare precedent in the literature (*61*), cleanly gave the desired C2 benzoate **35**. Introduction of the C13 hydroxy group proceeded through an oxidation and reduction (*39*) sequence with excellent regio- and stereoselectivity. Finally, attachment of the side chain by the method of Ojima (*31,32*) and Holton (*28-30*) followed by desilylation gave Taxol **1** (*62*).

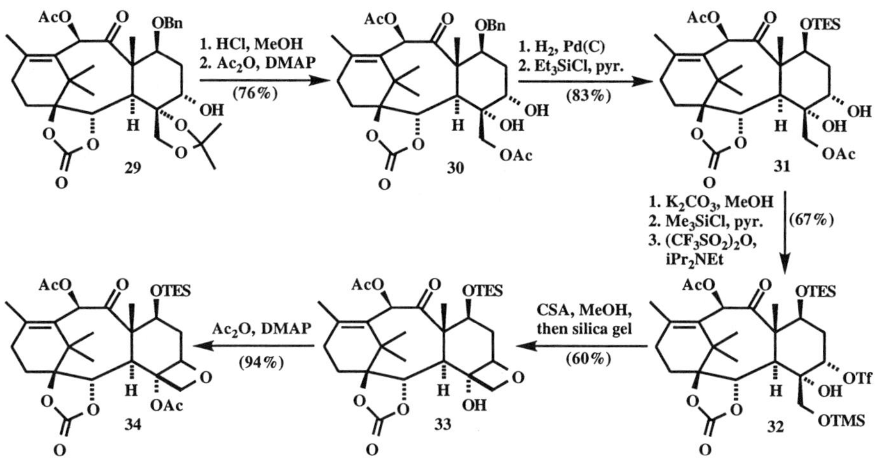

Figure 7. Synthesis of Taxol's D-Ring.

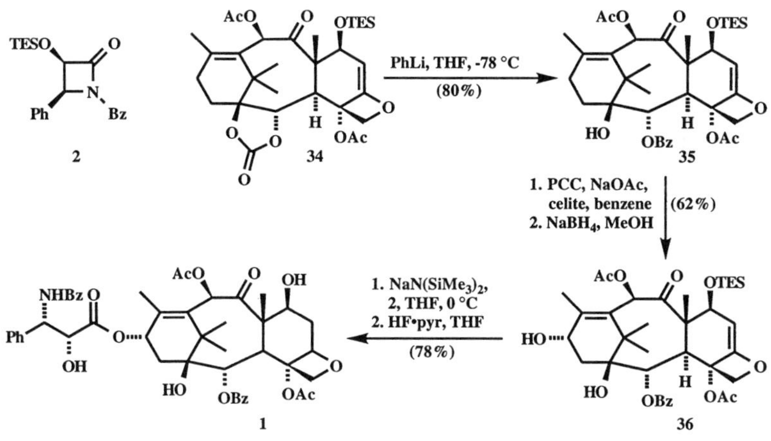

Figure 8. Attachment of Taxol's Side Chain.

Conclusion

In looking back upon this project one must admit mixed feelings about the present results. Certainly the vindication of one's planning by the completion of a total synthesis is gratifying. The meeting of a challenge such as the one put by Taxol and the new strategies developed in the process provided a sense of accomplishment. Unfortunately, the length of the route and the modest yield of some crucial steps preclude the practical application of this method to the production of Taxol. However, the ready access to derivatives of Taxol that this route provides opens new areas for exploration in Taxol's structure activity relationships. Additionally, future improvements in the present synthesis may give a useful second generation synthesis. It is certain that similar feelings must be held by the only other group to complete Taxol's total synthesis to date (*63,64*).

Acknowledgments

It is a pleasure to acknowledge the contributions of our collaborators in the Taxol project: Z. Yang, J. J. Liu, H. Ueno, P. G. Nantermet, C. F. Claiborne, J. Renaud, E. A. Couladouros, K. Paulvannan, and E. J. Sorensen. We thank I. Ojima for a sample of ß-lactam **2a** and E. Bombardelli for a gift of 10-deacetylbaccatin III. Our deepest gratitude goes to Raj Chadha for X-ray crystallography and W. Wrasidlo for biological assays. This work was supported by the NIH, The Scripps Research Institute, and fellowships from Mitsubishi Kasei Corporation (H. U.), Rhone Poulenc Rorer (P. G. N.), The Office of Naval Research USA (R. K. G.) Glaxo, Inc. (C. F. C.), Mr. Richard Staley (C. F. C.), NSERC (J. R.), The Agricultural University of Athens (E. A. C.), R. W. Johnson - ACS Division of Organic Chemistry (E. J. S.) and grants from Merck Sharp and Dohme, Pfizer, Inc., and Schering Plough.

Literature Cited

1. Nicolaou, K. C.; Dai, W. M.; Guy, R. K. *Angew. Chem. Int. Ed. Engl.* **1994**, *33*, 15 - 44.
2. Kingston, D. G. I. *Fortschritte Chem. Org. Natur.* **1993**, *61*, 1 - 206.
3. Lavelle, F. *Curr. Op. Invest. Drugs*, **1993**, *2*, 627 - 635.
4. Wani, M. C.; Taylor, H. L.; Wall, M. E.; Coggen, P.; McPhail, A. T. *J. Am. Chem. Soc.* **1971**, *93*, 2325 - 2327.
5. Manfredi, J. J.; Horwitz, S. B. *Pharmac. Ther.* **1984**, *25*, 83 - 125.
6. Schiff, P. B.; Fant, J.; Horwitz, S. B. *Nature*, **1979**, *277*, 665 - 667.
7. Schiff, P. B.; Horwitz, S. B. *Proc. Natl. Acad. Sci. USA*, **1980**, *77*, 1561 - 1565.
8. Weiss, R. B.; Donehower, R. C.; Wiernik, P. H.; Ohnuma, T.; Gralla, R. J.; Trump, D. L.; Baker, J. R.; Van Echo, D. A.; Von Hoff, D. D.; Leyland-Jones, B. *J. Clin. Oncol.* **1990**, *8*, 1263 - 1268.
9. Brown, T.; Havlin, K.; Weiss, G.; Cagnola, J.; Koeller, J.; Kuhn, J.; Rizzo, J.; Craig, J.; Phillips, J.; Von Hoff, D. *J. Clin. Oncol.*, **1991**, *9*, 1704 - 1712.
10. McGuire, W. P.; Rowinsky, E. K.; Rosenshein, N. B.; Grumbine, F. C.; Ettinger, D. S.; Armstrong, D. K.; Donehower, R. C. *Ann. Int. Med.* **1989**, *111*, 273 - 279.
11. Einzig, A. I.; Wiernik, P. H.; Sasloff, J.; Garl, S.; Runowicz, C.; O'Hanlan, K. A.; Goldberg, G. *Proc. Am. Assoc. Cancer Res.*, **1990**, *31*, 187 (Abstract 1114).
12. Pazdur, R.; Ho, D. H.; Lassere, Y.; Bready, B.; Kvakoff, I. H.; Raber, M. N. *Proc. Am. Soc. Clin. Oncol.* **1992**, *11*, 111 (Abstract 265).
13. Caldos, C.; McGuire, W. P. *Sem. Oncol.*, **1993**, *20(4 Suppl. 3)*, 50 - 55.
14. Holmes, F. A.; Waters, R. J.; Theriault, R. I.; Forman, A. D.; Newton, L. K.; Raber, M. N.; Buzdar, A. U.; Frye, D. K.; Hortobagyi, G. N. *J. Natl. Cancer Inst. USA*, **1991**, *83*, 1797 - 1805.
15. Chang, A.; Kim, K.; Glick, J.; Anderson, T.; Karp, D.; Johnson, D. *J. Natl. Cancer Inst. USA*, **1993**, *85*, 388 - 394.
16. Murphey, W. K.; Winn, R. J.; Fossella, F. V.; Shin, D. M.; Hynes, H. E.; Gross, H. M.; Davila, E.; Leimert, J. T.; Dhinga, H. M.; Raber, M. N.; Krakoff, I. H.; Hong, W. K. *Proc. Am. Soc. Clin. Oncol.*, **1993**, *85*, 384 - 388.
17. Ettinger, D. S. *Sem. Oncol.* **1993**, *20(4 Suppl. 3)*, 46 - 49.
18. Einzig, A. I.; Hochster, H.; Wiernik, P. H.; Trump, D. L.; Dutcher, J. P.; Garowski, E.; Sasloff, J.; Smith, T. J. *Invest. New Drugs*, **1991**, *9*, 59 - 64.

19. Legha, S. S.; Ring, S.; Papadopoulos, N.; Raber, M. N.; Benjamin, R. *Cancer*, **1990**, *65*, 2478 - 2481.
20. Forastiere, A. A. *Sem. Oncol.* **1993**, *20(4 Suppl. 3)*, 56 - 60.
21. Rowinsky, E. K.; Eisenhauer, E. A.; Chaudhry, V.; Arbuck, S. G.; Donehower, R. C. *Sem. Oncol.* **1993**, *20(4 Suppl. 3)*, 1 - 15.
22. Seidman, A. D.; Norton, L.; Reichman, B. S.; Crown, J. P. A.; Yao, T. J.; Heelan, R.; Hakes, T. B.; Lebwohl, D. E.; Gilewski, T. A.; Surbane, A.; Currie, V.; Hudis, C. A.; Klecker, R.; Jamis-Dow, C.; Collins, J.; Quinlivan, S.; Berkery, R.; Toomasi, F.; Canetta, R.; Fisherman, J.; Arbuck, S. *Sem. Oncol.*, **1993**, *20(4 Suppl. 3)*, 40 - 45.
23. Rowinsky, E. K.; Gilbert, M.; McGuire, W. P.; Noe, D. A.; Growchow, L. B.; Forastiere, A.; Ettinger, D. S.; Lubejko, B. G.; Clark, B.; Sartorious, S. E.; Cornblath, D. R.; Hendricks, C. B.; Donehower, R. C. *J. Clin. Oncol.* **1991**, *9*, 1692 - 1703.
24. *Science*, **1993**, *259*, 181.
25. Boven, E.; Venema-Gaberscek, E.; Erkelens, C. A.; Bissery, M. C.; Pinedo, H. M. *Ann. Oncol.* **1993**, *4*, 321 - 324.
26. Borman, S. *Chem. Eng. News*, **1991**, Sept. 2, 11 - 18.
27. Hartzell, H. *The Yew Tree, A Thousand Whispers*, Hulogosi: Eugene OR USA, **1991**.
28. Holton, R. A. *Workshop on Taxol and Taxus*, **1991**.
29. Holton, R. A.; *Eur. Pat. Appl. EP400,971*, **1990**.
30. Holton, R. A.; *Chem. Abstr.* **1990**, *114*, 164568q.
31. Ojima, I.; Habus, I.; Zhao, M.; Georg, G. I.; Jayasinghe, L. R. *J. Org. Chem.*, **1991**, *56*, 1681 - 1683.
32. Ojima, I.; Habus, I.; Zhao, M.; Zucco, M.; Park, Y. H.; Son, C. M.; Brigaud, T. *Tetrahedron*, **1992**, 6985 - 7012.
33. McCormick, D. *Bio/Technology*, **1993**, *11*, 26.
34. Mangatel, L.; Adeline, M. -T.; Guénard, D.; Guéritte-Voegelein, Potier, P. *Tetrahedron*, **1989**, *45*, 4177 - 4190.
35. Kingston, D. G. I. *Pharmac. Ther.* **1991**, *52*, 1-34.
36. Ettovati, L.; Ahond, A.; Poupat, C.; Potier, P. *Tetrahedron*, **1991**, *47*, 9823 - 9838.
37. Magee, T. V.; Bornmann, W. G.; Isaccs, R. C. A.; Danishefdky, S. J. *J. Org. Chem.*, **1992**, *57*, 3274 - 3276.
38. Nicolaou, K. C.; Liu, J. J.; Hwang, C. -K.; Dai, W. -M.; Guy, R. K. *J. Chem. Soc. Chem. Commun.* **1994**, 295 - 296.
39. Harrison, J. W.; Scrowsten, R. M.; Lythgoe, B. *J. Chem. Soc. C*, **1966**, 1932 - 1945.
40. McMurry, J. E. *Chem. Rev.* **1989**, *89*, 1513 - 1524.
41. Lenoir, D. *Synthesis*, **1989**, 883 - 897.
42. Chamberlin, A. R.; Bloom, S. H. *Org. React.* **1990**, *39*, 1 - 83.
43. Tokyo-Kasei
44. Alkonyi, I.; Szabó, D. *Chem. Ber.* **1967**, *100*, 2773 - 2775.
45. Wiley, R. H.; Jarboe, C. H. *J. Am. Chem. Soc.* **1956**, *78*, 2398 - 2401.
46. Begley, M. J.; Jackson, C. B.; Pattenden, G. *Tetrahedron*, **1990**, *46*, 4907 - 4924.
47. Kende, A. S.; Johnson, S.; Sanfilippo, P.; Hodges, J. C.; Jungheim, L. N. *J. Am. Chem. Soc.* **1986**, *108*, 3513 - 3515.
48. McMurry, J. E.; Rico, J. G.; Lectka, T. C. *J. Org. Chem.* **1989**, *54*, 3748 - 3750.
49. McMurry, J. E., Rico, J. G. *Tetrahedron Lett.* **1989**, *30*, 1169 - 1172.
50. McMurry, J. E.; Rico, J. G.; Shin, Y. -N. *Tetrahedron Lett.* **1989**, *30*, 1173 - 1176.
51. Nicolaou, K. C.; Yang, Z.; Sorensen, E. J.; Nakada, M. *J. Chem. Soc. Chem. Commun.* **1993**, 1024 - 1026.

52. Nicolaou, K. C.; Hwang, C. K.; Sorensen, E. J.; Claiborne, C. F. *J. Chem. Soc. Chem. Commun.* **1992**, 1117 - 1118.
53. Madge, N. C.; Holmes, A. B. *J. Chem. Soc. Chem. Commun.* **1980**, 956 - 957.
54. Cusack, N. J.; Reese, C. B.; Risius, A. C.; Roozpeikar, B. *Tetrahedron*, **1976**, *32*, 2157 - 2162.
55. Narasaka, K.; Shimada, S.; Osada, K.; Iwasawa, N. *Synthesis*, **1991**, 1171 - 1173.
56. Burgi, H. B.; Dunitz, J. D.; Lehn, J. M.; Wipff, G.*Tetrahedron*, **1974**, *30*, 1563 - 1572.
57. Griffith, W. P.; Ley, S. V. *Aldrichimica Acta*, **1990**, *23*, 13 - 18.
58. Martin, S. F.; Daniel, D.; Cherney, R. J.; Liras, S. *J. Org. Chem.*, **1992**, *57*, 2523 - 2525.
59. Sharpless, K. B.; Verhoeven, T. R. *Aldrichimica Acta*, **1979**, *12*, 63 - 89.
60. Nicolaou, K. C.; Nantermet, P. G.; Ueno, H.; Guy, R. K. *J. Chem. Soc., Chem. Commun.*, **1994**, 295 - 296.
61. Wender, P. A.; Kogan, H.; Lee, H. Y.; Munger, J. D.; Wilhelm, R. S.; Williams, P. D. *J. Am. Chem. Soc.* **1989**, *111*, 8957 - 8958.
62. Nicolaou, K. C.; Yang, Z.; Liu, J. J.; Ueno, H.; Nantermet, P. G.; Guy, R. K.; Claiborne, C. F.; Renaud, J.; Couladouros, E. A.; K. Paulvannan, Sorensen, E. J. *Nature*, **1994**, *367*, 630 - 634.
63. Holton, R. A.; Somoza, C; Kim, H. B.; Liang, F.; Biediger, R. J.; Boatman, P. D.; Shindo, M.; Smith, C. C.; Kim, S.; Nadizadeh, H.; Suzuki, Y.; Tao, C.; Vu, P.; Tang, S.; Zhang, P.; Murthi, K. K.; Gentile, L. N.; Liu, J. H. *J. Am. Chem. Soc.* **1994**, *116*, 1597 - 1598.
64. Holton, R. A.; Kim, H. B.; Somoza, C; Liang, F.; Biediger, R. J.; Boatman, P. D.; Shindo, M.; Smith, C. C.; Kim, S.; Nadizadeh, H.; Suzuki, Y.; Tao, C.; Vu, P.; Tang, S.; Zhang, P.; Murthi, K. K.; Gentile, L. N.; Liu, J. H. *J. Am. Chem. Soc.* **1994**, *116*, 1598 - 1600.

RECEIVED August 19, 1994

Chapter 23

Toward the Design of a Convergent Practical Route to All Classes of Taxanes

Leo A. Paquette

Evans Chemical Laboratories, Ohio State University, Columbus, OH 43210

The taxane diterpenes constitute a structurally intricate class of compounds which has been accorded considerable attention in recent years because one of its members, taxol, is remarkably effective against advanced ovarian cancer. The varied complexity and functional group arrays present in the various taxanes represent a major challenge to synthetic chemists. The present goal is to utilize (+)-camphor, an abundant enantiopure product of the camphor tree, as the key building block in a convergent approach to the several types of taxanes, as well as to suitable analogs not available by semi-synthesis or direct structural modification. The pathway under development is capable of rapid construction of the complete carbocyclic framework of the taxanes, with exceptional control of the bridgehead oxidation level.

Since the discovery by Wani and Wall of the antineoplastic agent taxol (**1**) in 1971 (*1*) and the subsequent demonstration by Horwitz that **1** functions as a mitotic spindle poison capable of inhibiting microtubule depolymerization (*2*), intense interest has arisen within the international synthetic organic chemistry community in the *de novo* construction of **1** and its congeners (*3-7*). Spurred on by the exciting inhibitory properties of taxol against ovarian, lung, and breast cancer and by the original scarce supply of this drug, researchers have responded admirably to the challenge of realizing the step-by-step assembly of the complex structural features inherent in taxane systems. Two syntheses of taxol have recently been reported (*8, 9*). Several years ago, a means for preparing the unnatural (-)-enantiomer of taxusin (**2**) was detailed (*10*).

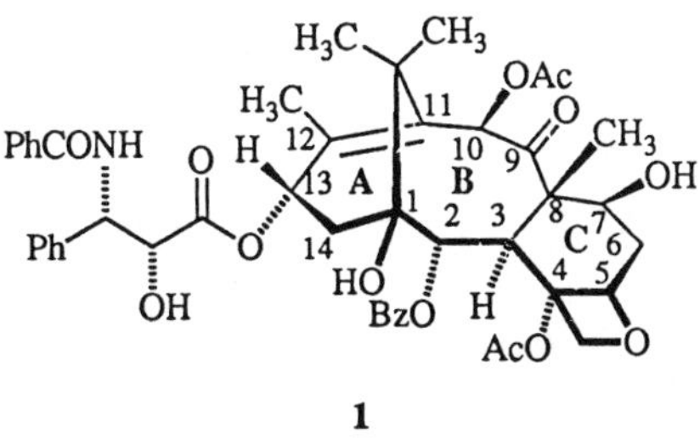

1

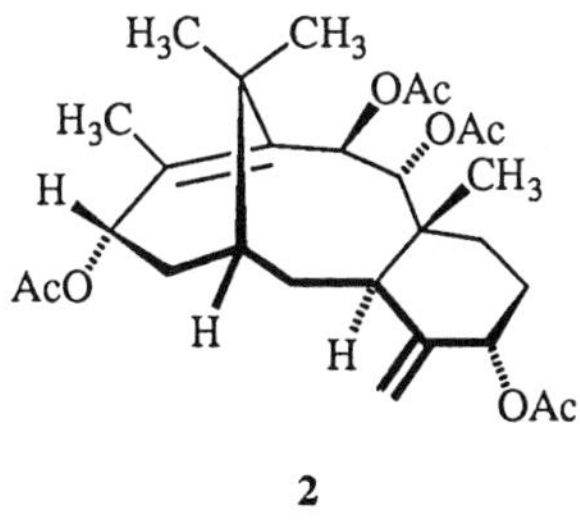

2

NOTE: Paclitaxel is the generic name for Taxol, which is now a registered trademark.

0097–6156/95/0583–0313$08.00/0

Formulation of the Synthetic Plan

Specific attention is called to the fact that **1** and **2** carry different levels of oxidation at C-1, a feature which broadly segregates the two sub-classes of taxanes. Although structural modifications of taxol have been investigated to some extent (*11-13*), 1-deoxytaxol is as yet unknown and seemingly unattainable by degradation because of the onset of framework rearrangement (*11, 14*). At the time that we began our work, the most attractive opportunity appeared to reside in the development of a practical (*viz.*, 25 steps or less), convergent protocol that would allow enantioselective synthetic entry into either sub-class of the taxane family on demand. The tactical plans were formulated with (+)-(*R*)-camphor in mind as a suitable starting material having the proper absolute configuration. An efficient means for transforming camphor into the optically pure bicyclic ketone **3** had already been documented (*15*). Our first objective was to determine specifically those features of the [3.3] sigmatropic rearrangement transition state which would be manifested by **4**, produced by coupling with a cyclohexenyl anion from the more accessible endo surface of the carbonyl (Scheme 1).

Scheme 1

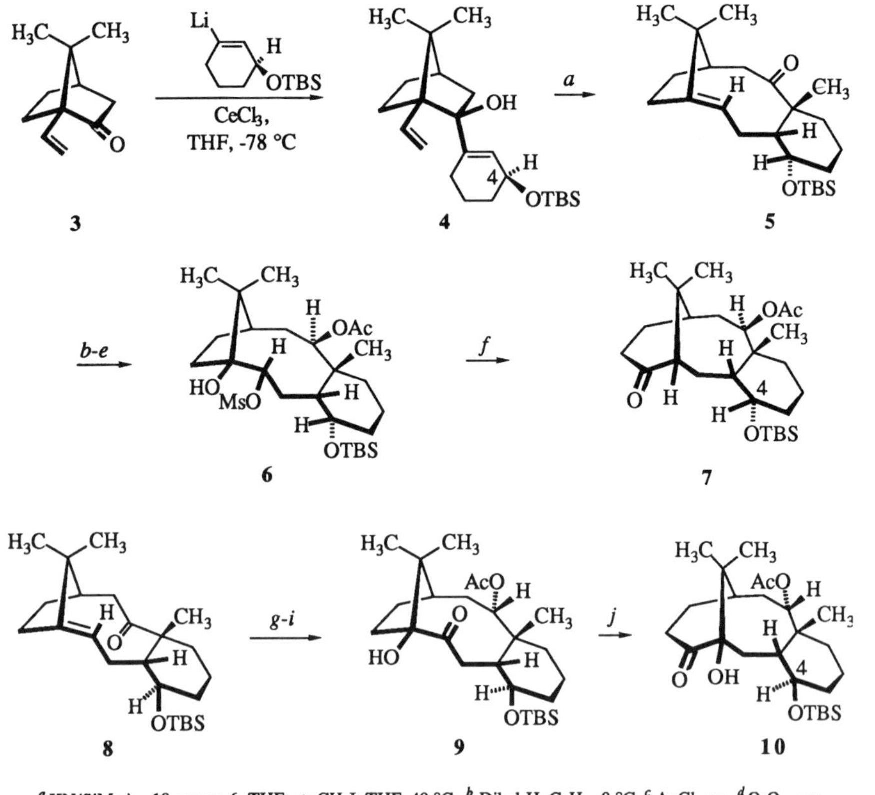

[a] $KN(SiMe_3)_2$, 18-crown-6, THF, rt; CH_3I; THF, 40 °C. [b] Dibal-H, C_6H_6, 0 °C. [c] AcCl, py. [d] OsO_4, py; $NaHSO_3$. [e] MsCl, py. [f] Et_2AlCl, CH_2Cl_2, -78 °C $\rightarrow$ 25 °C. [g] OsO_4, py; $LiAlH_4$. [h] Ac_2O, py; $Bu_4N^+ F^-$. [i] Swern. [j] $(i\text{-PrO})_3Al$, C_6H_6, Δ.

As matters have turned out, the endo-chair arrangement is uniquely utilized by **4** and related carbinols, thereby making ketones with stereochemistry defined as in **5** and **8** easily accessible (*16-18*).

These isomerizations happen to be remarkably atropselective as well. In the anionic oxy-Cope rearrangement originating from **4** only the "carbonyl down" conformer is produced. Warming solutions of this tricyclic ketone in tetrahydrofuran or benzene promotes conversion to the thermodynamically more stable "carbonyl up" modification **5**. By the simple expedient of inverting configuration at C-4 in **4, 8** becomes both thermodynamically and kinetically favored. Since the substitution plan in ring C lies at the root cause of these effects, purposeful control of the spatial orientation of these molecules is genuinely feasible. As a consequence, stereodirected chemical manipulation as at C-9 in **6** (β-OAc) and **9** (α-OAc) is made entirely possible (*17*).

Expedient Elaboration of Taxane Frameworks

The tantalizing objective of avoiding any vestiges of transannular bonding in constrained (*E*)-5-cyclononenones of this general type was quickly resolved (*19, 20*) and the four-step conversions of **5** into **6** and **8** into **9** soon were accomplished. At this point, it can be recognized that our intent was to implement a classical pinacol-like rearrangement in order to progress from **6** to members of the taxusin class exemplified by **7**. After some experimentation, it was determined that diethylaluminum chloride was a utilitarian Lewis-acidic promoter of the desired reaction (*21, 22*). Application of the α-ketol concept to **9** was equally successful. When heated with aluminum isopropoxide in benzene, **9** was transformed in highly regioselective fashion into **10**

Scheme 2

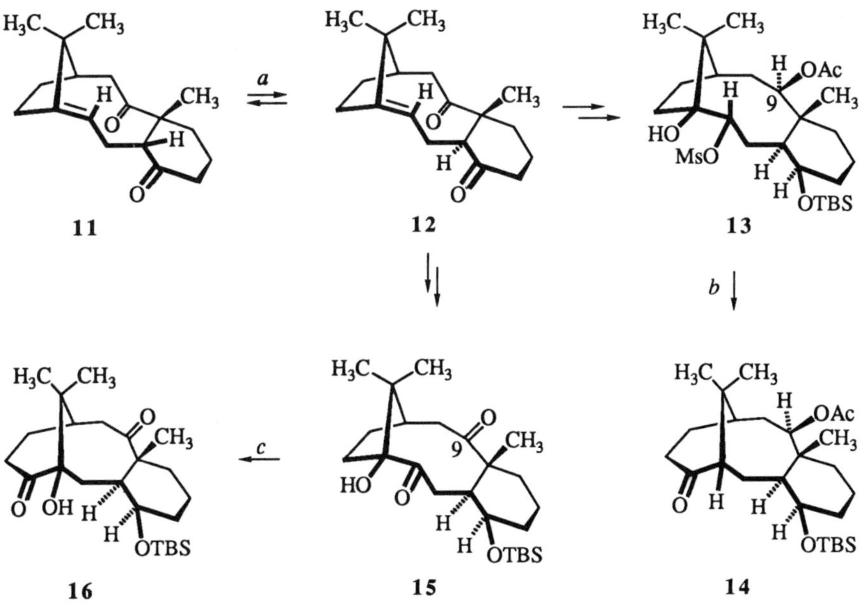

[a] NaOMe, MeOH, Δ. [b] Et_2AlCl, CH_2Cl_2, -78 °C → -40 °C. [c] $(t\text{-BuO})_3Al$, C_6H_6, Δ.

(78% isolated). Since the latter process is equilibrium controlled, some advance knowledge of the relative thermodynamic stabilities of the isomers would quite obviously be useful. MM2 calculations were noted by us to be very reliable in this regard, being completely in line with a wide range of experimental observations (*22, 23*).

The most important of these computational evaluations were associated with the trans B/C analogs of the above compounds because of their more closely aligned stereochemical relationship to **1** and **2**. In the event, diketone **11** was epimerized to **12**, and this intermediate was transformed along lines similar to those already detailed into **13** and **15** (Scheme 2). Evaluation of the global energy minima for the **13/14** and **15/16** isomer pairs suggested that the isomerizations would be exothermic only if C-9 were tetrahedral in **13** and trigonal in **14**, but not otherwise (*23*). In line with these predictions, the less strained **16** predominated totally once its equilibration with **15** had progressed to completion. Importantly, **16** was arrived at in only *ten* laboratory operations from **3**.

The Advance toward Taxusin

Although **16** is clearly underfunctionalized for expedient arrival at taxol, the substitution plan in **14** appeared conducive to the preparation of taxusin (**2**). With the acquisition of triketone **17** (*20*), a campaign was initiated to effect its possible conversion into **2**. The triad of carbonyl groups could easily be distinguished as a consequence of their quite different latent reactivities as in **18** (Scheme 3) (*24*). Steric factors resident in **18** allow for regiocontrolled oxidation to be implemented on the α-

Scheme 3

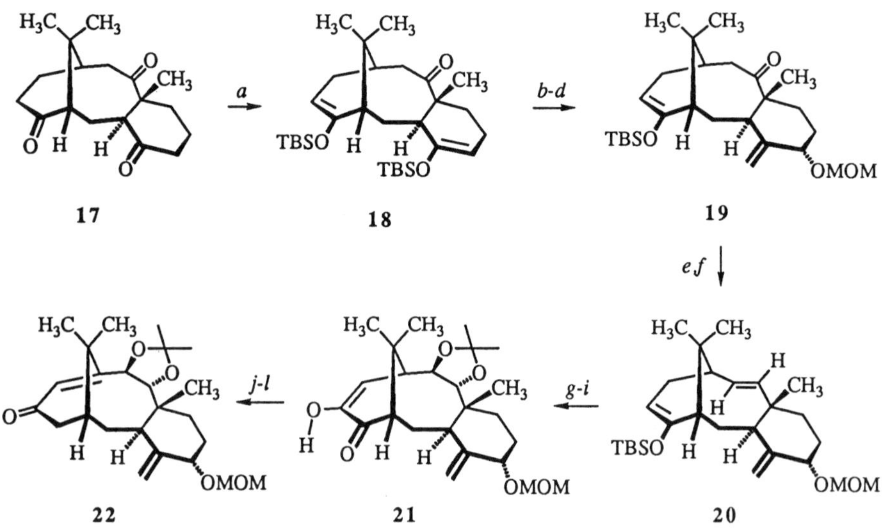

[a] TBSOTf, Et_3N, CH_2Cl_2. [b] , acetone, CH_2Cl_2; K_2CO_3, MeOH, rt. [c] MOMCl, $(i\text{-}Pr)_2NEt$. [d] $Ph_3P{=}CH_2$, THF, rt. [e] Dibal-H, C_6H_6. [f] Martin sulfurane, C_6H_6, rt. [g] OsO_4, py, 0 °C; $(CH_3)_2C(OCH_3)_2$, (TsOH). [h] TBAF, THF. [i] $KN(SiMe_3)_2$; O_2. [j] $LiAlH_4$, ether; PhCOCl, py, DMAP. [k] LDA, PhSeCl; H_2O_2. [l] SmI_2.

face of ring C. Subsequent MOM protection and Wittig olefination provided a sound basis for attaining **19** (Zhao, M., unpublished data). We then focused on reducing **19** with Dibal-H in benzene at 8°C (*24*), dehydrating the β-alcohol with Martin's sulfurane to generate trans olefin **20**, and osmylating this intermediate to set the proper trans diol arrangement at C-9 and C-10. Following this, the neighboring hydroxyl groups could be temporarily protected as the acetonide.

When the reduction of **19** was performed instead in hexane solution at -78 °C, nucleophilic attack at the carbonyl carbon was slowed significantly. This provides an opportunity to the >C=O---$AlHR_2$ complex for conformational reorientation to the "carbonyl down" direction prior to hydride delivery with resultant conversion to the α-carbinol. Dehydration of this epimer with Martin's sulfurane afforded the trans cycloalkene atropisomeric to **20**. This molecular asymmetry arises because of steric inhibition to internal rotation. When heated in C_6D_6 at 67 °C, **20** was irreversibly converted into its atropisomer with a $t_{1/2}$ of 45 min. Thus, the thermal stability of **20** is more than adequate to allow for its individual synthetic utilization.

Once rings B and C had been completely functionalized, carbonyl transposition in ring A was made entirely feasible by taking advantage of the facility with which enolates of C-14 ketones undergo air oxidation at C-13 (Zhao, M., unpublished data). As illustrated specifically for **21**, the α-diketones so produced are strongly enolized in the C12-C13 direction away from the bridgehead position. This central structural property allows for regiocontrolled reduction of the C-14 carbonyl with lithium aluminum hydride while the adjacent center is protected from attack because of enolization. Subsequent introduction of the bridgehead double bond and chemospecific removal of the α-benzoyloxy substituent via samarium diiodide reduction has presented no logistical problem. We have as yet not completed meaningful studies concerning

Scheme 4

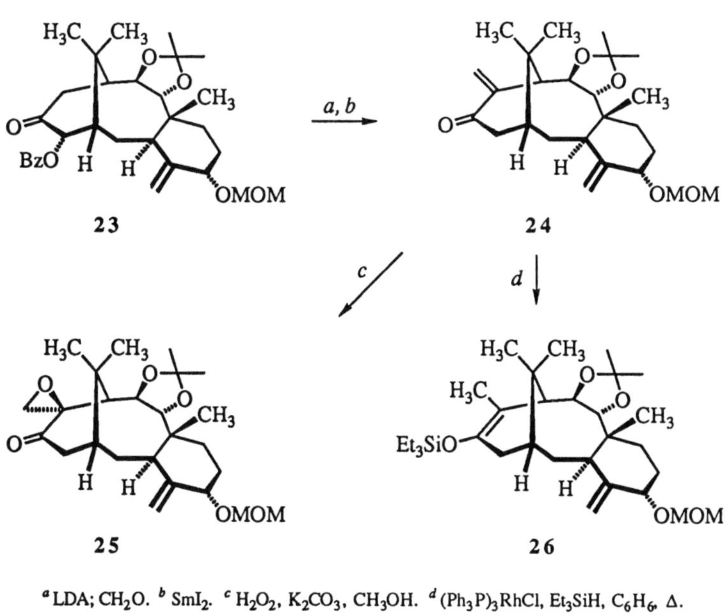

[a] LDA; CH_2O. [b] SmI_2. [c] H_2O_2, K_2CO_3, CH_3OH. [d] $(Ph_3P)_3RhCl$, Et_3SiH, C_6H_6, Δ.

introduction of the C-12 methyl group, but plan to do so in the near future. Suffice it to say that early efforts aimed at more advanced A-ring functionalization for the purpose of generating taxusin analogs have yielded encouraging results. The situation at this point, depicted in Scheme 4, may be summarized by conversion of the predescribed intermediate **23** into **24-26** and related compounds.

Enhancing the Oxygenation Level of Ring B

In light of the preceding accomplishments, we have moved on to address the several issues associated with stereocontrolled introduction of the various functionalized centers present in taxol. This is, of course, of particular importance if **1** is to be reached efficiently. Since the C-2 benzoate of **1** is disposed α, a first requirement is that our synthetic approach lend itself readily to this task. So reliable is the stereoselectivity of the oxy-Cope rearrangement that the geometry of the vinyl ether double bond in **27** and **28** is translated with complete configurational fidelity into α or β stereochemistry in the respective end products **29** and **30** (Scheme 5). The starting ketones are directly available from **3** via ozonolysis and chemoselective Wittig olefination (*25*).

Since the level of oxygenation present in ring B of taxol is still more dense than that contained in **30**, the convergency of this methodology had to be increased still more. As shown in Scheme 6, this obstacle was in the end surmounted once we recognized that β,γ-unsaturated ketone **28** could be oxidized by means of Davis' oxaziridine reagent to generate principally the exo alcohol **31**. Notwithstanding the favorable stereochemical features of this process, we have found that no need exists to

Scheme 5

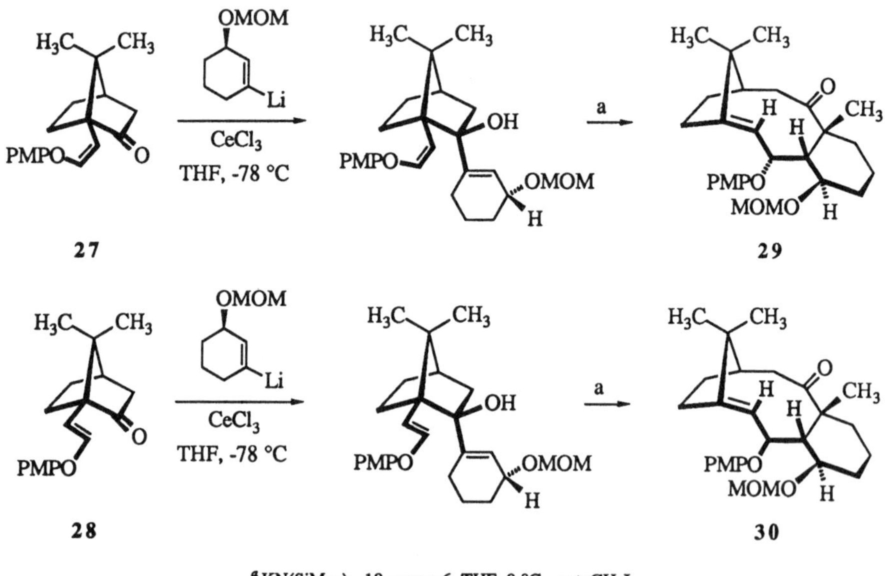

^{a} $KN(SiMe_3)_2$, 18-crown-6, THF, 0 °C → rt; CH_3I.

separate the endo diastereomer at this point or after MOM protection because the minor endo epimer does not react with the cyclohexenyllithium reagent (*26*). A very important feature of this chemistry is the fact that the presence of the OMOM substituent drastically reduces the acidity of the α-proton in the camphor derivative to a level such that recourse to cerium reagents is no longer required. Significantly, the additional oxygen substituent does not inhibit the oxy-Cope rearrangement, thereby allowing for the C-2, C-9, C-10 oxygen pattern in taxol to be quickly established. Product **32** has not as yet been advanced to bridge migration. Our more immediate objectives were to establish the feasibility of A-ring oxygenation in bridgehead hydroxyl-substituted *trans*-tricyclo[9.3.1.0^{3,8}]pentadecanones and to develop a promising means for completing C ring assembly. Both of these efforts have been crowned with partial success.

Scheme 6

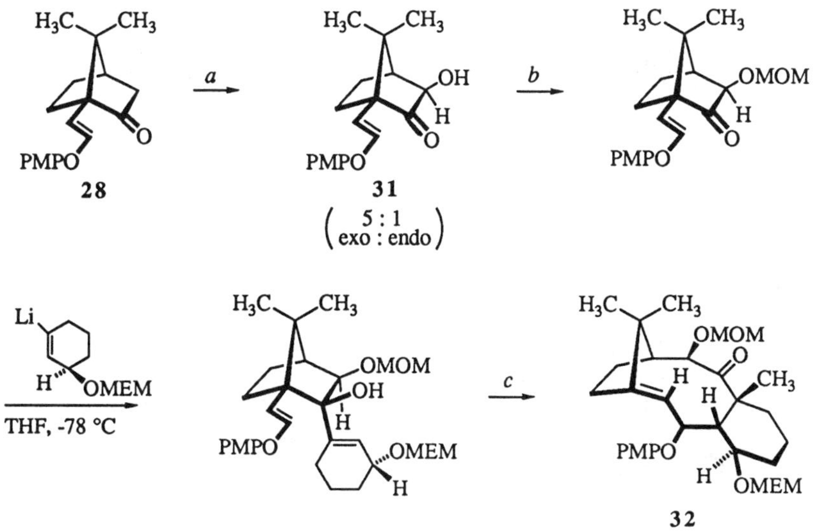

[a] KN(SiMe$_3$)$_2$, THF, -78 °C; N-sulfonyloxaziridine (PhSO$_2$, Ph). [b] MOMCl, (i-Pr)$_2$NEt, CH$_2$Cl$_2$. [c] KH, 18-crown-6, THF, 0 °C; MeI.

Ring A Oxygenation Studies

Once the tertiary hydroxyl group in **16** had been silylated, it was immediately apparent that diketone **33** underwent kinetically controlled deprotonation at C-13 to the virtual exclusion of C-10. As desirable as this was, transannular aldolization followed (see **34**) and had to be circumvented (*27*). The use of potassium hexamethyldisilazide at low temperature in the presence of chlorotrimethylsilane was found to enable conversion to crystalline **35** in good yield (83%, Scheme 7). Following reduction of the C-9 carbonyl and protection of the resulting α-alcohol as in **36**, carbonyl transposition in ring A was accomplished as shown to produce **39**. Despite severe steric congestion on the undersurface of the A ring in **36**, oxygenation occurs from the

Scheme 7

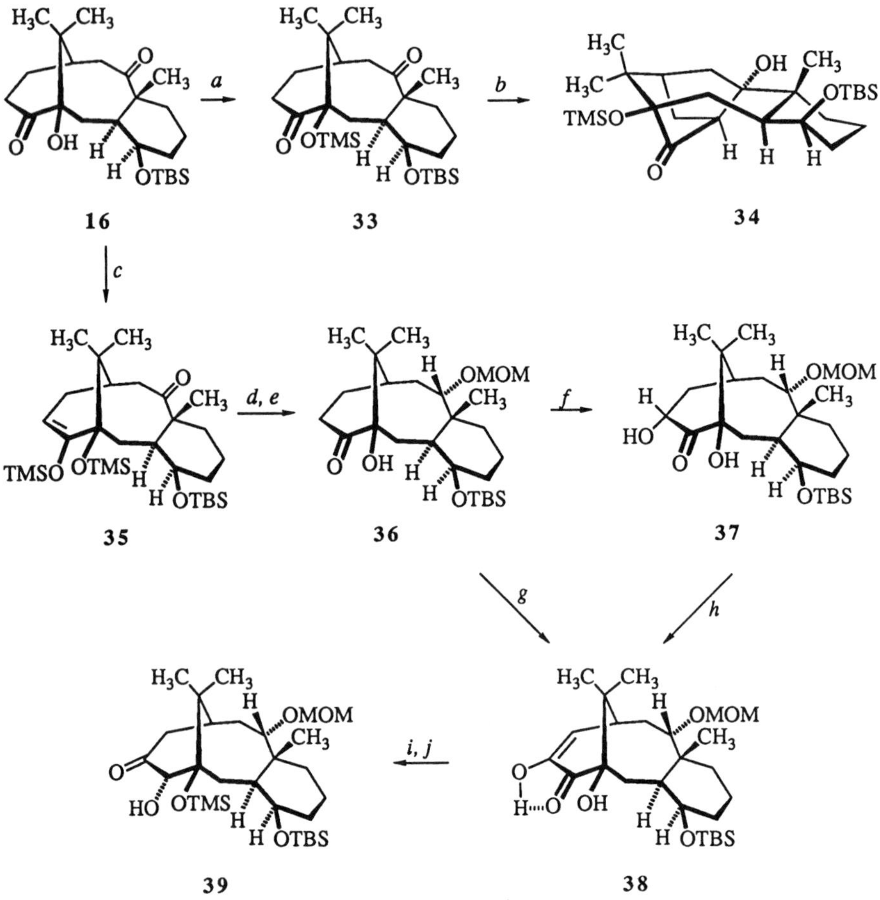

[a] Me_3SiCl, THF; $NaN(SiMe_3)_2$ (1.2 equiv), -78 °C. [b] $KN(SiMe_3)_2$, THF, -78 °C. [c] Me_3SiCl, THF; $KN(SiMe_3)_2$, -78 °C. [d] Dibal-H, C_6H_6. [e] MOMCl, $(i\text{-}Pr)_2NEt$, H_3O^+. [f] $KN(SiMe_3)_2$, THF, -78 °C, 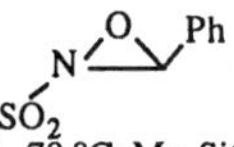 .
[g] As in *f*, except -78 °C → 0 °C. [h] Dess-Martin periodinane, CH_2Cl_2. [i] $KN(SiMe_3)_2$, THF, -78 °C; Me_3SiCl.
[j] $LiAlH_4$, Et_2O, -78 °C → rt.

α direction. Remarkably, the resultant α-hydroxy ketone **37** could be epimerized to the 13β isomer. Although **37** is prone to autoxidation, periodinane oxidation has proven more reliable and is the preferred means for securing **38**. The ultimate plan for **39** and its analogs is to complete A ring construction along lines comparable to those already implemented satisfactorily in the taxusin effort.

Introduction of the Oxetane Ring or its Functional Equivalent

In a rather bold and perhaps overoptimistic move to carry the functionalized oxetane ring present in taxol through the entire synthetic sequence, vinyl bromide **40** was prepared in racemic form, shown to undergo halogen-metal exchange efficiently, and incorporated smoothly into a variety of oxy-Cope presursors (*28*). If the ensuing charge-accelerated [3,3] sigmatropic rearrangement proceeds readily at or below 0 °C, the sensitive oxetane ring survives the strongly basic environment without event. The conversion of **41** to **42** in 80% yield illustrates the feasibility of our objectives (Scheme 8). However, neither of the two diastereomers **43** or **44** formed by addition of the cerate of **40** to **28** proved to be adequately reactive in this temperature range. For these systems, room temperature was required to promote structural reorganization. These somewhat more harsh conditions lead instead to kinetically favored aromatization. Heating of either alcohol results in operation of a thermal ene reaction instead.

Scheme 8

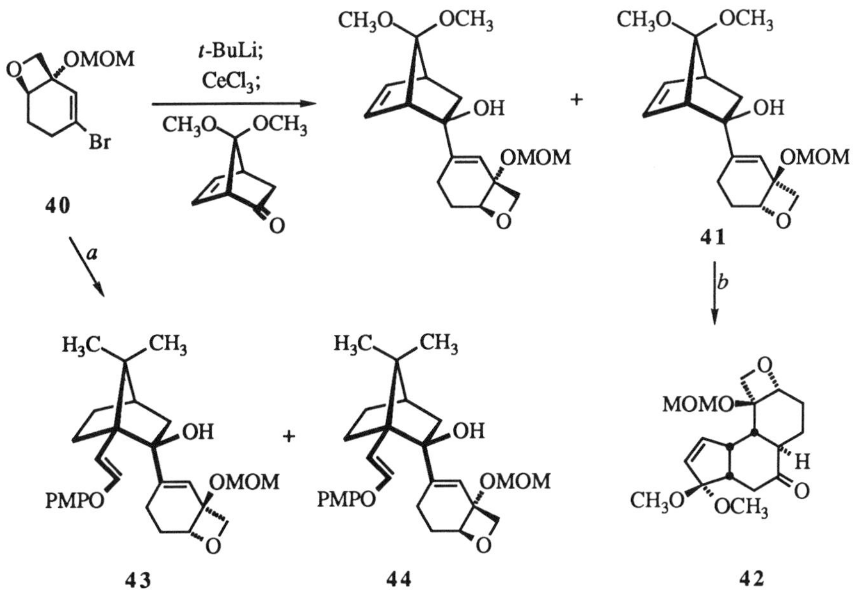

[a] *t*-BuLi; $CeCl_3$; **28**. [b] $KN(SiMe_3)_2$, 18-crown-6, THF, 0 °C.

A structurally less strained building block, the functionalized acetonide **45**, is equally receptive to transmetalation, adds to camphor-derived β,γ-unsaturated ketones without complication, and once present this preoxetane functionality is stable to the oxy-Cope reaction conditions (Scheme 9). As a consequence, this oxetane equivalent may be suited to our goals as it is expected to survive the requisite bridge migration (*28*).

Scheme 9

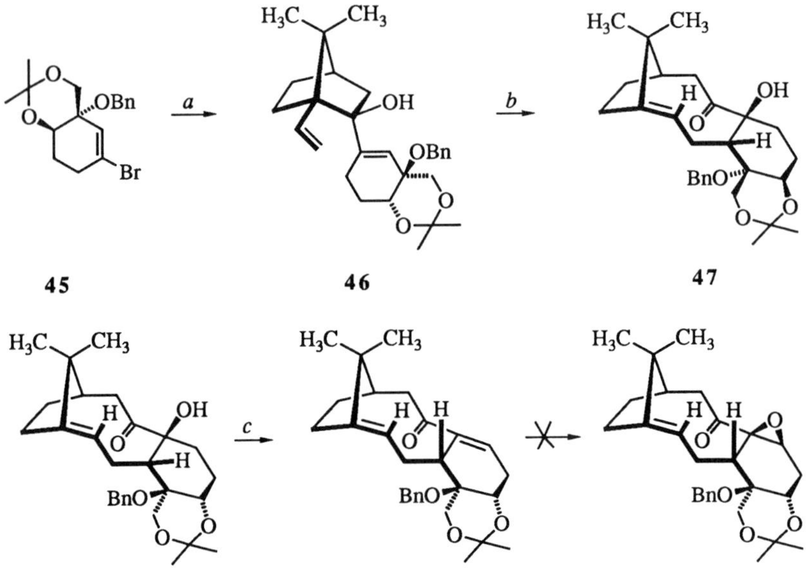

45 46 47

48 49 50

[a] *n*-BuLi; $CeCl_3$; 3. [b] $KN(SiMe_3)_2$, 18-crown-6, O_2, C_6H_6. [c] CH_3SO_2Cl, Et_3N, CH_2Cl_2.

The Issue of C-7 Oxygenation

When **46** and its diastereomer were subjected to anionic oxy-Cope rearrangement in the presence of potassium hexamethyldisilazide and then to air, the intermediate enolates experienced spontaneous α-oxygenation (*29*) to give **47** and **48**. These α-hydroxy ketones were examined for their ability to experience dehydration. This transformation was possible only for **48**, which underwent regioselective conversion to **49** spontaneously during mesylation. Surprisingly, **47** was totally resistant to the loss of water. Subsequent determination that **49** could not be epoxidized under a variety of conditions caused us to turn to a more viable tactic for introduction of the C-7 hydroxyl substituent.

From among the several attractive alternatives, the possibility of deploying ketals constituted of a 2-halo-2-cyclohexenone core has been evaluated. Pilot studies performed with the well-known building block **51** was a forerunner to the utilization of enantiomerically pure **52**, which is available from *D*-(-)-quinic acid (*30*). Relevantly, **52** is so constituted that it possesses a complementary sidechain that is to serve ultimately as the carbon atom of the oxetane ring. In a model study involving **51**, addition to ketone **53** gave alcohol **54** (88%), which was converted into its potassium salt in the usual manner (Scheme 10). To our amazement, the efficacious rearrangement to **55** was complete within 30 min at -40 °C. This discovery, which represented by far the most rapid [3.3] sigmatropic isomerization seen to that time in this series, foreshadowed comparable rate increases for related two-step processes

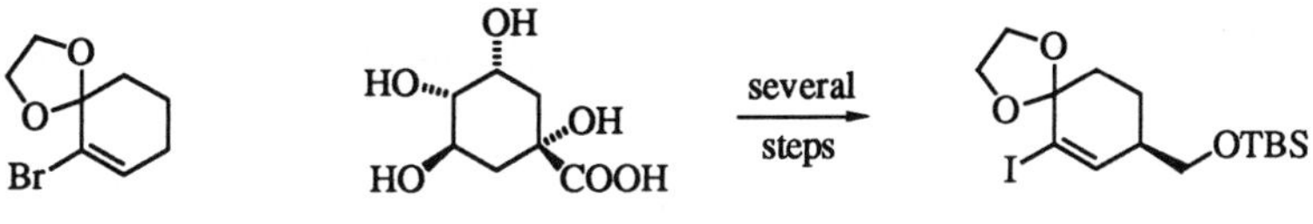

51 **52**

involving other norbornanones. The two-step sequence leading from **54** to **56** has also been implemented. Quite heartening has been our ability to produce **58** from **57** under extremely mild conditions. Since the precursor to **57** is optically pure, the carbinol having the proper absolute configuration at several key centers is formed exclusively. The highly selective nature of this chemistry and the overall brevity of the sequence are obvious and will continue to be exploited.

Scheme 10

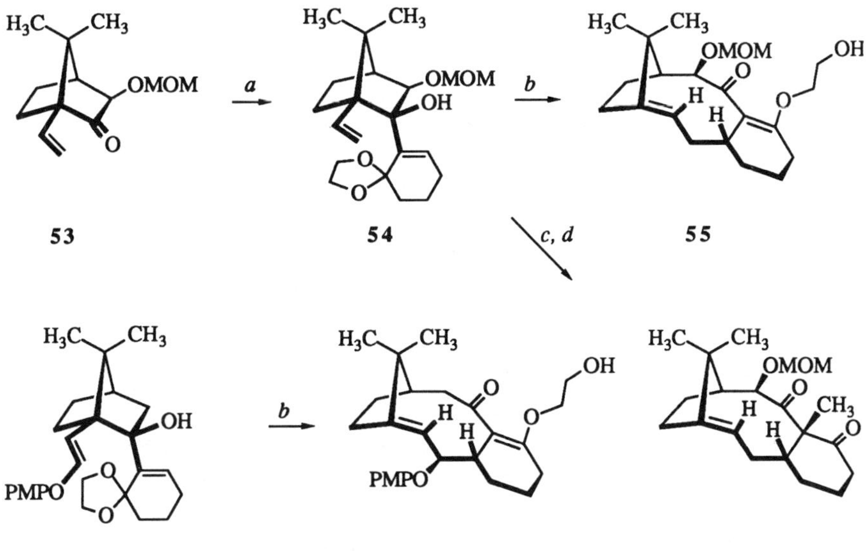

53 **54** **55** **57** **58** **56**

[a] *t*-BuLi on **51**, THF, -78 °C; add **53**. [b] $KN(SiMe_3)_2$, 18-crown-6, THF, -50 °C. [c] PPTS, acetone. [d] KO*t*-Bu, DMSO; MeI

The richness of the substitution level in **56** and **58** requires modest enhancement. A C-2 hydroxyl needs to be incorporated and epimerization of C-3 is mandatory. This stereochemical modification can presumably be realized by unmasking and oxidizing the C-2 hydroxyl. Should this course of action be followed, then the initial stereochemical disposition of the C-O bond is of no real consequence. As a result, the *E*- and *Z*-isomers of the vinyl ether precursor can be used in combination in order to maximize efficiency. The implementation of bridge migration in advance of setting trans B/C ring junction stereochemistry warrants exploration because a greater thermodynamic driving force may underlie the conversion to the natural taxane skeleton. Note that the epimerization will not be beset with potential complications stemming from β-elimination were **52** to be utilized. Reaction economy would be best served if the oxetane ring was subsequently introduced by one of several established options (*8,9,28,31*). The A-ring substitution plan to be adopted will be closely allied to that developed in the taxusin series.

Concurrent with implementation of the above tactics, other stratagems having as their objective a significant reduction in the number of steps associated with this convergent route to taxol will be studied. We are optimistic that the discoveries awaiting us in the course of this streamlining shall serve us proficiently in our quest of the fascinating taxane diterpenes.

Acknowledgments

This overview of our taxane synthesis program has been made possible by the efforts of a small, select group of students whose skill and devotion to chemistry have enabled us to come far. Their names are cited in the references. Collectively, we express our heartfelt thanks to the Bristol-Myers Squibb Company for the requisite financial support. The advances that we have made were facilitated at various times by Professor Robin Rogers (X-ray crystallography), Dr. Dirk Friedrich (NOE and COSY NMR studies), and Dr. Eugene Hickey and Mr. Scott Edmondson (molecular mechanics calculations). We thank them for their valuable input.

Literature Cited

(1) Wani, M. C.; Taylor, H. L.; Wall, M. E.; Coggon, P.; McPhail, A. T. *J. Am. Chem. Soc.* **1971**, *93*, 2325-2327.
(2) Schiff, P. B.; Fant, J.; Horwitz, S. B. *Nature* **1979**, *277*, 665-667.
(3) Suffness, M.; Cordell, G. A. In *The Alkaloids. Chemistry and Pharmacology*; Brossi, A., Ed.; Academic Press: New York, 1985; Vol. 25; pp 3-355.
(4) Kingston, D. G. I. *Pharmacol. Ther.* **1991**, *52*, 1-34.
(5) Swindell, C. S. *Org. Prep. Proced. Int.* **1991**, *23*, 465-543.
(6) Paquette, L. A. In *Studies in Natural Products Chemistry*; Rahman, A. U.; Basha, F. Z., Eds.; Elsevier: Amsterdam, 1992; Vol. 11, pp 3-69.
(7) Kingston, D. G. I.; Molinero, A. A.; Rimoldi, J. M. *Prog. Chem. Org. Nat. Prod.* **1993**, *61*, 1-206.
(8) Nicolaou, K. C., et al *Nature* **1994**, *367*, 630-634.
(9) Holton, R. A.; et al. *J. Am. Chem. Soc.* **1994**, *116*, 1597-1600.
(10) Holton, R. A.; Juo, R. R.; Kim, H. B.; Williams, A. D.; Harusawa, S.; Lowenthal, R. E.; Yogai, S. *J. Am. Chem. Soc.* **1988**, *110*, 6558-6560.
(11) Samaranayake, G.; Magri, N. F.; Jitraugsri, C.; Kingston, D. G. I. *J. Org. Chem.* **1991**, *56*, 5114-5119.
(12) Chen, S.-H.; Wei, J.-M.; Vyas, D. M.; Doyle, T. W.; Farina, V. *Tetrahedron Lett.* **1993**, *34*, 6845-6848 and earlier references cited therein.
(13) Klein, L. L. *Tetrahedron Lett.* **1993**, *34*, 2047-2050.
(14) Chen, S.-H; Huang, S.; Wei, J.; Farina, V. *Tetrahedron* **1993**, *49*, 2805-2828.
(15) Fischer, N.; Opitz, G. *Org. Synth., Collect. Vol. V* **1973**, 877-879.
(16) Paquette, L. A.; Pegg, N. A.; Toops, D.; Maynard, G. D.; Rogers, R. D. *J. Am. Chem. Soc.* **1990**, *112*, 277-283.
(17) Paquette, L. A.; Combrink, K. D.; Elmore, S. W.; Rogers, R. D. *J. Am. Chem. Soc.* **1991**, *113*, 1335-1344.
(18) Elmore, S. W.; Paquette, L. A. *Tetrahedron Lett.* **1991**, *32*, 319-322.

(19) Pegg, N. A.; Paquette, L. A. *J. Org. Chem.* **1991**, *56*, 2461-2468.
(20) Paquette, L. A.; Zhao, M.; Friedrich, D. *Tetrahedron Lett.* **1992**, *33*, 7311-7314.
(21) Elmore, S. W.; Combrink, K. D.; Paquette, L. A. *Tetrahedron Lett.* **1991**, *32*, 6679-6682.
(22) Paquette, L. A.; Elmore, S. W.; Combrink, K. D.; Hickey, E. R.; Rogers, R. D. *Helv. Chim. Acta* **1992**, *75*, 1755-1771.
(23) Paquette, L. A.; Combrink, K. D.; Elmore, S. W.; Zhao, M. *Helv. Chim. Acta* **1992**, *75*, 1772-1791.
(24) Paquette, L. A.; Zhao, M. *J. Am. Chem. Soc.* **1993**, *115*, 354-356.
(25) Paquette, L. A.; Huber, S. K.; Thompson, R. C. *J. Org. Chem.* **1993**, *58*, 6874-6882.
(26) Elmore, S. W. Ph.D. Dissertation, The Ohio State University, 1993.
(27) Elmore, S. W.; Paquette, L. A. *J. Org. Chem.* **1993**, *58*, 4963-4970.
(28) Paquette, L. A.; Thompson, R. C. *J. Org. Chem.* **1993**, *58*, 4952-4962.
(29) Paquette, L. A.; DeRussy, D. T.; Pegg, N. A.; Taylor, R. T.; Zydowsky, T. M. *J. Org. Chem.* **1989**, *54*, 4576-4581.
(30) Paquette, L. A.; Su, Z.; Bailey, S. submitted for publication.
(31) Magee, T. V.; Bornmann, W. G.; Issacs, R. C. A.; Danishefsky, S. J. *J. Org. Chem.* **1992**, *57*, 3274-3276.

RECEIVED August 23, 1994

Chapter 24

The Pinene Path to Taxanes

Genesis and Evolution of a Strategy for Synthesis

Paul A. Wender, Neil F. Badham, Simon P. Conway, Paul E. Floreancig, Timothy E. Glass, Jonathan B. Houze, Nancy E. Krauss, Daesung Lee, Daniel G. Marquess, Paul L. McGrane, Wei Meng, Thomas P. Mucciaro, Michel Mühlebach, Michael G. Natchus, Takeshi Ohkuma, Bernd Peschke, David B. Rawlins, Anthony J. Shuker, Jim C. Sutton, Richard E. Taylor, Katsuhiko Tomooka, and Ludger A. Wessjohann

Department of Chemistry, Stanford University, Stanford, CA 94305

"Clinical trials of taxol suggest that it may be one of the most promising anticancer drugs discovered during the last 10 years."
Rustin Howard, What's Happening in Chemistry, the American Chemical Society, 1992, p. 8.

"Without it [taxol], I don't think I'd be here now."
Audrey Avansino, the first patient at Stanford University Medical Center to receive taxol

Taxol is a recent and striking example of the continuing impact of natural products chemistry on science and human health and a reminder of the significant return that is realized through investment in basic research. First isolated in the sixties from the bark of the Pacific Yew by chemists Wall, Wani, and their colleagues working in collaboration with the National Cancer Institute (*1*), taxol was found to possess a novel polycyclic structure (**1**) punctuated by a central eight-membered ring. Interest in the chemotherapeutic potential of this compound was stimulated by early cytotoxicity studies and greatly amplified by the subsequent proposal by Horwitz and coworkers that taxol operates through a new molecular mode of action involving the facilitated assembly and stabilization of microtubules (*2*). Clinical trials of taxol in the eighties proved to be sufficiently promising to warrant its subsequent approval for use in the treatment of ovarian cancer and, more generally, to encourage further studies on its efficacy in the treatment of other cancer types (*3*). A curiosity some thirty years ago, the natural product taxol is now a promising chemotherapeutic agent and a significant lead in the development of new treatments for cancer.

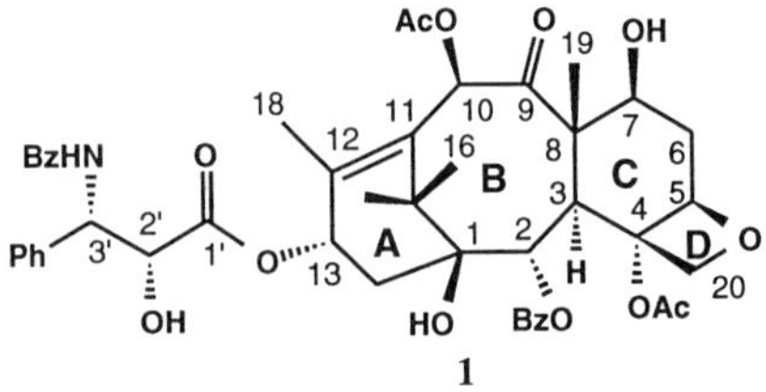

1

NOTE: Paclitaxel is the generic name for Taxol, which is now a registered trademark.

0097–6156/95/0583–0326$08.00/0

This chapter provides an overview of research conducted in the principal author's laboratory directed simultaneously at the synthesis of taxol and its analogs, the elucidation of its molecular mode of action, and the design of superior agents. Due to space constraints the emphasis of this overview is placed on the genesis and development of our synthesis program including some of its recent directions.

Taxol and the Transition Metal Catalyzed [4+4] Cycloaddition Strategy

Our initial interest in taxol was stimulated in the late seventies by its novel biological activity and structure, particularly its substitutionally complex eight-membered ring. At the time, relatively little attention had been given to the problem of eight-membered ring synthesis (*4*). This situation changed rapidly, however, as new structures were reported and as the medicinal potential of the taxanes grew. Prompted by these developments and the growing need for general methodology in this area, we sought to develop an approach to the synthesis of eight-membered rings that would address the taxane problem but would, in addition, be sufficiently flexible to be applicable to the synthesis of other targets incorporating eight-membered rings. Of the four classes of methods for ring formation (i.e., closure of an acyclic precursor, ring expansion/contraction, fragmentation, and cycloaddition), our attention was drawn initially to a cycloaddition process since it was the least developed approach to eight-membered rings and it offered the utility long appreciated and exploited in cycloaddition routes to other ring systems.

There are four classes of two bond connective cycloadditions for the synthesis of eight-membered rings: [1+7], [2+6], [3+5], and [4+4]. While precedent for the last is found in the seminal and impressive studies of Reed, Wilke, and others on transition metal mediated diene-diene cycloadditions (*5*), the [4+4] cycloaddition had not found use in complex molecule synthesis at the outset of our studies due in part to problems associated with poor selectivity and reactivity. For example, while the cyclodimerization of butadiene is of industrial importance, cycloadditions of even simple methyl-substituted butadienes are found to proceed more slowly, if at all, and often provide complex mixtures of regio- and stereoisomeric products (*6*). Crossed cycloadditions involving differently substituted dienes produce even more complex product mixtures.

To circumvent the problems of diminished reactivity and selectivity found in the *intermolecular* cycloaddition of substituted dienes, an initial investigation was conducted in our laboratory on the viability of the *intramolecular* [4+4] cycloaddition (*7*). While slowed by difficulties in finding a suitable catalyst, this study eventually provided the basis for an effective and general route to bicyclic systems incorporating an eight-membered ring. Pertinent to the goal of developing a route to taxanes was the finding (Scheme 1) that the cycloaddition of **2a** provided **3a** in high stereoselectivity (>65:1) and yield (84%). This adduct possesses the trans-fused ring juncture of taxol and an ester appendage corresponding to its C20 group. Given the steric problems encountered with increased diene substitution in the intermolecular reactions, it is noteworthy that further substitution of the diene in this intramolecular reaction has a negligible effect. The methyl substituted diene **2b** gives **3b** with even better selectivity (*ca.* 100%) and in higher yield (92%) than is found for **2a**. This cycloaddition methodology has also been used to produce the AB bicyclic core of the taxanes as illustrated (Scheme 2) in the type II cycloaddition of bis-diene **4** which gives cycloadduct **5** in 52% yield (*8*).

Scheme 1

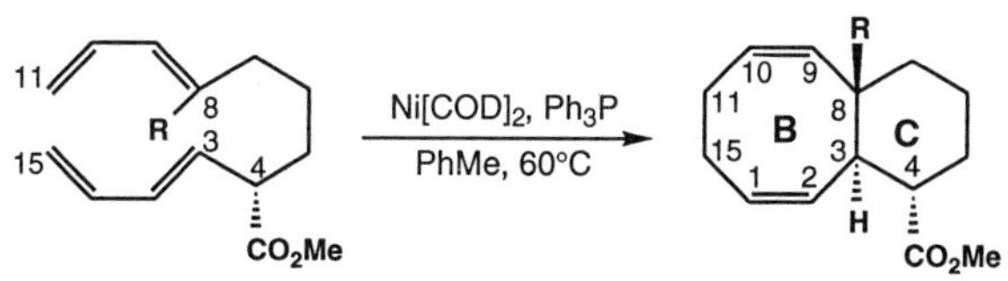

2a : R=H **3a : R=H (84%)**

2b : R=Me **3b : R=Me (92%)**

Scheme 2

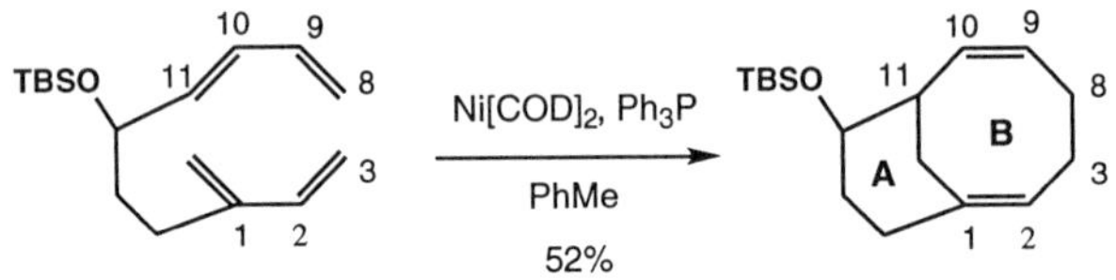

4 5

With the demonstration of the viability of this [4+4] cycloaddition methodology, numerous conceptually novel routes to taxanes became plausible. For example, the BC bicyclization route to **3b** lends itself nicely to introduction of the B-ring functionality of taxol (Scheme 3) since all four sites of oxygenation in taxol appear in this cycloadduct as alkene carbons which can be selectively oxidized. Thus, the C9-C10 double bond of cycloadduct **3b** can be readily epoxidized to give ester **6** (*9*). Subsequent hydrolysis of the ester and halolactonization provides lactone **7**, allowing for introduction of the C2 carbon-oxygen bond and versatile functionality at C1. Alternatively, catalytic osmylation of **3b** gives diol **8** which upon oxidation allows for alkylative introduction of the A-ring carbons at C11. Overall, this methodology shows great promise for the elaboration of taxanes and represents a useful general tool for the synthesis of complex molecules incorporating eight-membered rings (*10*).

Scheme 3

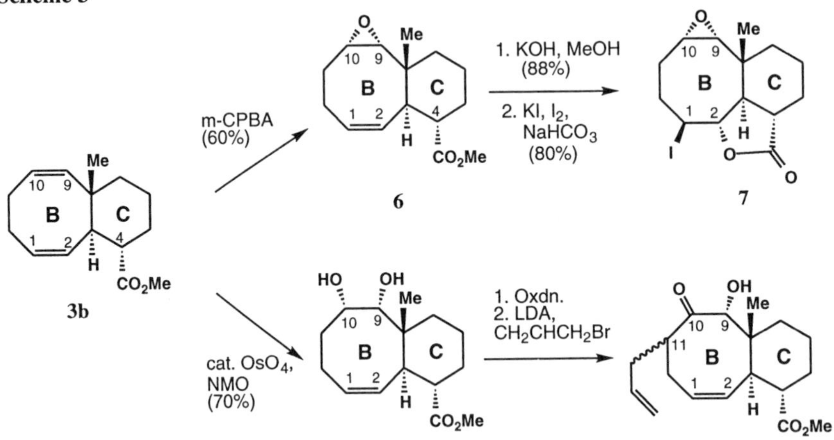

3b 6 7 8 9

Clinical Trials and the Emergence of the Pinene Path to Taxanes

While studies on the cycloaddition approach to taxanes continued in our laboratory, the strategic focus of our research was expanded in 1989 when the chemotherapeutic potential of taxol became more apparent from highly promising results of clinical trials conducted with ovarian cancer patients (*3*). At this point, our research priorities expanded from the development of general methodology to include the more specific development of a practical synthesis of taxol and its analogs, the use of this synthetic expertise in the elucidation of the molecular mode of action of taxol, and the use of this mode of action information in the design and development of superior analogs. Given that a long synthesis would neither address the taxane supply problem nor service mode of action studies, for which rapid access to systematically varied analogs is essential, an overriding emphasis was placed on developing a plan that would deliver the taxane core in under 10 steps and credible analogs as well as taxol itself in under approximately 25 steps. In retrospect, the demands of practicality and brevity proved to be as significant in the development of our strategy as any chemical consideration.

Connectivity Analysis and Strategic Considerations. The design of our strategy for the synthesis of taxanes was influenced at the time by a connectivity analysis that we and others had been independently developing (*11*). In this analysis, atoms of a target structure are treated as points and bonds as lines, thereby converting the structural symbolism of organic compounds into a graphical representation. As is the case for chemically biased retrosynthetic analysis, this graph based approach seeks to reveal how a complex target graph (e.g., taxol) could be assembled from a simpler precursor graph (a chemical intermediate) through the disconnections of lines (bonds) between points (atoms). The attractiveness of connectivity analysis is that it serves to identify strategies based on *what is possible* irrespective of the current state of synthetic methodology, i.e., *what is known* or more specifically *what one knows*. Thus, all options are identified and each can be prioritized according to its capacity to simplify the target.

Of special significance is the fact that connectivity analysis is exhaustive and easily executed. Thus, the taxane carbotricyclic core can be simplified in (only!) 153 ways: 17 one line (bond) and 136 two line (bond) disconnective approaches, 27 of which are given in Table I.

Table I. Two Bond (Line) Disconnections Corresponding to Convergent Approaches to the ABC Ring System of the Taxanes

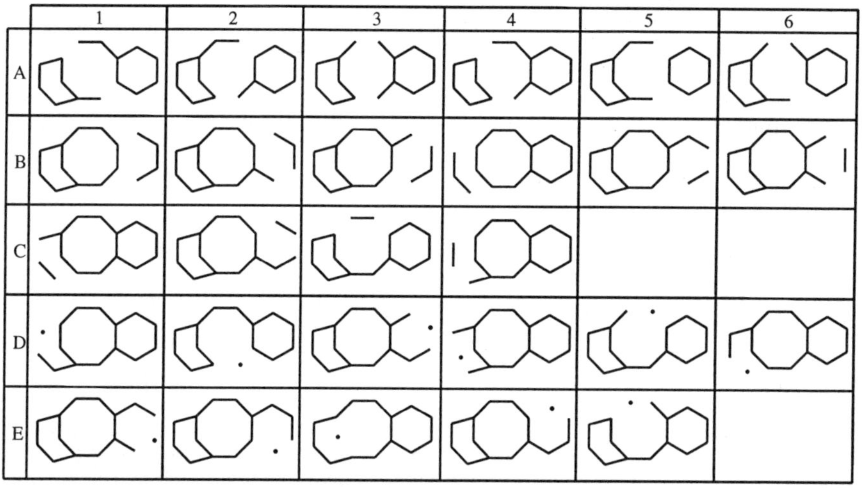

While not restricted to one and two line (bond) disconnections, this analysis emphasizes these two types since most methodology allows for the formation of only one or two bonds (lines) in a given reaction type. To convert these graphs (points and lines) into chemical opportunities (atoms and bonds), one need only add functionality that would allow for bond formation to occur between two or more unconnected atoms. Each graph can thus have many chemical translations, although these typically involve complementary reactivities associated with the major reactive intermediates (e.g., an electrophilic center combining with a nucleophilic center, a radical with an acceptor, a carbene with a π- or σ-bond, a carbon-metal bond with a π-system, etc) and pericyclic reactions. Scheme 4 illustrates how disconnection A3, for example, can be modified to exploit symmetry and then enabled chemically to produce an exceptionally concise but unexploited route to the taxane ABC ring system (*12*).

Scheme 4

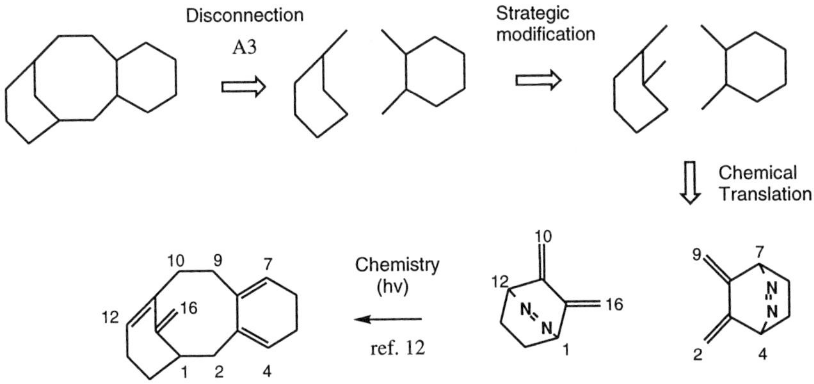

The value of connectivity analysis in this and other contexts derives in part from its utility in defining what is possible and thereby allowing one to begin to consider and prioritize *all* strategic options. As this is applied to taxane synthesis design, it becomes apparent that of the 136 two bond disconnections, only 27 produce two fragments (i.e., correspond to intermolecular processes in the bond forming direction). These represent convergent approaches (Table I). Notwithstanding some perceptions, a convergent synthesis does not necessarily reduce the number of steps in a synthesis plan, but rather it reduces only the longest linear sequence (of steps). In the final analysis, all steps in a synthesis figure in the overall effort and cost. The main advantages of convergency are that the syntheses of two or more fragments of the target can be shortened and developed simultaneously, thereby increasing material throughput, facilitating correction of a chemical problem such as an uncooperative protecting group, and providing more rapid access to fragment variations as required for example when a synthesis is used to generate structurally diverse analogs. Generally, the benefits of convergency are greatest for disconnections of a target that produce precursors of comparable size. For the carbotricyclic core of taxol, connectivity analysis thus reveals that only 6 of the 27 convergent approaches produce fragments of comparable size (A1-A6). It is noteworthy that all six involve disconnections of the eight-membered B-ring and all six have figured heavily in current strategies for taxane synthesis, including the pinene path (*4*).

The Pinene Path. Further direction in formulating a strategy based on connectivity analysis derives from the recognition that the six maximally simplifying convergent disconnections (i.e., A1-A6) all lead to precursors containing six-membered rings. This consideration, coupled with the general value of introducing stereochemistry as early in a synthesis as possible and of starting with commercial materials incorporating as many of the target atoms as possible, led to an examination of commercially available, enantiomerically enriched compounds based on six-membered rings. For reasons associated with the difficulty of assembling quaternary centers, this search (*13*) converged on compounds possessing a geminal dimethyl functionality such as α-ionone and eventually on what proved to be a superb candidate, the monoterpene α-pinene (**10**: Scheme 5). As a starting material for the synthesis of taxol, α-pinene is seen to have many attributes. It possesses 10 of the 20 atoms of the carbotricyclic core of taxol. It is available in either enantiomeric form and from a practical point of view it is inexpensive, being the major component of the industrial solvent turpentine. One could not ignore the additional aesthetic attractiveness of using the abundant constituent of one tree to produce the trace component of another.

Scheme 5

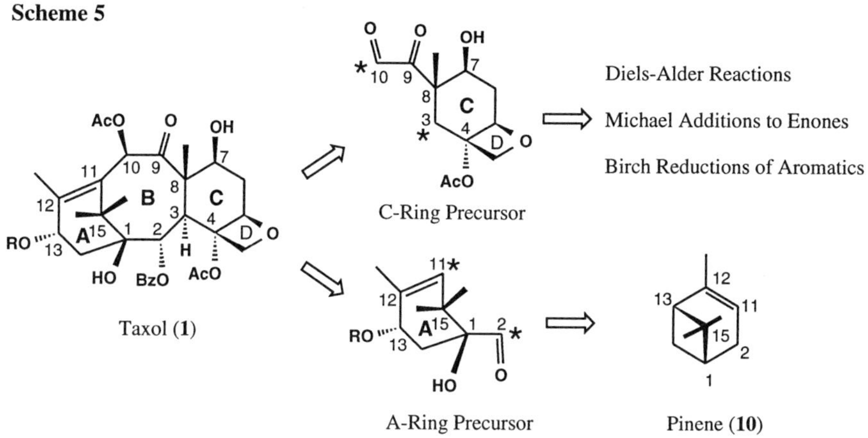

Notwithstanding the apparent advantages of α-pinene, its relationship to taxol and specifically the A-ring subunit of taxol is not immediately obvious. However, another striking example of the return on investment in basic research is found here as the relationship between pinene and the taxanes draws on basic research of the fifties and sixties. Specifically, α-pinene was shown (*14*) to be readily oxidized to verbenone (**11**: Scheme 6). Secondly, it was demonstrated in the late fifties (*15*) that verbenone can be photoisomerized to chrysanthenone (**12**), thereby suggesting how the connectivity of atoms in pinene could be changed to that required for the A-ring of taxol. Finally, it was demonstrated in the early sixties that chrysanthenone can be selectively epoxidized and that the resultant product (**13**) can be fragmented to the hydroxy acid **15** (*16*), a compound exhibiting a striking similarity to the A-ring of taxol.

Consideration of these earlier studies in connection with the present synthetic objectives led to the formulation of a remarkably concise route to the carbotricyclic core of taxanes (Scheme 7). In this plan, pinene (**10**) would be oxidized to verbenone and the latter alkylated with a C-ring precursor to give **16**, a process that would establish the C10-C11 bond of the B-ring. The resultant product would then be isomerized to the chrysanthenone derivative **17**. Epoxidation of the latter would set the stage for nucleophilic addition of C-3 to the carbonyl of the chrysanthenone subunit, thereby establishing the C2-C3 bond of the B-ring and producing a species (**19**) similar to that (**14**) involved in the fragmentation of the epoxide of chrysanthenone (*16*). Subsequent fragmentation would produce the taxane tricyclic core (**20**). In all, from commercially available α-pinene, the ABC-ring system of the taxanes would be assembled in the correct enantiomeric form in as few as five steps. Due to its potential to accommodate variations in the C-ring precursor and its brevity, this strategy was expected to provide the basis for an exceptionally concise route to a wide range of analogs of taxol as needed for mode of action studies and a potentially practical synthesis of taxol itself.

Scheme 6

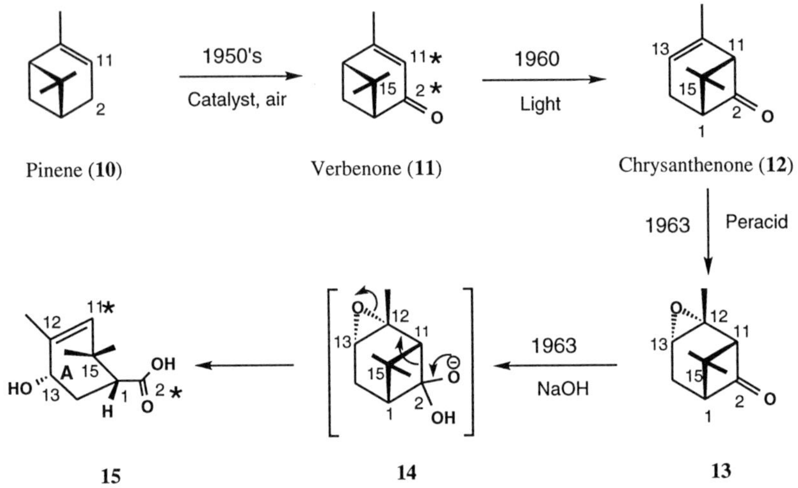

Scheme 7

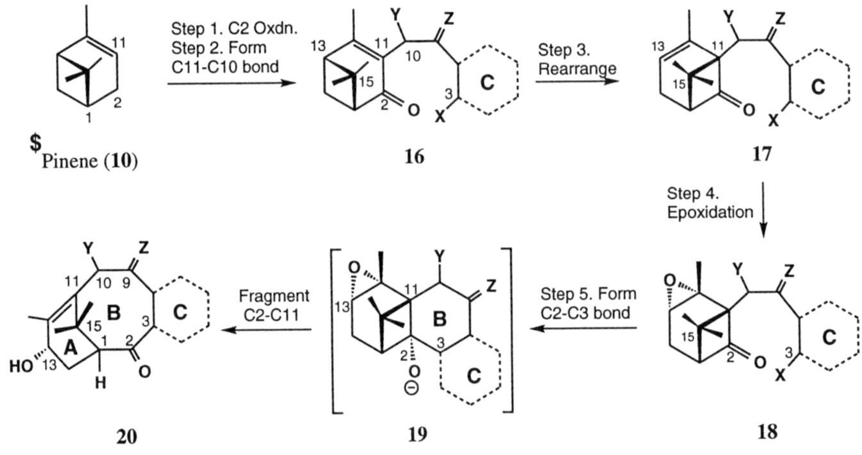

Feasibility of the Pinene Path

Six Steps to the Carbotricyclic Core of the Taxanes. Evaluation of the above strategy (*17*) started with the alkylation of verbenone (**11**) (Scheme 8) with dibromide **21** whose aromatic subunit was selected to serve as a potential taxol CD-ring precursor (*18*). Photoinduced rearrangement of the resultant product (**22**) proceeded uneventfully to give the chrysanthenone derivative **23** in 85% yield. In accord with previous mechanistic studies (*15*), some racemization was observed in this process. While **23** could be epoxidized with m-CPBA (81%), efforts to form the B-ring by lithium halogen exchange followed by 1,2-addition to the bridging carbonyl provided mostly the product of simple protonation at C3. The aryllithium intermediate was apparently formed but its addition to the carbonyl was inhibited. This result is consistent with the expectations that addition to the α-face of the carbonyl would incur strain arising from the formation of a trans-fusion between a four- and six-membered ring while β-face addition would be sterically encumbered by the bridging methyl group. Computer modeling suggested that addition to the carbonyl β-face would be less encumbered if the epoxide were replaced with an alkene. In accord with this analysis, carbonyl addition to the chrysanthenone derivative **23** proceeded readily, affording the tetracyclic product **24** in 67% yield.

Scheme 8

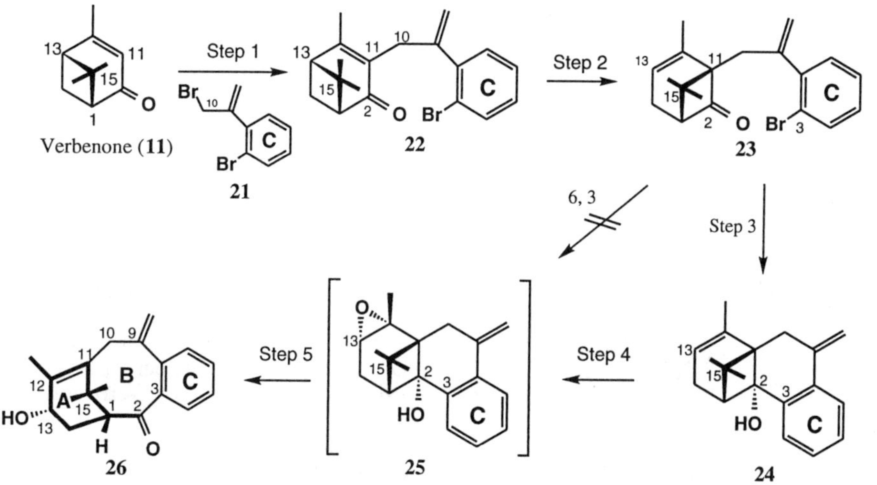

1) KO-*t*-Bu, DME, -78°C; then **21** (48-60%); 2) hv, Pyrex, C_6H_{12} (85%); 3) *t*-BuLi, TMEDA, THF -78°C to 25°C (67%); 4) *t*-BuOOH, Ti(O-*i*-Pr)$_4$, CH_2Cl_2, 0°C (70%); 5) DABCO, MeCN, reflux (80%); 6) *m*-CBPA, NaOBz, CH_2Cl_2, 0°C (81%);

It is noteworthy that the internal nucleophilic addition leading to formation of the C2-C3 bond is also sensitive to steric effects and trajectory requirements imposed by the nucleophilic C3 center and the C-ring. Alkenyl and aryl anions work well in this process, while enolates are less effective, notwithstanding their enormous strategic potential. For example, enantiomerically enriched ketones such as **27** (Scheme 9) are readily available and easily attached to the chrysanthenone subunit (*13*). However, when the alkylated chrysanthenone **28** was treated with base, no evidence for 1,2-addition leading to the formation of **29** was observed (*19*). This result is attributable in part to the different trajectories for C2-C3 bond formation expected for the enolate and aryllithium species, the greater steric interaction associated with the development of a bonding relationship between the C2 carbonyl and the enolate π-system relative to the sp^2-hybridized carbanionic site of the aryllithium and the reversibility of the enolate condensation. In addition, further steric congestion would arise in the enolate addition due to interaction of the sp^3-hybridized center at C8 with the bridging methyl group.

Scheme 9

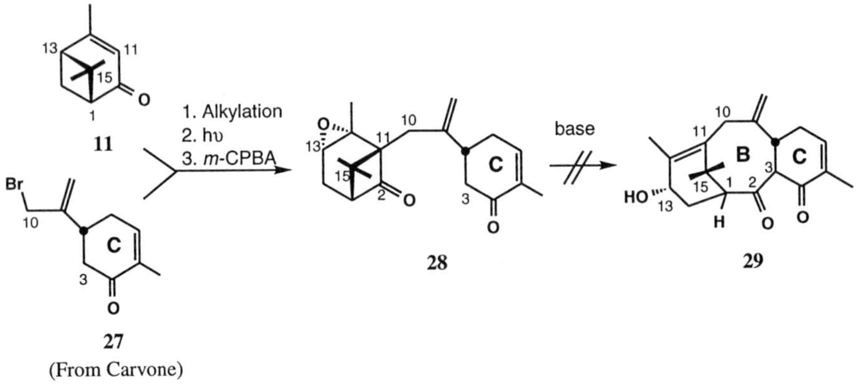

The reluctance of the epoxide of **23** (Scheme 8) to undergo internal carbonyl addition did not significantly alter our original plan since by resequencing the addition and epoxidation steps the desired fragmentation precursor was obtained. Accordingly, **23** was first converted to the tetracyclic **24**. While the subsequent epoxidation step was initially problematic, producing rearrangement products when **24** was treated with mCBPA, it was eventually achieved with t-BuOOH/ Ti(O-i-Pr)$_4$. Fragmentation of resultant epoxide (**25**) was accomplished with DABCO, providing the carbotricycle **26** as two atropisomers (*20*) in 80% yield (16% overall yield and 5 steps from verbenone).

Introduction of the C1 Alcohol. At this point in the synthesis, attention was directed at introduction of the C1 oxygen, the one taxol feature that pinene lacked and a functionality required for establishing a general synthesis of taxanes. Examination of computer generated models of the fragmentation product **26** (Scheme 10: **26'** and **26"**) revealed that this limitation could be overcome through enolization of the C2 ketone which is possible only in the conformer corresponding to **26"** (H-C1-C2-O dihedral angle of ca. 100°). However, NMR analysis indicated that this was in fact the minor isomer and that it did not readily interconvert into the major isomer at room temperature. For example, when the C13 protected ketone **27** was treated with LDA at 0°C and the reaction mixture quenched with acetic acid-d$_4$, only starting material was obtained. Fortunately, it was found that exchange of the C1 hydrogen for deuterium could be achieved by treatment of **27** with NaOMe in methanol-d$_4$/THF-d$_8$ at 60°C. More importantly, when ketone **27** was treated with KO-*t*-Bu and oxygen in DMSO/THF at 60°C, the desired C1 hydroxylated product **28** was obtained in 80% yield.

C2 Stereochemistry. The next goal of this strategy was to set stereochemistry at C2. As suggested by enolization and molecular modeling studies, ketone **28** (Scheme 11) assumes a preferred conformation in which the α-face of the carbonyl group is exposed for hydride addition, a process that would lead to the undesired C2 stereochemistry. This was indeed observed. When ketone **28** was reduced with LiAlH$_4$, the β-C2-OH stereoisomer **29** was obtained. On the other hand, under thermodynamic control, reduction of this ketone gave the desired α-C2-OH stereoisomer **30** in 60% yield.

Scheme 10

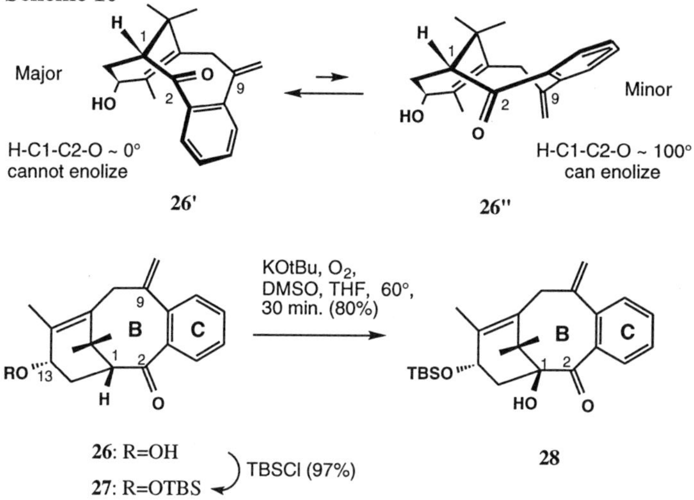

Scheme 11

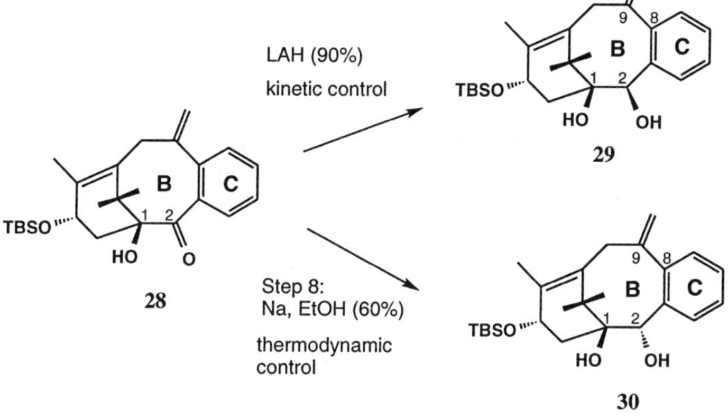

C-Ring Studies. At this point in our strategy, every carbon of α-pinene had been mapped into exactly the correct position and stereochemical and functional state required for taxol. What remained was elaboration of the C-ring. Toward this end, the C1,C2 diol was protected as an acetonide, a form of protection that was expected to conformationally bias the molecule for subsequent control of C8 stereochemistry (Scheme 12: see **33**). The exo-olefin at C9, having served well as a ketone protecting group, was oxidatively cleaved to afford ketone **31**. When this ketone was treated with potassium in ammonia followed by methyl iodide, three products were obtained: the 1,2-reduction product **32a**, its methyl ether **32b**, and the desired alkylated ketone **34**. With this demonstration of the viability of this uniquely short route to taxanes, our efforts turned to versions of this plan that offered even greater flexibility for taxane synthesis. This inaugural test of the pinene path to taxol established an efficient and general protocol for the translation of pinene into an A-ring precursor of the taxanes with provision for attachment to a C-ring precursor, introduction of the C1 oxygen and C9 carbonyl, and control of the C1, C2, and C13 stereochemistry.

Scheme 12

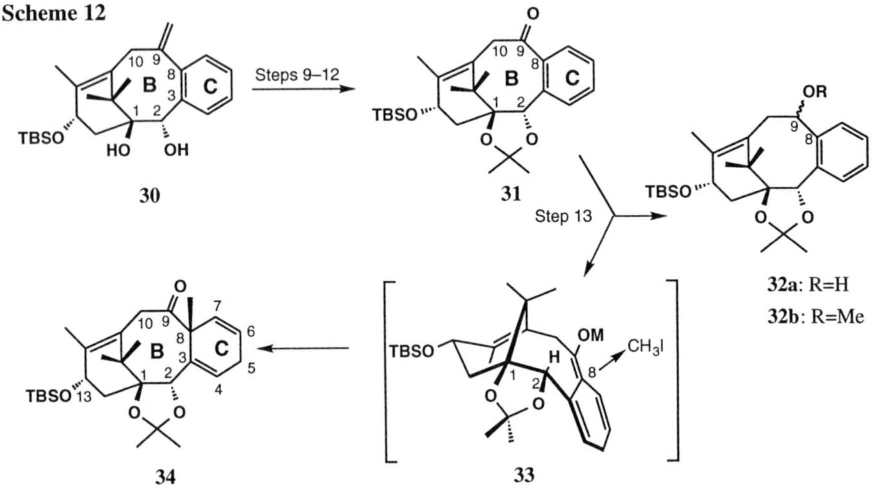

9) *p*-TSA, 2,2-dimethoxypropane, reflux (99%); 10) 9-BBN, THF, 60°C (83%); 11) Swern oxidation (60-80%); 12) KO-*t*-Bu, O_2, THF (60%); 13) K/NH_3, *t*-BuOH, THF; MeI (see text).

Methoxy-Aryl C-Ring Precursors. The pinene path strategy, established initially with a simple arene serving as a potential C-ring precursor, has been found to readily accommodate other C-ring precursor variations. Due to space constraints these options cannot be detailed but they proceed along a path closely paralleling the original viability study. For example, through Birch reduction and further functionalization of a methoxy substituted arene, the C-ring diene **35** (Scheme 13) can be prepared in 19 steps from pinene (*21*). Reaction of the carbonate subunit with phenyllithium allows for introduction of the C2 benzoate in **36** (*22*). The methoxy group serves in this plan to provide direct access to C4 oxidized intermediates. Moreover, by virtue of its C3 stereochemistry, enone **36** undergoes reductive alkylation from the more accessible β-face to provide ketone **37**, a flexible intermediate for the synthesis of a wide range of taxol analogs including the biologically active 7,10-dideoxytaxanes (*23*).

Scheme 13

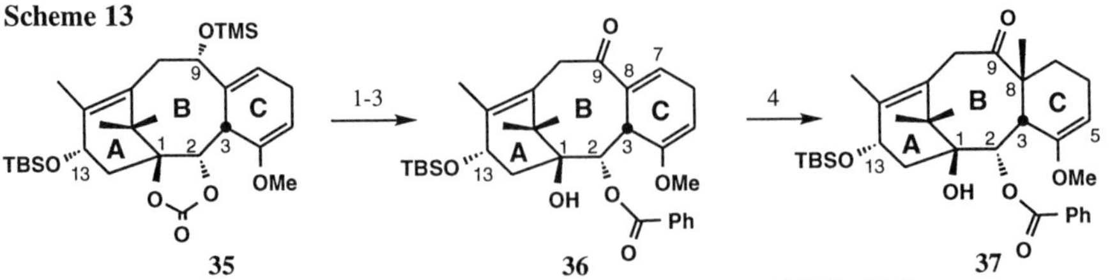

1) PhLi; 2) HOAc, MeOH; 3) Dess-Martin periodinane oxidation; 4) Li/NH_3; MeI

Non-Aromatic C-Ring Precursors. The pinene path strategy also accommodates non-aromatic precursors such as bromoenones (Scheme 14). For example, **39**, produced in 5 steps from verbenone, can be readily converted to the tricycle **40** in only 4 additional steps (*24*). C2-C3 bond formation in this case is achieved through carbanionic addition to a C2-carbonyl, the carbanion arising from a vinyl bromide rather than the aryl bromide employed with aromatic C-ring precursors. Subsequent oxidation, epoxidation and fragmentation gives enone **40**, possessing the tricarbocyclic core of the taxanes and functionality for the elaboration of taxol and its analogs. By accommodating more functionalized CD-ring precursors, this approach serves to decrease the number of post coupling operations needed to complete the synthesis.

Scheme 14

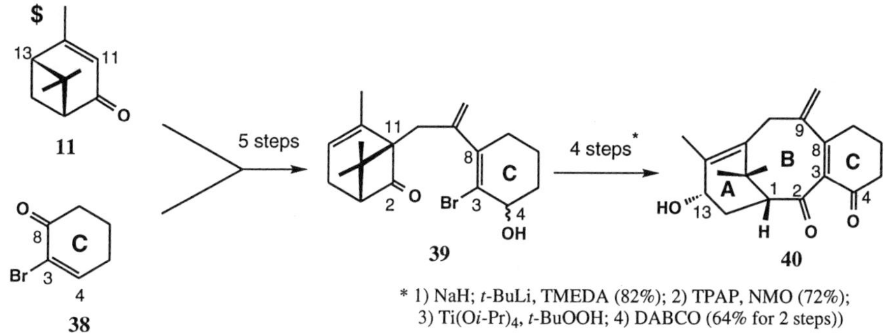

* 1) NaH; *t*-BuLi, TMEDA (82%); 2) TPAP, NMO (72%); 3) Ti(O*i*-Pr)$_4$, *t*-BuOOH; 4) DABCO (64% for 2 steps))

Scheme 15

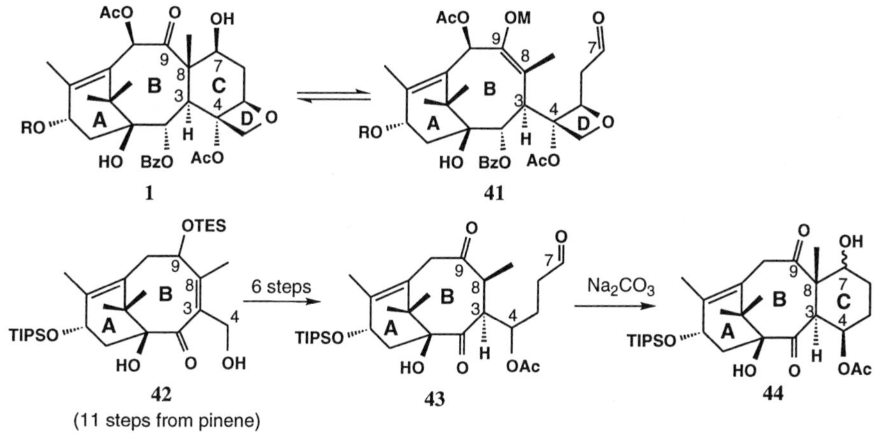

C-Ring Elaboration through an Aldol Closure. Finally, it is particularly noteworthy that the pinene path used above with pre-formed C-rings can also be used to produce AB-ring systems possessing an appendage at C3 which can subsequently be used to form the C-ring. This approach is based on a consideration of the mechanism of C7 epimerization for taxol and its derivatives (*25*) which presumably involves a retro aldol-aldol process (Scheme 15: **1** to **41**). We have found that similar aldol closures of bicyclic systems derived from the pinene path strategy can be readily accomplished (*26*). As illustrated for the conversion of ketoaldehyde **43** to the tricycle **44**, this process has proven to be a particularly effective method for C-ring construction, providing access to a versatile taxane precursor in only 18 steps overall from commercially available pinene.

In summary, our ongoing studies have resulted in a number of fundamental methodological and strategic advances relevant to taxol research and the more general problem of eight-membered ring synthesis. The intramolecular [4+4] cycloaddition that precipitated our initial involvement in this area has developed into a general strategy level reaction for the synthesis of eight-membered rings. While further theoretical, mechanistic, and synthetic studies (*10, 27*) are needed to exploit the enormous potential of this process, it has proven to be highly useful in the synthesis of advanced taxane precursors and in the total synthesis of other complex molecules incorporating an eight-membered ring. The pinene path strategy provides a uniquely short route to the carbotricyclic core of the taxanes in correct enantiomeric form and a potentially concise route to taxol itself. This strategy allows for the construction of the A- and B-rings of the taxanes with provision for control of stereochemistry and functionality. Its flexibility is indicated by the range of C-ring precursors that can be incorporated into the basic plan including aromatic and non-aromatic rings and a C3 appendage which can be used to form the C-ring through an aldol reaction. This strategy has already been used to make analogs of taxol and should prove useful for the synthesis of taxol itself.

Acknowledgments

The research outlined in this review is the product of an exceptionally creative and dedicated group whose support has been provided by the National Science Foundation (CHE-9321676), the National Institutes of Health (CA31845), and the Bristol-Myers Squibb Company. We are particularly indebted to Drs. S. Broder, B. Chabner, and especially M. Suffness of the NIH and Dr. T. Doyle of BMS for their support and encouragement and to Qi Gao of BMS for assistance with X-ray analyses.

Literature Cited

1. Wani, M. C.; Taylor, H. L.; Wall, M. E.; Coggon, P.; McPhail, A. T. *J. Am. Chem. Soc.* **1971**, *93*, 2325.
2. a) Schiff, P. B.; Fant, J.; Horwitz, S. B. *Nature (London)* **1979**, *277*, 665; b) Horwitz, S. B. *Trends Pharmacol. Sci.* **1992**, *13*, 134.
3. a) Slichenmyer, W. J.; Von Hoff, D. D. *J. Clin. Pharmacol* **1990**, *30*, 770; b) Slichenmyer, W. J.; Von Hoff, D. D. *Anti Cancer Drugs* **1991**, *2*, 519; c) Rowinsky, E.K.; Cazenave, L.A.; Donehower, R.C. *J. Natl. Cancer Inst.* **1990**, *82*, 1247; d) Suffness, M. *Ann. Rep. Med. Chem.* **1993**, *28*, 305.
4. For reviews on the taxanes and the general problem of eight-membered ring synthesis, see: a) Suffness, M.; Cordell, G.A. *The Alkaloids* **1985**, *25*, 1; b) Suffness, M. *Rep. Med. Chem.* **1993**, *28*, 305; c) Blechert, S.; Guénard, D. *The Alkaloids* **1990**, *39*, 195; d) Chen, H.; Huang, S.; Wei, J.; Farina, V. *Tetrahedron* **1993**, *49*, 2805; e) Guénard, D.; Guéritte-Voegelein, F.; Potier, P. *Acc. Chem. Res.* **1993**, *26*, 160; f) Khan, N. U.-D.; Parveen, N. *J. Sci. Ind. Res.* **1987**, *46*, 512; g) Kingston, D. G. I.; Molinero, A. A.; Rimoldi, J. M. *Progress in the Chemistry of Organic Natural Products*; Springer-Verlag: New York, 1993; Vol. 61, pp 3; h) Miller, R.W. *J. Nat. Prod.* **1980**, *43*, 425; i) Kingston, D. G. I. *Pharmacol. Ther.* **1991**, *52*, 1; j) Lythgoe, B. *The Alkaloids* **1968**, *10*, 597; k) Morelli, I. *Fitoterapia* **1976**, *47*, 31; l) Paquette, L. A. In *Studies in Natural Products Chemistry*; A.-u.-Rahman, Ed.; Elsevier: 1992; Vol. 11; pp 3; m) Swindell, C. S. *Org. Prep. Proc. Intl.* **1991**, *23*, 465; m) Nicolaou, K.C.; Dai, W.-M.; Guy, R.K. *Angew. Chem. Int. Ed. Engl.* **1994**, *33*, 15; n) Wender, P.A.; Natchus, M.G.; Shuker, A.J. *The Taxanes* Suffness, M., Ed., *in press*. For reviews pertinent to cycooctane syntheses, see: o) Ihle, N. C. Ph.D. Thesis, Stanford University, 1988; p) Feldman, K. S.; Come, J. H.; Kosmider, B. J.; Smith, P. M.; Rotella, D. P.; Wu, M.-J. *J. Org. Chem.* **1989**, *54*, 592; q) Petasis, N. A.; Patane, M. A. *Tetrahedron* **1992**, *48*, 5757. For a discussion of issues relating to eight-membered ring formation, see: r) Mundy, B. *Concepts of Organic Chemistry*; Marcel Dekker: New York, 1979, pp 50-57; s) Illuminati, G.; Mandolini, L. *Acc. Chem. Res.* **1981**, *14*, 95. For the synthesis of taxol, see: t) Holton, R.A.; Kim, H.-B.; Samosa, C.; Liang, F.; Blediger, R.J.; Boatman, P.D.; Shindo, M.; Smith, C.C.; Kim, S.; Nadizadeh, H.; Suzuki, Y.; Tao, C.; Vu, P.; Tang, S.; Zhang, P.; Murthi, K.K.; Gentile, L.N.; Liu, J.H. *J. Am. Chem. Soc.* **1994**, *116*, 1597, 1599; u) Nicolaou, K.C.; Yang, Z.; Liu, J.J.; Ueno, H.;

Nantermet, P.G.; Guy, R.K.; Claiborne, C.F.; Renaud, J.; Couladouros, E.A.; Paulvannan, K.; Sorensen, E.J. *Nature* **1994**, *367*, 630.

5. Jolly, P.W.; Wilke, G. *The Organic Chemistry of Nickel, vol. II* **1975**, Academic Press, New York, pp 133-212.
6. Jolly, P.W. *Comp. Organomet. Chem.* **1982**, *56.4*, 671
7. In order to facilitate comparisons, the taxane numbering system is used for all structures in this chapter. Wender, P. A.; Ihle, N. C. *J. Am. Chem. Soc.* **1986**, *108*, 4678; Wender, P. A.; Ihle, N. C. *Tetrahedron Lett.* **1987**, *28*, 2451.
8. Wender, P. A.; Snapper, M. L. *Tetrahedron Lett.* **1987**, *28*, 2221; Wender, P. A.; Tebbe, M. J. *Synthesis* **1991**, 1089.
9. Snapper, M. L. Ph.D Thesis, Stanford University, 1990.
10. Wender, P.A.; Ihle, N.C.; Correia, C.R.D. *J. Am. Chem. Soc.* **1988**, *110*, 5904.
11. Wender, P.A.; Miller, B.L. In *Organic Synthesis, Theory and Applications*; T. Hudlicky, Ed.; JAI Press: 1993; Vol. 2; pp 27-66; b) Mehta, G.; Barone, R.; Azario, P.; Barberis, F.; Arbelot, M.; Chanon, M. *Tetrahedron* **1992**, *48*, 8953 and references cited therein.
12. Roth, W. R.; Erker, G. *Angew. Chem. Int. Ed. Engl.* **1973**, *12*, 503.
13. Mucciaro, T.P. Ph.D. Thesis, Stanford University, 1992.
14. Moore, R.N.; Fisher, G.S. *J. Am. Chem. Soc.* **1956**, *78*, 4362; Kizlink, J.; Hronec, M.; Cvengrosova, Z.; Kuruc, L.; Kriz, M.; Oblozinsky, A.; Ilavsky, J. *Czech CS* **1986**, 634 (*Chem. Abstr.* **1989**, *111*, 214768r).
15. a) Hurst, J.J.; Whitham, G.H. *Proc. Chem. Soc.* **1959**, 160; b) Hurst, J.J.; Whitham, G.H. *J. Chem. Soc.* **1960**, 2864; c) Erman, W.F. *J. Am. Chem. Soc.* **1967**, *89*, 3828; Erman, W.F.; Kretschmar, H.C. *J. Am. Chem. Soc.* **1967**, *89*, 3842; Erman, W.F. *J. Am. Chem. Soc.* **1969**, *91*, 779.
16. Chretien-Bessiere, Y.; Retamar, J.-A. *Bull. Soc. Chim. Fr.* **1963**, *30*, 884.
17. Wender, P. A.; Mucciaro, T. P. *J. Am. Chem. Soc.* **1992**, *114*, 5878.
18. Wender, P. A.; Rawlins, D. B. *Tetrahedron*, **1992**, *48*, 7033.
19. Mucciaro, T. P.; Sutton, J. unpublished results, Stanford University.
20. Shea, K.J.; Higby, R.G.; Gilman, J.W. *Tetrahedron Lett.* **1990**, 1221 and references cited therein.
21. Studies pertinent to the methoxy arene route have been conducted by Glass, T.; Krauss, N.; Lee, D.; Mühlebach, M.; Peschke, B.; Rawlins, D.; Tomooka, K. (Stanford University).
22. Wender, P.A.; Kogen, H.; Lee, H.Y.; Munger, J.D.; Wilhelm, R.S.; Williams, P.D. *J. Am. Chem. Soc.* **1989**, *111*, 8957; see also ref. 4t and 4u.
23. Klein, L.L.; Yeung, C.M.; Li, L.; Plattner, J.J. *Tetrahedron Lett.* **1994**, *27*, 4707 and references cited therein.
24. Studies pertinent to the bromoenone strategy have been conducted by Floreancig, P.; Glass, T.; Natchus, M.; Ohkuma, T.; Shuker, A.; Sutton, J.; and Wessjohann, L. (Stanford University).
25. Kingston, D.G.I.; Samaranayake, G.; Ivey, C.A. *J. Nat. Prod.* **1990**, *53*, 801; Miller, R.W.; Powell, R.G.; Smith, C.R.; Clardy, J.; Arnold, E. *J. Org. Chem.* **1981**, *46*, 1469.
26. Studies pertinent to the aldol closure strategy have been performed by Badham, N.; Conway, S.; Floreancig, P.; Houze, J.; Krauss, N.; Lee, D.; Marquess, D.; McGrane, P.; Meng, W.; Natchus, M.; Shuker, A.; Sutton, J.; Taylor, R. (Stanford University).
27. Gugelchuk, M.M.; Houk, K.N. *J. Am. Chem. Soc.* **1994**, *116*, 330.

RECEIVED August 19, 1994

Author Index

Affiliation Index

Subject Index

C

D

Production: Charlotte McNaughton
Indexing: Deborah H. Steiner
Acquisition: Barbara Pralle
Cover design: Alan Kahan

Printed and bound by Maple Press, York, PA